Lecture Notes on Data Engineering and Communications Technologies

Volume 69

The aim of the book series is to present cutting edge engineering approaches to data technologies and communications. It will publish latest advances on the engineering task of building and deploying distributed, scalable and reliable data infrastructures and communication systems.

The series will have a prominent applied focus on data technologies and communications with aim to promote the bridging from fundamental research on data science and networking to data engineering and communications that lead to industry products, business knowledge and standardisation.

Indexed by SCOPUS, INSPEC, EI Compendex.

All books published in the series are submitted for consideration in Web of Science.

More information about this series at http://www.springer.com/series/15362

Dmytro Ageyev · Tamara Radivilova ·
Natalia Kryvinska
Editors

Data-Centric Business and Applications

ICT Systems—Theory, Radio-Electronics, Information Technologies and Cybersecurity

Editors
Dmytro Ageyev
Department of Infocommunication
Engineering V. V. Popovsky
Faculty of Infocommunication
Kharkiv National University of Radio
Electronics
Kharkiv, Ukraine

Tamara Radivilova
Department of Infocommunication
Engineering V. V. Popovsky
Faculty of Infocommunication
Kharkiv National University of Radio
Electronics
Kharkiv, Ukraine

Natalia Kryvinska
Department of e-Business
Faculty of Business
Economics and Statistics
University of Vienna
Vienna, Austria

ISSN 2367-4512 ISSN 2367-4520 (electronic)
Lecture Notes on Data Engineering and Communications Technologies
ISBN 978-3-030-71891-6 ISBN 978-3-030-71892-3 (eBook)
https://doi.org/10.1007/978-3-030-71892-3

This Springer imprint is published by the registered company Springer Nature Switzerland AG
The registered company address is: Gewerbestrasse 11, 6330 Cham, Switzerland

Preface

With this volume, we analyze challenges and opportunities for infocommunication systems usage, taking in account theory, radio-electronics, information technologies and security aspects.

Explicitly, starting with the first chapter "Development and Research of Active Queue Management Method on Interfaces of Telecommunication Networks Routers" authored by Oleksandr Lemeshko, Tetiana Lebedenko, Mykola Holoveshko presented the results of the development and further research of the method of active queue management at the interfaces of telecommunication networks routers. The research was conducted as part of a full-scale experiment based on the laboratory of Cisco Systems. To evaluate the effectiveness of the proposed method of active queue management, we compared the results of its work (according to the main indicators of quality of service) with technological solutions of congestion management mechanism—WFQ and congestion avoidance mechanism—WRED, which were automatically configured on the router interfaces. In particular, the obtained results were compared with the average packet delay and the packet loss probability. Such characteristics of flows and queues as a bandwidth of the router interface; the number of packet flows arriving at its input, the value of their classes and intensities; the number of queues formed on the interface, and their classification were changing. According to the results of the laboratory experiment, recommendations were developed on the practical application of the proposed solutions in modern and promising telecommunication networks. It was found that the algorithmic implementation in practice of the proposed models and methods can be the basis for promising queue management mechanisms and interface bandwidth in order to improve the quality of service in telecommunication networks as a whole. It was also established that the recommended area of the practical application of the proposed method of active queue management is a high-load area (over 80–85%) and congestion of interfaces of telecommunication network routers, especially in conditions of increased dynamics of their state change.

The second chapter "Improving the Structural Reliability of Mobile Radio Networks Based on the AD-Hoc Algorithms," devoted the description of specific features of Ad-Hoc technology implementation. The possibility of significant increase of the radio communications stability without introducing structural redundancy into the system is shown. For this, it is proposed to use mobile stations deployed on the terrain in non-stationary circumstances and operating in the retransmission mode. The use of repeaters is considered as a tool to improve the structural reliability of Ad-Hoc network, as well as a tool for reducing intra-system interference. The feasibility assessment of using automatic transmitter's power adjustment for radio stations within a decentralized mobile ad-hoc network is performed. The measure of intra-system interference-level reduction is estimated at the receiving point based on the assumption of subscribers' random location on the terrain.

The third chapter ("Van der Pol Oscillators Based on Transistor Structures with Negative Differential Resistance for Infocommunication System Facilities") proposes new circuits of Van der Pol oscillators based on nonlinear and reactive properties of transistor structures with negative differential resistance to be employed in facilities of infocommunication systems. Mathematical models of voltage-controlled Van der Pol oscillators, which are based on nonlinear and reactive properties of transistor structures with negative differential resistance and generate electric oscillation with regular dynamics, have been improved. Influence of additive white noise on dynamics of electric oscillations from the Van der Pol oscillators is investigated. Results of theoretical calculations, numerical simulations and experimental studies of dynamics of the electric oscillation from the microwave Van der Pol oscillators in the frequency ranges 860..910 MHz and 1,8..2,1 GHz are presented. Application of the reactive properties of transistor structures with negative differential resistance provides the parameter readjustment expansion for the self-oscillating systems of Van der Pol oscillators up to 20..30% with operation modes being stable. The effective relative operational frequency readjustment is 5.5% for the range 860..910 MHz and 9.88% for the range 1.8..2.1 GHz.

The chapter "Study of the Influence of Changing Signal Propagation Conditions in the Communication Channel on Bit Error Rate" focuses on the transmission of the information signal through the communication channel accompanied by the addition of additive white Gaussian noise, industrial interference, atmospheric noise, etc. In addition, the signal may have an additional frequency and phase shift caused by the movement of the receiver concerning the transmitter. The article is devoted to the study of the effect of the listed conditions on the errors number dependence in the communication channel on the signal-to-noise ratio (SNR). It also explores the possibilities of reducing the effect of signal propagation conditions in a communication channel by using symbolic synchronization, which is based on a phase-locked loop (PLL). The early-late-time and Gardner synchronization error detectors are investigated. The early-late-time synchronization error detector is 1.5 dB more efficient than the Gardner detector at low SNR and has a simpler implementation scheme. An energy-efficiency study of a coherent digital communication system with QPSK modulation at the phase shift in the propagation medium is performed. Increasing of the phase shift from 0 to 40° decreases the energy efficiency by 3 dB at low SNR. The energy efficiency of a non-coherent digital QPSK modulation system is reduced by 10 dB at the phase shift of 30° in the propagation medium. Adding a symbolic synchronization circle compensates for the rotation of the signal constellation. Increasing of the phase shift in the propagation medium to 45° for a coherent communication system leads to reduction of the energy efficiency by 2 dB. Frequency shifting has a significant impact on the energy efficiency of the communication system. The energy efficiency of the digital communication system decreases by 10 dB when 0.1 Hz frequency offset occurs and symbolic synchronization is missing. Symbol synchronization circuit increases the energy efficiency by 7 dB with a frequency shift of 0.1 Hz. When the value of frequency offset increases, the energy efficiency of coherent communication expands. The efficiency of coherent digital communication is increased by 24 dB with the introduction of frequency shift at 2 Hz.

The authors of chapter "Quality Assessment of Measuring the Coordinates of Airborne Objects with a Secondary Surveillance Radar" show that the principle of constructing aircraft responders and SSRs as a whole predetermined the low quality of information support for the systems under consideration under the influence of intra-system and deliberate interference. A brief description of the tasks solved by the considered information tools is given, as well as quality of information support integral indicator, the quality of which can be the probability of information support, which is defined as the product of the probability of detecting the airborne object of the requester, the probability of correctly receiving on-board information and the probability of combining flight and coordinate information. The effect of deliberate and unintended (impulse and fluctuation) interference on the quality of the assessment of measuring the range and azimuth of an air object by the considered information tool is evaluated. Based on the assessment of the influence of destabilizing factors, it is shown that in order to obtain higher accuracy in the range and azimuth measuring in the SSR, it is necessary to ensure a responder availability coefficient close to unity and high probabilities of detecting single pulses of response signals. It is shown that improving the quality of SSR information support can be achieved by searching for methods to reduce the influence of intentional and unintentional interference on the aircraft responder readiness coefficient, which is possible by changing the principle of service of request signals.

The sixth chapter ("Pulse and Multifrequency Van der Pol Generators Based on Transistor Structures with Negative Differential Resistance for Infocommunication System Facilities") proposes new circuits of Van der Pol generators based on nonlinear and reactive properties of transistor structures with negative differential resistance, which operate in relaxation and quasi-periodic modes when parameters of their self-oscillating systems are controlled in a wide range. New analytical relations are obtained, which describe parameters of sawtooth-shaped and rectangular voltage pulses of the relaxation-type Van der Pol generators, built on bipolar and field-effect transistor structures with negative differential resistance. A new multifrequency Van der Pol generator of quasi-periodic electric oscillations based on a field-effect transistor structure with negative differential resistance is developed, and its mathematical model is proposed. There are obtained analytical rations for calculating the stationary oscillation amplitude in single-frequency and multifrequency modes, the critical off-tuning between single-frequency and multifrequency modes, as well as the lower and upper cutoff frequencies of the operating frequency band. Phase portraits, time diagrams and amplitude spectra of electric oscillations of the

Van der Pol generators in oscillatory, relaxation and quasi-periodic modes are studied. Influence of parameters of the Van der Pol generators' self-oscillating on generated oscillation dynamics is estimated.

The authors of chapter "The Method of Redistributing Traffic in Mobile Network" proposed to use the sector analysis method for optimizing the load distribution between base stations when predicting the coverage areas of base stations, in addition to using the frequency-spatial planning method, when forecasting service areas of base stations. The technology of cellular systems is changing at such a speed that 4G networks have not yet had time to fully deploy, as 5G is already being introduced. The fourth generation is characterized by LTE-advanced technology, which implies an intelligent network with self-training and partial adjustment of its parameters. The distribution functions of the radio resource of the cellular communication network of this standard lie at the base stations. However, clear control algorithms for such networks have not yet been developed. As part of situationally adaptive planning of radio resources in radio communication systems, a method is proposed for determining the optimal coverage areas of base stations depending on the distribution of subscribers according to billing data. To this end, in addition to the statistics of the base stations for servicing the load, enrich it with billing system data.

The authors of chapter "Development of System for Registration and Monitoring of UAVs Using 5G Cellular Networks" aimed to analyze the current state of the UAV registry in Ukraine, to identify the main weak points and to develop an architecture for the UAV registration and monitoring system. The following tasks were solved to achieve the goal: analysis of the current state of the UAV registry in Ukraine; identifying the main weaknesses of UAV registration and monitoring in Ukraine and determining the directions for their solutions; development of a UAV registration and monitoring method; development of a model for the registration and monitoring of UAVs using 5G cellular networks; development of software for UAVs' registration and monitoring. As a result of the work, a system was developed that allows registering UAVs when they are turned on and monitoring the current coordinates of the UAVs which are using this system.

The authors of chapter "Complex Tools for Surge Process Analysis and Hardware Disturbance Protection" show the necessity of additional protection measures from a direct lightning hit or voltage induction by a remote discharge with determining protective devices and charts of their settings. The protection of information and communication systems and the components of their electronic devices is particularly dependent on the thunderstorms in the environment. Methods of calculation and protecting schemes from impulsive overvoltage recommended by international standards of IEC are implemented in Ukraine's standardization. The critical analysis of existing protectively switching devices is lead. The surge protection devices of low-voltage electric networks are compared to combine with spark protection devices at defective contact connections. The analysis of initial signals for the reception of the information on the occurrence of defective contact connections is lead, and limits of sensitivity necessary for the operation of such devices are certain. Therefore, it is proposed to apply means of measurements against impulse overvoltage in the protected circuits. The proposed means for measuring control detections and fix impulse deviations from the sinusoidal form of network voltage. On this basis, the application schemes of control devices for impulsive overvoltage are complemented, which substantially reduces the risk of fires and electronic device damage from thunderstorm origin. Implementation in the field of operation and maintenance of electronic and telecommunication equipment is proposed.

The main goal of the chapter authored by Olexander Belej and Tamara Lohutova "Development of Evaluation Templates for the Protection System of Wireless Sensor Network" is to formalize ontologies based on templates using software tools. With its help, the protection system in sensor wireless networks is analyzed and the viability of the proposed approach is noted. The implementation of the model of multi-agent systems is described; and the simulation results for the protection system are analyzed. Also, we describe the implementation of the proposed estimation algorithms for the coefficient of deviation of the request to the database and the implementation results for sensor wireless networks. Particular attention is paid to defense agents, who calculate the coefficient of deviation in the work of the ward component and identify a possible attack. Adaptive algorithms for the operation of protection agents are proposed for estimating the coefficients of deviations of requests to the database based on statistics and using a neural network.

The chapter "Studying of Useful Signal Impact on Convergence Parameters of the Gradient Signal Processing Algorithm for Adaptive Antenna Arrays that Obviates Reference Signal Presence" discusses the place and role of signal processing algorithms in adaptive antenna arrays that are analyzed with respect to their application in mobile telecommunication systems. It is established that both in literature and in practice insufficient attention is paid to search for and design of such simple adaptation algorithms that do not require presence of the reference signal. A simple algorithm of antenna array adaptation was earlier deduced that operates in the absence of the reference signal. Its successful testing was carried out under assumptions that the useful signal in the adaptive antenna array does not have an impact on the adaptation process. This assumption is valid for radar applications with short duration signals but not necessarily applicable for telecommunication systems with continuous ones. To address these limitations, influences of the useful signal on the algorithm convergence parameters are carried out. Computer simulation of adaptive antenna arrays with the designed algorithm with account of useful signal impact in different situations is performed. It is shown that with some adjustments of the constant that determines the adaptation step, the impact of continuous useful signals on the designed gradient algorithm convergence can be greatly mitigated.

In the chapter called "Interference Immunity Assessment Identification Friend or Foe Systems," based on the consideration of the place and role of Identification friend or foe (IFF) systems in the airspace control system, it is shown that the principle of building modern IFF systems in the form of an asynchronous network for transmitting request and response signals and the implementation of the principle of servicing request signals based on an open single-channel queuing system with refuses, as well as the use of primitive coding of request signals and response signals do not allow an acceptable level of information support for the air control space and air traffic control in conditions of significant intensities of intra-systemic, as well as deliberate correlated and uncorrelated interference. A general description of the considering information systems is given, and a brief description of the signals used in IFF systems is given. Based on the presentation of IFF systems in the form of two-channel systems for transmitting request and response signals, the noise immunity of aircraft responders is evaluated under the action of request signals and intentional as well as unintended (intra-system), correlated and uncorrelated interference in the request channel, which made it possible to evaluate the noise immunity of the entire IFF system in the form estimates of the probability of detection of airborne objects by the considering system.

The authors of chapter "A Stand for Diagnosing the Durability of Infocommunication Equipment to the Effects of Powerful Ultra-Wideband Electromagnetic Pulses and Laser Radiation" develop of a laboratory stand for diagnosing the durability to ultra-wideband electromagnetic impulses and laser radiation of infocommunication equipment at the stage of research. Powerful microwave nanosecond pulses parameters on the generator line design features dependence analysis. The shaper design parameters have been optimized according to the criterion of achieving maximum generator power. The main attention was paid to the study of illumination by laser radiation of optoelectronic devices, taking into account the peculiarities of their design and lens adjustment modes, in which there is a significant loss of the information component obtained in real time.

In the next chapter "Method for Planning Storage Area Network Based on Fiber to the Home Technology" is considered analytical models for determining the length of an optical cable based on symmetric graphical models with one-way and two-way cable laying topology in an urban area with a base of potential customers of a subscriber access network uniformly distributed over the square area, which allow optimizing the economic costs of deploying an access network. An access network model has been developed, which allows to consider the selection of access technology, operating and capital costs. An optimization procedure has been developed that is aimed at minimizing the objective function according to the criterion of the cost of an FTTH network deploying, taking into account the costs of purchasing network elements and deploying street optical cable infrastructure. An optimization problem has been formulated and solved that allows minimizing the cost of a passive optical (PON) access network, considering the number of optical splitters, floor splitters, and ONU subscriber units. The solution to this problem is presented in general form, which allows to adapt it for any set of these network elements. The developed methodology for calculating the cost of the designed access network includes the total capital and operating expenditures for the purchasing, installation and maintenance of network elements presented in approximate prices. The calculation of expenditures on a typical cable structure is shown on a specific example. The presented methodology also allows to take into account the costs of deploying cable infrastructure and linear structures.

The chapter authored by Volodymyr Vasylyshyn "Estimation of Signal Parameters Using SSA and Linear Transformation of Covariance Matrix or Data Matrix" discusses the joint application of singular spectrum analysis (SSA) approach (basic variant or adaptive variant) and linear transformation of extended data matrix obtained after SSA technique or corresponding covariance matrix is proposed for improvement of performance of the signal parameter estimation. The unitary transformation which reduces the computational load and improves the performance of spectral analysis performed by subspace-based techniques is used. Performance improvement can be explained by forward–backward averaging effect that has a place when performing the unitary transformation. This averaging effectively doubles the number of samples. The proposed approach can be characterized by reduced computational load such as the computations with real-valued numbers are performed after unitary transformation. Unitary Root-MUSIC is mainly used for simulation. The Unitary ESPRIT is obtained for the problem of frequency estimation. The possible applications of considered approach in the communication systems (including channel estimation, speech processing, automatic modulation classification, and so on) are considered. Simulation results confirm the improvement of performance when using proposed approach.

The chapter "Method of Creating a Passive Optical Network Monitoring System" authored by Liubov Tokar and Yana Krasnozheniuk discussed the necessity of monitoring a passive optical network (PON) and analyzed the architecture and topologies of PONs, the equipment management tools for creating a monitoring system. The protocol and control information database were selected using the MIB-I, MIB-II, RMON MIB standards. It is proved that the selection of SNMP for management is conditioned by its simplicity and efficiency, as well as the ability to unifiedly control equipment of various manufacturers. SNMP features are analyzed. The general procedure for creating a monitoring system is formulated. It is shown that for high-quality network monitoring it is necessary to implement periodic polling of all available OLTs and tracking important indicators: detecting ONUs on OLTs that are not included in the ONU database or registered simultaneously on two OLTs. Tools and technologies for creating a PON monitoring system (client and server technologies) are analyzed. Algorithms have been developed that allow processing and outputting OLT data structured according to PON specifics: an algorithm for adding a new OLT and obtaining information about ONUs; an algorithm for displaying information about ONUs in real time; and an algorithm for displaying information about client devices in real time. It is shown that in the process of monitoring a PON network, an important role is played by data obtained in real time—on demand. A PON monitoring system database has been developed with the allocation of the necessary set of domain objects. Its entities and relationships have been determined; and a database scheme as well as code examples has been presented.

The object of next chapter "Statistical Analysis and Optimization of Telecommunications Company Operating Business Processes" is the operational business processes of a telecommunications company. Authors describe the correspondences between the processes, and the elements of the organizational structure that allow controlling the correctness of the processes, their correspondence to real goals, objectives and performers. This used to analyze the implementation of business processes. The use of cluster analysis as a classification method is proposed. As a result, the entire set of processes is divided into several clusters based on the indicator of the excess of the work execution time, which makes it possible to make individual decisions on the optimization of work from each cluster. The overall runtime of business processes by network planning methods is adjusted. As a result, the manager has the opportunity to assess how productive and optimal the operational activities of this company are and to make the most profitable solution option in the field of telecommunications services management. The proposed approach allows improving the efficiency of the company's operations based on the optimization of business processes.

The authors of last chapter ("Probabilistic Method Proactive Change Management in Telecommunication Projects") Viktor Morozov and Olena Kalnichenko propose the use of a comprehensive proactive approach to managing such projects in the existing turbulent environment, which is characterized by super-complex influences of various environmental factors. The basis of this approach is the coherent perception of the processes of interaction of the "product–project organization" system, which is formed during the project implementation and is a temporary factor of interaction with the external environment. Such interaction is illustrated by the information impacts that are realized through four categories of processes: project management processes, product management processes, stakeholder management processes and processes for managing interaction with the external environment. At the same time, the proposed approach focuses on the possibilities for forming the desired actions for managing change through proactive measures.

Kharkiv, Ukraine
Kharkiv, Ukraine
Vienna, Austria

Dmytro Ageyev
dmytro.aheiev@nure.ua
Tamara Radivilova
tamara.radivilova@nure.ua
Natalia Kryvinska
natalia.kryvinska@univie.ac.at

Contents

Development and Research of Active Queue Management Method on Interfaces of Telecommunication Networks Routers

Oleksandr Lemeshko, Tetiana Lebedenko, and Mykola Holoveshko

Abstract The paper presents the results of the development and further research of the method of active queue management at the interfaces of telecommunication networks routers. The research was conducted as part of a full-scale experiment based on the laboratory of Cisco Systems. To evaluate the effectiveness of the proposed method of active queue management, we compared the results of its work (according to the main indicators of quality of service) with technological solutions of congestion management mechanism—WFQ and congestion avoidance mechanism—WRED, which were automatically configured on the router interfaces. In particular, the obtained results were compared with the average packet delay and the packet loss probability. Such characteristics of flows and queues as a bandwidth of the router interface; the number of packet flows arriving at its input, the value of their classes and intensities; the number of queues formed on the interface, and their classification were changing. The D-ITG package was used as a load testing package for generating and further analyzing network traffic. The results of the research confirmed the effectiveness of the proposed method in terms of improving the average packet delay—from 12–17 to 22–25% for high-priority flows (EF, AF41-43), and from 8–12 to 16–19%—low priority flows (AF11-13). Using the active queue management method allowed to reduce the chance of packet loss by 7–12% for high-priority flows (EF, AF41-43), and by 10–17%—low priority flows (AF11-13). Also, according to the results of the laboratory experiment, recommendations were developed on the practical application of the proposed solutions in modern and promising telecommunication networks (TCNs). It was found that the algorithmic implementation in practice of the proposed models and methods can be the basis for promising queue management mechanisms and interface bandwidth in order to improve the quality of service in telecommunication networks as a whole. It was also

O. Lemeshko (✉) · T. Lebedenko · M. Holoveshko
Kharkiv National University of Radio Electronics, Nauky Ave. 14, Kharkiv 61166, Ukraine
e-mail: oleksandr.lemeshko.ua@ieee.org

Ivan Kozhedub Kharkiv National Air Force University, Klochkivs'ka St. 228, Kharkiv 61045, Ukraine

T. Lebedenko
e-mail: tetiana.lebedenko@nure.ua

D. Ageyev et al. (eds.), *Data-Centric Business and Applications*, Lecture Notes on Data Engineering and Communications Technologies 69,
https://doi.org/10.1007/978-3-030-71892-3_1

established that the recommended area of the practical application of the proposed method of active queue management is a high-load area (over 80–85%) and congestion of interfaces of telecommunication network routers, especially in conditions of increased dynamics of their state change.

Keywords Active queue management · Congestion management · Congestion avoidance · Resource allocation · Laboratory experiment · Quality of service

1 Introduction

An important step in improving existing or developing new methods for queue management on interfaces of routers of telecommunication networks (TCNs) is the organization and conducting of their experimental study. It is known that the use of analytical modeling, as a rule, is not time consuming and it allows to study network processes for a wide range of source data: traffic characteristics, interface parameters, service disciplines, ets [1–4]. However, to increase the adequacy and accuracy of the research results, analytical calculations must be added during a full-scale (laboratory) experiment on real network equipment [5–7].

Conducting a laboratory (full-scale) experiment is based on solving the following important tasks:

(1) development of the general scheme of carrying out the experiment in active management of queues on interfaces of routers of telecommunication networks and the description of its structure;
(2) review and selection of means for load testing and analyzing the network traffic;
(3) description of a network fragment and the order of configuring network interfaces for solving queue management problems;
(4) analysis of the results obtained during the laboratory experiment;
(5) development and substantiation of recommendations on the areas of the most effective practical application of the studied methods for queue management.

In this paper, the object of study is the method of active queue management on the interfaces of telecommunication networks routers, which is described in [8–16]. The basis of this method are the following models:

(1) a model of optimal aggregation and allocation of packet flows among queues;
(2) a model of balanced allocation of telecommunication networks router interface bandwidth among the formed class queues;
(3) a model of active queue management on the interfaces of telecommunication networks routers, which will be described below.

2 Model of Optimal Aggregation and Allocation of Packet Flows Among Queues of Interfaces of Telecommunication Networks Routers

As shown in [8–10, 12, 14], the model of optimal aggregation and allocation of packet flows among queues of interfaces of telecommunication networks routers is based on the solution of the optimization problem of linear integer programming with the optimality criterion:

$$\min_{x} F,\ F = \sum_{i=1}^{N} \sum_{j=1}^{M} h_{i,j}^{x} x_{i,j}, \tag{1}$$

where

$$x_{i,j} \in \{0, 1\} \tag{2}$$

are the control variables, each of which characterizes the part of the ith flow of packets sent for servicing to the jth queue; N denotes the total number of packet flows coming to the selected interface in accordance with the content of the routing table on the TCNs router; M is the number of organized queues on the router interface; $h_{i,j}^{x}$ is the conditional cost (metric) of servicing packets of the i th flow by the j th queue [8–10, 14]:

$$h_{i,j}^{x} = \left(k_i^f - k_j^q\right)^2 + 1. \tag{3}$$

According to the physical content of the problem being solved, the metric $h_{i,j}^{x}$ will be minimal, i.e. equal to one, in case of matching of classes k_i^f and k_j^q. The larger the difference in the values of the flow and queue classes is, the larger the metric (3) will be, which will restrain the direction of the flows of different classes to one class queue during the solution of optimization problem (1) in the presence of constraints (2).

3 Model of Balanced Interface Bandwidth Allocation Among Formed Class Queues

In the studied queue management method, after solving the problem of aggregation and allocation of packet flows among queues, the bandwidth of the TCNs router interface is allocated among them. To do this, we introduce a set of control variables b_j, each of which determines a part of interface bandwidth allocated for servicing of the j th queue [8–10, 14]. In order to adequately allocate the total bandwidth of the interface b among the queues created on it, the following constraints are imposed on the introduced control variables:

$$b_j \geq 0, \quad \sum_{j=1}^{M} b_j \leq b, \quad \left(j = \overline{1, M}\right). \tag{4}$$

To ensure the balanced allocation of the interface bandwidth among the class queues formed on it, the structure of the model introduces conditions to prevent overloading of queues by the bandwidth allocated to them [14]:

$h_j^\alpha \sum_{i=1}^{M} a_i x_{i,j} \leq \alpha b_j,$

where h_j^α denotes class coefficient, which is introduced to consider the classes of queues when balancing the allocation of the interface bandwidth among queues:

$$h_j^\alpha = 1 + \frac{k_j^q}{K \cdot D}, \left(j = \overline{1, M}\right), \tag{5}$$

where $D \geq 1$ is the rationing factor, by which the differentiation of router interface bandwidth allocation is performed among the class queues organized on it. The higher the value D, the less the queue class k_j^q will affect the amount of interface bandwidth allocated to it; $a_i\left(i = \overline{1, N}\right)$ is the average intensity of the th flow, which is measured in packets per second (1/s); α is control variable that is quantitatively related to the upper dynamically controlled threshold for using queues according to the bandwidth of the interface and which satisfies the condition $0 < \alpha \leq 1$.

To further simplify the algorithmic-software implementation of the computational solutions obtained in practice, it is proposed to reduce the above-formulated nonlinear optimization problem with constraints (4)–(5) to a linear programming class problem without losing the adequacy and accuracy of the final results. Thus, it is proposed to replace the control variable α:

$$\alpha^* = \frac{1}{\alpha}, \tag{6}$$

where α^* is an additionally introduced control variable, which is inversely proportional to the upper threshold of the load of interface queues (α) and satisfies the condition:

$$\alpha^* > 0. \tag{7}$$

Then the condition of preventing queue congestion with a balanced allocation of interface bandwidth between them will take the form:

$$h_j^{\alpha}\alpha^* \sum_{i=1}^{N} a_i x_{i,j} \leq b_j, \quad \left(j = \overline{1, M}\right). \tag{8}$$

Then the model of the balanced allocation of the interface bandwidth among the formed class queues will be based on the solution of the optimization problem of linear programming with the optimality criterion

$$\max_{b,\,\alpha*} \alpha^*, \tag{9}$$

in the presence of constraints (4)–(8), when $x_{i,j}$ are known values obtained during the preliminary solution of problem (1)–(3).

4 Model of Active Queue Management on Interfaces of Telecommunication Networks Routers

In the case of interface overload, when the bandwidth of the interface is not enough to service the flows coming to it, it is impossible to ensure the fulfillment of the condition (8). Then, instead of the model (4)–(9), it is necessary to use the model of active queue management on the interfaces of telecommunication networks routers [12, 14], which focuses on the agreed solution of problems of interface bandwidth allocation and preventive (early) limiting of the intensity of flows incoming to the input of the same interface of the telecommunication network router.

To solve these problems by analogy with the model of balanced interface bandwidth allocation, we introduce a set of control variables b_j $\left(j = \overline{1, M}\right)$, which are subject to the conditions (4). In order to implement preventive (early) limiting of the intensity of flows incoming to the input of the TCNs router interface, we determine a set of control variables y_i, that characterize a part of the ith $\left(i = \overline{1, N}\right)$ packet flow that received a denial of service on the TCNs router interface when solving the active queue management problem.

In fact, the variables y_i will determine the probability of dropping the packets of the ith flow from the queue formed on the interface and will be subject to the following conditions [14, 17]:

$$0 \leq y_i \leq 1, \quad \left(i = \overline{1, N}\right). \tag{10}$$

To ensure the controllability of the process of active queue management, it is necessary to satisfy the system of constraints [12, 14, 18]:

$$\sum_{i=1}^{N} a_i x_{i,j}(1 - y_i) \leq b_j, \quad \left(j = \overline{1, M}\right). \tag{11}$$

Based on the formulated conditions and constraints (4), (10)–(11), the optimization problem for active queue management is represented by the optimality criterion [14]:

$$\min_{b,y} P, \quad P = \sum_{j=1}^{M} h_j^b b_j + \sum_{i=1}^{N} h_i^y a_i y_i, \tag{12}$$

where h_j^b is the metric that characterizes the conditional cost of allocating a unit of bandwidth of the router interface to the jth queue; h_i^y is the metric that determines the conditional cost of denial of service to packets of the ith flow.

The use of optimality criterion (12) is associated, firstly, with minimizing the use of router interface bandwidth based on its optimal allocation among the formed queues, and secondly, with minimizing possible denials of service caused by preventive limitation of packet flow intensity. Metrics of direction h_j^b and h_i^y depend on the corresponding values of queue and flow classes.

5 Development and General Description of Laboratory Experiment Scheme

To organize a laboratory experiment for studying the above method of active queue management, a research technique using network equipment manufactured by Cisco Systems and a load testing package for generation and further analysis of network traffic was used [19–25]. As part of the laboratory experiment, queue and bandwidth management processes were studied on a high-speed Fast Ethernet interface of the Cisco Systems router of 2801 Series with the bandwidth of 10/100 Mbps. The traffic sending side was the terminal PC1, the receiving side was terminal PC2. The interaction of the end stations was carried out through a section of the network, which was represented by the router Router1. Standard personal computers connected to Router1 via Fast Ethernet technology were used as endpoints. An example of the technological scheme of connection is shown in Fig. 1.

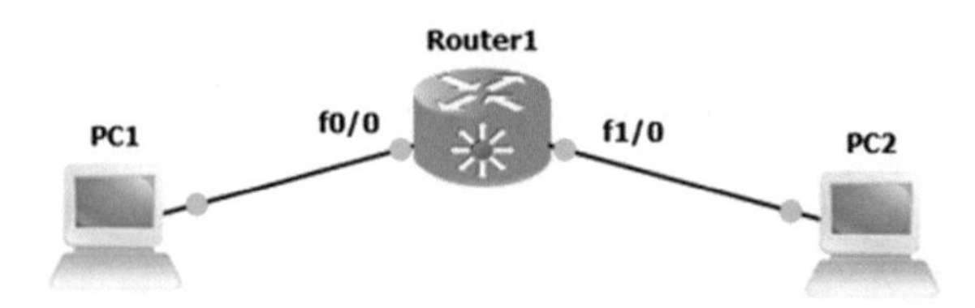

Fig. 1 A variant of the scheme for the router and terminals connection during the laboratory experiment

```
Router(config)#policy-map dscp-based
Router(config-pmap)#class class1
Router(config-pmap-c)#bandwidth 40
Router(config-pmap-c)#class class2
Router(config-pmap-c)#bandwidth 3500
Router(config-pmap-c)#exit
```

(a)

```
Router(config)#class-map class1
Router(config-cmap)#match ip dscp af13
Router(config-cmap)#exit
Router(config)#class-map class2
Router(config-cmap)#match ip dscp ef
Router(config-cmap)#exit
Router(config)#
```

(b)

Fig. 2 An example of configuring the overload management mechanism (class-based weighted fair queueing, CBWFQ)

In addition, the class commands were used to configure congestion management mechanisms, such as the Class-Based Weighted Fair Queuing (CBWFQ) (Fig. 2). In the framework of this example, two classes were created, class1 and class2, the classification criteria of which were the DSCP field codes in the header of the IP packet [19, 26–28]. Packet flows with the field code DSCP AF13 corresponded to class1, while those with DSCP EF corresponded to class2. The bandwidth command explicitly specified the bandwidth of the interface for each class, allocated for each of the queues formed on it (Fig. 2a)—class1 and class2 received 40 kbps and 3500 kbps, respectively.

Figure 2b shows an example of classification and aggregation of packet flows according to the value of the field code DSCP AF13 and EF during the creation of two service classes—class1 and class2 given in Fig. 2a.

After the configurations have been made for generating traffic and controlling the quantitative values of the main indicators of service quality at the end stations of PC1 and PC2 subscribers, it is necessary to use the load testing package [20, 26–28].

Table 1 shows a brief description and comparison of load testing packages that are most commonly used in solving traffic management problems [24, 25].

In the general case, load testing packages are simulation tools, the distinguishing features of which are licensing conditions, the number of supported functions, applicability to different types of operating systems, the use of different types of protocols and the ability to configure them. Today, the following packages are widely used for traffic generation and analysis in modern TCNs: IxChariot, IPerf, IPerf3, NetPerf, D-ITG (Distributed Internet Traffic Generator), MTools, UDP-Generator, Mgen, IP Traffic, Rude/Crude.

The result of the analysis showed that the most correct, easy to use and illustrative in the study of queue management processes on the interfaces of telecommunications networks routers are IxChariot and D-ITG load testing packages. These packages support a large number of traffic parameters. In particular, the D-ITG package is able to generate traffic with different types of load distribution (uniform, deterministic, exponential, Pareto, Cauchy, normal, Poisson, gamma distribution, Weibull); support network layer protocols (IPv4 and IPv6), transport layer protocols (TCP, UDP, ICMP, DCCP, SCTP) and application layer protocols (VoIP (G.711, G.723, G.729), DNS, Telnet); work on Linux, Windows, MAC OS X (Leopard), FreeBSD, Android, iOS operating systems; provide high packet rate between a sender and a receiver (up to 612

Table 1 Comparison and characteristics of load testing packages

Brief characteristics	Load testing package			
	IxChariot	IPerf	NetPerf	D-ITG
Operational system	Linux, Windows MAC OS Solaris Android iOS	Linux Windows MAC OS	Linux Windows MAC OS	Linux Windows MAC OS X (Leopard) FreeBSD, Android, iOS
Interface view	Graphic interface	Command line	Command line	Command line/graphic interface
Supported protocols	TCP, UDP, RDP, VoIP, IPv4, IPv6	TCP, UDP, IPv4, IPv6	TCP, UDP, SCTP, DLPI, IPv4, IPv6	TCP, UDP, ICMP, DCCP, SCTP, VoIP, IPv4, IPv6
QoS-indicators	Bandwidth, jitter, packet loss, packet delays, MOS, R-factor	Bandwidth, jitter, packet loss	Bandwidth, jitter, packet loss	Bandwidth, jitter, bitrate, packet loss, RTT, packet delays
Manual	Yes	Yes	Yes	Yes
Free software	No	No	No	No

Mbps). In addition, D-ITG allows to measure and analyze the bitrate, packet rate, one way delay (OWD), round trip time (RTT), as well as determine the throughput, jitter and packet loss [24–26]. An additional advantage of D-ITG is also the availability of free software.

Using the D-ITG load testing package, the ITGSend subprogram was started on the sender side, in which both the number of packet flows and their characteristics were set using control parameters (keys) (Fig. 3).

According to the example shown in Fig. 4, the ITGSend packet sender has the following flow characteristics:

- the address of the receiver side (-a 192.168.1.107);
- the packet rate (pack/s) (-C 40);
- the size of each packet (Byte) (-c 100);
- the type of transport layer protocol (-T UDP);
- the duration of the experiment generation (-t 15,000);

```
C:\D-ITG-2>ITGSend.exe -a 192.168.1.107 -C 40 -c 100 -T UDP -t 15000 -l sender.log -x receiver.log
ITGSend version 2.8.1 (r1023)
Compile-time options:
Started sending packets of flow ID: 1
Finished sending packets of flow ID: 1
```

Fig. 3 An example of configuring the ITGSend subprogram of packet sender

```
-----------------------------------------------------------
Total time                  =     14.976000 s
Total packets               =           600
Minimum delay               =      0.000000 s
Maximum delay               =      0.002000 s
Average delay               =      0.000283 s
Average jitter              =      0.000416 s
Delay standard deviation    =      0.000483 s
Bytes received              =         60000
Average bitrate             =     32.051282 Kbit/s
Average packet rate         =     40.064103 pkt/s
Packets dropped             =             0 (0.00 %)
Average loss-burst size     =      0.000000 pkt
-----------------------------------------------------------

___________________________________________________________
****************  TOTAL RESULTS   ******************
___________________________________________________________
Number of flows             =             1
Total time                  =     14.976000 s
Total packets               =           600
Minimum delay               =      0.000000 s
Maximum delay               =      0.002000 s
Average delay               =      0.000283 s
Average jitter              =      0.000416 s
Delay standard deviation    =      0.000483 s
Bytes received              =         60000
Average bitrate             =     32.051282 Kbit/s
Average packet rate         =     40.064103 pkt/s
Packets dropped             =             0 (0.00 %)
Average loss-burst size     =             0 pkt
Error lines                 =             0
-----------------------------------------------------------
```

Fig. 4 An example of displaying basic QoS indicators using the D-ITG package

- the names of the files, in which the received data will be entered, will be created both on the sender side (-l sender.log) and on the receiver side
- (-x receiver.log).

On the receiving side with the help of subprograms ITGRecv and ITGDec the results of the laboratory experiment were processed and formed in accordance with the characteristics of the flows shown in Fig. 3.

Example of decoding a log file (receiver.log) displaying the values of the total number of transmitted packets, the time of the experiment, as well as the main indicators of QoS (values of minimum, maximum and average packet delay, jitter, average bit and packet rates, and percentage of packet loss) are presented in Fig. 4.

Figure 5 shows an example of traffic generation with the characteristics of VoIP:

- the type of codecs used (G. 711.2 and G. 729.2);
- the number of generated flows (one flow for VoIP G. 711.2 traffic and one flow for VoIP G. 729.2 traffic);
- the receiving port for the flow G. 711.2—8001;
- the receiving port for the flow G. G. 729.2—8002;
- the packet rate (50 1/s);
- the VAD sub-option (enables VoIP voice activity detection).

```
C:\D-ITG-2>ITGSend.exe script_file.txt
ITGSend version 2.8.1 (r1023)
Compile-time options:
Voice Codec: G.711
Framesize: 80.00
Samples:    2
Packets per sec.:   50
Voice Codec: G.729
Framesize: 10.00
Samples:    2
Packets per sec.:   50
VAD: Si
VAD: Si
Started sending packets of flow ID: 1
Started sending packets of flow ID: 2
Finished sending packets of flow ID: 1

Finished sending packets of flow ID: 2
```

Fig. 5 An example of configuring the ITGSend subprogram of packet sender when generating traffic with VoIP characteristics

Due to the multi-flow mode, the script described in the text file script_file.txt was formed to generate traffic. Similar to the example shown in Fig. 4, to display the values of the main quality of service indicators, the created log file was decoded (Fig. 6).

```
----------------------------------------------------------
Total time               =       9.980000 s
Total packets            =            500
Minimum delay            =       0.000000 s
Maximum delay            =       0.002000 s
Average delay            =       0.000444 s
Average jitter           =       0.000613 s
Delay standard deviation =       0.000647 s
Bytes received           =          16000
Average bitrate          =      12.825651 Kbit/s
Average packet rate      =      50.100200 pkt/s
Packets dropped          =              0 (0.00 %)
Average loss-burst size  =       0.000000 pkt
----------------------------------------------------------
----------------------------------------------------------
Flow number: 1
From 192.168.1.107:62636
To    192.168.1.107:8001
----------------------------------------------------------
Total time               =       9.980000 s
Total packets            =            500
Minimum delay            =       0.002000 s
Maximum delay            =       0.001000 s
Average delay            =       0.000522 s
Average jitter           =       0.000591 s
Delay standard deviation =       0.000661 s
Bytes received           =          86000
Average bitrate          =      68.937876 Kbit/s
Average packet rate      =      50.100200 pkt/s
Packets dropped          =              0 (0.00 %)
Average loss-burst size  =       0.000000 pkt
```

Fig. 6 An example of displaying the main QoS indicators using the D-ITG package when generating traffic with VoIP characteristics

6 Results of Experimental Research and Assessment of Effectiveness of the Investigated Active Queue Management Method

On the basis of the considered method, a number of laboratory experiments regarding the research of active queue management processes on the interface of the TCNs router was carried out. To assess the effectiveness of the proposed models, the results of their performance were compared (according to the main quality of service indicators) with technological solutions of congestion management mechanisms—WFQ, and congestion avoidance—WRED, which were configured on router interfaces automatically. In particular, the obtained results were compared in terms of the average packet delay and the probability of packet loss. The following were subject to change: bandwidth of the interface; the number of packet flows received at its input; values of flow classes and intensities; the number of queues formed on the interface and their classification [14, 29–33].

For example, suppose that a TCNs router interface received aggregated traffic, in which the number of packet flows varied from one to twenty. Flow differentiation was performed based on the content of the differentiated DSCP field code in the header of the IP packet. VoIP traffic flows were then matched with EF field code (priority), interactive video (video conferencing) and audio/video on demand flows—AF41, critical data flows—AF31, data transmission flows—AF13. Each flow was serviced within separately formed queues.

Figures 7, 8, 9 and 10 show the results of comparisons of solutions obtained during the laboratory experiment in the impact of router interface load on the average packet delay for the proposed model of balanced allocation of the TCN interface bandwidth (1)–(9), and for WFQ/WRED mechanisms configured on the router interface "by default". Interface load (ρ) was defined as the ratio of the total packet load on the

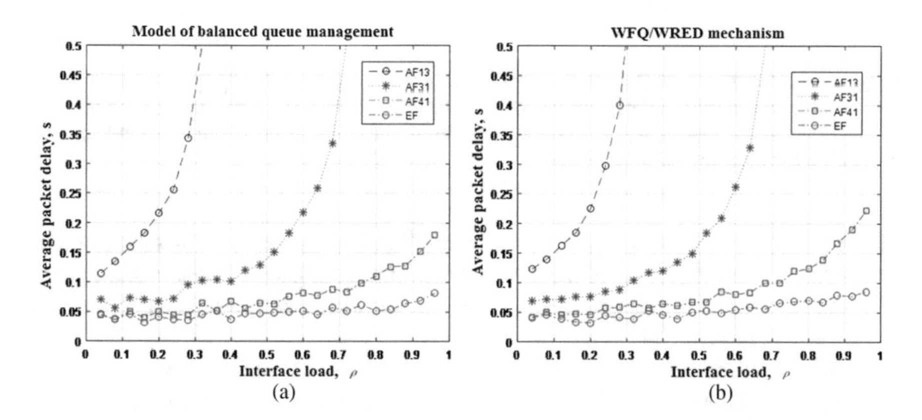

Fig. 7 Analysis of the impact of interface load on the average packet delay when generating symmetric traffic with different DSCP field values for the balanced queue management model (**a**) and the WFQ/WRED mechanism (**b**)

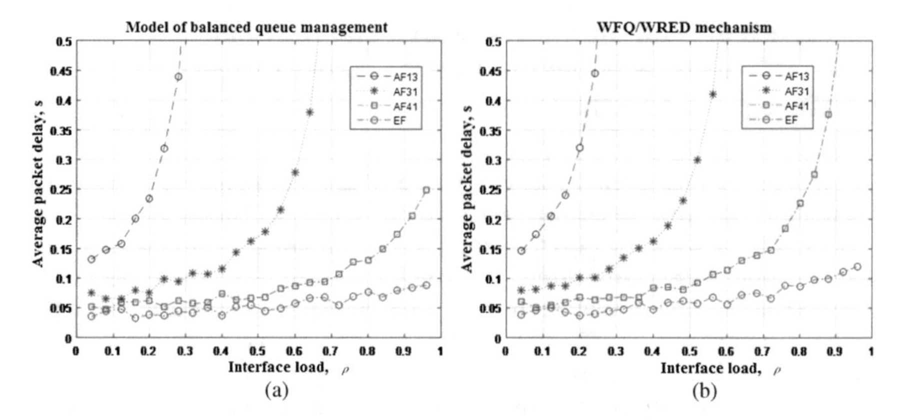

Fig. 8 Analysis of the impact of interface load on the average packet delay when generating asymmetric traffic with different values of the DSCP field code and increasing of the number of low-priority flows AF13 ($N_{AF13} = 8$) for the balanced queue management model (**a**) and WFQ/WRED mechanism (**b**)

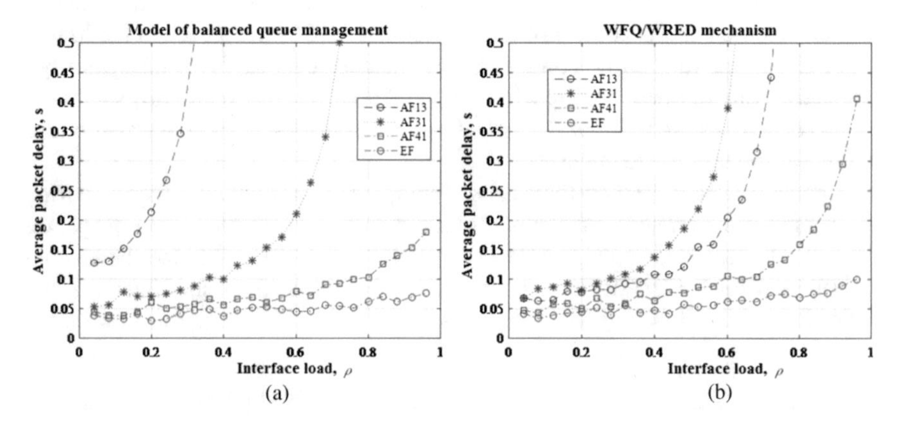

Fig. 9 Analysis of the impact of interface load on the average packet delay when generating asymmetric traffic with different values of the DSCP field code and increasing of the number of low-priority AF13 ($N_{AF13} = 15$) flows for the balanced queue management model (**a**) and WFQ/WRED mechanism (**b**)

TCN router interface to its bandwidth. Thus, Fig. 7 shows the results of the study, when the traffic structure is symmetric, i.e. the number of packet flows coming to the class queues formed on the telecommunication network router interface is the same: four flows per queue [14, 29].

Analysis of the obtained solutions regarding the impact of interface load on the average packet delay (Fig. 7) showed that in the conditions of generating symmetric traffic with different values of the DSCP field code both solutions work adequately, allocating to each queue the interface bandwidth according to the value of its class. However, the proposed balanced queue management model (in contrast to the WFQ

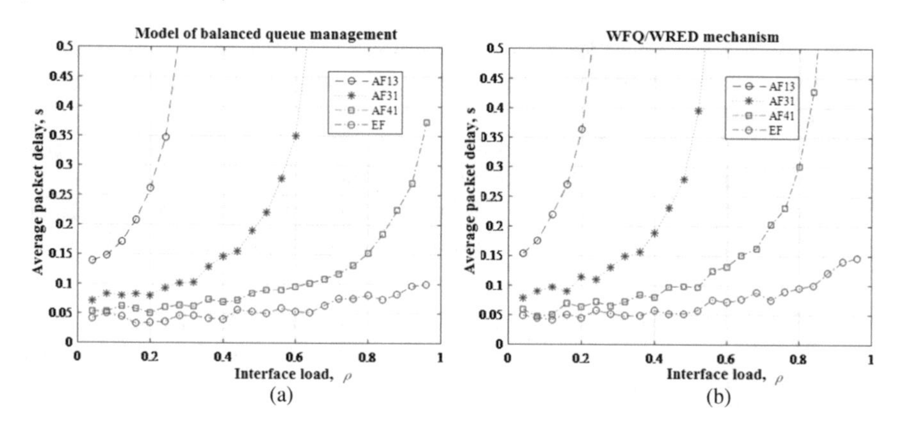

Fig. 10 Analysis of the impact of interface load on the average packet delay when generating asymmetric traffic with different values of the DSCP field code and increasing of the number of high-priority EF ($N_{EF} = 8$) flows for the balanced queue management model (**a**) and the WFQ/WRED mechanism (**b**)

mechanism) allows, if necessary, to adjust the influence of the queue class on the level of QoS due to the conditions and constraints introduced into the model structure (5) taking into account the rationing factor *D*.

Figure 8 shows the results of the analysis of the interface load impact on the average packet delay for the conditions under which the priority queues generated on the router interface receive traffic with an asymmetric structure for the following output data:

- $M = 4$—the total number of queues organized on the interface;
- $N_{AF13} = 8$—the number of packet flows with the field code DSCP AF13;
- $N_{AF31} = 4$—the number of packet flows with the field code DSCP AF31;
- $N_{AF41} = 4$—the number of packet flows with the field code DSCP AF41;
- $N_{EF} = 4$—the number of packet flows with the field code DSCP EF;
- $b = 100$ 1/s—the bandwidth of the TCN router interface.

Figure 9 shows the results of the study of the case in which the traffic of the asymmetric structure has the following initial data:

- $M = 4$—the total number of queues organized on the interface;
- $N_{AF13} = 15$—the number of packet flows with the field code DSCP AF13;
- $N_{AF31} = 4$—the number of packet flows with the field code DSCP AF31;
- $N_{AF41} = 4$—the number of packet flows with the field code DSCP AF41;
- $N_{EF} = 4$—the number of packet flows with the field code DSCP EF;
- $b = 100$ 1/s—the bandwidth of the TCN router interface.

It can be concluded from Figs. 8 and 9 that with increasing traffic asymmetry, which manifested itself in an increase in the number of low-priority flows compared to higher-priority ones, the use of the WFQ mechanism led to a significant reallocation of link resource. The bandwidth of the router interface was allocated more intensively to low-priority flows, which led to the lack of link resource for priority queues (flows). This was reflected in a significant increase in the average delay of packets with EF and AF41 priorities.

The use of the proposed model of balanced interface bandwidth allocation (1)–(9) allowed a more rational allocation of link resource among flows and queues of different classes and priorities. This fact made it possible to improve the average packet delay for high-priority flows EF, AF41-AF43 compared to using the WFQ mechanism:

- by 12–17%—in the area of average load ($0.55 \le \rho \le 0.8$) of the router interface;
- by 22–25%—in the area of high ($0.85 \le \rho \le 0.95$) and critical ($\rho \ge 0.96$) load of the router interface.

To analyze the impact of interface load on the average packet delay, the case was studied (Fig. 10), in which the priority queues formed on the interface of the telecommunication router received traffic with the following asymmetric structure:

- $M = 4$—the total number of queues organized on the interface;
- $N_{AF13} = 4$—the number of packet flows with the field code DSCP AF13;
- $N_{AF31} = 4$—the number of packet flows with the field code DSCP AF31;
- $N_{AF41} = 4$—the number of packet flows with the field code DSCP AF41;
- $N_{EF} = 8$—the number of packet flows with the field code DSCP EF;
- $b = 100$ 1/s—the bandwidth of the TCN router interface.

As we can see in Fig. 10, in the case of increasing the number of high-priority EF flows, the WFQ mechanism significantly limits the link resource allocated for low-priority queues. Therefore, by analogy with the examples shown in Figs. 8 and 9, to ensure the differentiation of servicing packet flows with different QoS requirements, it is also advisable to use the proposed model (1)–(9).

The results of the study (Fig. 10) allowed to confirm the effectiveness of the model of balanced allocation of the TCN router interface bandwidth (1)–(9) proposed within the method of active queue management from the point of view of improving the average packet delay for low-priority flows AF11-AF13:

- by 8–12%—in the area of average ($0.55 \le \rho \le 0.8$) load of the TCN router interface;
- by 16–19%—in the area of high ($0.85 \le \rho \le 0.95$) and critical ($\rho \ge 0.96$) load of the TCN router interface.

During the experimental study we also compared the solutions obtained during the laboratory experiment in the impact of the TCN router interface load on the probability of packet loss for the proposed model of active queue management (1)–(4), (10)–(12) and for WFQ/WRED mechanisms configured on the router interface "by default".

The results of performed studies on the impact of TCN router interface load on the probability of packet loss when generating symmetric traffic with different values of the DSCP field code show that both of these approaches work adequately, and the nature of their results is similar to the comparison given in Fig. 7. The analysis of the impact of the interface load on the probability of packet loss during the generation of asymmetric traffic is shown in Fig. 11.

In the case of generating asymmetric traffic and provided that the number of low-priority packet flows AF13 increases similarly to the example shown in Fig. 8, the use of the model of active queue management (Fig. 11) allows to ensure a smoother nature of denials of the service for high-priority packets. That is, denials of service will apply to all packet flows, but to a greater extent to those from low-priority queues and to a lesser extent to those from high-priority queues.

Figure 12 shows the results of the study for the example when the number of high-priority packet flows has been increased in the structure of asymmetric traffic.

According to the results of the study (Figs. 11 and 12) it was found that the use of the proposed model of active queue management (1)–(4), (10)–(12) has reduced the probability of packet loss:

- by 7–12%—for high-priority packet flows EF, AF41-AF43;
- by 10–17%—for low-priority packet flows AF11-AF13, depending on the source of asymmetry in the structure of network traffic.

It was found that the recommended area for application the proposed models of active queue management and balanced allocation of the TCN interface bandwidth is the area of high ($0.85 \leq \rho \leq 0.95$) and critical ($\rho \geq 0.96$) load of network interfaces, especially in conditions of increased dynamics of their state change. In the area of low ($0 \leq \rho \leq 0.5$) load of the router interface, all compared solutions provided almost the same QoS level.

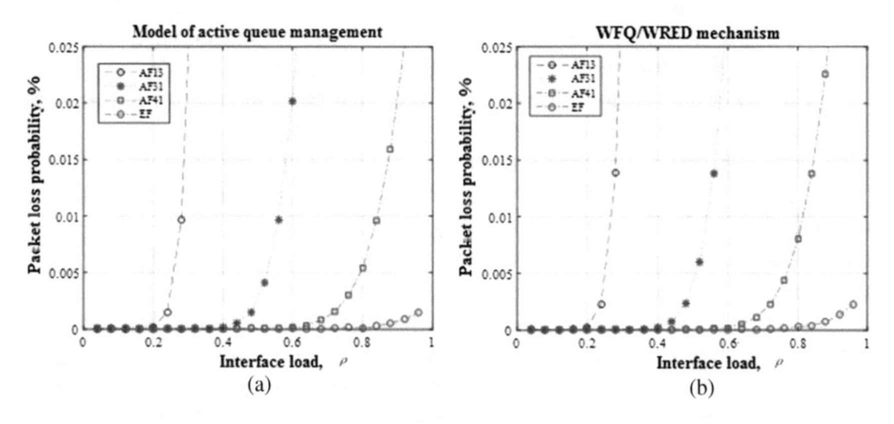

Fig. 11 Analysis of the impact of interface load on the probability of packet loss when generating asymmetric traffic with different values of the DSCP field code and increasing of the number of low-priority AF13 flows for the active queue management model (**a**) and WFQ/WRED mechanism (**b**)

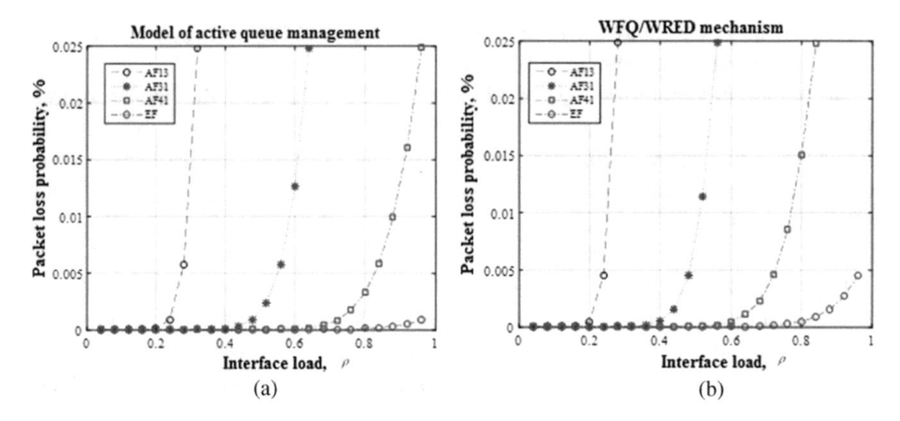

Fig. 12 Analysis of the impact of interface load on the probability of packet loss when generating asymmetric traffic with different values of the DSCP field code and increasing of the number of high-priority EF flows for the active queue management model (**a**) and the WFQ/WRED mechanism (**b**)

7 Recommendations for Practical Implementation of the Method of Active Queue Management in Telecommunication Networks

Based on the results of the research, within the framework of laboratory experiment, recommendations for the practical implementation of the studied method of active queue management on the interfaces of routers of telecommunication networks were developed. Its practical algorithmic realization can be the basis of promising mechanisms for managing queues and bandwidth of interfaces in order to improve the quality of service in TCNs as a whole.

Figure 13 presents a functional diagram of the TCN router interfaces, where each of the physical interfaces is connected to the hardware queue and software queue systems. The hardware queues always use the FIFO (Tx-ring) mechanism by default. The configuration of system queues depends on the mechanisms of queue scheduling (WFQ, CBWFQ, LLQ) and active queue management (WRED). These mechanisms start working at the moment when the hardware queue is filled to its maximum capacity, and the router interface begins to be overloaded. Then hardware queue signals to the software queue that any scheduling policies, such as WFQ/LLQ/CBWFQ, and active queue management, such as RED/WRED, that have been configured on the router interface before, must be enabled.

Therefore, given that the process of queuing on the interfaces of TCN routers is directly related to their overload or high load (more than 85–95%), the recommended area of implementation of the obtained solutions is their application on the TCN routers interfaces, in which there is lack of link resource. Thus, the model of balanced

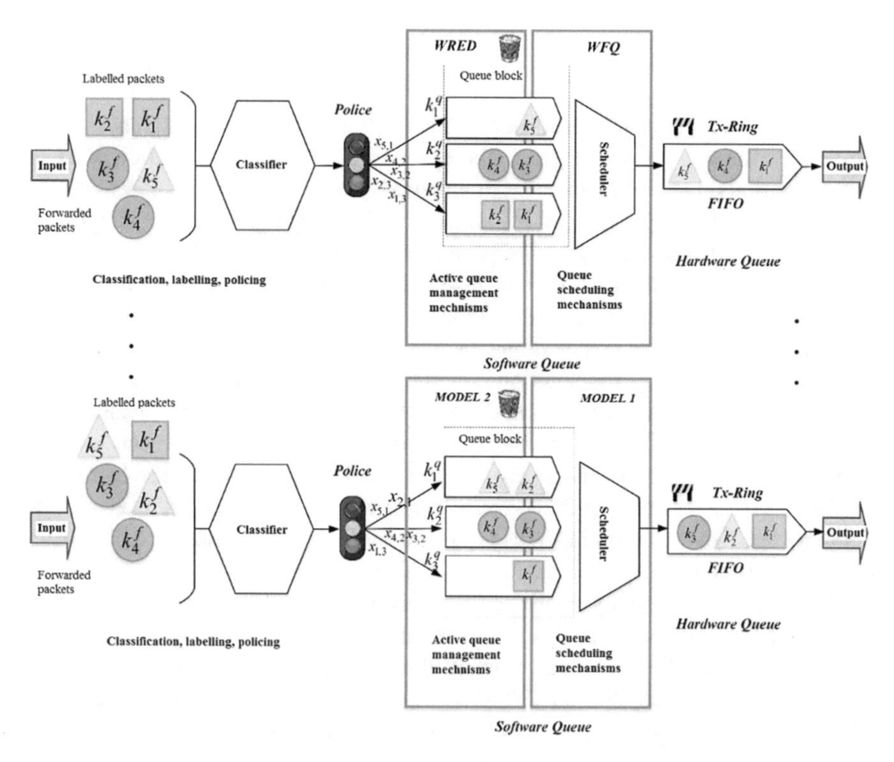

Fig. 13 Principles of realization of the proposed method and models of queue management on interfaces of telecommunication networks routers

allocation of the TCN router interface bandwidth (1)–(9) is recommended to be used instead of the classic queuing mechanisms FIFO, CQ, PQ, WFQ, etc. This will help avoid their overload by bandwidth due to more flexible and rational allocation of link resources.

In case of interface overload, it is advisable to use the model of active queue management (1)–(4), (10)–(12) instead of the RED/WRED mechanisms configured "by default". The use of this model allows to limit in advance the intensity of the flow that causes interface overload. That is, in the event of router interface overload, denial of service will have a balanced nature: denials will affect all packet flows, but to a greater extent those from low-class queues and to a lesser extent those from high-class queues. This will ensure a smoother nature for denial of service of packet flows instead of abrupt.

On low-loaded interfaces or during their low load (up to 50–60%) it is advisable to use the simplest queuing mechanism—FIFO. In addition, in the arae of medium loads with a symmetrical structure of network traffic, the WFQ mechanism can be used.

8 Conclusions

An experimental study of the adequacy and assessment of the effectiveness of the proposed solutions for active queue management and balanced allocation of TCNs router interfaces bandwidth in terms of key quality of service indicators was carried on the basis of the Cisco Systems laboratory. For load testing, generating and further analyzing of network traffic, the D-ITG package was used. To test the control solutions for the implementation of link resource allocation functions obtained when applying the model of balanced allocation of TCNs routers bandwidth (1)–(9) they were compared with WFQ scheduling mechanisms by average packet delay (Figs. 7, 8, 9 and 10). To verify the solutions for the realization of active queue management functions obtained when using the model of active queue management of TCN interfaces (1)–(4), (10)–(12) they were compared with AQM WRED mechanisms in terms of packet loss probability (Figs. 11, 12 and 13). All studies were carried out on Cisco Systems routers of 2801 Series, when the bandwidth of the interface, the number of packet flows coming to its input, the values of their classes and intensities, the number of queues formed and their classification were subject to change.

The results of the study confirmed the effectiveness of the model of balanced interface bandwidth allocation in terms of improving the average packet delay—from 12–17 to 22–25% for high-priority flows (EF, AF41-43) and from 8–12 to 16–19% for low priority flows (AF11-13). The use of the active queue management model reduced the probability of packet loss by 7–12% for high-priority flows (EF, AF41-43) and by 10–17% for low-priority flows (AF11-13).

It is established that the recommended area for practical application of the proposed models of queue and bandwidth management includes areas of high load (more than 80–85%) and overload of TCNs routers interfaces, especially in conditions of increased dynamics in their state change.

References

1. Rao DS (2012) Queue management and quality of service (QoS) in the internet: a novel approach for flow protection for providing better than best-effort service in the internet. LAP LAMBERT Academic Publishing
2. Tan L (2017) Resource allocation and performance optimization in communication networks and the internet. CRC Press
3. White R, Banks E (2018) Computer networking problems and solutions: an innovative approach to building resilient, modern networks, 1st edn. Addison-Wesley Professional
4. Smelyakov K, Chupryna A, Hvozdiev M, Sandrkin D, Martovytskyi V (2019) Comparative efficiency analysis of gradational correction models of highly lighted image. In: 2019 IEEE international scientific-practical conference problems of infocommunications, science and technology (PIC S&T). Kyiv, Ukraine, pp 8–11. https://doi.org/10.1109/PICST47496.2019.9061356
5. Tipper D (2014) Resilient network design: challenges and future directions. Telecommun Syst 56(1):5–16. https://doi.org/10.1007/s11235-013-9815-x

6. Muscariello L, Carofiglio G, Papalini M (2018) System and method to facilitate robust traffic load balancing and remote adaptive active queue management in an information-centric networking environment: U.S. Patent Application No. 15/658,628
7. Lemeshko OV, Ali AS, Semenyaka MV (2012) Results of the dynamic flow-based queue balancing model research. In: 2012 IEEE international conference on modern problem of radio engineering, telecommunications and computer science, Lviv-Slavske, Ukraine, pp 318–319
8. Lebedenko T, Simonenko A, Fouad Abdul Razzaq Arif A (2016) Queue management model on the network routers using optimal flows aggregation. In: 2016 IEEE 13th international conference modern problems of radio engineering, telecommunications, and computer science (TCSET), Lviv, Ukraine, pp 605–608. https://doi.org/10.1109/TCSET.2016.7452129
9. Lemeshko O, Lebedenko T, Yeremenko O, Simonenko O (2018) Mathematical model of queue management with flows aggregation and bandwidth allocation. In: International conference on theory and applications of fuzzy systems and soft computing, Springer, Cham, pp 165–176. https://doi.org/10.1007/978-3-319-91008-6_17
10. Lebedenko T (2019) Method of scheduling and active queues management on routers interfaces of telecommunication networks. Innov Technolo Sci Solut Ind 2(8):54–61. https://doi.org/10.30837/2522-9818.2019.8.054
11. Irazabal M, Lopez-Aguilera E, Demirkol I (2019) Active queue management as quality of service enabler for 5G networks. In: 2019 IEEE European conference on networks and communications (EuCNC'2019), Valencia, Spain, pp 421–426. https://doi.org/10.1109/EuCNC.2019.8802027
12. Lemeshko O, Lebedenko T, Al-Dulaimi A (2019) Improvement of the method of balanced queue management on the routers interfaces of the telecommunication networks. In: 2019 IEEE 3rd international conference advanced information and communication technologies (AICT), Lviv, Ukraine, pp 170–175. https://doi.org/10.1109/AIACT.2019.8847749
13. Chitra K, Padamavathi G (2010) Classification and performance of AQM-based schemes for congestion avoidance. Int J Comput Sci Inf Secur 8(1):331–340
14. Lemeshko O, Lebedenko T, Nevzorova O, Snihurov A, Mersni A, Al-Dulaimi A (2019) Development of the balanced queue management scheme with optimal aggregation of flows and bandwidth allocation. In: 2019 IEEE the15th international conference the experience of designing and application of CAD system in microelectronic (CADSM), Polyana-Svalyava, Ukraine, pp 1–4. https://doi.org/10.1109/CADSM.2019.8779246
15. Okokpujie KO, Chukwu EC, Noma-Osaghae E, Okokpujie IP (2018) novel active queue management scheme for routers in wireless networks. Int J Commun Antenna Propag (IRECAP) 8(1):53–61. https://doi.org/10.15866/irecap.v8i1.13408
16. Lebedenko T, Kholodkova A, Al-Dulaimi A (2018) Linear-quadratic model of optimal queue management on interface of telecommunication network router. In: 2018 IEEE international conference on information and telecommunication technologies and radio electronics (UkrMiCo), Odessa, Ukraine, pp 1–4. https://doi.org/10.1109/UkrMiCo43733.2018.9047602
17. Rao SS (2019) Engineering optimization: theory and practice, 5th edn. Wiley
18. Lemeshko O, Semenyaka M, Simonenko O (2013) Researching and designing of the dynamic adaptive queue balancing method on telecommunication network routers. In: 2013 IEEE 12th international conference on the experience of designing and application of cad systems in microelectronics (CADSM), Polyana Svalyava, Ukraine, pp 204–207
19. An architectural framework for support of quality of service in packet networks, ITU-T Recommendation Y.1291. ITU-T (2004)
20. Network performance objectives for IP-based services. ITU-T Recommendation Y. 1541. ITU-T (2011)
21. Ramachendra GA, Reshma B, Ali ahammed GF (2014) Performance comparison of ModRED AQM with round robin scheduling. Int J Adv Res Comput Eng Technol 3(7):2560–2566
22. Yeremenko O, Lebedenko T, Mersni A (2018) Features of dynamic modeling of routers operation modes in simulink. In: 2018 IEEE fifth international scientific-practical conference problems of infocommunications. Science and Technology (PIC S&T), Kharkiv, Ukraine, pp 520–524. https://doi.org/10.1109/INFOCOMMST.2018.8632061

23. Fei Z, Xing C, Li N (2015) QoE-driven resource allocation for mobile ip services in wireless network. Sci China Inf Sci 58(1):1–10. https://doi.org/10.1007/s11432-014-5163-z
24. Avallone S, Guadagno S, Emma D, Pescapè A, Ventre G (2004) D-ITG distributed internet traffic generator. In: 2004 IEEE first international conference quantitative evaluation of systems (QEST'2004), Enschede, The Netherlands, pp 316–317
25. Kolahi S, Narayan S, Nguyen DDT, Sunarto Y (2011) Performance monitoring of various network traffic generators. In: the13th international conference computer modelling and simulation (UKSim-AMSS), Cambridge, United Kingdom, pp 501–506. https://doi.org/10.1109/UKSIM.2011.102
26. Stallings W (2016) Foundations of modern networking: SDN, NFV, QoE, IoT, and Cloud, 1st edn. Pearson Education Inc.
27. Varma S (2015) Internet congestion control. Morgan Kaufmann
28. Haghighi A, Mishev D (2016) Delayed and network queues. Wiley
29. Yeremenko O, Lebedenko T, Vavenko T, Semenyaka M (2015) Investigation of queue utilization on network routers by the use of dynamic models. In: 2015 IEEE second international scientific-practical conference problems of infocommunications science and technology (PIC S&T), Kharkiv, Ukraine, pp 46–49. https://doi.org/10.1109/INFOCOMMST.2015.7357265
30. Ramakrishna BB, Prashant A, Shrinivasa DMD (2012) A survey on new load based active queue management mechanisms. Int J Sci Eng Res 3(10):1–4
31. Romanov O, Mankivskyi V (2019) Optimal traffic distribution based on the sectoral model of loading network elements. In: 2019 IEEE international scientific-practical conference problems of infocommunications, science and technology (PIC S&T), Kyiv, Ukraine, pp 683–688. https://doi.org/10.1109/PICST47496.2019.9061296
32. John J, Balan R (2017) Priority queuing technique promoting deadline sensitive data transfers in router based heterogeneous networks. Int J Appl Eng Res 12(15):4899–4903
33. Romanov O, Nesterenko M, Mankivsky V (2016) Application of the regression model of the coefficient of use of channels for forming the plan of load distribution in the network. In: Bulletin of NTUU "KPI". Radio Engineering Series, Radio apparatus construction, No 67, pp 34–42

Improving the Structural Reliability of Mobile Radio Networks Based on the Ad-Hoc Algorithms

Leonid Uryvsky, Alina Moshynska, and Serhiy Osypchuk

Abstract The paper is devoted the description of specific features of Ad-Hoc technology implementation. The possibility of significant increase of the radio communications stability without introducing structural redundancy into the system is shown. For this, it is proposed to use mobile stations deployed on the terrain in non-stationary circumstances and operating in the retransmission mode. The use of repeaters is considered as a tool to improve the structural reliability of Ad-Hoc network, as well as a tool for reducing intra-system interference. The feasibility assessment of using automatic transmitter's power adjustment for radio stations within a decentralized mobile Ad-Hoc network is performed. The measure of intra-system interference level reduction is estimated at the receiving point based on the assumption of subscribers' random location on the terrain. It is concluded that the combination of automatic transmitter power adjustment for subscriber stations using an auxiliary repeater-station can improve the electromagnetic compatibility conditions of radio equipment in communication systems with mobile objects. This measure is an effective way to economical use of the radio spectrum allocated to communication systems with moving objects. It is important to emphasize that the operation conditions of many mobile radio communications systems do not always allow us to achieve improvements in the characteristics of stability and communication quality by generally accepted organizational and technical means. Therefore, the search for effective ways to improve radio systems operating in adverse conditions remains as an interesting and important task. In this article, the attempt has been made to show new possibilities that retransmission of signals gives precisely in such mobile systems taking into account their features.

Keywords Ad-Hoc technology · The retransmission mode · Structural reliability · Intra-system interference · Automatic transmitter's power adjustment

---

L. Uryvsky (✉) · A. Moshynska · S. Osypchuk
Department of Telecommunication Systems, Igor Sikorsky Kyiv Polytechnic Institute, Industrialnyi Lane, 2 (Campus 30), Kyiv 03056, Ukraine
e-mail: leonid_uic@ukr.net

D. Ageyev et al. (eds.), *Data-Centric Business and Applications*, Lecture Notes on Data Engineering and Communications Technologies 69,
https://doi.org/10.1007/978-3-030-71892-3_2

1 Introduction

Nowadays, a decentralized type of mobile systems is classified as Ad-Hoc networks—radio networks where randomly fixed and mobile users establish fully decentralized control over the network without base stations or backbone nodes [1–4].

They are commonly used as sensor networks in cases of special operations in emergency situations in poorly populated areas, search and rescue operations, when stationary infrastructure is removed in case of earthquakes, hurricanes and fires. Such systems have more specific name—MANET (Mobile Ad-Hoc Networks): radio networks with randomly placed mobile subscribers establish fully decentralized control over the network without base stations or backbone nodes, creating peer-to-peer networks which have rapidly changing topology with random node allocation [5–7].

The dynamic structure of MANET networks raises the question of the reliability of mutual communication between users in such network.

Currently, there is a lack of studies dedicated to mobile radio communications systems which operate on rough terrain and are used outside populated areas (in the rescue service, in agriculture, in the forest industry, etc.) [8]. The analysis of the operating conditions of radio equipment used in systems of this kind, as well as those measures that may contribute to the improvement of their performance characteristics, is not presented in sufficient depth.

This article has the following structure:

- Mobile radio features and functional requirements
- Repeaters usage principles
- The system structural reliability using repeaters;
- Measures on improving the electromagnetic environment and using efficiently the radio spectrum.

2 Features and Conditions of Functioning Radio Equipment with Ad-Hoc Technology

Ad-Hoc Radio Equipment (AHRE) is a set of distributed on the aria nodes (stationary and mobile) used for processing of the management messages from officials, collecting data from the environment and for transferring them to the control center by the retransmission between nodes.

The data received by a node data are transferred to one or several gateways straight or via relay line.

Communication systems require their previous installation and adjustment, which is often difficult to implement for the conditions of monitoring objects of critical infrastructure, in areas of natural disaster (man-made disasters). In such circumstances, the possible solution is the use of systems built on the MANET (Mobile

Ad-Hoc Networks) principle, which allows creating of a radio network capable of self-organization and node adaptation to operate in conditions that cannot be predicted during the design process.

Installation simplicity, cost-effectiveness and high efficiency of the independently self-organized repeater radio systems facilitate their wide application in various fields: industry, environmental protection, emergency services, etc.

Radio devices operate in automatic or semi-automatic mode, so nodes should be able to make decisions on managing nodal and network resources.

Ad-Hoc network devices may be heterogeneous and consist of a set of nodes (and, accordingly, subnets) of different types: stationary, mobile, air, etc., providing large area coverage.

An important feature of AHRE networks is limited resources: power, radio frequency, computing, etc.

Systems similar to AHRE, have the following basic requirements [5–7]:

- monitoring the specified objects (zones) at the given time;
- transferring various types of traffic (data, video, voice messages) with a given quality;
- automation of network management processes;
- provide adaptive and distributed operation of the network with the possibility of its self-organization;
- making real or near to real time decisions;
- network characteristics optimization (maximization of its lifetime, minimal traffic load of network with service information, coverage area maximization, nodes energy efficiency, minimization of node transmission power, number of active nodes minimization, topology and connectivity of active nodes optimization, etc.) towards achieving the desired degree of stability.

System should be considered as stable, when it has the possibility to perform the tasks under the influence of interfering factors.

Since the systems under consideration are systems with a random structure which changes in time, then for them quantified index of stability (security, at the same time) can be characterized by comparing the probability of a P_C event, which is related to the possibility of organizing an arbitrary pair of subscriber's mutual exchange channel.

A model representing a system with a variable topology is a random graph the structure of which is described by the set of nodes $\{1, 2, \ldots, N\}$ and the set of simplistic situations $\{\xi ij, i, j = 1,2, \ldots, n\}$. The ξij denotes an event consisting of the presence of a directed edge between nodes i and j. If ξij is independent equivalence events, then $p\{\xi ij\}$ corresponds to the probability that any node i in a graph is connected by an edge with some given node j, and

$$p\{\xi_{ij}\} = 1/n. \tag{1}$$

For a real system, the condition (1) means that the probability of a direct exchange channel between a given subscriber's communication device (SCD) and any SCD system is the same for all SCDs.

If δ_i is the number of nodes for which there is a path to them from the node i, then equality

$$\gamma = M\{\delta_i / n\} \tag{2}$$

corresponds to the probability that an arbitrary node of a graph is associated with any other node of this graph in the absence of restrictions on the number of intermediate nodes. Then for a system in which the amount of relay is not limited, $Pc = \gamma$.

In real circumstances, the number of retransmissions is limited. Therefore, in the modeling graph, it is expedient to use the index γ(m), that corresponds to the probability of the connection of two arbitrary nodes connected by no more than m ($m \geq 0$) intermediate nodes. In this case, $Pc = \gamma(m)$.

The value of $Pc = \gamma(0)$ corresponds to the connectivity index of a system in which only direct links between objects are allowed, that is, without retransmitting SCDs.

If we denote $P(k)$ as the probability that some node is distant from an arbitrarily chosen node of a random graph exactly for k edges, under condition (1), then the parameter $\gamma(m)$ can be expressed through this probability in the following way:

$$\gamma(m) = \sum_{k=1}^{m-1} P(k), \tag{3}$$

Mathematical relation between the above arguments may be found based on known procedure [9]:

$$\gamma = 1 - e^{-\alpha\gamma}, \tag{4}$$

$$\alpha = \frac{-\ln(1-\gamma)}{\gamma}, \tag{5}$$

where α is the average number of nodes which have a direct connection to an arbitrary node of the graph.

Determining the network connectivity index, as the probability of having a path for the transfer of information between any nodes of the system, it is assumed that each node of a random graph has α edges for connecting to adjacent nodes.

Consequently, with known values of α and n, the system connectivity can be estimated by assuming an unlimited number of relays—in accordance with (1) and (4).

Features of systems with AHRE are influenced by natural factors (propagation mechanism radio waves of the working range; terrain and the presence of all sorts of obstacles; dominant interference) as well as organizational and technical factors (nature and density of station placement on the terrain; structure of the organized relations; additional capabilities embedded in the equipment to improve system performance).

Systems with AHRE deployed outside human settlements, on rough terrain are characterized by a well-defined set (combination) of such factors.

For the systems under further consideration, this set includes the following key components.

The terrain on which the related systems of geological expeditions, forestry brigades, agricultural units, etc. perform can be very different and take into account its influence in general terms is impossible.

It is difficult to take into account the effect of natural interference because in the microwave range the main mechanism for the propagation of electromagnetic energy from the moving stations is the surface waves. Such obstacles as uneven terrain, forest, etc., are often considerable compared to the wavelength. In this case beyond the obstacle a "dead zone" arises, in which the signal field cannot be determined using radiation theory. To systematize the study of the operating conditions of mobile radio stations on natural terrain, a method of modeling typical terrains is proposed in Sect. 3.

The predominant type of natural interference is cosmic noises, representing a very stable in time fluctuation process. The effect of atmospheric noise can be neglected, because at frequencies above 100 MHz, this interference has little effect on the radio communication systems operation.

The noise level of the receiver, among other factors, depends on the operating range of the stations.

The influence of mutual (system) interference is associated with organizational and technical problems, which will be discussed below.

Placement of stations on the terrain significantly affects the stability of the mobile radio system, that is, its ability to maintain its performance under all influencing factors. Assuming that at any time the subscriber stations of the system need to communicate with each other, being at different distant from each other (not exceeding the limiting distance of communication $\boldsymbol{r}_{\max}$) in theoretical and engineering calculations the placement of stations is described through the distribution functions or the probability density of the distances at which the communication is carried out. Wherein the distribution functions of the useful signal voltage and mutual interference voltage at each point of reception depends on the actual distribution function of the distance between subscriber stations on the terrain.

Accurate data on the statistics of mutual deletions of stations in systems for various purposes are not available. Therefore, in necessary cases, the characteristics of the mobile subscribers distribution on the terrain are set.

The following assumptions are useful for practical use:

- all distances between subscribers in the interval $\boldsymbol{\rho} = \boldsymbol{r}_{\min} \ldots \boldsymbol{r}_{\max}$ are equally likely.

The probability density of mutual distances between stations in this case is equal to:

$$\omega(\rho) = \frac{1}{\mathrm{r}_{\max} - \mathrm{r}_{\min}}; \tag{6}$$

- subscribers are evenly distributed on the ground. In this case, it is fair.

$$\omega(\rho) = \frac{2\rho}{r_{max}^2 - r_{min}^2}; \tag{7}$$

- some distance $\check{R}$ is most likely in a given interval of distances; in this case, it is possible to use the assumption of the Rayleigh distribution of mutual deletions of subscribers:

$$\omega(\rho) = \frac{\rho}{\sigma^2} e^{-\frac{\rho}{2\sigma^2}} \tag{8}$$

where $\check{R} = \sigma$ is distribution mode (argument corresponding to the maximum distribution density).

Smetric about Ř looks density distribution:

$$\omega(\rho) = \sqrt{\frac{2}{\pi} \frac{\rho^2}{\sigma^2}} e^{-\frac{\rho^2}{2\sigma^2}} \tag{9}$$

In (9) $\check{R} = \sigma$, where σ is the standard deviion of the random variable ρ.

Among the most used additional capabilities of the station equipment contributing to the improvement of the radio communication system performance as a whole, includes automatic retransmission, automatic adjustment of the radiation power, etc.

Due to these features, the mutual interference level in the system is reduced, the communication range is increased or the level of the useful signal at the point of reception is increased, the reliability of communication and its resistance to various throwing factors are increased too.

Thus, the Ad-Hoc communication systems characterized by the features described in this section, are of interest both in terms of prospects and practical value, being in some cases as the only means of communication in the specified conditions of functioning, as from the standpoint of the complexity of solving engineering problems arising in the development and operation of such systems.

3 Principles of Repeaters Use in Ad-Hoc Networks

Most mobile communications systems include repeaters as an integral part of the system.

In such systems, among intermediate stations there is no the main or central one. All these stations are equivalent, that is, their functions as repeaters are not differentiated by any attribute. Therefore, as a rule, systems with a core network of equivalent stations (not necessarily stationary) are called integrated. The advantage of such systems is the possibility of establishing a connection between any pair of subscribers on any of many paths which are different in number and composition of intermediate stations—repeaters participating in each chosen path.

In this systems, the principle of organization of communication "every with each" (or "peer-to-peer") is used, when communication between any pair of subscribers can be established only in one way, and only this pair of subscribers participates in the process of establishing communication. The implementation of this principle provides additional opportunities for establishing communication between subscribers, if it is impossible to provide direct communication between them.

Under certain conditions of providing communication, serviceable technical means and subscribers' readiness for information exchange, conditions for direct communication may not be fulfilled for the following reasons:

(a) the distance between subscribers exceeds the permissible $\boldsymbol{r_{max}}$, due to the technical characteristics of the equipment

$$r_{\max}^2 = \frac{k \cdot \sqrt{P_{\max}}}{U_{\min}}. \tag{10}$$

where

k coefficient taking into account the characteristics of the antenna devices, as well as the attenuation of the radio wave into the propagation medium (air) over a smooth surface;
$P_{\max}$ the maximum power of the transmitter;
$U_{\min}$ minimum voltage detected by the receiver (sensitivity).

(b) the presence of natural obstacles on the terrain breaks the conditions of direct visibility, and the diffracting radio waves do not provide the required signal level at the subscriber's receiver input;
c) the conditions of the electromagnetic environment do not allow the normal reception of signals of the subscriber who transfers the information.

Traditionally, in cases where providing a direct connection is hampered by the first of these reasons, signals through stations located between the leading subscribers are used.

In the absence of additional, support stations the transfer can only be facilitated by an exchange-free station equipped with appropriate devices (for example, a relay unit) subscriber station system.

If it is known, the law of distribution of subscribers on the terrain, given, for example, through the probability density $\boldsymbol{\omega(\rho,\varphi)}$ of mutual distances between subscribers by the azimuth $\boldsymbol{\varphi}$, then you can determine the probability $\boldsymbol{p_R}$ that the station will be able to provide retransmission, that is, it is in the communication zone of two subscribers, the removal of R between which satisfies the condition:

$$r_{\max} \leq R \leq 2 \cdot r_{\max} .$$

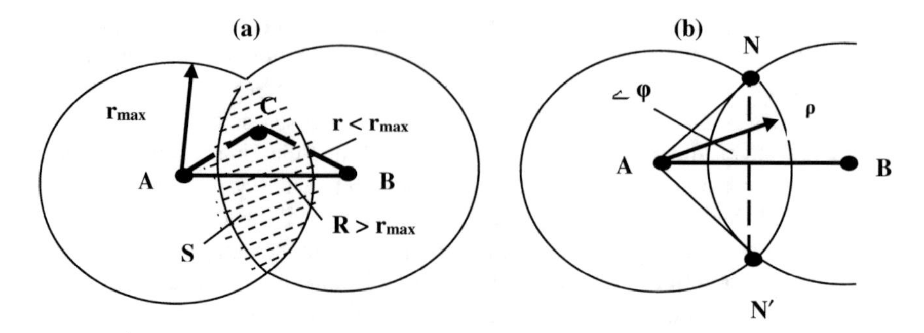

Fig. 1 The geometric model for determining the location of a station (C) that is capable of retransmission

To find the probability p_R, we use constructions on the plain surface (Fig. 1).

Taking into consideration that the main type of antennas installed on mobile objects is a whip antenna, in the subsequent antenna patterns of mobile stations in the horizontal plane are displayed by a circle of the corresponding radius.

From Fig. 1a it is shown that the repeater station (point C) should be located in the hatches area.

The seeking probability is determined as an integral over a given area of the known distribution function of mutual deletions of subscribers (in polar coordinates):

$$p_r = \int\limits_S \omega(\rho, \phi)\, d\rho\, d\phi\,. \tag{11}$$

Taking the point A as the center of the system of polar coordinates (Fig. 1b), it is easy to determine that for the case shown in Fig. 1a, the limits of change in the phase φ of the vector $\boldsymbol{\rho}$: $\left\{-\arccos\left(\frac{R}{2r_{\max}}\right); +\arccos\left(\frac{R}{2r_{\max}}\right)\right\}$.

Modulus change $\rho = |\rho|$ occurs within:

$\left\{R\cos\varphi - \sqrt{r_{\max}^2 - R^2\sin^2\varphi}; r_{\max}\right\}$.

Then the formula (11) is converted to the form:

$$p_{r0} = \int\limits_{-\arccos\left(\frac{R}{2r_{\max}}\right)}^{\arccos\left(\frac{R}{2r_{\max}}\right)} \left[\int\limits_{R\cos\varphi-\sqrt{r_{\max}-R^2\sin^2\varphi}}^{r_{\max}} \omega(\rho,\varphi)d\rho\right] d\varphi \tag{12}$$

Formula (12) is right when (Fig. 1b).

$$\angle NAN\prime \leq \frac{\pi}{2} \tag{13}$$

that is, when segment ***AN*** does not intersect the arc ***N–N′***.

Condition (13) is satisfied for

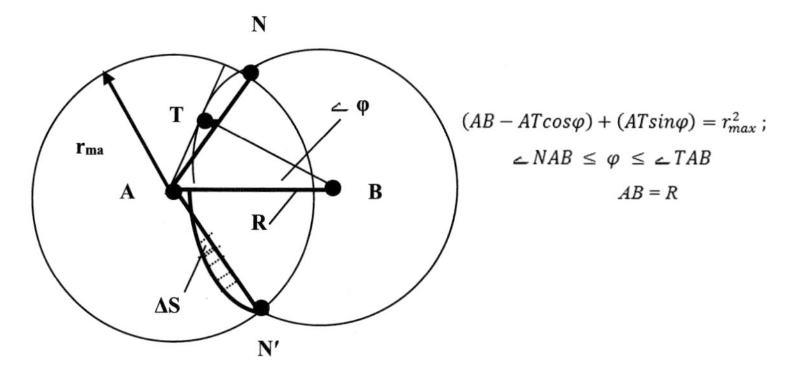

Fig. 2 The geometric model for determining the location of a station that is capable of retransmission when subscribers come close to each other

$$\sqrt{2} \cdot r_{\max} \leq R \leq 2 \cdot r_{\max} . \tag{14}$$

$$r_{\max} \leq R \leq \sqrt{2} \cdot r_{\max} , \tag{15}$$

when $\angle NAN\prime > \frac{\pi}{2}$, then

$$p_r = p_{r0} + \Delta p_r , \tag{16}$$

where (Fig. 2)

$$\Delta p_R = 2 \cdot \int\limits_{(\Delta S)} \omega(\rho, \phi)\, d\rho\, d\phi \ .$$

From Fig. 2 it follows that in formula (16) the limits of change in phase $\boldsymbol{\varphi}$.

$$\left\{ \arccos\left(\frac{R}{2r_{\max}}\right); \arcsin\left(\frac{r_{\max}}{R}\right) \right\},$$

and the limits of variation of the argument $\boldsymbol{\rho}$ are defined as the roots of the equation: $(R - x\cos\varphi)^2 + (x\sin\varphi)^2 = r_{\max}^2$,

where $\boldsymbol{x}$—length of the segment connecting the point A with the intersection point of the segment AT with the circle centered at the point B.

Seeing

$$x_{1,2} = R\cos\varphi \mp \sqrt{r_{\max}^2 - R^2\sin^2\varphi},$$

then

$$\Delta p_r = 2 \int\limits_{\arccos\left(\frac{R}{2r_{\max}}\right)}^{\arcsin\left(\frac{r_{\max}}{R}\right)} \left[\int\limits_{R\cos\varphi - \sqrt{r_{\max} - R^2 \sin^2\varphi}}^{R\cos\varphi + \sqrt{r_{\max} - R^2 \sin^2\varphi}} \omega(\rho, \varphi) d\rho \right] d\varphi \qquad (17)$$

In the particular case when subscribers are evenly distributed on the terrain, the probability $\boldsymbol{p_R}$ is determined by the ratio of the area of a circle of radius $\boldsymbol{r}_{\max}$ to the area of the figure shown in Fig. 1a by hatching:

$$p_R = \frac{S}{\pi \cdot r_{\max}^2} = \frac{2 \cdot \left[r_{\max}^2 \cdot \left(\arccos \frac{R}{2r_{\max}} - 0,5 \cdot \sin\left(2 \arccos \frac{R}{r_{\max}}\right) \right) \right]}{\pi \cdot r_{\max}^2}. \qquad (18)$$

The formula in square brackets of the numerator is the area of the segment, shown by dashed lines, cut off by the straight line AN' from the circle with center B and radius $\boldsymbol{r}_{\max}$.

Finally

$$p_R = \frac{2}{\pi} \cdot \left[\arccos \frac{R}{2 \cdot r_{\max}} - 0,5 \cdot \sin\left(2 \cdot \arccos \frac{R}{2 \cdot r_{\max}}\right) \right]. \qquad (19)$$

So, the inculcation in subscriber stations devices that retransmit signals, will allow temporarily using free from the information exchange station as a repeater, and thus avoiding the structural redundancy of the system. The probability that such a repeater station will be in the communication zone of interested subscribers can be determined by the formulas (12), (16), (17).

Below it will be shown that the implementation of this measure can also significantly decrease the influence of other factors interfering with the establishment and maintenance of the connection, mentioned at the beginning of this section.

4 Improving Structural Reliability by Mean of Repeaters

Often, natural obstacles cause communication breakdowns in the process of moving objects, as well as in the parking lot, even if the distance between subscribers is small. Then, with a certain probability, communication can be provided through a subscriber station which is located in the communication zone of two subscribers and switching to retransmission mode.

Thus, changing the structure of the information load passage can increase the reliability of the information system as a whole.

The apparatus of the graph theory is a convenient tool for studying systems with repeaters. In the traditional formulation of the question, a random planar graph in the form of a lattice, the nodes of which are displayed by the subscriber stations, and

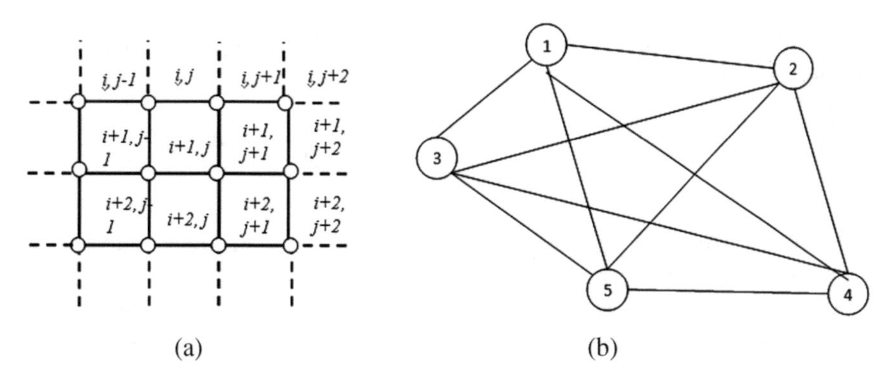

Fig. 3 Planar graph in the form of a lattice (**a**) and in the form of a fully-connected graph (**b**)

the branches connecting the nodes are the radio links between stations (for example, as in Fig. 3a becomes the object of research in this case. The reliability of a system represented by such a graph is characterized by the connectivity of its elements, that is, the probability of a path connecting two arbitrary nodes (passing in the general case through other nodes). In this regard, the problem of structural reliability of an information system with transponders is not new. However, in this case, this task has its own characteristics.

First of all, the implementation of the principle of mutual linking of the AHRE "ever with each" requires using of adequate quality model in form of a fully-connected graph (Fig. 3b), but not in form of a lattice (Fig. 3a).

The number of paths from node $\boldsymbol{i}$ to node $\boldsymbol{j}$ in indirect complete (fully-connected) graph N equals

$$S_N = 1 + (N-2) + (N-2)\cdot(N-3) + \cdots + (N-2) \cdot (N-3)\cdot \ldots (N-M-1), \quad (20)$$

where $\boldsymbol{M} = N - 2$ is maximum allowed number of intermediate nodes.

In Sect. 2, using the formula (4), the probability of the existence of a path between any two SCDs is determined, when in any accidental network structure each object has an average of α related objects in the absence of restrictions on the number of retransmissions, then:

$$\gamma = 1 - e^{-\alpha\gamma}.$$

The exact calculation shows that even with the value of the indicator $\alpha = 2$, the probability of the existing path is $\gamma = 0.8$, and with $\alpha = 3$ $\gamma = 0.94$.

At the same time, it is impossible to avoid restrictions on the number of retransmissions in the real system, and, therefore, on the structures of possible paths between the nodes in the graph.

Provided that the number of retransmissions should not be greater than ***m***, the number of paths from the node ***i*** to the node ***j*** decreases:

$$S_N = 1 + (N-2) + (N-2)\cdot(N-3) + \cdots + (N-2) \\ \cdot (N-3)\cdot \ldots (N-m-1). \tag{21}$$

It follows from (21) that with the increase in the number of retransmissions ***m*** the value of $\boldsymbol{S_{N(m)}}$ increases, and consequently, the probability of the connection of the selected pair of nodes increases (with non-zero probabilistic weights of the graph branches), but the procedures for constructing the message transmission paths becomes complicated.

The ratio (4) can be refined for the case of a limited number of retransmissions in a network with random structure.

We denote by $P(k)$ the probability that this node was found at the ***k***-th step for the first time since the beginning of the procedure for finding adjacent nodes (this probability coincides with the probability $p(k)$ to distance this node from an arbitrary node exactly on the ***k*** branches), wherein

$$P(0) = p(0) = \frac{1}{n}, \quad q(0) = 1 - \frac{1}{n}$$

For ***k***-th step is valid

$$P(k) = p(k)\cdot\left[\prod_{s=0}^{k-1} q(s)\right]. \tag{22}$$

On the other hand

$$P(k) = p(k)\cdot\left[1 - \sum_{s=0}^{k-1} P(s)\right]. \tag{23}$$

The object in square brackets (22) and (23) defines the probability that this node was not encountered on all previous $(k-1)$ steps.

It can be shown that the parameter $\gamma(m) = \mathrm{P}(m)$, in contrast to the index γ (4), and can be calculated by the relation:

$$P(m) = \left[1 - \sum_{s=0}^{m-1} P(s)\right]\cdot(1 - e^{-\alpha\cdot P(m-1)} \tag{24}$$

Consequently, with the known values of α and n, the system's connectivity can be evaluated by assuming an unlimited number of r retransmissions—in accordance with (4), and in the case of number of r retransmissions not exceeding m—according to (24).

There is an opportunity to supplement the outlined approach with additional indicators of structural stability of the system.

Thus, in the graph of the type shown in Fig. 3b, we denote by $\boldsymbol{p_{ij}}$ the probability that between the subscriber stations represented graphs by the nodes $\boldsymbol{i}$ and $\boldsymbol{j}$ there is a direct radio channel. The probability of the opposite event is denoted by $\boldsymbol{q_{ij}}$. The probability that a station that is represented by a node $\boldsymbol{S}$ can act as a repeater is denoted by $\boldsymbol{p_S}$.

Then, the probability of having at least one path between the nodes $\boldsymbol{i}$ and $\boldsymbol{j}$ corresponding to the radio link is equal to

$$Pij = 1 - q_{ij} \cdot \left[\prod_{s=1}^{N-2} (1 - p_{is} \cdot p_s \cdot p_{sj}) \right] \tag{25}$$

In the general, the values of $\boldsymbol{p_{iS}}$ and $\boldsymbol{p_S}$ can be given on the basis of the data of field tests or determined either analytically or according to the results of the test of a statistical model.

According to the statement of the problem which was formulated at the beginning of the section, with sufficient accuracy the probability of $\boldsymbol{p_{iS}}$ equals to the probability of the presence of direct visibility between two subscriber stations operating in a territory with known relief. The refined probability value of $\boldsymbol{p_{Si}}$ can be achieved by taking into account the influence of the interference fading of the signal at the receiving point, the change of the operating heights of the antennas by moving objects due to the fluctuations of the body of the object and the antennas themselves and other factors.

Statistical methods are the most acceptable to determine $\boldsymbol{p_{Si}}$.

Within the bound of the task, the correspondence of the obtained probabilistic characteristics to the real conditions depends on how well the model of the statistical experiment and especially the model of the terrain have been chosen.

There is a reference to some variants of simulation of inequalities of the earth's surface of varying complexity in [10].

A relatively simple version of simulation of typical reliefs can be proposed. The terrain is represented in the form of a plane divided into many adjoining cells. Using the statistical characteristics of a particular type of terrain, the random law determines the positions of the cells corresponding to the tops of the unevenness of the relief. Each top is fitted with some height and angle of inclination of the rays, using the values that cells in the vicinity of the cell—" elevations" are assigned to "excess" above the original plane level. The conjugation of all the irregularities of such a relief is achieved at the highest value of the heights assigned to each cell.

For a relief characterized as a hilly plain, statistics on the number of tops $\boldsymbol{n}$ in the area of 10×10 km^2, the angles of slopes inclination $\boldsymbol{\alpha}$ and the relative heights of individual tops $\boldsymbol{\Delta H} > 20$ m were compiled on topographic maps. The results are presented in the form of the histograms in Fig. 4a–c, where the function of these characteristics is the frequency **P** of observations of various values of $\boldsymbol{n}$, $\boldsymbol{\alpha}$, $\boldsymbol{\Delta H}$.

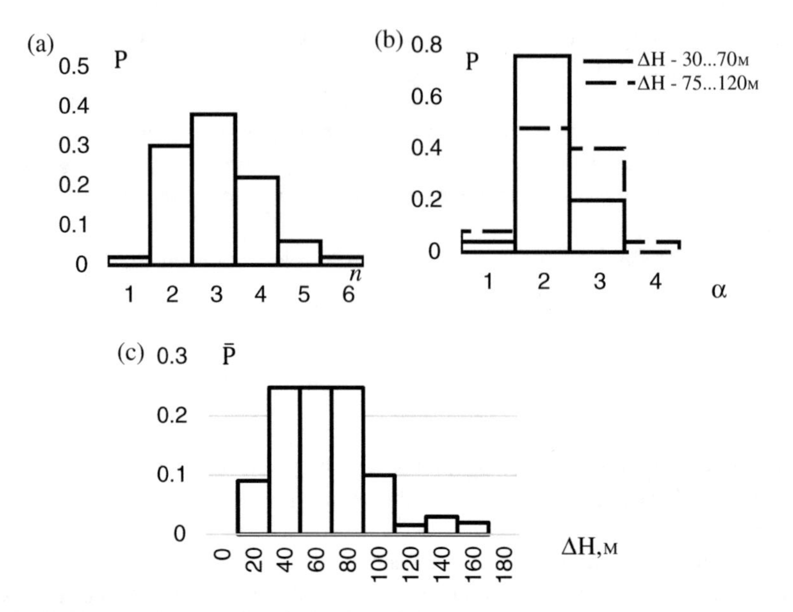

Fig. 4 Histograms of probability distribution of the number of tops ***n*** (**a**), an angles of slopes inclination α (**b**) and relative heights of individual tops ΔH (**c**)

When the actual situation is displayed, within each considered ***i***-th component of graph, the probability $\boldsymbol{p_{is}}$ and $\boldsymbol{p_S}$ can be different for different S = 1,*N* − 2 (for example, it is a function of mutual removal of objects).

However, it is possible to qualitatively estimate the effect of the use of a retransmission compared to the case of direct ties only with a simplified assumption, namely: the probability of the presence of a connection between two arbitrary subscriber stations displayed by the nodes of the graph is constant and equal to $\boldsymbol{p_i}$, the probability of the possibility of using any node as an intermediate also constant and equal to $\boldsymbol{p_S}$.

Then

$$P_{ij} = 1 - q_1 \cdot [1 - p_1^2 \cdot p_2]^{N-2} \tag{26}$$

The results of the calculations $\boldsymbol{p_{ij}}$ using the formula (26) for $\boldsymbol{N}$ = 1 … 20, $\boldsymbol{p_1}$ = 0.1 … 0.7 and $\boldsymbol{p_2}$ = 0.1 … 1.0 are shown in Fig. 5a, where it is evident that the reliability of communication establishment increases with an increase in the number of possible roundabouts (with an increase in ***N***). However, the velocity of the probability $\boldsymbol{p_{ij}}$ for small $\boldsymbol{p_1}$ grows almost linearly; for $\boldsymbol{p_1}$, close to 1.0—faster, quickly reaching "saturation": $\boldsymbol{p_{ij}}$ = 1 − 0 [***N***].

Let's use the win indicator ***Q***, which is to be determined through the ratio of probabilities of nodes (***i***, ***j***) connectivity for the retransmissions and without their:

$$Q = \frac{P_{ij}(p_1, p_2, N)}{p_1} \tag{27}$$

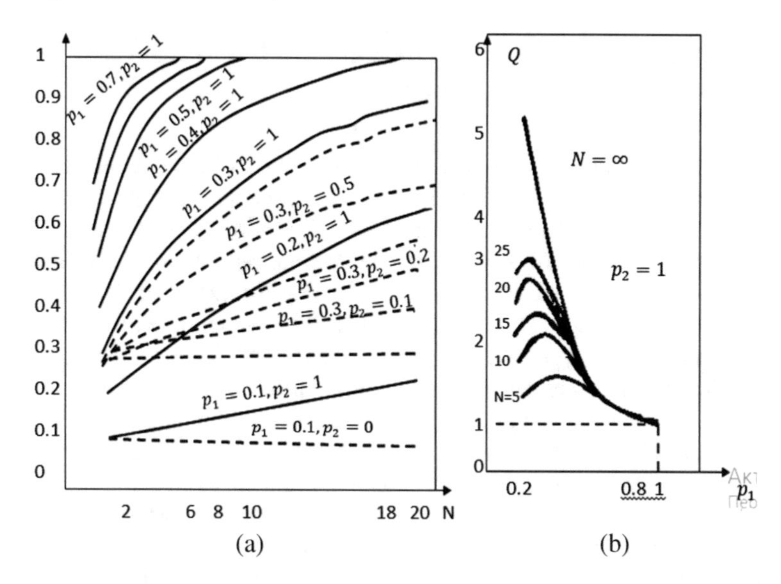

Fig. 5 The probability p_{ij} of the presence of the connection between two arbitrary network objects (**a**) and the indicator Q from the use of retransmissions (**b**)

The dependency $Q = (p_1, N)$ with $p_2 = 1.0$ is presented in Fig. 5b.

In Fig. 5b it is shown that the value of Q for N ∞, when p_{ij} 1.0, and $Q \rightarrow 1/p_1$.

The dependency $Q = (p_1, 1, \infty)$ has no extremum, when N are finite the area of p_l appears, for which the profit from retranslation is maximum. This area consistently covers $p_1 = 0.15 \ldots 0.5$ with a change in N in a fairly wide range.

Hence the conclusion is made on the expediency of using the proposed method to increase the structural reliability of the system in those cases where the probabilities of direct links between subscriber stations in the system are far from the limit. In the absence of conditions for direct communication between subscribers, a single retransmission does not significantly improve the characteristics of the connectivity of system elements; with the stable work of direct links, the proposed measure is superfluous.

Thus, providing a retransmission mode in subscriber stations deployed *i* on rough terrain, it is possible to improve the stability characteristics of communication systems with mobile objects without introducing structural redundancy into them. At the same time, the high reliability of the system is achieved only with a high degree of automation of the processes of establishing direct and by-pass communications. Otherwise, those features that give additional use of the retransmission mode will not be realized.

5 Measures to Improve Electromagnetic Facilities and Economic Use of Radio Spectrum

It is known that among the possible reasons for the lack of mutual communication between subscribers in systems that implement the principle "every with each", the influence of the parameters of the electromagnetic environment on the stability of individual directions plays a significant role. We are talking about the impact on the receiving devices of subscriber stations, leading the exchange of information, interference of natural and artificial origin, as well as mutual (system) interference.

For the first of these types of interference is characteristic, on the one hand, their identical effect on the receivers of all subscriber stations operating in a certain territory in the same frequency range. On the other hand, the effect of these interferences does not depend on the internal state of the system (on the number of emitting stations, their mutual position, structures of organized connections, etc.).

This cannot be said about the mutual interference arising in the system during its operation. The influence of these interferences on the conditions for providing communications in the near radio communication systems of many correspondents in a limited area is especially noticeable.

The known way to reduce significantly the level of mutual interference is the use of automatic transmitter's power adjustment (ATPA) of subscriber stations.

As a measure of quantifying the benefits of systems with adjustable radiation power compared to systems with a constant transmitter power, the coefficient β can be used, that shows how many times the average values of interfering voltages at the receiving point are reduced when using ATPA:

$$\beta = \frac{m(U)_{reg.}}{m(U)_{unreg.}} \tag{28}$$

The initial dependence that connects the voltage of the signal at the receiver input U, the transmitter power P and the distance r between them is the formula of Vvedensky [6]:

$$U = \frac{k\sqrt{P}}{r^2} \tag{29}$$

where $\boldsymbol{k}$—coefficient taking into account the influence of the propagation medium and the properties of the antennas.

This formula displays the propagation of USW signals near a smooth surface for a wide range of distances $\boldsymbol{r}_{\mathbf{min}} \leq \boldsymbol{r} \leq \boldsymbol{r}_{\mathbf{max}}$, где $\boldsymbol{r}_{\mathbf{min}} = (5 \ldots 3) * \boldsymbol{\lambda}$, a $\boldsymbol{r}_{\mathbf{max}}$ is limited by the curvature of the Earth's surface.

The method of determining the coefficient β can be traced by the example of a system in which subscribers are evenly distributed on the terrain according to (7).

It is known, the probability density of one from two random values x and y, related by the functional dependence $\boldsymbol{y} = \boldsymbol{f}(\boldsymbol{x})$ can be determined through another known density:

$$\omega(x) = \omega(y)\frac{dy}{dx} \tag{30}$$

With the known density $\boldsymbol{w}(\boldsymbol{\rho})$ it is easy to determine the probability density of interfering voltages $\boldsymbol{w}(\boldsymbol{U})$ in systems without power control at $\boldsymbol{P} = \boldsymbol{P}_{\text{макс}}$:

$$\omega(U) = \frac{k\sqrt{P_{\max}}}{r_{\max}^2 - r_{\min}^2}\frac{1}{U^2}, \tag{31}$$

where $\boldsymbol{k}$—is the coefficient taking into account the characteristics of antenna devices, as well as the attenuation of the radio wave in the propagation medium (air) over a smooth surface.

The average voltage at the point of reception:

$$m(U) = \int\limits_{U_{\min}}^{U_{\max}} U\omega(U)dU = \frac{k\sqrt{P_{\max}}}{r_{\max}^2 - r_{\min}^2}\ln\frac{U_{\max}}{U_{\min}} \tag{32}$$

where

$$r_{\min} = \sqrt{\frac{k\sqrt{P_{\max}}}{U_{\max}}\omega(y)\frac{dy}{dx}} \tag{33}$$

For systems with adjustable radiation power, the power $\boldsymbol{P}$ of each transmitter varies, being a function of the distance between subscribers, so as to create a voltage $\boldsymbol{U}_{\min}$ at the input of the subscriber receiver.

With restrictions $\boldsymbol{r}_{\min} \le \boldsymbol{\rho} \le \boldsymbol{r}_{\max}$ and $\boldsymbol{P}_{\min} \le \boldsymbol{P} \le \boldsymbol{P}_{\max}$, it is convenient to express the average value of the mutual interference voltage at the reception point in terms of the probability density of the transmitter power $\boldsymbol{\psi}(\boldsymbol{P})$. In this case, the formula (28) can be reduced to the following form:

$$\beta = \frac{\sqrt{P_{\max}}}{\int_{P_{\min}}^{P_{\max}}\sqrt{P}\psi(P)dP}, \tag{34}$$

From formulas (31), (30) in consideration (29) and (33) follows

$$\psi(\rho) = \frac{1}{2\sqrt{P}\left(\sqrt{P_{\max}} - \sqrt{P_{\min}}\right)} \tag{35}$$

Substituting (35) into (34) we have

$$\beta = \frac{2\sqrt{P_{\max}}\left(\sqrt{P_{\max}} - \sqrt{P_{\min}}\right)}{\int_{P_{\min}}^{P_{\max}} dP} = \frac{2\sqrt{P_{\max}}}{\sqrt{P_{\max}} - \sqrt{P_{\min}}} \cong 2,\ P_{\max} \gg P_{\min} \tag{36}$$

The obtained values of $\boldsymbol{\beta}$ and the distribution function of interfering voltages $\boldsymbol{F}(\boldsymbol{U})$ for systems with an ATPA, where the probability densities of mutual distances of subscribers correspond to (3) and (4), are summarized in Table 1.

From the formulas for $\boldsymbol{F}(\boldsymbol{U})$ it follows that with $\boldsymbol{U} = \boldsymbol{U}_{\mathbf{min}}$ for each type of distribution $\mathbf{F}(\mathbf{U}_{\mathbf{min}}) \sim \mathbf{1}/\boldsymbol{\beta}$.

This means that the number of transmitters the signals of which at the receiving point exceed the receiver's sensitivity $\boldsymbol{U}_{\mathbf{min}}$ in systems with adjustable power is $\boldsymbol{\beta}$ times less than the total number of transmitters evenly distributed around the reception point in radius $\boldsymbol{r}_{\mathbf{max}}$.

The effectiveness of the automatic transmitted power adjustment (ATPA) use can be improved by implementing structural changes in the information load transmission path, by using retransmission through a free subscriber station system.

If a certain station $\boldsymbol{B}$ (Fig. 6a) receives signals from station $\boldsymbol{A}$ operating on a whip antenna, the area in which station A inrferes with other stations not less than $\boldsymbol{U}_{\mathbf{min}}$ can be approximated by a circle of radius $\boldsymbol{AB} = \boldsymbol{R}$.

If communication in the same direction is carried out through a repeater $\boldsymbol{T}$ (Fig. 6b), then with the ATPA of the initiator station and the repeater stations, the area where mutual interference will occur can be reduced (the interference zone is shown by dimming). An additional effect is always achieved when the repeater is located on a straight line connecting two terminal stations. In general, the existence of this effect depends on the position of the repeater (Fig. 6c, d).

From the geometric interpretations of the problem, it follows that the gain from using retransmission in combination with the ATPA is achieved when the ratio $\boldsymbol{Y}_{\boldsymbol{S}}$ of the area of the figure formed by intersecting the radiation patterns of stations $\boldsymbol{A}$ and $\boldsymbol{T}(\boldsymbol{S}_{\Sigma})$ to the area of a circle of radius $\boldsymbol{R}(\boldsymbol{S}_0)$ does not exceeds 1, that way.

$$Y_S = \frac{S_\Sigma}{S_0} \leq 1, \tag{37}$$

By analyzing constructions similar to those shown in (Fig. 6a–d), it was determined that for all $\boldsymbol{AT} = \boldsymbol{R1}$ and $\boldsymbol{TB} = \boldsymbol{R2}$ within a circle with radius $\boldsymbol{R}$, the following relations are true:

Table 1 Distortion voltage distribution functions

Distribution functions	β	$F(U)$
$\omega(\rho) = \frac{1}{r_{\max} - r_{\min}}$	3	$\frac{1}{3}\frac{U_{\min}}{U_{\max}}\left[\left(\frac{U_{\max}}{U} - 1\right) - 2\left(1 - \sqrt{\frac{U}{U_{\max}}}\right)\right]$
$\omega(\rho) = \frac{2\rho}{r_{\max}^2 - r_{\min}^2}$	2	$\frac{1}{2}\frac{U_{\min}}{U_{\max}}\left[\left(\frac{U_{\max}}{U} - 1\right) - \left(1 - \frac{U}{U_{\max}}\right)\right]$

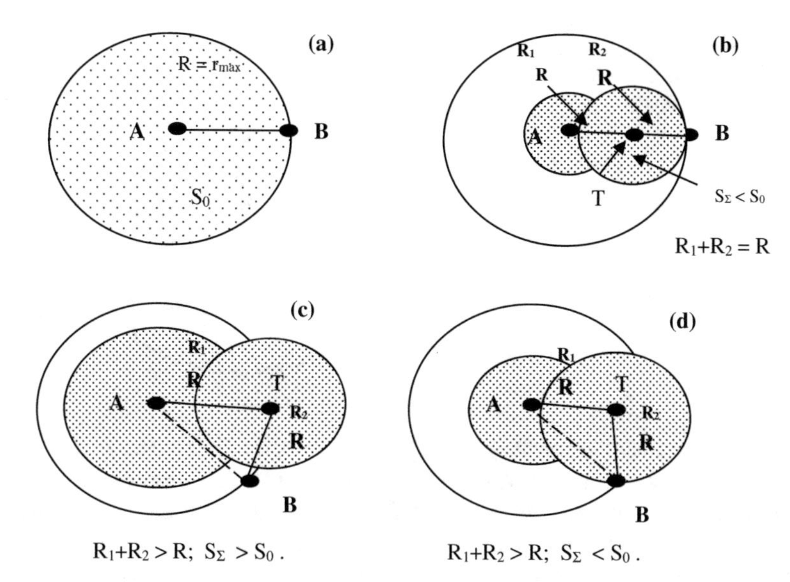

Fig. 6 The options for the relative position of subscribers and repeater

- $R_2 = \sqrt{(R - R_1\cos\alpha) + R_2\sin^2\alpha}$. где α = ▶TAB.
- при $0 < R_2 \le R\sqrt{2}$:

$$S_\Sigma = \pi R_1^2 + \pi R_2^2 + R_1^2\left\{\arccos\left(1 - \frac{R_2^2}{2R_1^2}\right) - \frac{1}{2}\sin\left[2\arccos\left(1 - \frac{R_2^2}{2R_1^2}\right)\right]\right\} - R_2^2\left\{\arccos\left(\frac{R_2}{2R_1}\right) - \frac{1}{2}\sin\left[2arccos\left(\frac{R_2}{2R_1}\right)\right]\right\};$$

- при $R\sqrt{2} < R_2 < 2R_1$:

$$S_\Sigma = \pi R_2^2 + R_1^2\left\{\arccos\left(\frac{R_2^2}{2R_1^2} - 1\right) - \frac{1}{2}\sin\left[2\arccos\left(\frac{R_2^2}{2R_1^2} - 1\right)\right]\right\} - R_2^2\left\{\arccos\left(\frac{R_2}{2R_1}\right) - \frac{1}{2}\sin\left[2\arccos\left(\frac{R_2}{2R_1}\right)\right]\right\};$$

- при $R_2 \ge 2R_1$: $S_\Sigma = \pi R^2$.

As a result of solving, using a computer, a set of transcendental equations of the form:

$$\boldsymbol{S}_\Sigma/\boldsymbol{S}_0 = \boldsymbol{Y}_S(\boldsymbol{Y}_S \le 1)\text{ for }\boldsymbol{R}_1 = 0\ldots\boldsymbol{R}\text{ and }\boldsymbol{\alpha} = -90^\circ\ldots 90^\circ$$

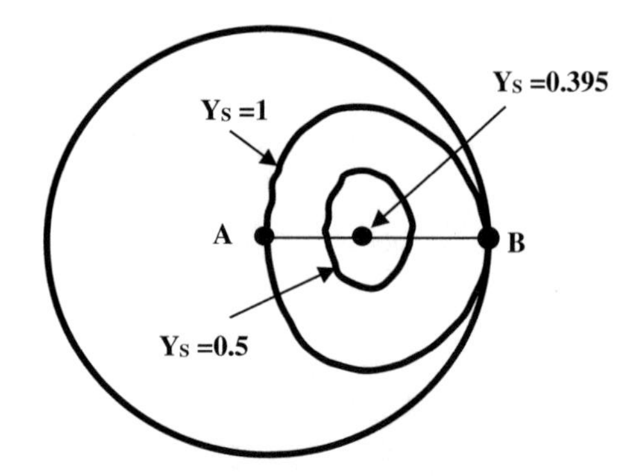

Fig. 7 Geometrical place of points for which spatial gain is achieved using ATPA

was determined the locus of the points for which relation (37) is valid. The type of figure obtained, shown in Fig. 7, is not amenable to analytical description.

From the data presented in Fig. 7, it follows that the maximum gain that can be achieved with the implementation of the proposed measure to improve the electromagnetic environment is equal $\mathbf{1/}\boldsymbol{Y_s} = \mathbf{1/0.395} \approx \mathbf{2.5}$ and can be reached when placing the repeater on the direct link connecting the end stations, at a distance from the transmitting subscriber at a distance $0.43\boldsymbol{R}$.

In the case of an even distribution of subscribers on the terrain, the ratio of the area of the figure, which is valid for formula (37), to the area $\boldsymbol{S}_0 = \boldsymbol{\pi R}^2$ characterizes the probability that at least one station will appear on the terrain, the use of which as a relay of messages between stations remote by distance $\boldsymbol{R}$, gives an additional gain in the above sense. The value of this probability is 0.305 for this law of subscriber station placement.

The reduction of the area in which mutual interferences from radiating stations of the system are created allows using economically the spectrum of frequencies allocated for communication, that is, it makes it possible to reuse each emission frequency with a smaller protective territorial interval.

6 Conclusion

In order to solve problems of heterogeneous networks management with mobile nodes, fundamentally new approaches to the construction of a data management network system are proposed. It is proposed the network architecture which uses AHRE (Ad-Hoc Radio Equipment) that has similar features to the MANET (Mobile Ad-Hoc Networks) principle and allows creating a radio network capable of self-organizing and adapting nodes to operating conditions that cannot be predicted during the design process.

Physical security management requires adaptation to rapid changes in the parameters of the radio channel: mobility (frequency maintenance and adaptation to the channel attenuation at the physical level); adaptation of the channel using correction codes; change of transmission power (in order to minimize interference, minimize the delay, maximize bandwidth, etc.).

It is proposed the model and index of stability of radio systems which is based on the probability of the possibility of organizing an arbitrary pair of subscribers of a channel of mutual exchange in a system with a random structure. Accordingly, the model representing a system with variable topology is a graph with a random structure.

It is shown that application of the automatic retransmission algorithm in subscriber radio stations provides increase of structural reliability of the network and its survivability. In the case of relay number in 1–3, the probability of having a message transmission path between any two subscriber devices in the system in the absence of the destructive effect increases in 2.5–3 times compared with the version of the system without retransmission. With an adverse effect on the system of destructive factors, the probability of preserving the transmission of messages in the system is higher in 1.5–2.5 times.

The use of more than three retransmissions is not feasible, since the additional survival benefit does not exceed 2–5%.

The use of retransmissions gives the greatest gain in structural reliability (1.5–2 times) with the reliability of lines of direct radio communication $p = 0.2 \ldots 0.55$. For $p = 0 \ldots 0.1$ and $p = 0.8 \ldots 0.99$ the gain from the use of retransmissions decreases to 0.05–0.1.

The combination of automatic transmitter power adjustment for subscriber stations using an auxiliary repeater-station can improve the electromagnetic compatibility conditions of radio equipment in communication systems with mobile objects. This measure is an effective way to economical use of the radio spectrum allocated to communication systems with moving objects.

The operation conditions of many mobile radio communications systems do not always allow us to achieve improvements in the characteristics of stability and communication quality by generally accepted organizational and technical means. Therefore, the search for effective ways to improve radio systems operating in adverse conditions remains as an interesting and important task. In this article, the attempt has been made to show new possibilities that retransmission of signals gives precisely in such mobile systems taking into account their features.

References

1. Features and uses of an ad hoc wireless network. https://www.lifewire.com/what-is-an-ad-hoc-wireless-network-2377409
2. Bondarenko O, Ageyev D, Mohammed O (2019) Optimization model for 5G network planning. In: 2019 IEEE 15th international conference on the experience of designing and application of CAD systems (CADSM). IEEE, pp 1–4. https://doi.org/10.1109/CADSM.2019.8779298
3. Kryvinska N (2010) Converged network service architecture: a platform for integrated services delivery and interworking. Electronic business series, vol 2. International Academic Publishers, Peter Lang Publishing Group

4. Ilchenko ME, Moshinskaya AV, Urywsky LA (2011) Levels separation and merging in the OSI reference model for information–telecommunication systems. Cybern Syst Anal 47:598. https://doi.org/10.1007/s10559-011-9340-4
5. Bunin SG (2013) UWB and ad hoc networks are future of telecommunications. Wireless Ukraine 13–14(1–2):38–42
6. Hepko IA, Oleynik VF (ed), Gull YD, Bondarenko AV (2009) Modern wireless networks: state and prospects of development. ECMO, Kyiv
7. Urywsky L, Shmigel B (2014) Analyze the productivity of mobile communication system with nano- and pico-cellular. Sci Prod Collect Sci Notes Ukrainian Res Inst Commun 4:5–10
8. Ilchenko M, Urywsky L (2010) Aspects of system analysis in the applied information theory for telecommunications. Cybern Syst Anal 46(5):737–743. https://doi.org/10.1007/s10559-010-9255-5
9. Ore O (1962) Theory of graphs. American Mathematical Society
10. Burshtynska KhV (2002) Kolokatsiia z umovamy u tsyfrovomu modeliuvanni reliefu. Heodeziia, Kartohrafiia I Aerofotoznimannia. 62:103–110

Van der Pol Oscillators Based on Transistor Structures with Negative Differential Resistance for Infocommunication System Facilities

Andriy Semenov, Olena Semenova, Oleksandr Osadchuk, Iaroslav Osadchuk, Serhii Baraban, Andrii Rudyk, Andrii Safonyk, and Oleksandr Voznyak

Abstract The paper proposes new circuits of Van der Pol oscillators based on nonlinear and reactive properties of transistor structures with negative differential resistance to be employed in facilities of infocommunication systems. Mathematical models of voltage-controlled Van der Pol oscillators, which are based on nonlinear and reactive properties of transistor structures with negative differential resistance, and generate electric oscillation with regular dynamics, are improved. Influence of additive white noise on dynamics of electric oscillations of the Van der Pol oscillators is investigated. Results of theoretical calculations, numerical simulations and experimental studies of dynamics of the electric oscillation of the microwave Van der Pol oscillators in the frequency ranges 880…910 MHz and 1.8…2.1 GHz are presented. Application of the reactive properties of transistor structures with negative differential resistance provides the parameter retuning expansion for the self-oscillating systems of Van der Pol oscillators up to 20…30% with operation modes being stable. The effective relative operational frequency retuning is 3.4% for the range of 880…910 MHz and 9.88% for the range of 1.8…2.1 GHz.

A. Semenov (✉) · O. Semenova · O. Osadchuk · I. Osadchuk · S. Baraban
Vinnytsia National Technical University, Khmelnytske Shose 95, Vinnytsia, Ukraine
e-mail: semenov.a.o@vntu.edu.ua

O. Semenova
e-mail: semenova.o.o@vntu.edu.ua

S. Baraban
e-mail: serg@politex.org.ua

A. Rudyk · A. Safonyk
National University of Water and Environmental Engineering, 11 Soborna st, Rivne, Ukraine
e-mail: a.v.rudyk@nuwm.edu.ua

A. Safonyk
e-mail: a.p.safonyk@nuwm.edu.ua

O. Voznyak
Vinnytsia National Agrarian University, 3 Soniachna St, Vinnitsia, Ukraine

D. Ageyev et al. (eds.), *Data-Centric Business and Applications*, Lecture Notes on Data Engineering and Communications Technologies 69,
https://doi.org/10.1007/978-3-030-71892-3_3

Keywords Van der Pol oscillator · Transistor structure · Negative differential resistance · Phase portrait · Electric oscillations · White noise · Amplitude spectrum · Voltage control

1 Introduction

Infocommunication technologies are developing in two interrelated directions – hardware and software. Nowadays, such areas of software infocommunication systems as converged network service architecture [1–5], parametric synthesis of networks with self-similar traffic [6–12], fractal and multifractal analisys of data stream [13–17], managing the process of servicing hybrid telecommunications services [18–21] are developing very fast.

The current state of hardware is mostly determined by the latest advances in the semiconductor device engineering. This approach leads to the fact that someday the limit of technological capabilities of the physical implementation of semiconductor devices will be reached. Therefore, recently more and more attention is paid to properties of transistor structures with negative differential resistance.

A promising direction of engineering is the development and investigation of devices on transistor structures with negative differential resistance compatible with microelectronic technology for generating and forming signals with regular and chaotic dynamics for infocommunication systems [22–25]. The development and application of the devices on transistor structures with negative differential resistance for generating and forming signals have been carried out since the early 1980s [26–31]. Theoretical and practical aspects of applying devices with chaotic signal dynamics in infocommunication systems have been developed [32–38]. Currently, the development and implementation of devices for generating and forming signals with regular and chaotic dynamics based on transistor structures with negative differential resistance is an urgent scientific and technical task [39–46].

Modern oscillators for facilities of infocommunication systems must have a wide operating frequency band and sufficient power. Also, devices for generating and forming signals for infocommunication systems must be quite efficient in terms of energy consumption. Such oscillators must are intended to generate signals with regular and chaotic dynamics in a given frequency range and to have low out-of-band radiation [47–49]. In addition, the electric oscillators must be implemented as integrated circuits of silicon or silicon-germanium technologies.

The rapid growth of the modern information market has led to increased demand for high-performance, inexpensive and energy-efficient radio-frequency integrated circuits. The core of such energy-efficient radio frequency integrated circuits are voltage-controlled oscillators. The most common oscillators are those in the frequency bands of 900, 1800, 2000, 2400 and 5000 MHz. In facilities of infocommunication systems, oscillators with frequencies 900, 1800, 2000 MHz are used in frequency synthesizers for GSM and CDMA standards. Oscillators for frequency bands of 2400 and 5000 MHz are employed in direct conversion receivers for wireless LANs of the 802.11a/b/g standard.

Traditionally, microwave oscillators are built on the basis of two-transistor cross-connected structures, in particular of CMOS or HEMT technology [50, 51]. Varactors or MOS transistors are utilized as elements for controlling the frequency change voltage [52, 53]. However, this implementation is not suitable for monolithic integration of the oscillator with the radioreceiver due to a high sensitivity of the oscillator and the tuning nonlinearity. Therefore, in practice, a broadband implementation of the voltage-controlled oscillator with division into subbands is used. This can be implemented by one of two approaches: (1) by adding or removing a discrete amount of capacitance or inductance from a oscillating circuit of the oscillator; (2) by switching between oscillating circuits that are individually optimized and tuned for different frequency ranges.

In this study, the authors propose an approach for constructing voltage-controlled microwave oscillators based on nonlinear and reactive properties of transistor structures with negative differential resistance. These oscillators are able to generate electric oscillations with regular and chaotic dynamics. This paper presents results of theoretical and experimental studies of regular dynamics of electric oscillations in voltage-controlled oscillators on transistor structures with negative differential resistance.

2 The Theory of Van der Pol Oscillators on Transistor Structures with Negative Differential Resistance

An important and urgent task in the analysis of dynamic processes in oscillators of electric oscillations on transistor structures with negative differential resistance is an examination of phase portraits of their self-oscillating systems [39, 42, 43]. For an equivalent oscillator circuit (Fig. 1a) based on Kirchhoff's first law, we have obtained the equation [54, 55]

$$i_T(u) = C\frac{du}{dt} + \frac{u}{R} + \frac{1}{L}\int\limits_{-\infty}^{t} udt. \tag{1}$$

Having differentiated the Eq. (1) and reduced the similar, we obtained such a differential equation in normalized time $\tau = \omega_0 t$ (where $\omega_0 = 1\big/\sqrt{LC}$ is a frequency of generated oscillations) relative to the generated voltage [56]

$$\frac{d^2u}{d\tau^2} + \sqrt{\frac{L}{C}}\left(\frac{1}{R} - \frac{di_T(u)}{du}\right)\frac{du}{d\tau} + u = 0. \tag{2}$$

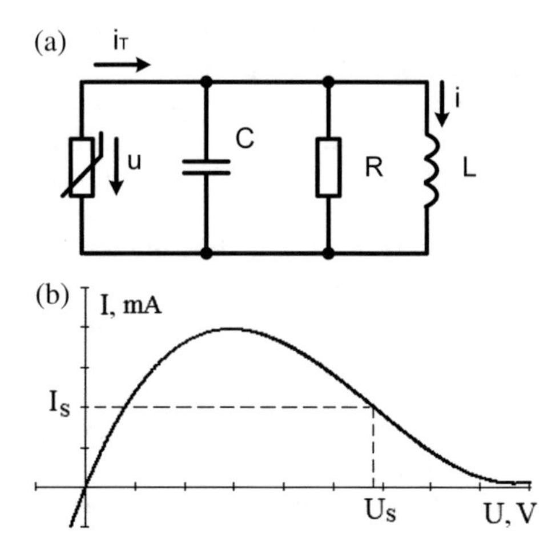

Fig. 1 Equivalent circuit (**a**) and approximated I–V curve of the transistor structure with negative differential resistance (**b**) for the oscillator of electric oscillation

The equation of approximation by the third degree polynomial for the static I–V curve of the transistor structure with negative differential resistance is [56]

$$i_T(u) = \left(I_S + gU_S - hU_S^3\right) - \left(g - 3hU_S^2\right)u - 3hU_S u^2 + hu^3, \tag{3}$$

where g, h are the approximation coefficients, U_S, I_S are the coordinates of the middle of the descending section at the I–V curve (the coordinated of the initial setting of the operating point on the descending section of the I–V curve in Fig. 1b).

According to the method [56], having substituted $i_T(u)$ from (3) to (2) and reduced the similar, the authors transformed differential Eq. (2) into the form of Van der Pol [54, 55] relative to the generated voltage in normalized time

$$\frac{d^2u}{d\tau^2} - \mu\left(1 - bu - qu^2\right)\frac{du}{d\tau} + u = 0, \tag{4}$$

where the coefficients taking into account approximation (3) are calculated by the following formulas

$$\mu = \sqrt{\frac{L}{C}}\left(g - 3hU_S^2 - \frac{1}{R}\right), b = -\sqrt{\frac{L}{C}}\frac{6hU_S}{\mu}[1/\mathrm{V}], q = \frac{3h}{\mu}\sqrt{\frac{L}{C}}[1/\mathrm{V}^2]. \tag{5}$$

The system of second-order differential equations in absolute variables with respect to normalized time is [56]

$$\begin{cases} \frac{dx_1}{d\tau} = x_2, \\ \frac{dx_2}{d\tau} = \mu\left(1 - bx_1 - qx_1^2\right)x_2 - x_1, \end{cases} \tag{6}$$

where $x_1 = u,\ x_2 = \frac{du}{d\tau}$ [56]. The condition of self-excitation is $\mu \geq 0$.

The results of studying dynamic processes in the Van der Pol oscillators on transistor structures with negative differential resistance are presented in Figs. 2, 3 and 4 (the normalized time is on the abscissa axis, the voltage in volts is on the ordinate axis) [56]. Shown in Figs. 2, 3 and 4 the simulation results are grouped in ascending order per decade of the parameter q, to visualize how the quadratic term of nonlinearity bx_1x_2 Eq. (6) impacts the dynamics of the generated voltage fluctuations.

3 Investigating the Parameters of Electric Oscillations During Broadband Retuning of Self-Oscillating Systems of Van der Pol Oscillators on Transistor Structures with Negative Differential Resistance

In the general case, the mathematical model of Van der Pol oscillators is a nonlinear inhomogeneous second-order differential equation [57]

$$\ddot{x} - w(x)\dot{x} + f(x) = 0, \tag{7}$$

where

$$f(x) = \frac{dU(x)}{dx}, \tag{8}$$

$U(x)$ is the potential function of the self-oscillating system, $w(x)$ is the nonlinear function that specifies the negative resistance of the oscillator. According to Eqs. (4) and (7)

$$\begin{aligned} w(x) &= -\mu\left(1 - bx - qx^2\right), \\ f(x) &= x. \end{aligned} \tag{9}$$

For the Van der Pol oscillator on transistor structures with negative differential resistance, the potential function with accuracy up to an integration constant is [57]

$$U(x) = \int f(x)dx = \frac{x^2}{2} + C. \tag{10}$$

Equation (7) in the standard form for second-order dynamical systems is [57]

$$\begin{cases} \dot{x} = y, \\ \dot{y} = w(x)y - f(x). \end{cases} \tag{11}$$

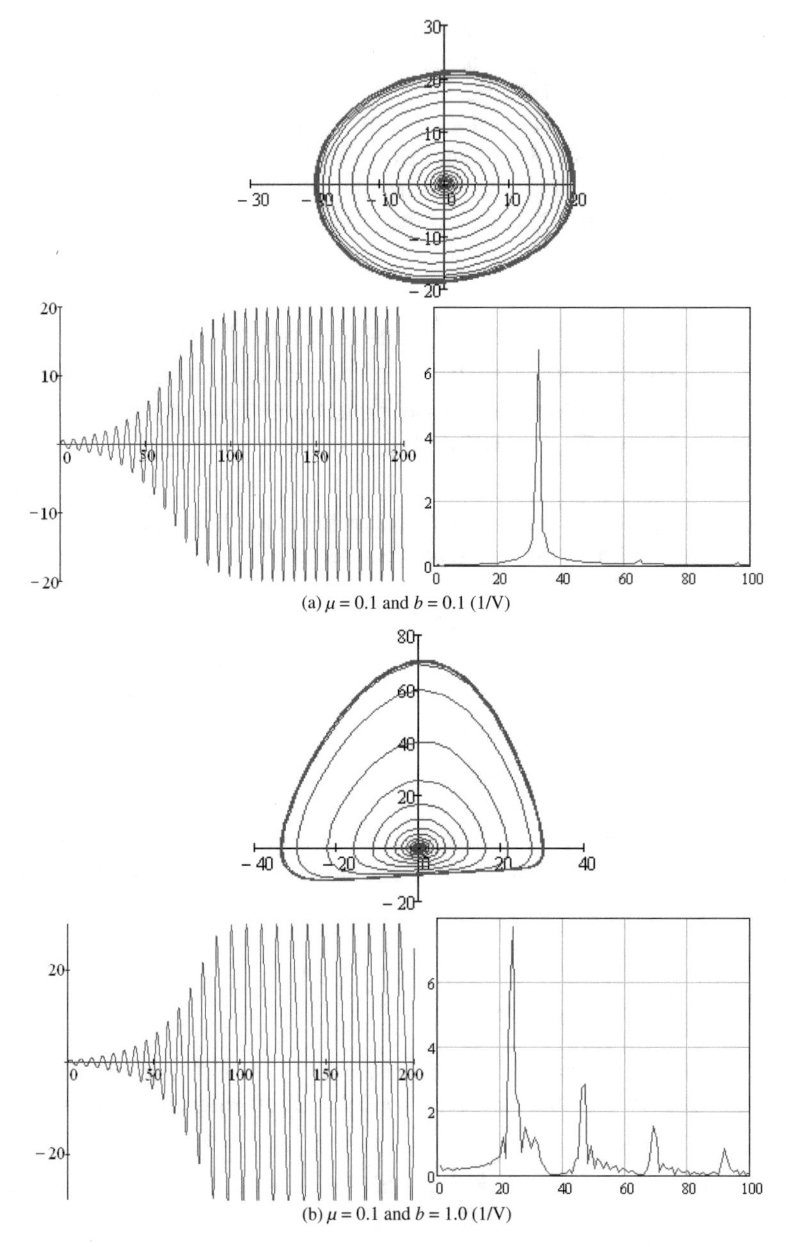

Fig. 2 Results of modelling the Van der Pol oscillators on transistor structures with negative differential resistance at $q = 0.01$ (1/V^2) and different values of the self-oscillating system (6) parameters: phase portraits of oscillators in the plane of dynamic variables x_1–x_2 (on the left), diagrams of oscillations of the generated voltage relative to the normalized time $\tau = \omega_0 t$ (in the center) and its amplitude spectra (on the right)

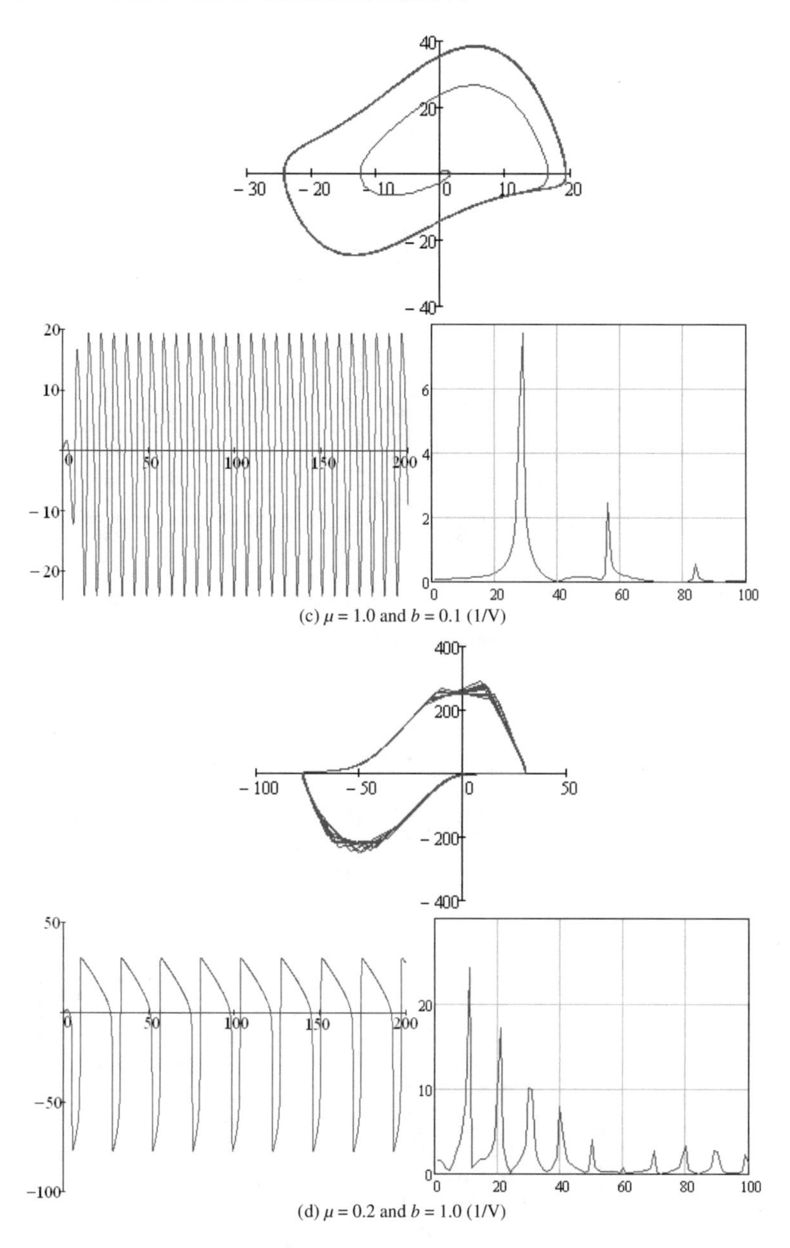

(c) $\mu = 1.0$ and $b = 0.1$ (1/V)

(d) $\mu = 0.2$ and $b = 1.0$ (1/V)

Fig. 2 (continued)

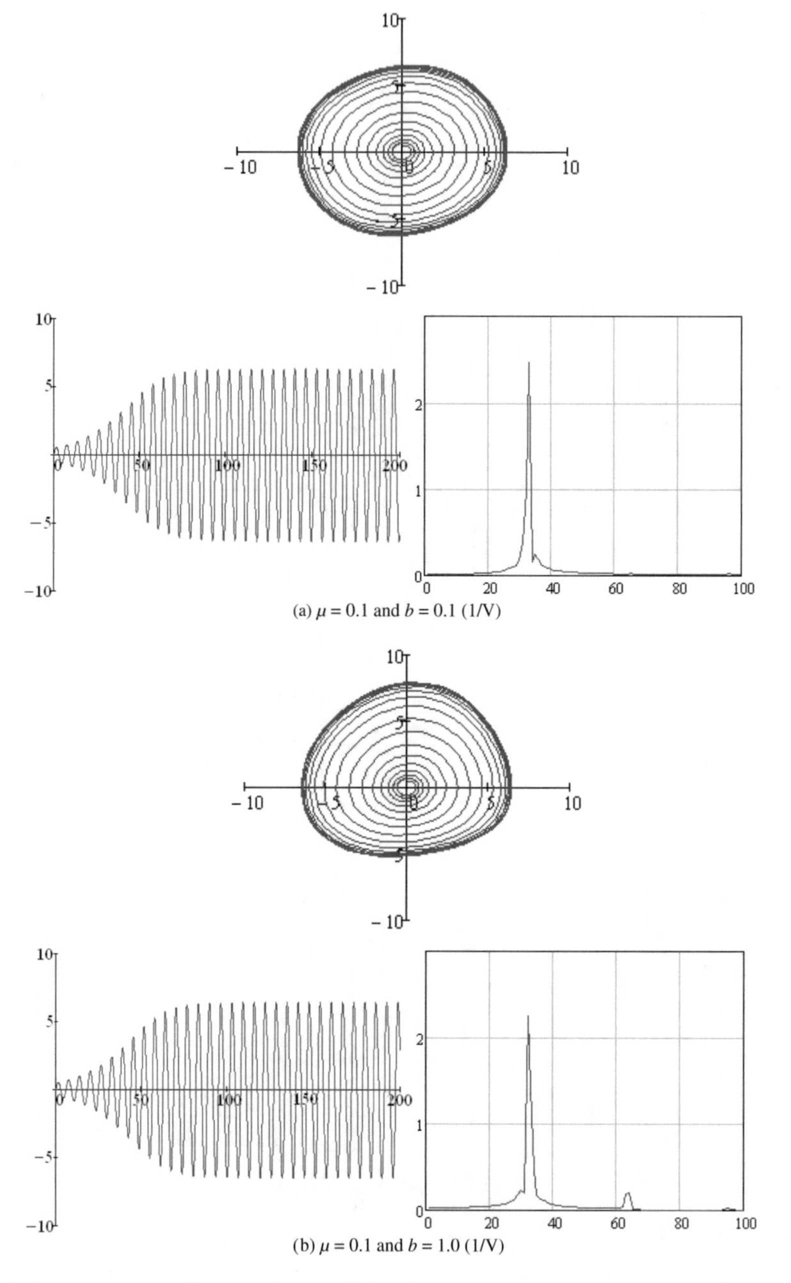

(a) $\mu = 0.1$ and $b = 0.1$ (1/V)

(b) $\mu = 0.1$ and $b = 1.0$ (1/V)

Fig. 3 Results of modelling the Van der Pol oscillators on transistor structures with negative differential resistance at $q = 0.1$ (1/V^2) and different values of the self-oscillating system (6) parameters: phase portraits of oscillators in the plane of dynamic variables x_1-x_2 (on the left), diagrams of oscillations of the generated voltage relative to the normalized time $\tau = \omega_0 t$ (in the center) and its amplitude spectra (on the right)

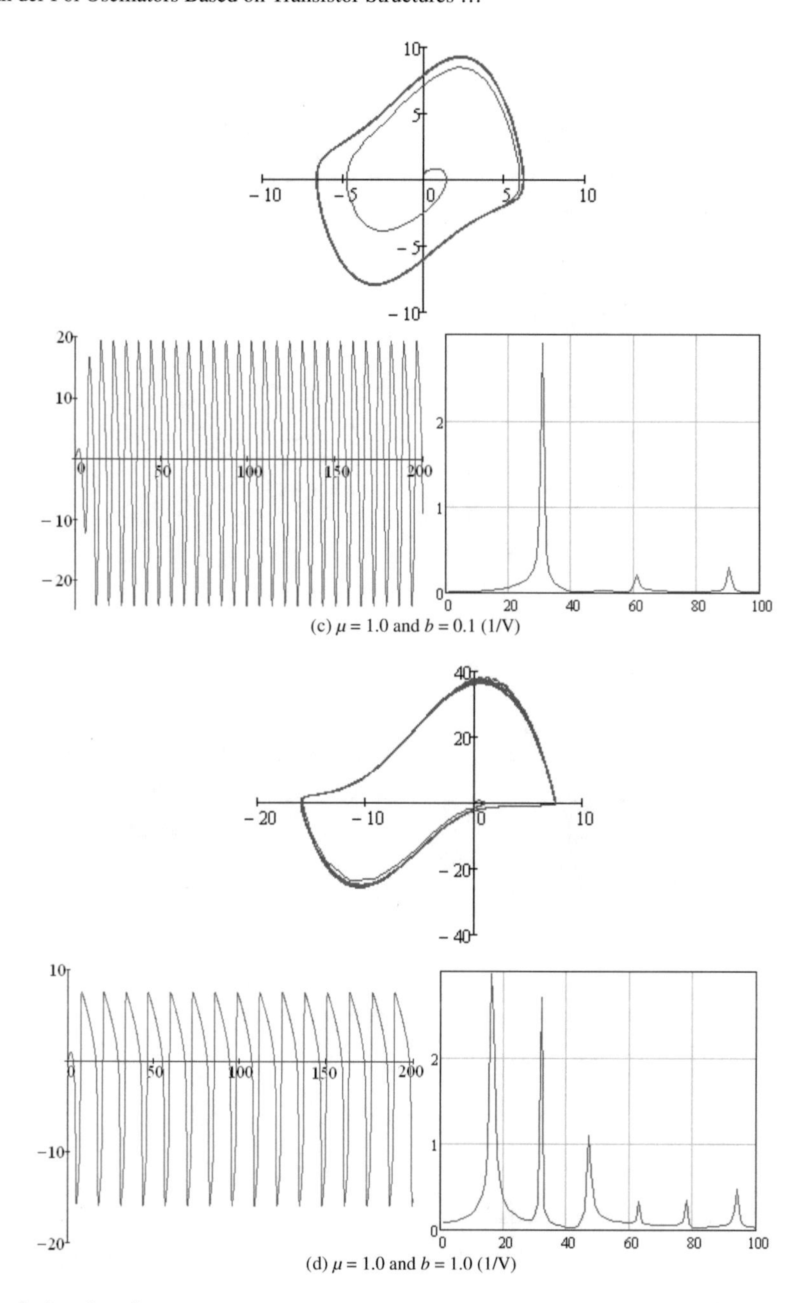

(c) $\mu = 1.0$ and $b = 0.1$ (1/V)

(d) $\mu = 1.0$ and $b = 1.0$ (1/V)

Fig. 3 (continued)

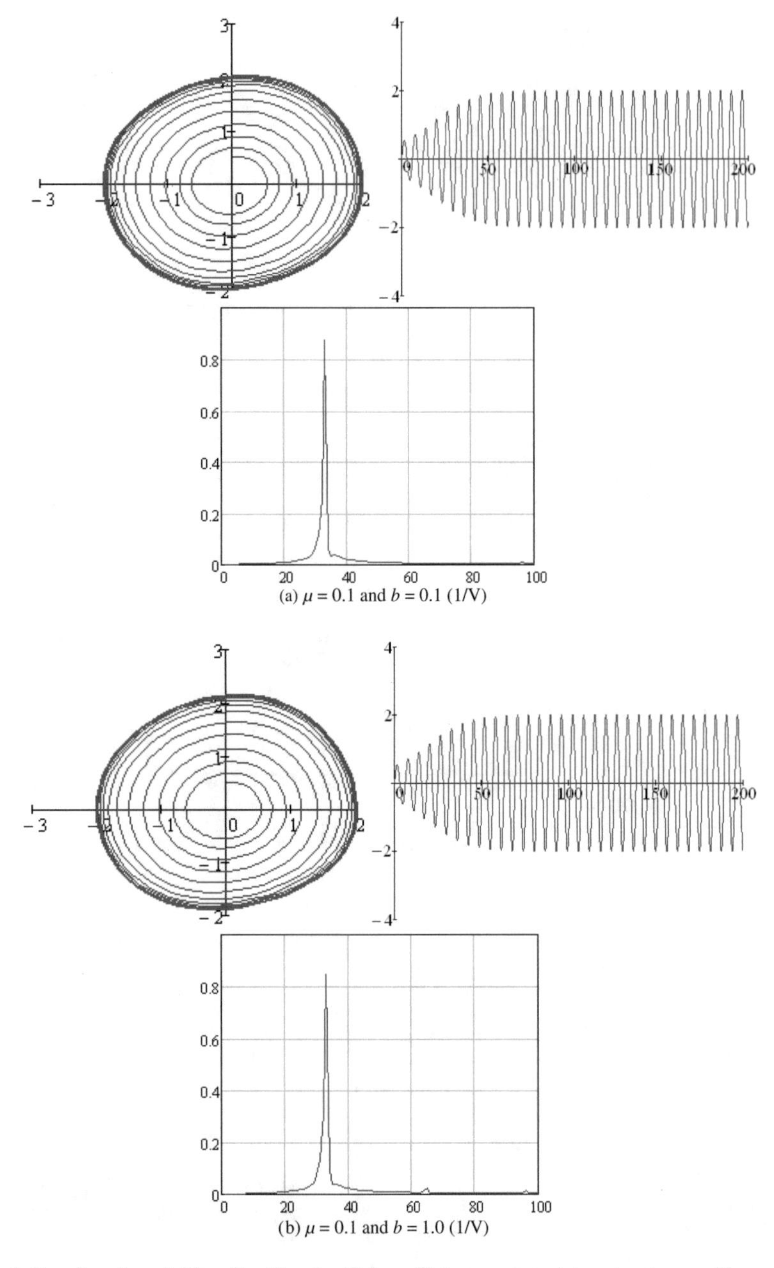

Fig. 4 Results of modelling the Van der Pol oscillators on transistor structures with negative differential resistance at $q = 1.0$ (1/V^2) and different values of the self-oscillating system (6) parameters: phase portraits of oscillators in the plane of dynamic variables x_1-x_2 (on the left), diagrams of oscillations of the generated voltage relative to the normalized time $\tau = \omega_0 t$ (in the center) and its amplitude-frequency spectra (on the right)

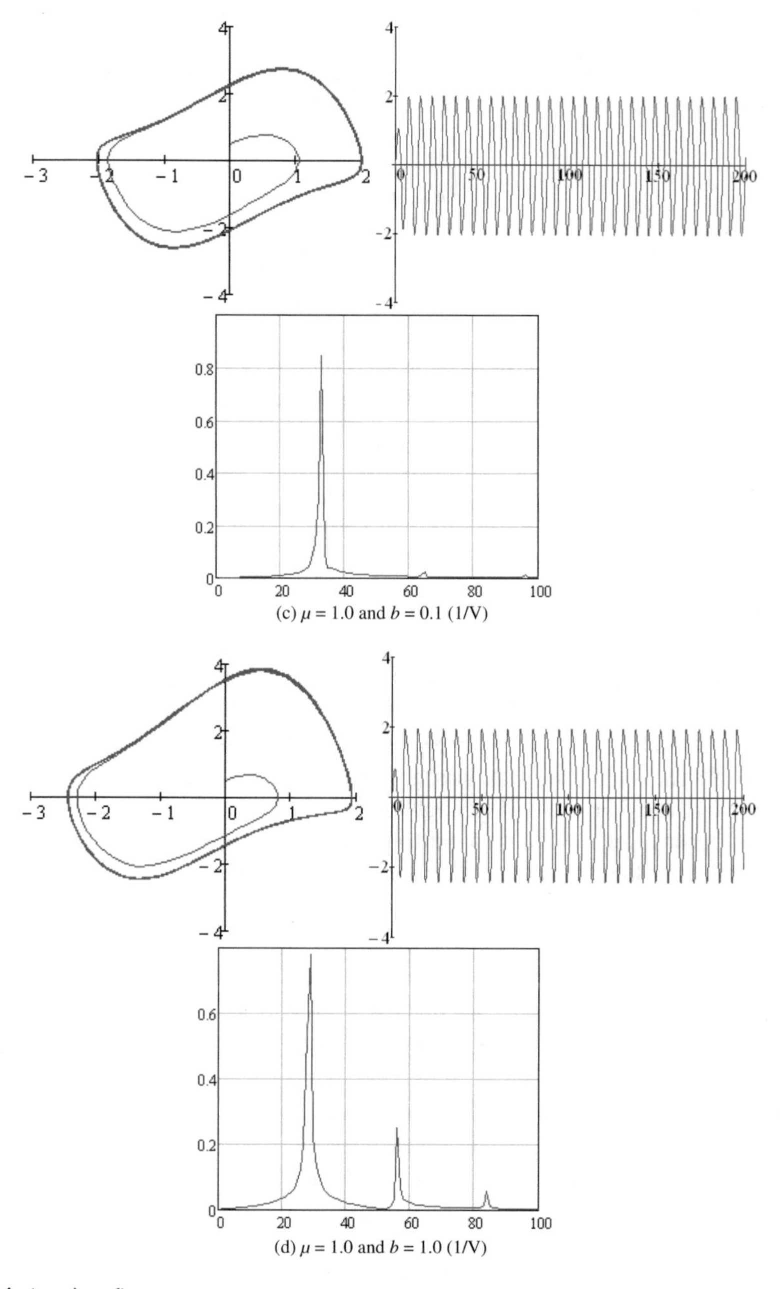

(c) $\mu = 1.0$ and $b = 0.1$ (1/V)

(d) $\mu = 1.0$ and $b = 1.0$ (1/V)

Fig. 4 (continued)

In [57], conditions of main bifurcations in the self-oscillating oscillator described by Eq. (7) with arbitrary dissipative and potential functions were investigated. Fixed points are determined from the condition that left parts of system of Eqs. (11) equal zero

$$\begin{aligned} y &= 0, \\ f(x) &= 0. \end{aligned} \tag{12}$$

It follows from (12) that the fixed points of the self-oscillating system correspond to the extremes of the potential function $U(x)$.

Having differentiated the right parts of system of Eq. (11) and calculated the corresponding partial derivatives, we have [57] a linearization matrix, a trace and the Jacobian of this matrix

$$M = \begin{pmatrix} 0 & 1 \\ f'(x) & w(x) \end{pmatrix}. \tag{13}$$

$$\xi = w(x), \;\; J = f'(x). \tag{14}$$

The coordinate is determined at the fixed point (12). Having Eq. (14) of trace ξ and jacobian J, the authors investigated the self-oscillating system (4)–(6) considering types of bifurcations. It was shown in [58] that in the soft mode of excitation of the oscillators on transistor structures with negative differential resistance described by Eq. (4) only the Andronov-Hopf bifurcation is possible. Other known bifurcations, in particular the saddle-node, Bogdanov-Takens or triple equilibrium those, were not observed because their existence requires the jacobian to be zero, i.e. $f'(x) = 0$, that is impossible for the studied self-oscillating system [57].

For the Andronov-Hopf bifurcation $\xi = 0$ under condition $J > 0$ [57], so

$$f(x) = 0, \;\; w(x) = 0, \;\; f'(x) > 0. \tag{15}$$

$$qx^2 + bx - 1 = 0. \tag{16}$$

The solution of the quadratic Eq. (16) is points

$$x_{1,2} = \frac{-b \pm \sqrt{b^2 + 4q}}{2q}. \tag{17}$$

Results of theoretical study obtained by the authors in [56] show that dynamic processes in the oscillators on transistor structures with negative differential resistance correspond to the classical Van der Pol oscillator. The quadratic term of nonlinearity bx_1x_2 in the second equation of the system (6) does not change the generation modes.

To detect imbalance that occurs in self-oscillating systems of the Van der Pol oscillators, we convert Eq. (4) to a form [59]

$$\frac{d^2u}{d\tau^2} + u = \mu\left(qu^2 + bu - 1\right)\frac{du}{d\tau}, \tag{18}$$

where μ is the small parameter, u, $\frac{du}{d\tau}$, $\frac{d^2u}{d\tau^2}$ is the investigated time function of the generated voltage and its derivatives.

The left part of Eq. (18) is linear, and the right part is a nonlinear part of the self-oscillating system. At $\mu = 0$, we have $\frac{d^2u}{d\tau^2} + u = 0$, its solution is $u(\tau) = U_m \cos(\omega_0\tau + \varphi_0)$, where U_m is the amplitude of the generated voltage oscillation; $\omega = 1$ is the frequency and φ_0 its initial phase [59]. Magnitudes of the amplitude and phase of the generated sinusoidal voltage oscillations are determined from initial conditions.

The right part of the differential Eq. (18) determines nonlinear properties of an active element of the oscillator on a transistor structure with negative differential resistance. As it can be seen in Figs. 2, 3 and 4, the mode of operation of the self-oscillating system (oscillatory or relaxation), the frequency of self-oscillations, the time of setting the stationary mode and the harmonic coefficient depend not only on the value of the small parameter μ, but also on the quadratic term of nonlinearity qu^2.

Figure 5 illustrates a diagram of imbalance that takes place in self-oscillating systems of the Van der Pol oscillators [59]. The authors have identified four main factors of self-oscillating systems that provide imbalance when retuning parameters

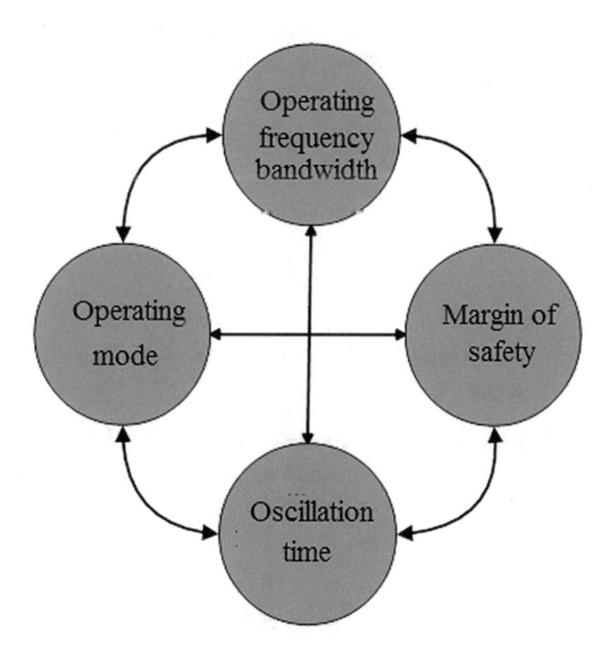

Fig. 5 Imbalance taking place at the broadband retuning of self-oscillating systems in the Van der Pol oscillators on transistor structures with negative differential resistance

of the self-oscillating system of the Van der Pol oscillators on transistor structures with negative differential resistance: (1) operating mode (oscillatory or relaxation); (2) margin of safety (large or small); (3) frequency range width (wide or narrow); (4) time of stationary oscillations installation (small or large).

These imbalances can be explained by their physical content. The mode of operation changes from oscillatory to relaxation and vice versa (Figs. 2c–d, 3c–d and 4c–d) when changing parameters of the self-oscillating system of the Van der Pol oscillator in a wide range. In the oscillatory mode, the emergence of quasi-harmonic oscillations with a low harmonic coefficient requires the maximum time of the stationary oscillations installation (Figs. 2a, 3a–b and 4a–b). Retuning of the generated oscillation frequency changes the operating mode and reduces the margin of safety (in the range of HF Fig. 2a, 3a and 4a and in the range of LF Fig. 2c). The classical approach to deal with these imbalances taking place at retuning parameters of the self-oscillating system of the Van der Pol oscillator is to increase its order. But this leads to nonlinear and chaotic oscillations.

In order to evaluate the dynamic processes in self-oscillating systems of autonomous and non-autonomous oscillators of electric oscillations, parameters of nonlinear systems must be calculate. Numerical values characterizing the dynamics of the self-oscillating system as invariant measure were obtained taking into account values of the Lyapunov exponents [60]. To analyze the stability and synchronization of self-oscillating systems, the method of maps of the Lyapunov exponents was widely used [60]. But this method is convenient to apply to discrete mappings of nonlinear dynamical systems [61]. In case of continuous nonlinear dynamical systems, the method of signature of spectra of the Lyapunov exponents is applyied [61].

The range of the Lyapunov exponents for the attractor of a nonlinear dynamical system, represented by autonomous differential equations, must satisfy two requirements [62]:

(1) the sum of all N Lyapunov exponents must be negative

$$\sum_{i=1}^{N} \lambda_i < 0, (\lambda_1 \geq \lambda_2 \geq \ldots \geq \lambda_N);$$

(2) an attractor that differs from a fixed point must have at least one zero Lyapunov exponent.

Depending on the dimension of dynamic systems of the Van der Pol oscillators bon transistor structures with negative differential resistance, there several cases of the signature of the spectrum of Lyapunov exponents [62].

For $N = 2$:

$(\lambda_1, \lambda_2) = (-, -)$—stable fixed point;
$(\lambda_1, \lambda_2) = (0, -)$—limit cycle;

For $N = 3$:

$(\lambda_1, \lambda_2, \lambda_3) = (-, -, -)$—attractive fixed point;
$(\lambda_1, \lambda_2, \lambda_3) = (0, -, -)$—limit cycle;
$(\lambda_1, \lambda_2, \lambda_3) = (0, 0, -)$—two-dimensional torus;
$(\lambda_1, \lambda_2, \lambda_3) = (+, 0, -)$—strange attractor (chaos).

For $N = 4$:

$(\lambda_1, \lambda_2, \lambda_3, \lambda_4) = (+, 0, -, -)$—strange attractor (chaos);
$(\lambda_1, \lambda_2, \lambda_3, \lambda_4) = (+, +, 0, -)$—strange attractor hyperchaos.

4 Investigation of Dynamic Processes in the Van der Pol Oscillators on Transistor Structures with Negative Differential Resistance in the Presence of Additive White Noise

Dynamic processes in real oscillators proceed under the action of internal and external random forces. The effect of noise on dynamic processes near bifurcation points should be studied in detail. Bifurcations in dynamical systems with noise are called stochastic [63]. The self-oscillation mode in the Van der Pol oscillators is associated with the Andronov-Hopf bifurcation. It can be soft (supercritical) [64] and hard (subcritical) [65]. The method of analysis of dynamic processes in the Van der Pol oscillators on transistor structures with negative differential resistance in the presence of additive white noise uses the numerical modelling of the non-autonomous stochastic equation of the Van der Pol oscillator in soft [63, 64] or hard [65] self-excitation modes.

An AC model of the quasi-harmonic oscillators on transistor structures with negative differential resistance when operating at a fixed frequency is a first-type parallel oscillating circuit with the following elements: $G(u)$ is the negative differential conductivity of the active element of the oscillator; G is the conductivity of active losses; $C(u)$ is the capacitance of the transistor structure depending on both the amplitude and the frequency; L is the equivalent inductance of the oscillating circuit [66].

The mathematical model of the oscillators is a nonlinear homogeneous differential equation of the second order, its form depends on the approximation functions of the negative conductivity $G(u)$ and the nonlinear capacitance $C(u)$ of the transistor structure. Noise properties of the oscillators on transistor structures with negative differential resistance were studied by introducing a white noise current source into the equivalent circuit [66]. This leads to the transformation of the nonlinear homogeneous differential equation to an inhomogeneous [66]

$$C(u)\frac{dv(t)}{dt} + [G - G(u)]v(t) + \frac{1}{L}\int v(t)dt = i_n(t), \tag{19}$$

where $u(t)$ is the voltage of stationary oscillation, $v(t)$ is the voltage of noise, $i_n(t)$ is the current of noise.

The theoretical study is aimed is to specify the parameters of the modified Van der Pol model at the presence of noise used for analyzing physical processes in the Van der Pol oscillators on transistor circuits and structures with negative differential resistance and studying phase portraits of the oscillators.

Noise is known to have a double effect on self-oscillation at output of the oscillators: it simulates the amplitude of the generated oscillations and introduces a randomly fluctuated phase shift, and therefore the solution of the differential Eq. (19) with respect to the noise voltage $v(t)$ in the general case is [66]

$$v(t) = v_0[1 + a(t)]\cos[\omega_0 t - \psi(t)], \tag{20}$$

where $a(t)$ describes amplitude modulation, $\psi(t)$ describes phase modulation, v_0 is the amplitude of natural oscillations in the absence of noise. Functions $a(t)$ and $\psi(t)$ are stationary stochastic processes. In the case of a high-quality oscillatory system of the oscillator on transistor structures with negative differential resistance, which takes place in practice, spectral components of the parts $a(t)$ and $\psi(t)$ are in the frequency range that is much lower than frequencies of their own oscillations.

Further analysis of the modified Van der Pol model in the presence of noise was performed using the software package of mathematical calculations MathCad by the methodology described in [67]. When the Van der Pol oscillator operates at a fixed frequency, the nonlinear properties of the capacitive component of the impedance of the transistor structure are negligibly small, so Eq. (19) in the normalized time $\tau = \omega_0 t$ will be as follows

$$\frac{d^2u}{d\tau^2} + \sqrt{\frac{L}{C}}\left(\frac{1}{R} - \frac{di_T(u)}{du}\right)\frac{du}{d\tau} + u = \sqrt{2D}n(t), \tag{21}$$

where $n(t)$ is the normalized source of white Gaussian noise, D is the noise intensity level (W/Hz). Parameters of the normalized source of white Gaussian noise [67]

$$\langle \mathrm{n(t)} \rangle = 0, \langle n(t)n(t-\tau) \rangle = \delta(\tau).$$

Having divided the variables in (21), the authors obtained a system of second-order differential equations [56]

$$\begin{cases} \frac{dx_1}{d\tau} = x_2, \\ \frac{dx_2}{d\tau} = \mu(1 - bx_1 - qx_1^2)x_2 - x_1 + \sqrt{2D}n(t), \end{cases} \tag{22}$$

where $x_1 = u$, $x_2 = \frac{du}{d\tau}$, $0 < D \leq 1$ [56]. Results of the dynamic processes examination in the oscillator in the presence of noise are shown in Figs. 6, 7 and 8 [56].

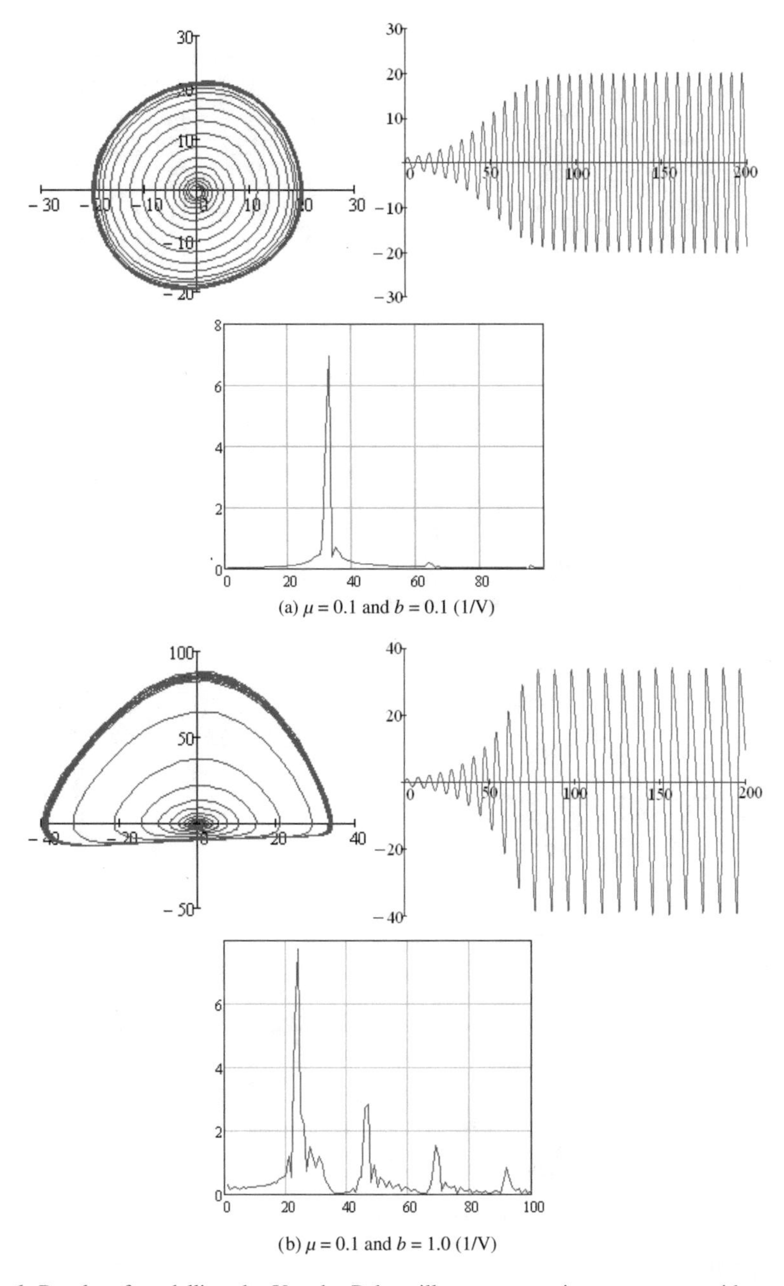

(a) $\mu = 0.1$ and $b = 0.1$ (1/V)

(b) $\mu = 0.1$ and $b = 1.0$ (1/V)

Fig. 6 Results of modelling the Van der Pol oscillators on transistor structures with negative differential resistance at $q = 0.01$ (1/V^2) and different values of the self-oscillating system (6) parameters in the presence of additive white Gaussian noise with intensity $D = 0.125$ (W/Hz)

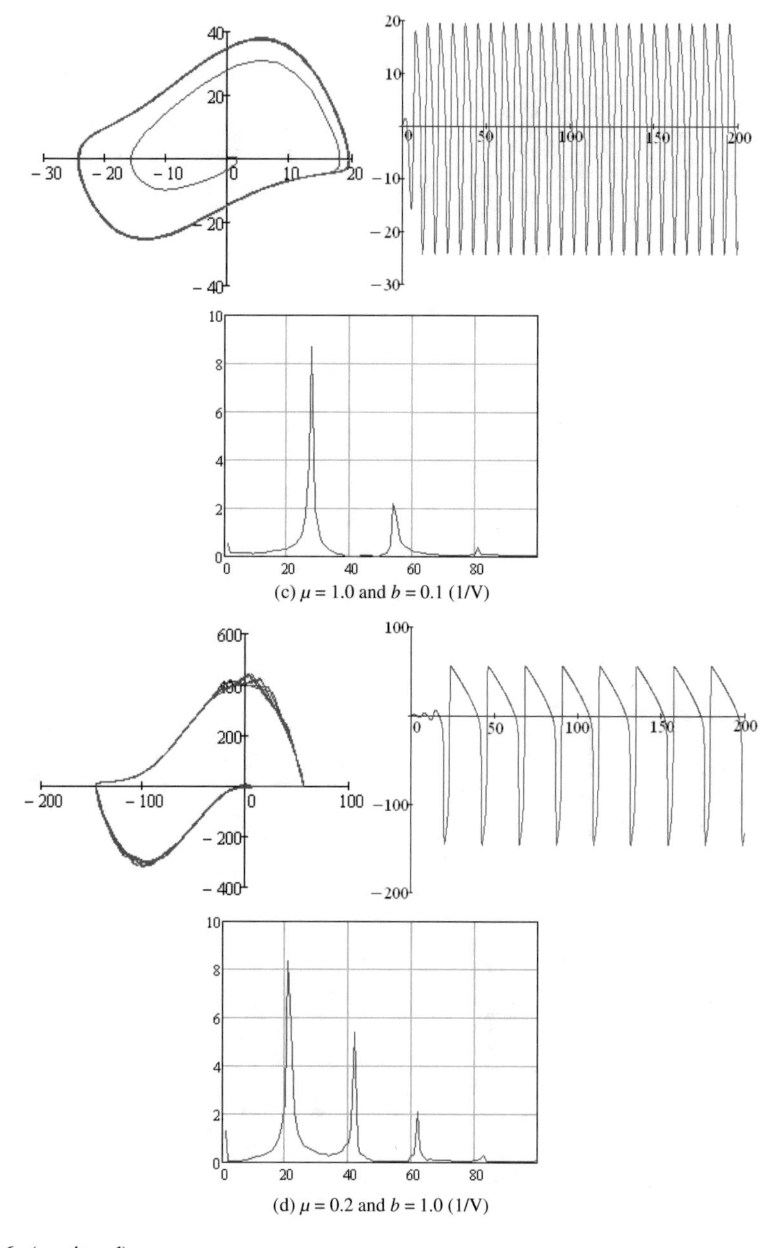

(c) $\mu = 1.0$ and $b = 0.1$ (1/V)

(d) $\mu = 0.2$ and $b = 1.0$ (1/V)

Fig. 6 (continued)

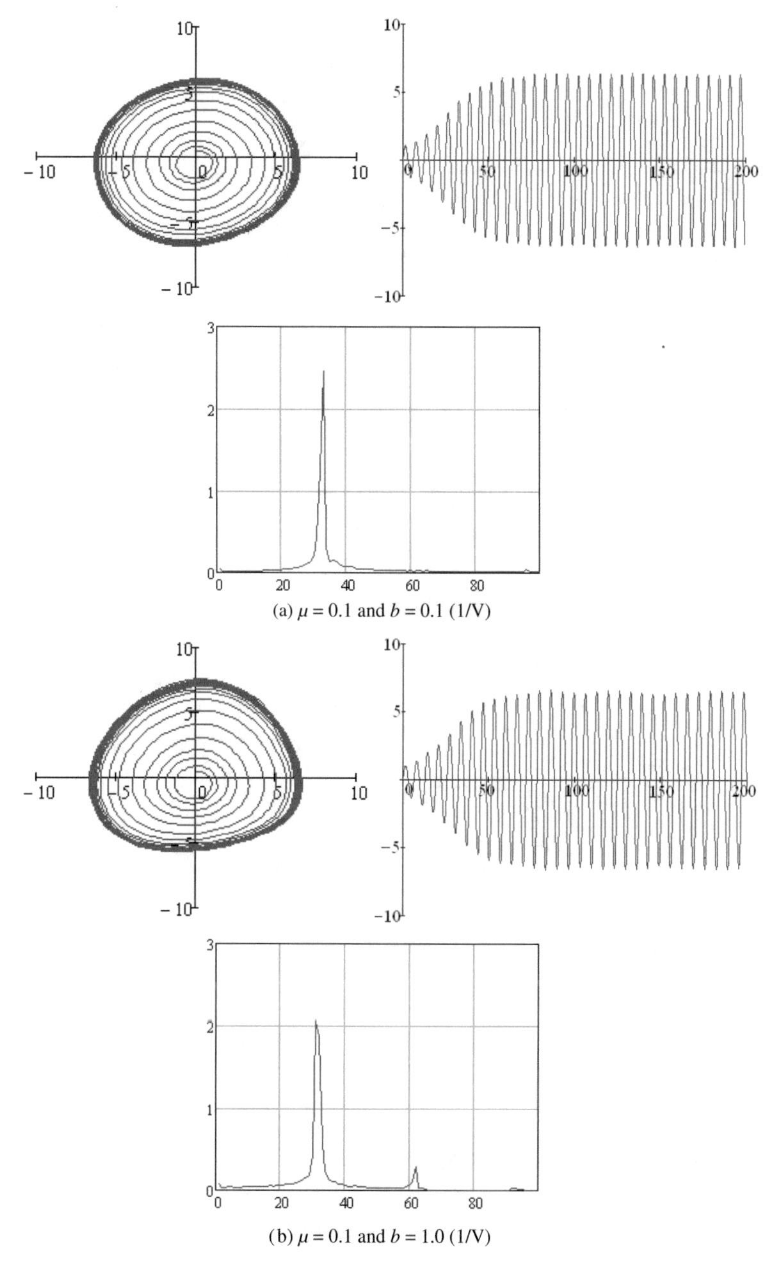

(a) $\mu = 0.1$ and $b = 0.1$ (1/V)

(b) $\mu = 0.1$ and $b = 1.0$ (1/V)

Fig. 7 Results of modelling the Van der Pol oscillators on transistor structures with negative differential resistance at $q = 0.1$ (1/V^2) and different values of the self-oscillating system (6) parameters in the presence of additive white Gaussian noise with intensity $D = 0.125$ (W/Hz)

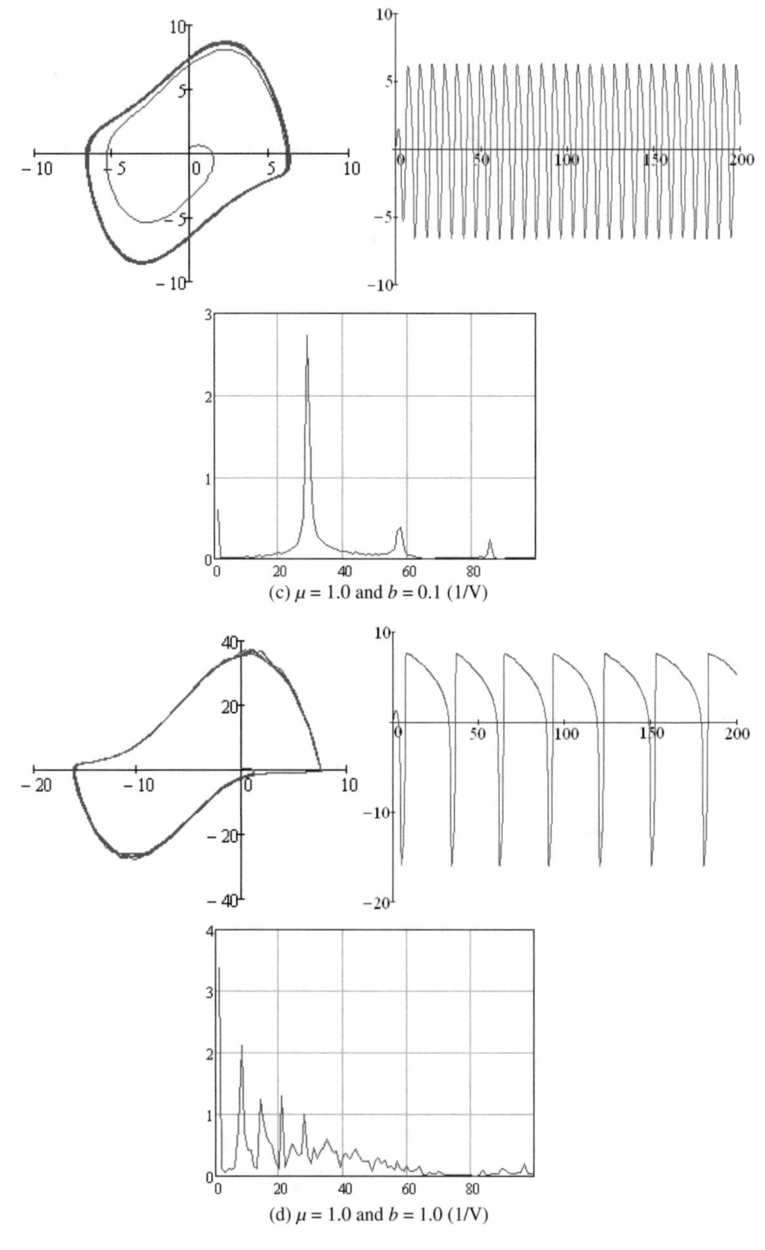

(c) $\mu = 1.0$ and $b = 0.1$ (1/V)

(d) $\mu = 1.0$ and $b = 1.0$ (1/V)

Fig. 7 (continued)

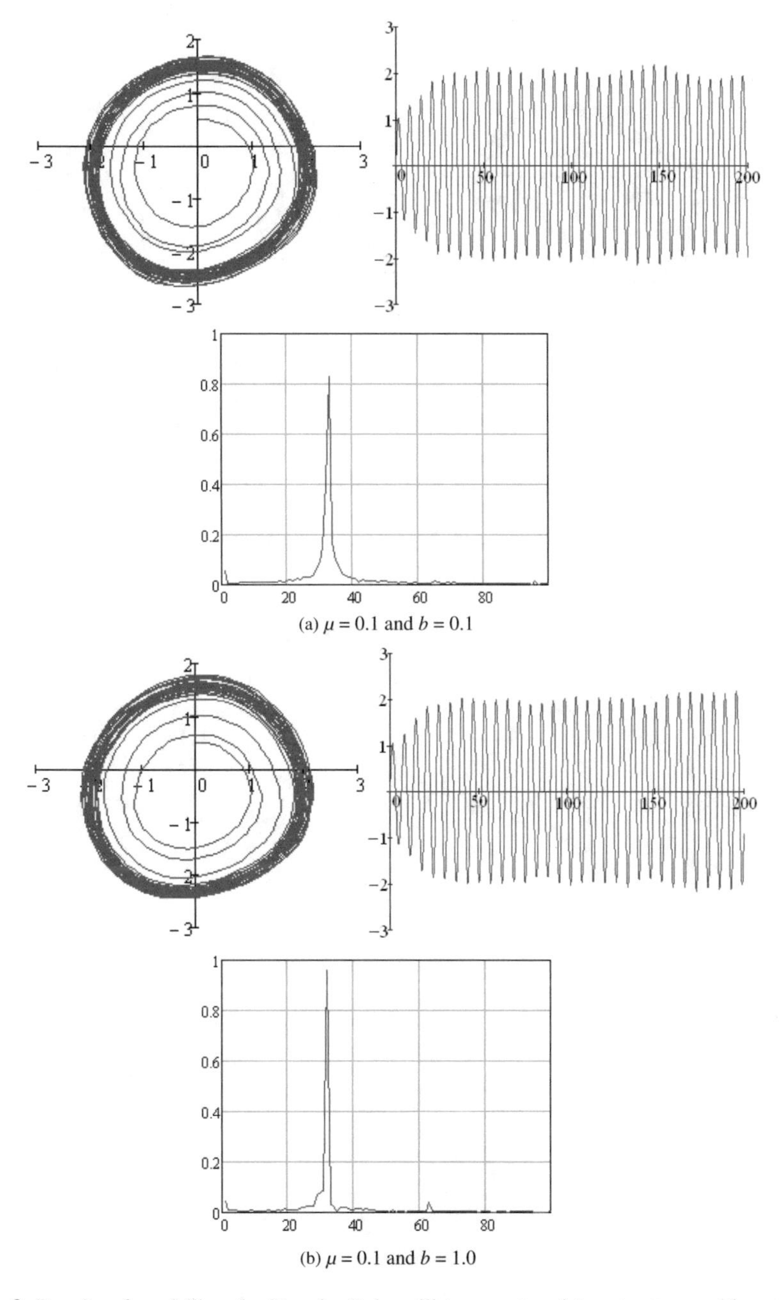

(a) $\mu = 0.1$ and $b = 0.1$

(b) $\mu = 0.1$ and $b = 1.0$

Fig. 8 Results of modelling the Van der Pol oscillators on transistor structures with negative differential resistance at $q = 1.0$ (1/V ^{2}) and different values of the self-oscillating system (6) parameters in the presence of additive white Gaussian noise with intensity $D = 0.125$ (W/Hz)

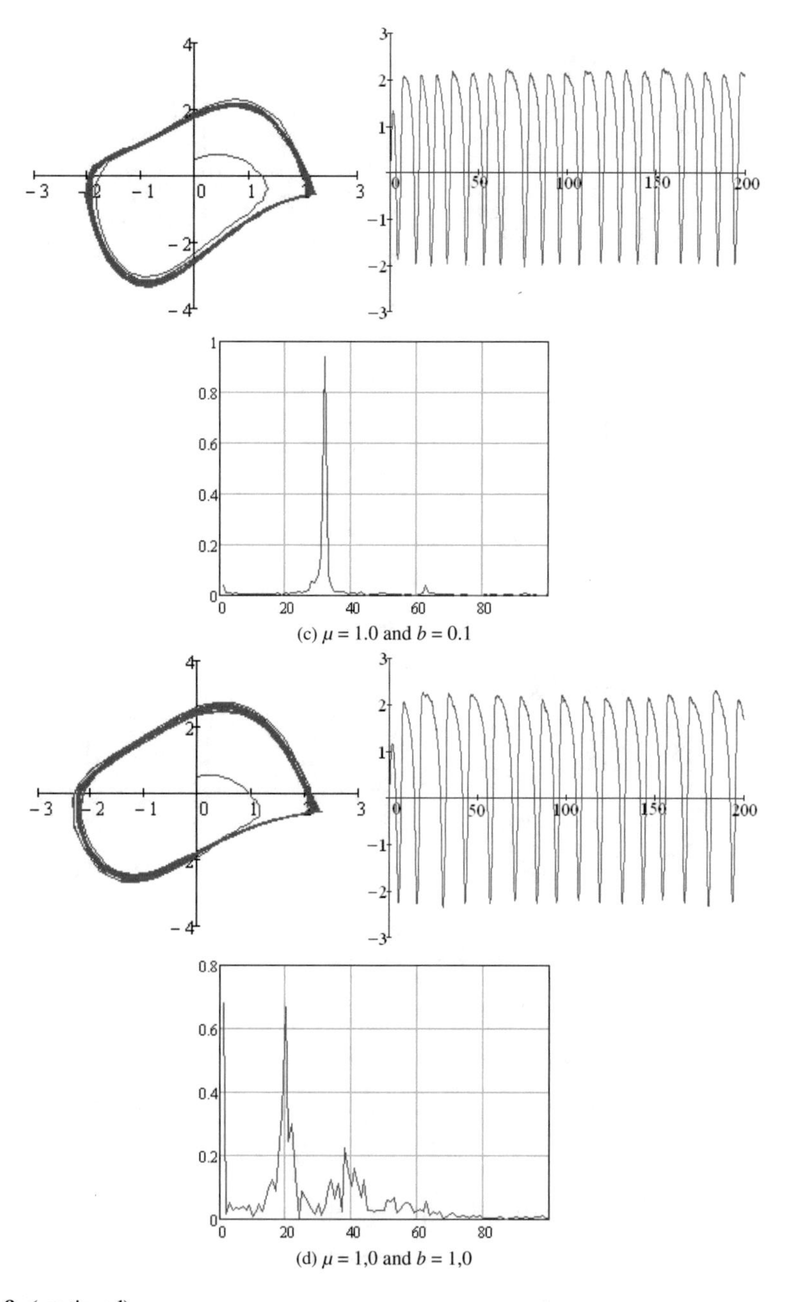

(c) $\mu = 1.0$ and $b = 0.1$

(d) $\mu = 1,0$ and $b = 1,0$

Fig. 8 (continued)

Figures 6, 7 and 8 present phase portraits of oscillators in the plane of dynamic variables $x_1 - x_2$, diagrams of the generated voltage oscillations relative to the normalized time $\tau = \omega_0 t$ and the amplitude spectra. Under the action of additive white Gaussian noise, amplitude modulation of oscillations and random fluctuations of the phase shift take place. Thus, the solution of the differential Eq. (6) for the voltage generated in the presense of noise in general case is [66]

$$u(t) = U_0[1 + a(t)]\cos[\omega_0 t - \psi(t)], \quad (23)$$

where $a(t)$ describes the amplitude modulation, $\psi(t)$ describes phase modulation, U_0 is the amplitude of stationary oscillations in the absence of noise. Functions $a(t)$ and $\psi(t)$ are stationary random processes in the transistor structures with negative differential resistance. In practice, the spectral components of a high-frequency oscillating system based on the capacitive effect of the transistor structure with negative differential resistance are much lower in the frequency range than the self-oscillation of the oscillator [66].

5 Voltage-Controlled Van der Pol Microwave Oscillators on Transistor Structures with Negative Differential Resistance

The voltage-controlled microwave oscillators on transistor structures with negative differential resistance were constructed according to the equivalent circuit in Fig. 1 and the mathematical model of Van der Pol (4)–(5). Previuos studies carried out by the authors [68–74] have shown that in the UHF and microwave ranges the reactive properties and the magnitude of the negative resistance of transistor structures provided conditiond for the single-frequency quasiharmonic oscillation excitation. Electrical diagrams of these oscillators are given in Fig. 9. The oscillating circuits of both oscillators are formed by a parallel connection of the inductor with the capacitive component of the corresponding transistor structure. To increase the operating frequency of the oscillations in both circuits, a dynamic power supply using the RC1 circuit is used. The capacitance C1 shunts the AC channel of the transistor VT2, while the DC resistor R limits the base current of the transistor VT1 and forms a static I–V curve of Λ-type [69]. Theoretical and experimental studies carried out by the authors show that frequency properties of the transistor VT1 must be considered when constructing these oscillator [68, 69]. The cutoff frequency of the transistor VT2 may be 1.5.2 times less than the lower frequency of the operating range of the oscillator. In the oscillator circuit based on HEMT (Fig. 9b), the control parameter is the amount of optical lighting power acting on the p-i-n photodiode VD1 [69].

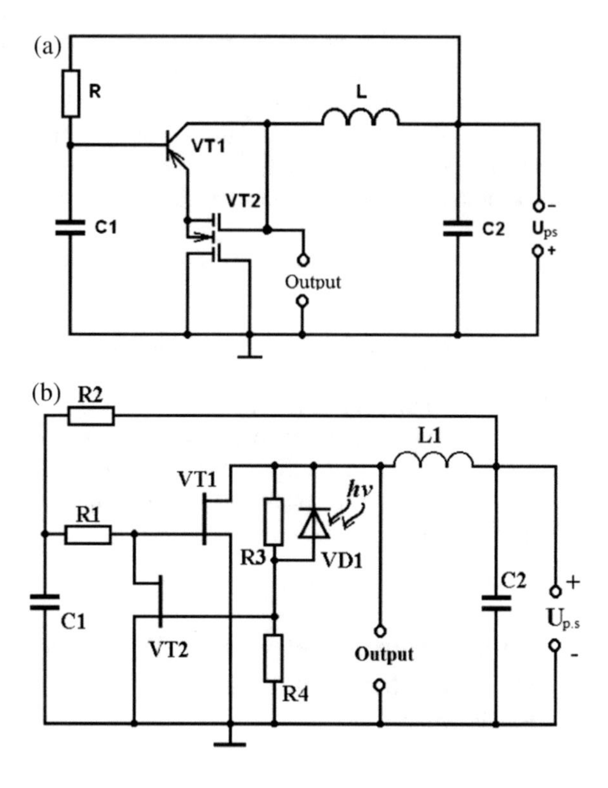

Fig. 9 Electrical circuits of the Van der Pol oscillators based on transistor structures with negative differential resistance: **a** BJT-MOS [72], **b** HEMT [69]

Figure 9a illustrates the electrical circuit of the electrically controlled UHF Van der Pol oscillator on a bipolar field-effect transistor structure with negative differential resistance [69]. Its oscillating circuit is formed by the inductor and the reactive component of the impedance of the transistor structure, consisting of bipolar and MOS transistors. The capacitive effect of the transistor structure takes place at electrodes of the bipolar transistor collector and of the MOS transistor drain. The phase-shifting circuit RC1 provides occurrence of a reactive capacitive component at electrodes of the bipolar transistor collector and of the MOS transistor drain. The capacitor C2 prevent generated signal's passing through the DC power supply $U_{p.s.}$ [68, 69, 72, 74].

Figure 9b shows a circuit diagram of the optically controlled UHF Van der Pol oscillator on the HEMT transistor structure with negative differential resistance [69]. The oscillating circuit is formed by the capacitive component of the impedance at drain electrodes of the transistor and the source of the transistor VT2 and the passive inductance L1. Circuit R2C1 increases the dynamic negative resistance of the HEMT transistor structure with negative differential resistance. The effect of optical radiation on the p-i-n photodiode VD1 leads to a change in the capacitive component of the impedance of the drain-source electrodes, transistors VT1 and VT2, which provides the adjustment of the generation frequency. Resistors R1, R3, R4 and the DC voltage

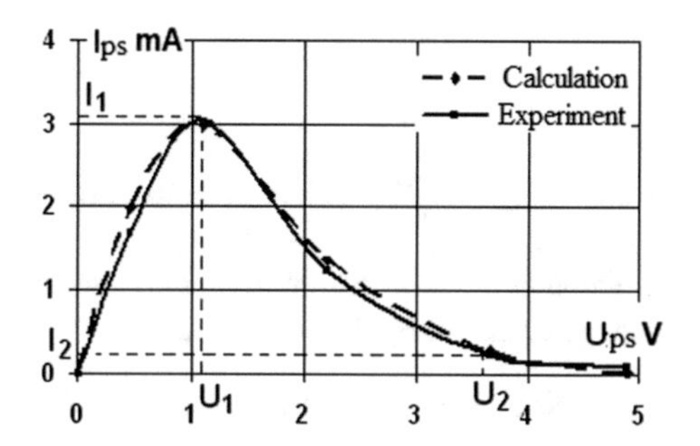

Fig. 10 Static I–V curve of the BJT-MOS transistor structure with negative differential resistance

source $U_{p.s.}$ provide the choice of the operating point on the descending part of the I–V curve of the HEMT structure [73].

The study of dynamic processes in the stationary mode of operation for the Van der Pol microwave oscillators was carried out by approximating the static I–V curves of transistor structures with negative differential resistance by semi-empirical equations using the hyperbolic tangent function:

- for the bipolar field-effect transistor structure with negative differential resistance

$$I_T(U_{p.s.}) = g_S U_{p.s.} + a(U_2 - U_{p.s.})^2 th\frac{eU_{p.s.}}{kT}, \tag{24}$$

where $a = \frac{I_1 - g_S U_1}{(U_2 - U_1) th \frac{1}{2}\frac{eU_1}{kT}}$; $g_S = \frac{I_2}{U_2}$; I_1, U_1, I_2 and U_2 are the coordinates of the end points of the descending section of the I–V curve (Fig. 10);

- for the HEMT transistor structure with negative differential resistance

$$I(U_{p.s.}) - \frac{U_{p.s}}{R_2 + R_3} + I_{C0}(1 - P1(U_{p.s.}))^2 (thM)^{-1} th\left[M\frac{U_{p.s.}/U_0}{1 - P1(U_{p.s.})}\right], \tag{25}$$

where

$$P1(U_{p.s.}) = \frac{U_{p.s.}}{U_0}\frac{R_2}{R_1 + R_2 + R_{VT2}} - SR_1\left[\frac{U_{p.s.}}{U_0}\frac{R_3}{R_2 + R_3} - \frac{bU_{p.s.}^2}{U_0}\left(\frac{R_3}{R_2 + R_3}\right)^2 - 1\right], \tag{26}$$

where I_{D0} is the drain current at $U_{GS} = 0$, $U_{DS} = U_0$; U_{GS}, U_{DS} is the voltage on electrodes, respectively, gate-source and drain-source; U_0 is the cut-off voltage. The parameter M is defined as

$$M = S_{\max}\frac{U_0}{I_{D0}}, \tag{27}$$

where $S_{\max} = \frac{dI_D}{dU_{DS}}$ is the steepness of the output characteristics of the HEMT transistor at $U_{DS} = U_{GS} = 0$.

Dependences of differential conductivities of transistor structures on the supply voltage obtained by the authors are:

- for bipolar-field-effect transistor structure

$$G_{\sim}(U_{p.s.}) = g_S - a(U_2 - U_{p.s.})\left[2th\frac{eU_{p.s.}}{kT} - \frac{e}{kT}(U_2 - U_{p.s.})ch^{-2}\frac{eU_{p.s.}}{kT}\right]; \tag{28}$$

- for the HEMT transistor structure

$$\begin{aligned} G_{\sim}^{(-)}\left(U_{p.s.}\right) &= \frac{1}{R2+R3} + \frac{I_{S0}}{U_0}(\tanh M)^{-1}\left[1 - P1\left(U_{p.s.}\right)\right]\times \\ &\times\left\{2\cdot S\cdot R1\cdot\left[\frac{R3}{R2+R3} - 2bU_{p.s.}\left(\frac{R3}{R2+R3}\right)^2\right]\cdot\tanh\left(M\frac{U_{p.s.}/U_0}{1-P1\left(U_{p.s.}\right)}\right) + \right. \\ &\left. + \frac{M\left[1-P1\left(U_{p.s.}\right)\right]}{\cosh^2\left[M\frac{U_{p.s.}/U_0}{1-P1\left(U_{p.s.}\right)}\right]}\cdot\frac{2\left[1-P1\left(U_{ж}\right)\right] - S\cdot R1 - 1}{\left[1-P1\left(U_{p.s.}\right)\right]^2}\right\}. \end{aligned} \tag{29}$$

For providing the stability, parameters of electrical circuits of the microwave Van der Pol oscillators on transistor structures with negative differential resistance, are calculated according to the equations of quasiharmonic electric oscillations generated in the stationary mode obtained by the authors [68–74].

Dependences of the first harmonic of the currents of transistor structures on the voltage amplitude in real time are as follows:

- for the bipolar-field-effect transistor structures

$$i(u) = \left(g_S + a\frac{eU_2^2}{kT} - \frac{3}{4}\frac{ae}{kT}\left(\frac{U_2^2}{3}\left(\frac{e}{kT}\right)^2 - 1\right)U^2\right)U\cos\omega_0 t + \ldots, \tag{30}$$

- for the HEMT transistor structures

$$i_T(u) = \left[\frac{1}{R_2 + R_3} + \frac{I_{D0}}{U_0} \frac{M}{thM} \left(P_2 - \frac{U^2}{U_0^2} \left(\frac{R_3}{R_2 + R_3} \right)^2 P_3 \right) \right] U \cos \tau + \ldots \quad (31)$$

where U is the amplitude of the generated signal. The coefficients P_2 and P_3 are determined from the ratios

$$P_2 = \left(1 + SR_1 + S^2 R_1^2\right)^2 - 3SR, \quad (32)$$

$$P_3 = \left(1 + SR_1 + S^2 R_1^2\right)\left(S^2 R_1^2 - 2bU_0(1 + SR_1)\right) + SR_1(bU_0 - 1), \quad (33)$$

where S is the steepness of the HEMT transfer characteristic in the gain mode, b is the coefficient of the approximation Eq. (34), which is determined from the static I–V curve of the HEMT

$$I_C(U_{GS}) = S\left(U_{GS} - U_0 - bU_{GS}^2\right), \quad (34)$$

where U_{GS} is the gain-source voltage.

Amplitude differential equations describing the oscillators in Fig. 9 are:

- for the bipolar-field-effect transistor structure

$$RC_1 \frac{dU}{dt} = \left[\left(g_S + a \frac{eU_2^2}{kT} \right) R - 1 \right] U - \frac{3}{4} \frac{ae}{kT} \left(\frac{U_2^2}{3} \left(\frac{e}{kT} \right)^2 - 1 \right) RU^3, \quad (35)$$

- for the HEMT transistor structure

$$R_2 C_1 \frac{dU}{dt} = \left(\frac{R_{eq}}{R_2 + R_3} + \frac{I_{D0} R}{U_0} \frac{M}{thM} P_2 - 1 \right) U - I_{D0} R \frac{M}{thM} \frac{U^3}{U_0^3} P_3 \left(\frac{R_3}{R_2 + R_3} \right)^2. \quad (36)$$

Conditions of self-excitation (balance of amplitudes) of the Van der Pol oscillators (Fig. 9) are determined by relations:

- for the bipolar-field-effect transistor structure

$$\left(g_S + a \frac{eU_2^2}{kT} \right) R > 1. \quad (37)$$

- for the HEMT transistor structure

$$\frac{R_{eq}}{R_2 + R_3} + \frac{I_{C0} R_{eq}}{U_0} \frac{M}{thM} P_2 > 1. \tag{38}$$

Amplitudes of the stationary oscillations of the Van der Pol microwave oscillators are:

- for the bipolar-field-effect transistor structure

$$U_{DT} = 2\sqrt{\left(g_S + a\frac{eU_2^2}{kT}\right)R - 1} / \sqrt{\frac{ae}{kT}\left(U_2^2\left(\frac{e}{kT}\right)^2 - 3\right)R}. \tag{39}$$

- for the HEMT transistor structure

$$U_{ST} = U_0\sqrt{\frac{R_{eq}}{R_2 + R_3} + \frac{I_{D0} R_{eq}}{U_0} \frac{M}{thM} P_2 - 1} / \sqrt{\frac{M}{thM} \frac{I_{D0} R_{eq}}{U_0} P_3\left(\frac{R_3}{R_2 + R_3}\right)^2}. \tag{40}$$

The dependence of the amplitude of quasiharmonic oscillations of the Van der Pol microwave oscillators in real time is

$$U(t) = U_0(\exp \gamma t) \Big/ \sqrt{1 + \left(U_0^2 / U_{ST}^2\right)(\exp 2\gamma t - 1)}, \tag{41}$$

where $U_0 = U(0)$ is the initial amplitude of the generated oscillations. The coefficient γ is determined from the ratios:

- for the bipolar-field-effect transistor structure

$$\gamma = \left(\left(g_S + a\frac{eU_2^2}{kT}\right)R - 1\right)/T, \tag{42}$$

- for the HEMT transistor structure

$$\gamma = \left(\frac{R_{eq}}{R_1 + R_2} + \frac{I_{D0} R_{eq}}{U_0} \frac{M}{thM} P2 - 1\right)/T \tag{43}$$

When the Van der Pol oscillator operates in the microwave range, the HEMT source currents are the sum of two components, 2D-electron gas and electrons AlGaAs [73]. Each of these components is described by the equation [69, 73]

$$I_{Ci} = qZn_{Si}(U_{GS}, U_i(x))\nu_i(x), \quad i = 1, 2, \tag{44}$$

where Z is the gain width; q is the electron charge; υ_i is the electron drift velocity; U_i is the simulation parameter. Analytically, the dependence of the drift velocity on the field strength, taking into account the effect of saturation is [69]

$$\nu_i(x) = \begin{cases} \frac{\mu_i E(x)}{1+E(x)/E_{Hi}}, & E(x) < E_{Hi}; \\ V_{Hi}, & E(x) \geq E_{Hi}; \end{cases} \tag{45}$$

where μ_i is the electron mobility in a weak field; E_{Hi} is the critical value of the field at which saturation occurs; V_{Hi} is the electron velocity at $E_i = E_{Hi}$.

The drain currents of the HEMT in the linear mode and the saturation mode can be described by Eqs. (46) and (47), respectively [69]

$$I_{\text{сл}_i} = \frac{A_i}{L + U_{\text{сл}} / E\text{н}_i}\left[B_i U_{\text{сл}} - \ln\cosh\frac{U_{GS} - U_{mi} - U_{\text{сл}}}{U_{1i}} + \ln\cosh\frac{U_{GS} - U_{mi}}{U_{1i}}\right]; \tag{46}$$

$$I_{\text{сн}_i} = \frac{A_i}{L\text{н}_i + U\text{н}_i / E\text{н}_i}\left[B_i U_{\text{сл}} - \ln\cosh\frac{U_{GS} - U_{mi} - U_{\text{сл}}}{U_{1i}} + \ln\cosh\frac{U_{GS} - U_{mi}}{U_{1i}}\right], \tag{47}$$

where

$$A_i = q\mu_i Zn_{0i}(1 - \alpha_i)U_{li} \tag{48}$$

and

$$B_i = \alpha_i/(1 - \alpha_i)U_{li}, \tag{49}$$

$U_{\text{н}i}$ is the potential of the HEMT channel at the saturation point, $L_{\text{н}}$ is the effective channel length in saturation and is equal to [69]

$$L\text{н}_i = L - \frac{2d\text{н}_i}{\pi}\sinh^{-1}\frac{\pi\left(U_{DS} - U\text{н}_i\right)}{2d\text{н}_i E\text{н}_i}. \tag{50}$$

Experimental results received by the authors are presented in Figs. 11 and 12 [69, 73]. Figure 11 depicts that the effective relative frequency retuning is 3.4% for the range of 880...910 MHz. Figure 12 depicts that the relative optical retuning of the generation frequency is 9.88%, and the electrical one 3.7% [69, 73]. In [68] results of the experimental study of microwave Van der Pol oscillators on bipolar static

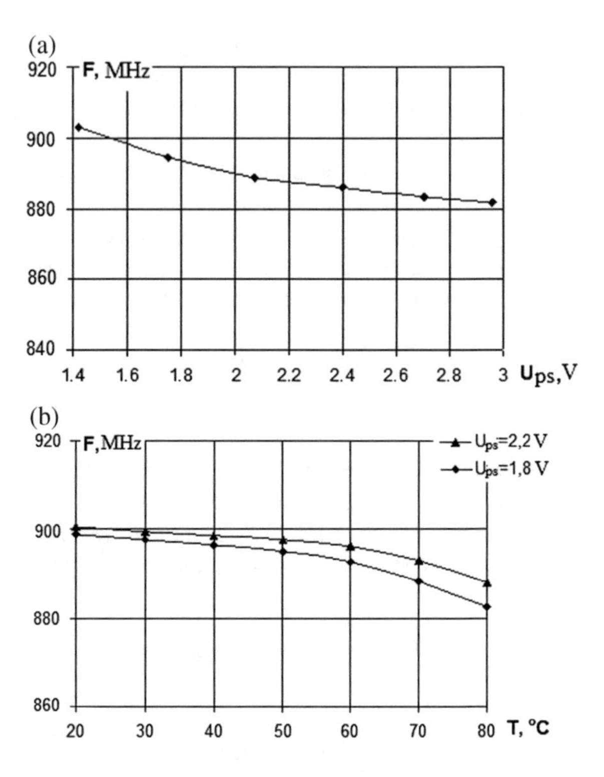

Fig. 11 Dependence of the generation frequency on the supply voltage (**a**) and dependence of the generation frequency on the temperature at different values of the supply voltage (**b**) for the microwave Van der Pol oscillator on the BJT-MOS transistor structure with negative differential resistance

inductance transistor structures with negative differential resistance are given. The results show generation oscillations in the frequency ranges from 770 to 910 MHz [68]. The relative electrical retuning of the generation frequency is 16.67%.

6 Conclusions

Application of nonlinear and reactive properties of the transistor structures with negative differential resistance provides elimination of imbalances in development of the Van der Pol oscillators. These imbalances are the impossibility to obtaine self-oscillations with low harmonic coefficient along with short time of the stationary self-oscillation fixing, and to save the stability margin while electric retuning in wide limits parameters of the self-oscillating system of the oscillator.

There were obtained new analytical ratios that determine the excitation conditions, the amplitude of stationary oscillations and the stability margin of the voltage-controlled microwave Van der Pol oscillators on BT-MDN and HEMT transistor structures with negative differential resistance. These equations were obtained on the basis of nonlinear approximation of static I–V curves of the transistor structures with negative differential resistance by semi-empirical equations.

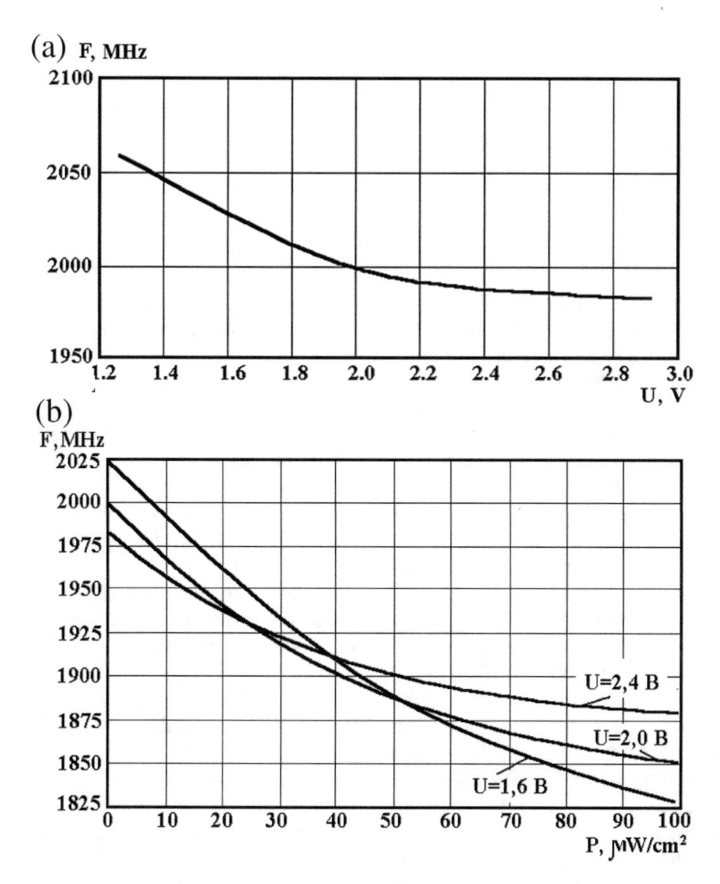

Fig. 12 Dependences of the frequency of generation for the microwave Van der Pol oscillator on the HEMT transistor structure with negative differential resistance on the supply voltage (**a**) and the power density of the radiation at different values of the supply voltage (**b**)

Phase portraits, time diagrams and amplitude spectra of electric oscillations in the Van der Pol microwave oscillators were obtained. The influence of the parameters of the self-oscillating system of the van der Pol oscillator on te dynamics of the generated oscillations in the presence and absence of additive white noise was estimated. It is shown that the quadratic nonlinearity of the self-oscillating system of the Van der Pol oscillator was shown to impact the amplitude and phase equations of stationary oscillations. But this quadratic nonlinearity does not change the dynamic mode of the Van der Pol oscillators and does not cause appearance of new bifurcations in the phase plane.

New circuits of the Van der Pol oscillators based on nonlinear and reactive properties of BJT-MOS and HEMT transistor structures with negative differential resistance for equipment of infocommunication systems were proposed. The oscillators were studies theoretically and experimentally. Mathematical models of the Van der Pol oscillators on transistor structures with negative differential resistance were developed. Results of numerical modeling and experimental study were considered. The convergence of the obtained results confirms the adequacy of the developed mathematical models.

References

1. Kryvinska N (2010) Converged network service architecture: a platform for integrated services delivery and interworking. Electronic Business series, vol 2. International Academic Publishers, Peter Lang Publishing Group
2. Kryvinska N (2008) An analytical approach for the modeling of real-time services over IP network. Math Comput Simul 79(4):980–990. https://doi.org/10.1016/j.matcom.2008.02.016
3. Kryvinska N (2004) Intelligent network analysis by closed queuing models. Telecommun Syst 27:85–98. https://doi.org/10.1023/B:TELS.0000032945.92937.8f
4. Kryvinska N, Zinterhof P, van Thanh D (2007) An analytical approach to the efficient real-time events/services handling in converged network environment. In: Enokido T, Barolli L, Takizawa M (eds) Network-based information systems. NBiS 2007. Lecture notes in computer science, vol 4658. Springer, Berlin, Heidelberg
5. Kryvinska N, Zinterhof P, van Thanh D (2007) New-emerging service-support model for converged multi-service network and its practical validation. In: First international conference on complex, intelligent and software intensive systems (CISIS'07). IEEE, pp 100–110. https://doi.org/10.1109/cisis.2007.40
6. Ageyev DV, Salah MT (2016) Parametric synthesis of overlay networks with self-similar traffic. Telecommun Radio Eng (English translation of Elektrosvyaz and Radiotekhnika) 75(14):1231–1241. https://doi.org/10.1615/TelecomRadEng.v75.i14.10
7. Ageyev D, Qasim N (2015) LTE EPS network with self-similar traffic modeling for performance analysis. In: Proceedings of the 2015 second international scientific-practical conference problems of infocommunications science and technology (PIC S&T). Kharkov, Ukraine, IEEE, 13–15 Oct 2015, pp 275–277. https://doi.org/10.1109/infocommst.2015.7357335
8. Ageyev DV, Evlash DV (2008) Multiservice telecommunication systems design with network's incoming self-similarity flow. In: Proceedings of the international conference on modern problems of radio engineering, telecommunications and computer science (TCSET 2008), Lviv-Slavsko, 19–23 Feb 2008, pp 403–405
9. Ageyev D et al (2018) Method of self-similar load balancing in network intrusion detection system. In: 2018 28th International conference radioelektronika (RADIOELEKTRONIKA), IEEE, 19–20 Apr 2018, pp 1–4. https://doi.org/10.1109/radioelek.2018.8376406
10. Ageyev D et al (2019) Infocommunication networks design with self-similar traffic. In: 2019 IEEE 15th international conference on the experience of designing and application of CAD systems (CADSM). IEEE, 26 Feb–2 Mar 2019, pp 24–27. https://doi.org/10.1109/cadsm.2019.8779314
11. Ageyev D et al (2018) Provision security in SDN/NFV. In: 2018 14th International conference on advanced trends in radioelecrtronics, telecommunications and computer engineering (TCSET), IEEE, 20–24 Feb 2018, pp 506–509. https://doi.org/10.1109/tcset.2018.8336252
12. Ageyev D et al (2018) Classification of existing virtualization methods used in telecommunication networks. In: Proceedings of the 2018 IEEE 9th international conference on dependable systems, services and technologies (DESSERT), IEEE, 24–27 May 2018, pp 83–86. https://doi.org/10.1109/dessert.2018.8409104
13. Radivilova T et al (2018) Decrypting SSL/TLS traffic for hidden threats detection. In: Proceedings of the 2018 IEEE 9th international conference on dependable systems, services and technologies (DESSERT), IEEE, 24–27 May 2018, pp 143–146. https://doi.org/10.1109/dessert.2018.8409116

14. Radivilova T, Kirichenko L, Ageiev D, Bulakh V (2020) The methods to improve quality of service by accounting secure parameters. In: Hu Z, Petoukhov S, Dychka I, He M (eds) Advances in computer science for engineering and education II. ICCSEEA 2019. Advances in intelligent systems and computing, vol 938. Springer, Cham
15. Radivilova T, Kirichenko L, Ivanisenko I (2016) Calculation of distributed system imbalance in condition of multifractal load. In: Proceedings of the third international scientific-practical conference problems of infocommunications science and technology, Kharkiv, Ukraine, 4–6 Oct 2016, pp 156–158. https://doi.org/10.1109/infocommst.2016.7905366
16. Ivanisenko I, Kirichenko L, Radivilova T (2016) Investigation of multifractal properties of additive data stream. In: Proceedings of the IEEE first international conference on data stream mining & processing, Lviv, Ukraine, 23–27 Aug 2016, pp 305–308. https://doi.org/10.1109/dsmp.2016.7583564
17. Ivanisenko I, Radivilova T (2015) The multifractal load balancing method. In: Proceedings of the second international scientific-practical conference problems of infocommunications science and technology, Kharkiv, Ukraine, 13–15 Oct 2015, pp 123–123. https://doi.org/10.1109/infocommst.2015.7357289
18. Skulysh M (2017) The method of resources involvement scheduling based on the long-term statistics ensuring quality and performance parameters. In: Proceedings of the 2017 international conference on information and telecommunication technologies and radio electronics, Odessa, Ukraine, 11–15 Sept 2017. https://doi.org/10.1109/ukrmico.2017.8095430
19. Globa L, Skulysh M, Reverchuk A (2014) Control strategy of the input stream on the online charging system in peak load moments. In: Proceedings of the 24th international crimean conference microwave & telecommunication technology, sevastopol, Ukraine, 7–13 Sept 2014. https://doi.org/10.1109/crmico.2014.6959409
20. Skulysh M, Romanov O (2018) The structure of a mobile provider network with network functions virtualization. In: Proceedings of the 14th international conference on advanced trends in radioelecrtronics, telecommunications and computer engineering, Lviv-Slavske, Ukraine, 20–24 Feb 2018. https://doi.org/10.1109/tcset.2018.8336370
21. Skulysh MA (2019) Managing the process of servicing hybrid telecommunications services. Quality control and interaction procedure of service subsystems. Adv Intell Syst Comput 889:244–256. https://doi.org/10.1007/978-3-030-03314-9_22
22. Kennedy MP, Rovatti R, Setti G (2000) Chaotic electronics in telecommunications. CRC Press London
23. Kushnir NY, Horlsy PP, Grygoryshyn AN (2005) Dynamic chaos in phase syncronization devices. In: 2005 15th International crimean conference microwave and telecommunication technology, CriMiCo'2005—conference proceedings, vol 1, pp 346–347. https://doi.org/10.1109/crmico.2005.1564936
24. Anishchenko VS, Vadivasova TE, Strelkova GI (2014) Deterministic nonlinear systems. Springer International Publishing, Switzerland, A Short Course
25. Galiuk SD, Kushnir MY, Politanskyi RL (2011) Communication with use of symbolic dynamics of chaotic systems. In: CriMiCo 2011—2011 21st international crimean conference: microwave and telecommunication technology, conference proceedings, pp 423–424
26. Chua LO, Yu J, Yu Y (1985) Bipolar—JFET—mosfet negative resistance devices. IEEE Trans Circuits Syst 32(1):46–61. https://doi.org/10.1109/TCS.1985.1085599
27. Kumar U (2000) Design of an indigenized negative resistance characteristics curve tracer. Active Passuve Elec Comp 23:13–23
28. Kumar U (2003) Simulation of a novel Bipolar-FET type-S negative resistance circuit. Active and Passive Elec Comp 26:129–132
29. O'Donoghue K, Kennedy MP, Forbes P, Qu M, Jones S (2005) A fast and simple implementation of chua's oscillator with cubic-like nonlinearity. Int J Bifurcat Chaos Appl Sci Eng 15(09):2959–2971. https://doi.org/10.1142/S0218127405013800
30. O'Donoghue K, Kennedy MP, Forbes P (2005) A fast and simple implementation of Chua's oscillator using a "cubic-like" Chua diode. In: Proceedings of the 2005 European conference on circuit theory and design, Cork, Ireland, 2 Sept 2005, vol 2, pp 1–3. https://doi.org/10.1109/ecctd.2005.1522998

31. Eltawil AM, Elwakil AS (1999) Low-voltage chaotic oscillator with an approximate cubic nonlinearity. Int J Electron Commun (AEU) 53(3):11–17
32. Kushnir M, Galiuk S, Rusyn V, Kosovan G, Vovchuk D (2014) Computer modeling of information properties of deterministic chaos. In: Proceedings of the 7th chaotic modeling and simulation international conference (CHAOS 2014), pp. 265–276
33. Rusyn V, Kushnir M, Galameiko O (2012) Hyperchaotic control by thresholding method. In: Proceedings of international conference on modern problem of radio engineering, telecommunications and computer science, Lviv-Slavske, Ukraine, 21–24 Feb 2012, p 67
34. Semenko AI, Bokla NI, Kushnir MY, Kosovan GV (2018) Features of creating based on chaos pseudo-random sequences. In: Proceedings 14th international conference on advanced trends in radioelectronics, telecommunications and computer engineering, TCSET 2018, Slavske, Ukraine, 20–24 Feb 2018, pp 1087–1090. https://doi.org/10.1109/tcset.2018.8336383
35. Semenov A, Osadchuk O, Semenova O et al (2018) Signal statistic and informational parameters of deterministic chaos transistor oscillators for infocommunication systems. In: 2018 International scientific-practical conference problems of infocommunications. Science and technology (PIC S&T), Kharkiv, Ukraine, 9–12 Oct 2018, pp 730–734. https://doi.org/10.1109/infocommst.2018.8632046
36. Kushnir M, Ivaniuk P, Vovchuk D, Galiuk S (2015) Information security of the chaotic communication systems. In: Proceedings of the 8th Chaotic Modeling and Simulation International Conference, CHAOS 2015, Paris, France, pp 441–452
37. Volos CK, Kyprianidis IM, Stouboulos IN (2006) Experimental demonstration of a chaotic cryptographic scheme. WSEAS Trans Circ Syst 5(11):1654–1661
38. Volos CK, Kyprianidis IM, Stouboulos IN (2006) Chaotic cryptosystem based on inverse duffing circuit. In: Proc of the 5th WSEAS international conference on non-linear analysis, non-linear systems and chaos, Bucharest, Romania, Oct 16–18, 2006, pp 92–97
39. Semenov A, Baraban S, Semenova O, Voznyak O, Vydmysh A, Yaroshenko L (2019) Statistical express control of the peak values of the differential-thermal analysis of solid materials. Solid State Phenom 291:28–41. https://doi.org/10.4028/www.scientific.net/SSP.291.28
40. Ulansky V, Raza A, Oun H (2019) Electronic circuit with controllable negative differential resistance and its applications. Electronics 8(4):409. https://doi.org/10.3390/electronics8040409
41. Ulansky VV, Ben Suleiman SF (2013) Negative differential resistance based voltage-controlled oscillator for VHF band. In: 2013 IEEE XXXIII international scientific conference electronics and nanotechnology (ELNANO), Kiev, 16–19 Apr 2013, pp 80–84. https://doi.org/10.1109/elnano.2013.6552016
42. Semenov AO, Baraban SV, Osadchuk OV et al (2020) Microelectronic pyroelectric measuring transducers. In: Tiginyanu I, Sontea V, Railean S (eds) 4th International conference on nanotechnologies and biomedical engineering, ICNBME 2019, Chisinau, Moldova, 18–21 Sept 2019. IFMBE proceedings, vol 77, pp 393–397. https://doi.org/10.1007/978-3-030-31866-6_72
43. Semenov A, Osadchuk O, Semenova O et al (2021) Research of dynamic processes in the deterministic chaos oscillator based on the colpitts scheme and optimization of its self-oscillatory system parameters. Lect Notes Data Eng Commun Technol 48:181–205. https://doi.org/10.1007/978-3-030-43070-2_10
44. Ulansky VV et al (2016) Optimization of NDR VCOs for microwave applications. In: 2016 IEEE 36th international conference on electronics and nanotechnology (ELNANO), Kiev, 19–21 Apr 2016, pp 353–357. https://doi.org/10.1109/elnano.2016.7493083
45. Tamaševičius A, Bumelienė S, Kirvaitis R et al (2009) Autonomous duffing-holmes type chaotic oscillator. Elektronika ir Elektrotechnika 5(93):43–46
46. Tamaševičiūtė E, Tamaševičius A, Mykolaitis G et al (2008) Analogue electrical circuit for simulation of the duffing-holmes equation. Nonlinear Anal Model Control 13(2):241–252
47. IEEE standard for information technology (2007) Telecommunications and information exchange between systems. Local and metropolitan area networks. Specific requirements. Part 15.4: wireless medium access control (MAC) and Physical Layer (PHY) Specifications for low-rate wireless personal area networks (WPANs); Amendment 1: Add Alternate PHYs

48. Revision of Part 15 of the commission's rules regarding ultra-wideband transmission systems, first report and order. Federal communications commission (FCC), ET Docket 98–153, FCC 02-48; Adopted: February 14, 2002; Released: April 22, 2002
49. Report from the commission to the european parliament and the council on the implementation of the radio spectrum policy programme, brussels, 22 Apr 2014. 13 p. https://eur-lex.europa.eu/LexUriServ/LexUriServ.do?uri=COM:2014:0228:FIN:EN:PDF. Accessed 15 Jan 2020
50. Khanna APS (2015) State of the art in microwave VCOs. Microwave J 58(5):22–42
51. Do TNT, Lai S, Hörberg M et al (2015) A MMIC GaN HEMT voltage-controlled-oscillator with high tuning linearity and low phase noise. In: 2015 IEEE compound semiconductor integrated circuit symposium (CSICS), 11–14 Oct 2015, New Orleans, LA, USA, pp 1–4. https://doi.org/10.1109/csics.2015.7314478
52. Kang S, Chien JC, Niknejad AM (2011) A 100 GHz phase-locked loop in 0.13 μm SiGe BiCMOS process. In: 2011 IEEE radio frequency integrated circuits symposium (RFIC), 5–7 June 2011, Baltimore, MD, USA, pp 1–4. https://doi.org/10.1109/rfic.2011.5940606
53. Kang S, Chien JC, Niknejad AM (2014) A W-Band low-noise PLL with a fundamental VCO in SiGe for millimeter-wave applications. IEEE Trans Microw Theory Tech 62(10):2390–2404. https://doi.org/10.1109/tmtt.2014.2345342
54. Gan KJ, Jiang ZJ, Chen YH, Yeh WK (2016) Application of NDR-based Van der Pol oscillator based on BiCMOS technology. J Electron Commun 16(1):189–198. https://doi.org/10.17654/EC016010189
55. Gan KJ, Jiang ZJ, Chan DY et al (2013) Design and application of Van Der Pol oscillator using NDR circuit. In: 2013 IEEE international symposium on next-generation electronics (ISNE), 25–26 Feb 2013, Kaohsiung, Taiwan, pp 329–332. https://doi.org/10.1109/isne.2013.6512358
56. Semenov A (2016) The Van der Pol's mathematical model of the voltage-controlled oscillator based on a transistor structure with negative resistance. In: Proceedings of the XIII international conference modern problems of radio engineering, telecommunications, and computer science, Lviv-Slavsko, Ukraine, 23–26 Feb 2016, pp 100–104. https://doi.org/10.1109/tcset.2016.745 1982
57. Kuznetsov AP et al (2010) Synchronization in tasks [Sinhronizatsiya v zadachah]. Publishing center Nauka, Saratov, Russia
58. Mathis W, Bremer J (2009) Modelling and design concepts for electronic oscillator and its synchronization. Open Cybern Syst J 3:47–60
59. Rybin YK (2014) Measuring signal generators. Springer, Dordrecht, Heidelberg, London, New York
60. Feudel U, Kuznetsov S, Pikovsky A (2006) Strange nonchaotic attractors. dynamics between order and chaos in quasiperiodically forced systems. In: World scientific series on nonlinear science, series a—vol 56, World Scientific, Singapore
61. Kuznetsov SP (2011) Hyperbolic Chaos: A Physicist's View. In: Luo ACJ, Afraimovich V (eds) Series: mathematical methods and modeling for complex phenomena. Higher Education Press, Beijing and Springer: Heidelberg, Dordrecht, London, New York
62. Kuznetsov SP (2006) Dinamicheskiy haos [Dynamical chaos], 2nd edn. Fizmatlit, Moscow, Russia
63. Emelianova YP, Kuznetsov AP, Turukina LV (2014) Quasi-periodic bifurcations and "amplitude death" in low-dimensional ensemble of van der Pol oscillators. Phys Lett A 378(3):153–157. https://doi.org/10.1016/j.physleta.2013.10.049
64. Kuznetsov AP, Seleznev EP, Stankevich NV (2012) Nonautonomous dynamics of coupled van der Pol oscillators in the regime of amplitude death. Commun Nonlinear Sci Numer Simul 17(9):3740–3746. https://doi.org/10.1016/j.cnsns.2012.01.019
65. Kuptsov PV, Kuptsova AV (2017) Radial and circular synchronization clusters in extended starlike network of van der Pol oscillators. Commun Nonlinear Sci Numer Simul 50:115–127. https://doi.org/10.1016/j.cnsns.2017.03.003
66. Buckingham NJ (1983) Noise in electronic devices and systems. Wiley, New York
67. Anishchenko VS et al (2005) Statistical properties of dynamical chaos. Phys Usp 48(2):151–166. https://doi.org/10.1070/PU2005v048n02ABEH002070

68. Osadchuk VS, Osadchuk AV, Semenov AA, Semenova EA (2010) Experimental research and simulation of microwave oscillator based on structure of static inductance transistor with negative resistance. In: 2010 20th International crimean conference "microwave & telecommunication technology", Sevastopol, Ukraine, 13–17 September 2010, pp 187–188. https://doi.org/10.1109/crmico.2010.5632543
69. Semenov A, Semenova O, Osadchuk O (2015) The UHF oscillators based on a HEMT structure with negative conductivity. In: 2015 International siberian conference on control and communications (SIBCON), Omsk, Russia, 21–23 May 2015, pp 1–4. https://doi.org/10.1109/sibcon.2015.7147215
70. Osadchuk AV, Semenov AA, Baraban SV et al (2013) Noncontact infrared thermometer based on a self-oscillating lambda type system for measuring human body's temperature. In: 2013 23rd International crimean conference "microwave & telecommunication technology", Sevastopol, Ukraine, 8–14 Sept 2013, pp 1069–1070
71. Semenov A (2017) Mathematical model of the microelectronic oscillator based on the BJT-MOSFET structure with negative differential resistance. In: 2017 IEEE 37th international conference on electronics and nanotechnology (ELNANO), Kiev, Ukraine, 18–20 Apr 2017, pp 146–151. https://doi.org/10.1109/elnano.2017.7939736
72. Semenov A (2016) Mathematical simulation of the chaotic oscillator based on a field-effect transistor structure with negative resistance. In: 2016 IEEE 36th international conference on electronics and nanotechnology (ELNANO), Kiev, Ukraine, 19–21 Apr 2016, pp 52–56. https://doi.org/10.1109/elnano.2016.7493008
73. Osadchuk VS, Osadchuk AV (2005) The magnetic controlled autogenerator superhigh frequencies. In: 2005 15th International crimean conference microwave & telecommunication technology, Sevastopol, Crimea, 12–16 Sept 2005, vol 2, pp 449–450. https://doi.org/10.1109/crmico.2005.1564986
74. Semenov A et al (2019) A deterministic chaos ring oscillator based on a MOS transistor structure with negative differential resistance. In: 2019 IEEE international scientific-practical conference problems of infocommunications, Science and technology (PIC S&T), Kyiv, Ukraine, 8–11 Oct 2019, pp 709–714. https://doi.org/10.1109/picst47496.2019.9061330

Study of the Influence of Changing Signal Propagation Conditions in the Communication Channel on Bit Error Rate

Juliy Boiko, Ilya Pyatin, Lesya Karpova, and Oleksander Eromenko

Abstract The transmission of the information signal through the communication channel is accompanied by the addition of additive white Gaussian noise, industrial interference, atmospheric noise, etc. In addition, the signal may have an additional frequency and phase shift caused by the movement of the receiver concerning the transmitter. The article is devoted to the study of the effect of the listed conditions on the errors number dependence in the communication channel on the signal-to-noise ratio. It also explores the possibilities of reducing the effect of signal propagation conditions in a communication channel by using symbolic synchronization, which is based on a phase-locked loop. The early-late-time and Gardner synchronization error detectors are investigated. The early-late-time synchronization error detector is 1.5 dB more efficient than the Gardner detector at low signal-to-noise ratio and has a simpler implementation scheme. An energy efficiency study of a coherent digital communication system with quadrature phase shift keying modulation at the phase shift in the propagation medium is performed. Increasing of the phase shift from 0 to 40° decreases the energy efficiency by 3 dB at low signal-to-noise ratio. The energy efficiency of a non-coherent digital quadrature phase shift keying modulation system is reduced by 10 dB at the phase shift of 30° in the propagation medium. Adding a symbolic synchronization circle compensates for the rotation of the signal constellation. Increasing of the phase shift in the propagation medium to 45° for a coherent communication system leads to reduction of the energy efficiency by 2 dB. Frequency shifting has a significant impact on the energy efficiency of the communication system. The energy efficiency of the digital communication system decreases by 10 dB when 0.1 Hz frequency offset occurs and symbolic synchronization is missing. Symbol synchronization circuit increases the energy efficiency by 7 dB with a frequency shift of 0.1 Hz. When the value of frequency offset increases, the energy efficiency of coherent communication expands. The efficiency of coherent digital communication is increased by 24 dB with the introduction of frequency shift at 2 Hz.

J. Boiko (✉) · I. Pyatin · L. Karpova · O. Eromenko
Khmelnytskyi National University, 11, Instytuts'ka str., Khmelnytskyi 29016, Ukraine

D. Ageyev et al. (eds.), *Data-Centric Business and Applications*, Lecture Notes on Data Engineering and Communications Technologies 69,
https://doi.org/10.1007/978-3-030-71892-3_4

Keywords Signal processing · Modulation · Synchronization · Error detector · Interpolator · Demodulator

1 Introduction

This paper is an extension of work originally presented in conference name "Synthesis of Ambiguity Functions for Complex Radar Signal Processing" that is published in 2019 IEEE International Scientific-Practical Conference Problems of Infocommunications, Science and Technology (PIC S&T) [1].

The proposed paper (additionally) presents the prospect of further development of the study is to take into account the multipath signal propagation through the communication channel, the use of multi-position types of modulation to accelerate the speed of information transmission and channel coding to noise immunity increase.

In modern systems of information transmission, it is necessary to provide clock synchronization of the transmitter and receiver. The use of independent reference generators leads to a difference in the sampling frequencies of the transmitter and receiver. The receiver must know where each character begins and ends. This information is required to know the appropriate integration interval for deciding on the meaning of a character. Symbol synchronization involves the generation of a portion of the transmitted signal in the receiver. These are rectangular oscillations that are synchronized with the speed of character transmission.

Recently, with the construction of digital communication systems, more and more functions that are traditionally performed by analog devices are being implemented on the basis of Digital Signal Processors (DSP). With the addition of DSP, the boundary between the analogue and digital segments of communication systems which are separated by Analog to Digital Converter (ADC) and moves relentlessly to the antenna.

Software-defined radio (SDR) includes analog components: antennas, prefilters, switches, preamplifiers and power amplifiers and digital components: GPP (General-purpose Processor); DSP; FPGA (Field-Programmable Gate Array); ASIC (Specialized Integrated Circuit). Great number of different functions can be implemented with these unified computing and control nodes, such as modulation, demodulation, filtering, coding. The ASIC scheme is not programmed but is specially designed for each application. It is characterized by low production costs, low power consumption and high productivity.

FPGAs are indispensable in software radios due to their high productivity and programmability. Their computing productivity is much higher than it is in DSP or GPP. This is due to the possibility of quasi-parallel data processing. DSP and GPP schemes only carry out successive but sometimes iterative data processing. The main advantages of DSP and GPP are their programmability and configuration restructuring.

In general, when designing a communication system, the following major factors that influence the signal when it spreads in a communication channel must be taken into account:

- influence of additive white Gaussian noise (thermal noise);
- frequency and phase signal offset, which may be caused by the mutual movement of the transmitter and receiver antenna or their heterodyne mismatch caused by instability of the reference frequency sources;
- signal delay in the communication channel which is caused by propagation from the transmitter antenna to the receiver antenna and in the feeder systems;
- fading caused by multipath signal propagation from transmitter to receiver, as well as by phenomena of refraction in the atmosphere and reverberation from different objects.

When modeling communication systems on a personal computer, particular attention should be given to signal delay in the communication channel. Whereas computer simulation can only be discrete in time so, a fractional delay in the communication channel must be created to properly research the receiver of synchronization systems.

2 Development of Schematic Solutions for Digital Signal Processing

2.1 Digital Processing of Radar Signals

When implementing the algorithm of estimating radar signals in the entire receiver bandwidth and taking into account the angular and amplitude fluctuations, the decisive criterion of efficiency is the quality of their transformation into digital form. At this processing stage all signal components that are outside of the known values are lost (maximum allowable deviation of frequency, amplitude, phase). In most modern radar systems analog-to-digital conversion is performed after the detector [2–4].

When implementing the autocorrelation estimation algorithm, it is necessary to apply a different approach in the construction of radar structures with digital signal processing, namely the application of ADC in the intermediate frequency path-to the detector. In this case, the received signal is sampled immediately after spatial selection in a sufficiently wide frequency band, amplifying and reducing the carrier frequency to a value that is convenient for further processing. In this embodiment, it is possible to perform all operations in digital processing including frequency-time signal processing [5–7] which can be implemented in one sounding, namely:

- optimal one- and multi-channel frequency-time filtering of signals;
- preliminary detection of useful signals by the results of one sounding;
- coordinates measurement corresponding to the detected signals;

- calculation of error signals by angular coordinates, range and speed for accompanying purposes, etc.

To implement the method which is proposed in this work, it is necessary to organize the operation of the signal processing device for digital processing in such a way as to ensure the fulfillment of both the listed tasks and additional ones, which are in the selection of echo-signals by the correlation features properties of their complex bypass, and therefore only quadratic ADC implementation or detection [8, 9].

This approach gives a definite system advantage:

- stability of characteristics throughout the range of operating conditions;
- the adapt ability to changing working conditions;
- device upgrading by modifying the software without changing the hardware,
- reduction of weight, dimensions and as a consequence, a significant reliability increase;
- easy configuration of equipment;
- reduced price compared with the analog variant due to the greater processability and low prices of the components in the wide production.

Signal processing device on the basis of a programmable processor consists of two interrelated parts as shown in Fig. 1, the device of gain and signal conversion and the actual programmable signal processor.

The main functions of the first unit are controlled amplification, digital signal processing, and analog-to-digital signal conversion. It should be emphasized that despite the fact that all the basic processing is carried out in the processor, the first block is a very important node, determining many important characteristics of the whole device such as sensitivity, dynamic range, maximum bandwidth of radar signals.

Figure 2 presents flowchart of the signal amplification and conversion channel.

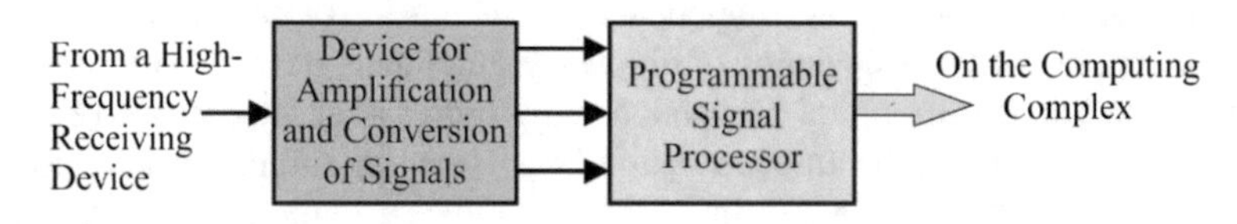

Fig. 1 Structural diagram of radar with digital signal processing to detection

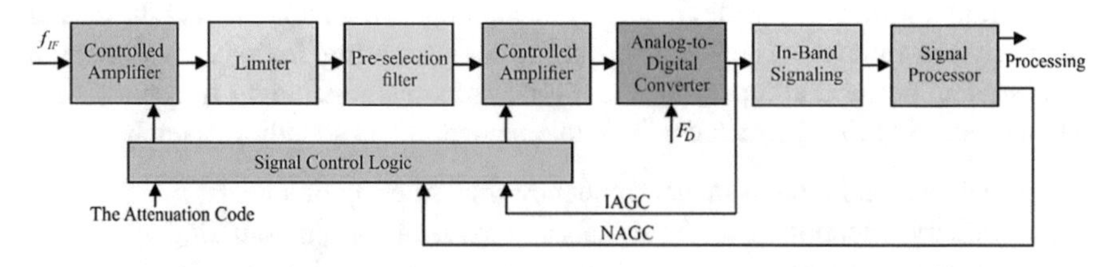

Fig. 2 Flowchart of the signal amplification and conversion channel: F_D—sampling rate; f_{IF}—intermediate frequency

The principles of the first adjustment as shown in Fig. 2, (Noise Automatic Gain Control-NAGC) are as follows. The timing diagram of the radar periodically (for example, once in a few minutes) highlights the special clock for the considered adjustment. No radar is emitted at this clock and the signal processing path operates at its own noise. In this case, the signal processor evaluates the value of the ratio σ_{in}/h (σ_{in}—rms input noise; h—quantization step) of the ADC source codes and calculates the necessary change in the gain.

The second adjustment is due to the fact that for the normal operation of the device the signal must be within the linear part of the amplitude characteristic (within the dynamic range of noise), for quasi-continuous signals gain tuning may be performed at the beginning of the sounding cycle.

In this case, the gain reduction command shall be made by a special scheme at the output of the ADC according to the results of source code analyzes for the pulse repetition period. This procedure is repeated cyclically during the first few repetition periods (instantaneous automatic gain control—IAGC).

Finally, the position can significantly improve the limiter before the pre-selection filter. The restriction level should correspond to the restriction level in the ADC. The transfer of the restriction to the point of the path to the filter allows the selection of the resulting harmonics significantly reduce their negative impact with the help of filter.

The second adjustment is due to the fact that for the normal operation of the device the signal must be within the linear part of the amplitude characteristic (within the dynamic range of noise), for quasi-continuous signals gain tuning may be performed at the beginning of the sounding cycle. In this case, the gain reduction command shall be made by a special scheme at the output of the ADC according to the results of source code analyzes for the pulse repetition period. This procedure is repeated cyclically during the first few repetition periods (Instantaneous Automatic Gain Control—IAGC). Finally, the position can significantly improve the limiter before the pre-selection filter. The restriction level should correspond to the restriction level in the ADC. The transfer of the restriction to the point of the path to the filter allows the selection of the resulting harmonics significantly reduce their negative impact with the help of filter.

The ADC is the most critical node of the device amplification and signal conversion, which limits both the frequency and dynamic characteristics of the device as a whole. The current state of ADC technology is presented in Table 1, which describes the characteristics of the best ADC integrated circuits available on the world market.

In addition to the bit rate and the maximum sampling rate, the table shows the dynamic range for its own noises and the dynamic range for the stray spectra (or harmonics), which are the main ADC parameters for the considered applications.

Table 1 shows that in the required interval of sampling frequencies and intermediate frequencies can be used commercially produced twelve-bit ADCs. For example, ADC A06640 can be used for $F_D = 40$ MHz, $f_{IF} = 30$ MHz. The dynamic noises range is about 70 dB and the harmonics are about 80 dB. It is possible to obtain a dynamic noises range of the entire device of about 90 dB with typical filter coefficients of the digital part (20–30 dB), which is sufficient for the case under consideration.

Table 1 Characteristics ADC integrated circuits

Type of Integrated circuits (ADC)	Bitrate	Sample rate, MHz	Dynamic range of, dB		Measurement frequency of dynamic range, MHz
			Noises	Harmonics	
AD9042	12	41	68	81	96
AD6640	12	65	67	79	32.4
AD9432	12	100	67	80	30
AD9240	14	10	78.5	90	1.0
AD9260	16	2.5	89.5	100	0.1

Although, the dynamic harmonic range does not improve with digital processing and remains at 80 dB. This is perfectly acceptable for pulse signal processing. However, for broadband signals this value lies at the lower limit of the allowable values [10].

The conversion of the analog signal to digital is accompanied by energy losses in SNR. It can be shown that if the RMS value of the external input noise σ_{in} is more than half of the quantization step h, then the losses p can be calculated by the formula:

$$p = 10lg\left[1 + \frac{k}{(\sigma_{in}/h)^2}\right] \tag{1}$$

where the coefficient k is calculated based on the certified value of the dynamic range D ADC relative to the noises:

$$k \approx \left[\frac{2^{r-1}}{10^{\frac{D}{20}}}\right]^2 \tag{2}$$

At $\sigma_{in} < h$ the losses sharply increase with σ_{in} decrease. On the other hand, the increase of σ_{in} reduces the dynamic range. The value of σ_{in}/h is compromise in the range of 1…2. For example, relation $\sigma_{in}/h = 1{,}5$ for ADC AD6040 provides a sufficient degree of linearization of the ADC amplitude characteristic, which corresponds to the loss values $p = 0.9$ dB. In signal processing, it is advisable to maintain relation σ_{in}/h at the ADC input ratio at a fixed optimum level. This can be achieved by special automatic gain control as shown in Fig. 2.

The signal is complex for a quadrature circuit of ADC, and therefore the spectrum $S(f)$ has the property of complex conjugation: $S(-f) = S^*(f)$, so for any of its harmonics with frequency f, there is a harmonic of the same amplitude with frequency minus f. In this case, the bandpass filter cuts two mirror-symmetric frequency bands from the signal spectrum in the positive and negative portions of the spectrum (Figs. 3 and 4). The complex conjugate of the spectrum part is shown by the dashed line. Other parts of the spectrum are muted by the bandpass filter.

Fig. 3 Signal spectrum position at the ADC input

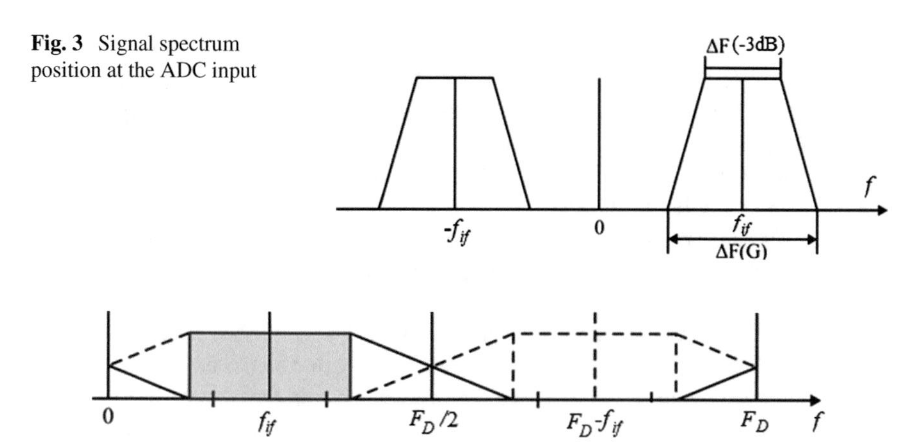

Fig. 4 Signal spectrum position after the ADC

Time sampling with F_D frequency leads to a well-known phenomenon of spectrum overlay, where the post-sampling signal spectrum $S_D(f)$ is connected to the signal spectrum $S(f)$ before sampling by the following ratio

$$S_D(f) = \sum_{m=-\infty}^{\infty} S(f + mF_D) \tag{3}$$

The estimate "underneath " for possible sampling rate will look:

$$F_D \geq (K_R + 1)\Delta F\ (-3\text{dB}) \tag{4}$$

where K_R is the coefficient of rectangularity of the filter.

Equality in expression (4) corresponds to location on the frequency axis of the straight and complex conjugated portions of the spectrum shown in Fig. 4. Part of the spectrum processed digitally is highlighted in the dark color.

The intermediate frequency cannot be selected arbitrarily. For example, if $f_{IF} = f_D/2$ then parts of the analyzed spectrum are completely overlapping and signal distortion occurs. Therefore, the intermediate frequency f_{IF} should be such that the straight and complex-connected parts of the spectrum do not overlap in the frequency range and can be interesting in the processing. The largest spacing along the frequency axis of the considered components of the spectrum occurs when choosing an intermediate frequency in the center of the first or second half of one of the segments of species frequencies $[mF_D, (m+1)F_D]$:

$$f_{IF} = \frac{(2m+1)F_D}{4} \tag{5}$$

where m is an arbitrary integer limited by the ADC frequency capabilities.

If the sampling rate is greater than the minimum required value, which is determined by the right-hand side of the expression 5, the value f_{IF} may vary slightly around the position 5 so that the filter slopes Fig. 3 do not fall into the treated area.

Formulas 4 and 5 determine the necessary ratios for choosing the frequencies F_D and f_{IF}. For example, for the signals of the Table 1 at $\Delta F(-3\text{ dB}) = 2.4$ MHz from (4) when $K_R = 1$ have got $F \geq 9.6$ MHz. Choose $F_D = 10$ MHz. Then from (4) have got $f_{IF} = (2\,m + 1) \cdot 2.5$ MHz, for example, can be chosen $f_{IF} = 7.5$ MHz or $f_{IF} = 32.5$ MHz.

The Spectrum Overlay Sampling Theorem allows to formulate the requirements for the frequency response filter of the previous selection in the capture band. The level of spectral components entering the processing area (grey rectangle in Fig. 4 should be such that they do not lead to a significant increase in the side lobes of the mutual uncertainty function after optimal filtration. This condition will obviously be fulfilled if the amount of damping corresponds to the side lobes level or more. For pulse signals, the value of the latter is usually in the range of 20–40 dB; thus, it is sufficient to select the filter blank in the range $G = 40 \ldots 50$ dB.

When selected F_D and f_{IF}, according to the ratios 5, another property is provided that is useful for processing quasi-continuous signals: all harmonics of the main lobes of the spectrum are superimposed on the main lobes, which weakens the requirements for linearity of the path.

2.2 The Study of Methods of Symbols Synchronization Recovering

A coherent receiver must have information about the exact symbol synchronization in wireless systems to perform true demodulation [11–13]. Symbols synchronization recovery systems are used to evaluate the ideal symbol sampling point. Consider a digital quadrature phase shift keying (QPSK) modulation system: the digital signal with QPSK modulation on the transmitter side is converted into an analog one in the Up-conversion block, and its Simulink model is shown in Fig. 5.

Time sampling period of message source is $T_{d1} = 2 \times 10^{-7}$ s. Carrier frequency is $f_c = 50$ MHz. On the receiver side the signal is again sampled at a frequency of $f_{d2} = 200$ MHz. Then the spectrum of the information signal is shifted to the zero intermediate frequency in the Down-conversion unit, the sampling rate is reduced and the signal filtering is consisted. The receiver contains a timing synchronizer to adjust sampling points on the transmitter and receiver side and to reduce bit error rate. An Error Rate Calculation block is used to determine the number of bit errors comparing transmitter bits and receiver demodulated bits.

The research of the dependence of the number of bit errors on the signal-to-noise ratio (SNR) for two types of synchronization error detector with fractional delay 0.1×10^{-7} s in the transmission channel was conducted.

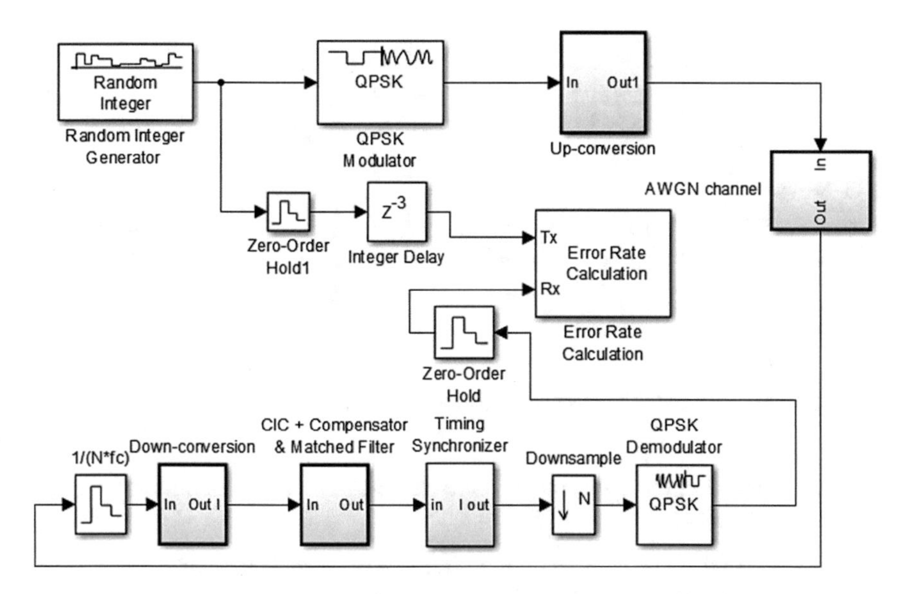

Fig. 5 A Simulink model of coherent digital communication system with QPSK modulation

The Simulink model of the early-late-time synchronization error detector is shown in Fig. 6.

Early and late character recovery is one of the simplest methods and is widely used in digital communications. This method's algorithm takes three samples, spaced by sample duration T_s, and each of them is within the current symbol duration T. Early and late samples are selected at $nT - T_s$ and $nT + T_s$ respectively. A synchronization error is the difference between late and early samples. On the basis of the synchronization error between late and early samples, the sample time of the next character is increased or delayed minimizing the synchronization error. The calculation of the synchronization error for I or Q is calculated as follows:

$$e = \{x[nT + T_s] - x[nT - T_s]\}x[nT] \tag{6}$$

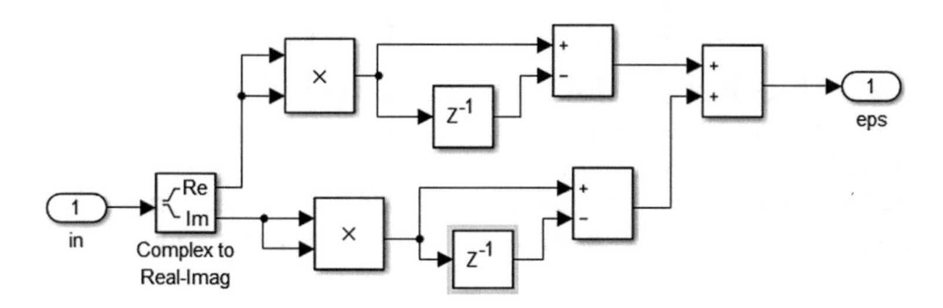

Fig. 6 A Simulink model of the early-late-time synchronization error detector

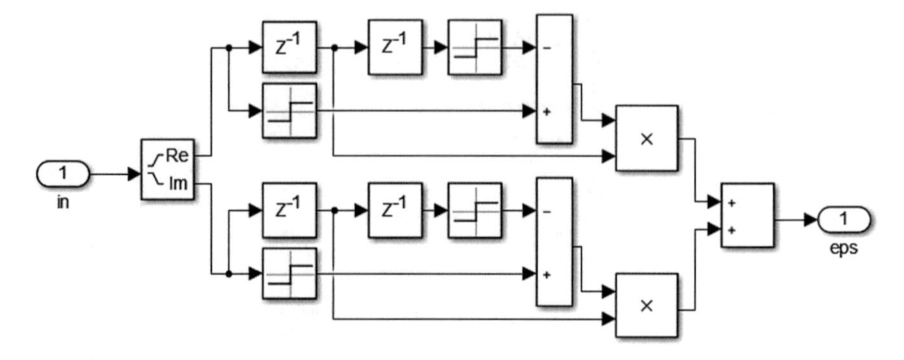

Fig. 7 A Simulink model of Gardner synchronization error detector

After calculating the error, the synchronization adjustment algorithm for I and Q-branch is running:

- if $e = 0$ then the next symbol does not require synchronization adjustment.
- if $e > 0$ then a time advance is required for the next symbol.
- if $e < 0$ then a time delay is required for the next symbol.

The Simulink model of Gardner synchronization error detector is shown in Fig. 7.

Gardner synchronization recovery algorithm requires two samples per symbol and the information about synchronization of the previous symbol to evaluate the synchronization error for the current symbol, as shown in Fig. 5 [14]. The calculation of the synchronization error for I or Q is:

$$e = \{x[nT] - x[(n-1)T]\}x[nT - T/2] \tag{7}$$

where T is the character duration.

After calculating the synchronization error, Gardner algorithm is executed by the algorithm:

- if $e = 0$ then the next character does not require synchronization adjustment.
- if $e < 0$ then the next character requires a time advance.
- if $e > 0$ then a time delay is required for the next symbol.

The dependence of the number of bit errors on the SNR of the synchronization error detector for a coherent digital communication system with QPSK modulation is shown in Fig. 8.

The early-late-time synchronization error detector is 1.5 dB more efficient than the Gardner detector at low SNR and it has a simpler implementation scheme.

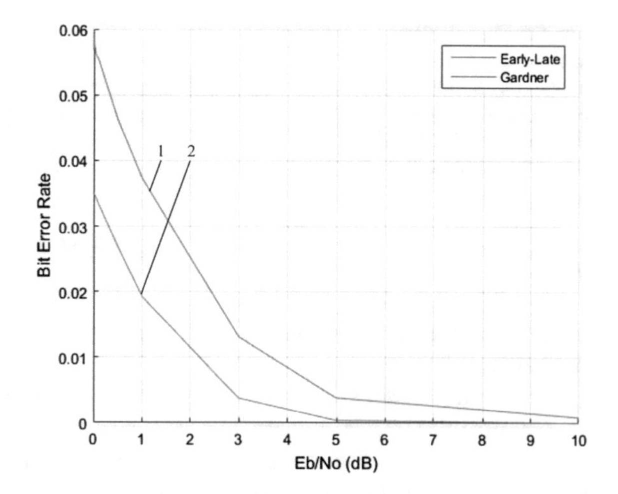

Fig. 8 Dependence of the number of bit errors on the SNR of the synchronization error detector for a coherent digital communication system with QPSK modulation (1 is a Gardner detector; 2 is an early-late detector)

2.3 *Exploring of the Symbol Synchronization Circuit*

The Simulink model of the symbol synchronization circuit is shown in Fig. 9. It was built similarly to the phase-locked loop (PLL) [15–20].

Farrow interpolator is required to perform a fractional delay at the less interval than the sampling period to adjust the exact sampling moments on the receiver and transmitter side [17, 21]. The interpolator and the synchronization error detector act as the phase detector of the PLL classic circuitry, and the Timing Control Unit acts as a controlled generator (direct digital frequency synthesizer). The interpolation-controlled unit provides the interpolator with a base point index as well as a fractional

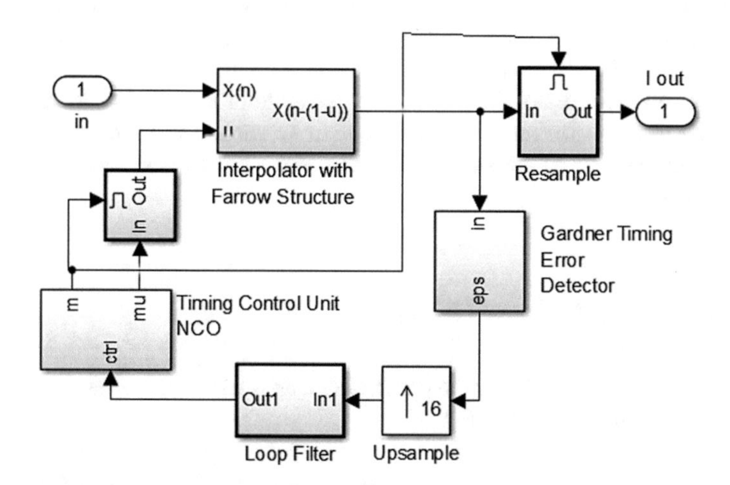

Fig. 9 The Simulink model of the phase-locked loop

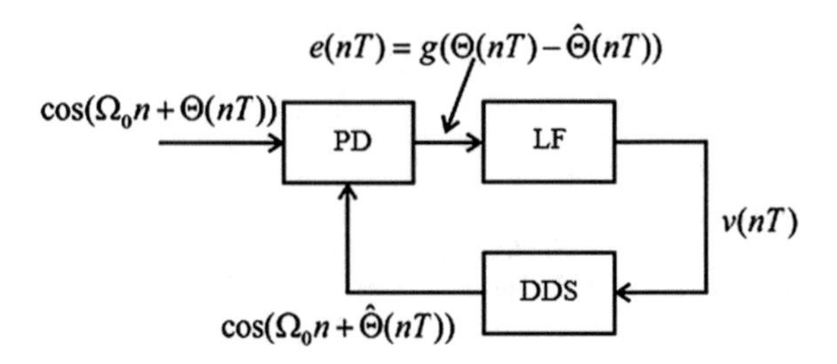

Fig. 10 The structural scheme of the phase-locked loop of the discrete-time (PD is the phase detector; LF is the loop filter; DDS is the direct digital frequency synthesizer)

interval and the base point index is the sampling index closest to the interpolation polynomic.

Structural scheme of the phase-locked loop of the discrete-time is shown in Fig. 10.

Samples of sinusoidal with frequency Ω and phase sampling $\Theta(nT)$ form the input to PLL [16, 17]. The discrete-time phase detector calculates the function of the phase difference between the DDS input and output. This phase difference is a phase error in discrete time. The phase error is filtered by a loop filter and entered into the DDS. The signal at the output of the DDS is given by the expression:

$$u_{DDS} = \cos\Big(\Omega n + \hat{\Theta}(nT)\Big) \tag{8}$$

where

$$\hat{\Theta}(nT) = K_0 \sum_{k=-\infty}^{n-1} v(kT).$$

Timing diagrams of signals at the outputs of the symbol synchronization circuit which Simulink model is shown in Fig. 9, corresponding to the structural blocks of the PLL system (Fig. 10), is shown in Fig. 11.

The controlled generator forms a controlling action for the phase detector which reduces the synchronization error at its output to zero. The synchronicity of the samples at the outputs of the structural blocks of the circuit is shown by the dotted lines in Fig. 11.

2.4 *Phase-Locked Loop (PLL)*

The PLL is designed to monitor the phase of the input harmonic signal [15, 16]. The schematic diagram is shown in Fig. 12.

The PLL system consists of three main components: a phase detector (PD), a loop filter (LF) and a controlled generator (CG). A sinusoidal signal is the input signal:

$$x(t) = U_m \cos(\omega_0 t + \Theta(t)) \tag{9}$$

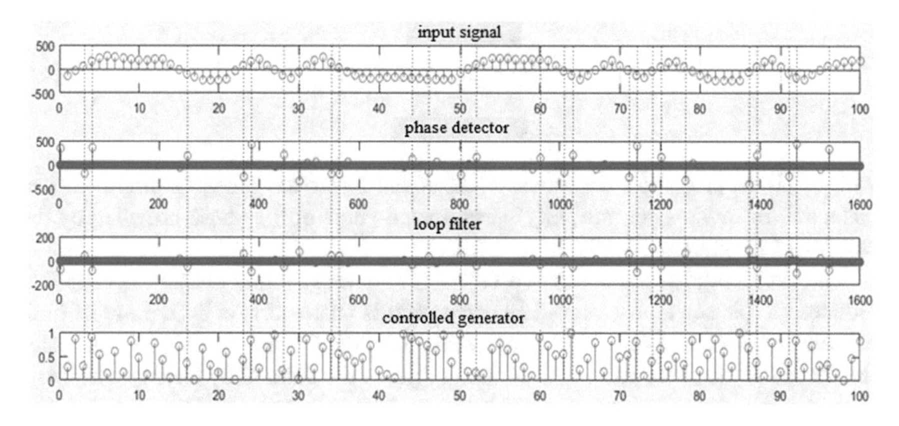

Fig. 11 Timing diagrams of signals at the outputs of the symbol synchronization circuit corresponding to the structural blocks of the PLL system

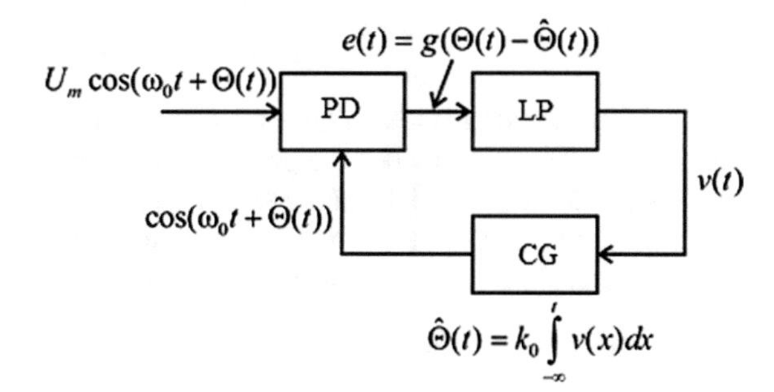

Fig. 12 Basic schematic diagram of a phase synchronization circuit (PD is a phase detector; LF is a loop filter; CG is a controlled generator)

At CG output the signal is determined by the expression:

$$y(t) = \cos(\omega_0 t + \hat{\Theta}(t)) \tag{10}$$

A phase detector is a device and its output are a function of the phase difference between two inputs [22–26]. Since the PLL input signal and the CG output signal form inputs for the phase detector, the output signal of the phase detector is determined by the expression: $g(\Theta(t)–\hat{\Theta}(t))$, and it can be written:

$$\Theta_e(t) = \Theta(t) - \hat{\Theta}(t)$$

is the phase error.

The phase error is filtered by a loop filter to obtain the control voltage $v(t)$ used to set the CG phase. CG output $y(t) = \cos(\omega_0 t + \hat{\Theta}(t))$ is related to input $v(t)$ by dependency:

$$\hat{\Theta}(t) = k_0 \int_{-\infty}^{t} v(x)dx \quad (11)$$

where k_0 is the proportionality constant called the controlled generator amplification, has a unit of radians per volt; $\hat{\Theta}(t)$ is estimated value of the phase provided by the controlled generator [27–30].

The circuit adjusts the control voltage $v(t)$ to obtain a phase estimation $\hat{\Theta}(t)$ that reduces the phase error to zero. PLL operation is characterized by the use of both phase error $\Theta_e(t)$ and estimated phase value from the CG output $\hat{\Theta}(t)$. The transfer functions for the phase error and CG output phase are defined by the expressions:

$$G_a(s) = \frac{\Theta_e(s)}{\Theta(s)} = \frac{s}{s + k_0 k_p F(s)} \quad (12)$$

$$H_a(s) = \frac{\hat{\Theta}(s)}{\Theta(s)} = \frac{k_0 k_p F(s)}{s + k_0 k_p F(s)} \quad (13)$$

So, PLL should provide a phase estimation that has zero phase error. The PLL must evaluate the phase when the phase error is zero.

This characteristic and the phase error transfer function will be used to determine the desired properties of the loop filter. The transfer function of phase estimation or the cycle transfer function is used to characterize the PLL performance [15].

2.5 Simulink Model of Farrow Interpolator with Third Degree Polynomial

Use a Lagrange polynomial for approximate representation of an analog signal $u(t)$ by its discrete values $u(n)$. The representation of the signal by the third degree polynomial is most widespread, it can be represented by:

$$u(t) = a_0 + a_1 u + a_2 u^2 + a_3 u^3 \quad (14)$$

By taking u out the brackets twice, the written expression can be converted as:

$$u(t) = u \cdot (u \cdot (u \cdot a_3 + a_2) + a_1) + a_0 \quad (15)$$

The recorded polynomial can be implemented in the form of the Simulink model shown in Fig. 13 [19, 21].

The given model is an integral part of the symbolic synchronization circuit, its Simulink model was shown in Fig. 9.

Simulink model of the loop filter (Fig. 8) is shown in Fig. 14.

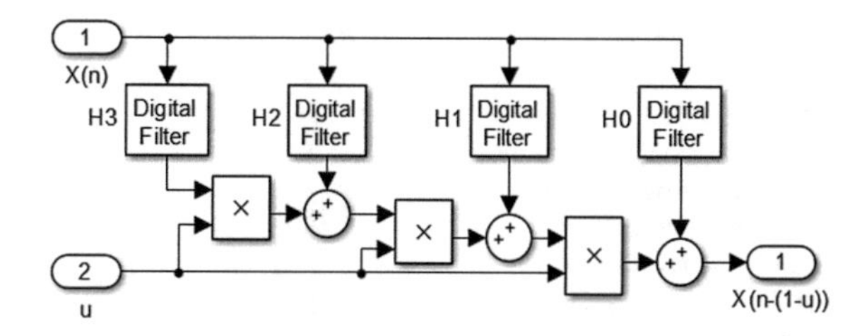

Fig. 13 Simulink model of Farrow interpolator with third degree polynomial

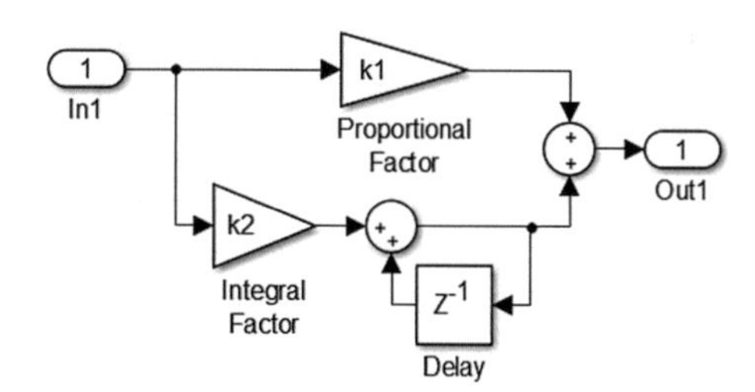

Fig. 14 Simulink loop filter model

A loop filter of proportional integration is used. The proportional gain K_1 and the integrator gain K_2 are calculated using the expressions:

$$K_1 = \frac{-4\zeta\Theta}{\left(1 + 2\zeta\Theta + \Theta^2\right)K_p} \tag{16}$$

$$K_2 = \frac{-4\Theta^2}{\left(1 + 2\zeta\Theta + \Theta^2\right)K_p} \tag{17}$$

where N, ζ, Θ, K_p correspond to the properties of the Samples Per Symbol, Damping Factor, Normalized Loop Bandwidth, and Detector Gain PLL respectively [31].

The filter transfer function looks like:

$$F(s) = k_1 + \frac{k_2}{s} \tag{18}$$

2.6 Dependence of Bit Error Rate on SNR for Different Values of Signal Phase Offset in Propagation Medium

When the input of the contour is different from the CG output by the phase difference $\Delta\Theta$, this is expressed by:

$$\Theta(t) = \Delta\Theta u(t) \tag{19}$$

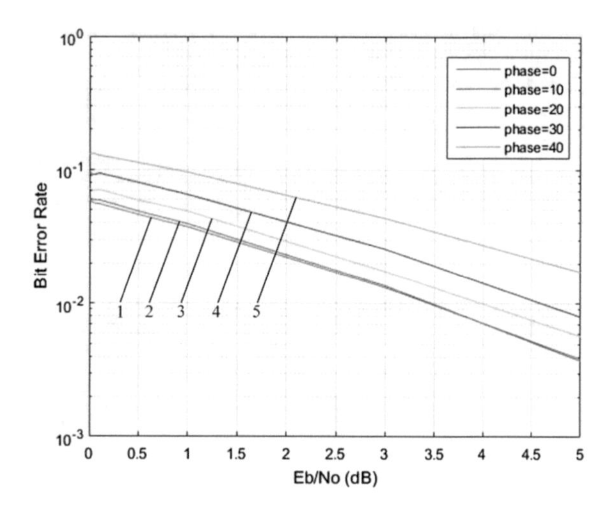

Fig. 15 Dependence of the number of bit errors on the SNR for different values of the signal phase shift in the propagation medium for a coherent digital communication system with QPSK modulation (1 is the absence of phase shift; 2 is the phase shift of 10°; 3 is the phase shift of 20°; 4 is the phase shift of 30°; 5 is the phase shift of 40°)

where $u(t)$ is a step function.

At time $t = 0$ a phase jump occurs. PLL outputs corresponding to the phase jump are called the PLL transition characteristic. The Laplace transform at the input is determined by the expression:

$$\Theta(s) = \frac{\Delta\Theta}{s} \tag{20}$$

The dependence of the number of bit errors [21, 22] on the SNR for different values of the signal phase shift in the propagation medium for a coherent digital QPSK modulation system is shown in Fig. 15 [32–34].

The study of an energy efficiency of a coherent digital communication system with QPSK modulation at the phase shift in the propagation medium is performed. Amplification of the phase shift from 0 to 40° decreases the energy efficiency by 3 dB at low SNR.

The dependence of the number of bit errors on the SNR at the phase shift in the propagation medium is shown in Fig. 16.

Energy efficiency of non-coherent digital communication system with QPSK modulation reduces by 10 dB at the occurrence of the phase shift of 30° in the propagation medium. Adding a symbol synchronization circle compensates for the rotation of the signal constellation. Amplification of the phase shift in the propagation medium to 45° for coherent communication systems leads to a reduction in energy efficiency by 2 dB.

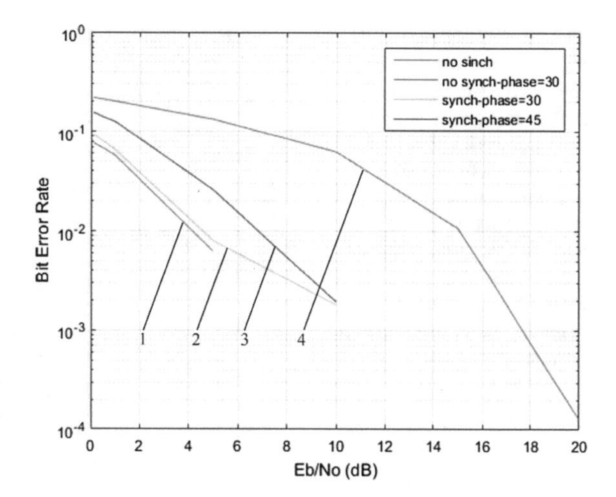

Fig. 16 Dependence of the number of bit errors on the SNR at the phase shift in the propagation medium (1 is the absence of symbol synchronization and the absence of phase shift; 2 is the presence of symbol synchronization and phase shift of 30°; 3 is the presence of symbol synchronization and phase shift of 45°; 4 is the absence of symbol synchronization and adding a phase shift of 30°)

2.7 *The Dependence of the Number of Bit Errors on the SNR at the Frequency Offset in the Propagation Medium*

Since the phase is the integral of frequency, the phase will change linearly at a constant input frequency shift. The Fourier transform of the integral of the frequency characteristic is Fourier transform of phase response. Since frequency response is the stage function, it can be expressed by:

$$\Theta(\omega) = \frac{\Delta\omega}{(j\omega)^2} \tag{21}$$

where $\Delta\omega$ is the value of the frequency jump, rad/s.

The voltage at the phase detector output of PLL range is determined by the expression:

$$e(t) = \frac{\Delta\omega}{K_0 F(0)} \tag{22}$$

where $F(0)$ is the Fourier transform of the impulse response of the loop filter for frequency $\omega = 0$.

The frequency offset can be recorded by the formula:

$$\cos((\omega_0 + \Delta\omega)t) = \cos(\omega_0 t + \Delta\omega t) \tag{23}$$

Denote

$$\Theta(e) = \Delta\omega t \cdot u(t) \tag{24}$$

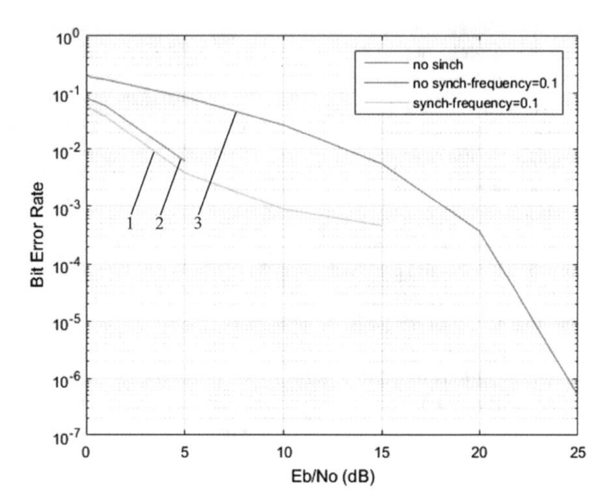

Fig. 17 Dependence of the bit errors number on the SNR at the frequency offset in the propagation medium (1 is the availability of circle symbol synchronization when frequency offset is 0.1 Hz; 2 is the lack of symbol synchronization and the absence of frequency offset; 3 is the lack of symbol synchronization and frequency offset is 0.1 Hz)

that is called a linearly variable input. The PLL outputs that correspond to the linearly variable entrance are called frequency linearly variable reviews. The Laplace transform can be written as:

$$\Theta(s) = \frac{\Delta\omega}{s^2} \tag{25}$$

The dependence of the number of bit errors on the SNR at the frequency offset in the propagation medium shown in Fig. 17.

Setting the frequency offset allows to explore the effects of Doppler shift the signal frequency through the relative movement of transmitter and receiver. The presence of frequency offset greatly affects the energy efficiency of the communication system. Energy efficiency of communication system decreases by 10 dB when the offset frequency is 0.1 Hz and the absence of symbol synchronization. The symbol synchronization circle increases the energy efficiency of 7 dB at the frequency offset of 0.1 Hz. When increasing the value of the frequency offset, the energy efficiency of coherent communication increases.

The properties of the phase error are used to determine the desired characteristics of the loop filter. Because the constant phase offset is modeled as a stage function for $\Theta(t)$ so, the starting point is the expression for the transient response of the phase error. The Laplace transform for the phase error of transient response is determined by the expression:

$$\Theta_e(s) = \frac{\Delta\Theta}{s + k_0 k_p F(s)} \tag{26}$$

If the loop filter has zero DC gain, the phase error for stage function at the input is equal to zero.

The Laplace transform of the phase error of linearly variable frequency is determined by the expression:

$$\Theta_e(s) = \frac{\Delta\omega}{s^2 + sk_0k_pF(s)} \tag{27}$$

If the loop filter has infinite DC gain, then the phase error for the linearly variable frequency is equal to zero in the steady state.

The loop filter should have zero DC gain to bring the steady-state phase error to zero at the phase shift and infinite DC gain to install the phase error to zero at the frequency offset. Loop filter that supports these conditions is a loop filter with transfer function of proportional integration:

$$F(s) = k_1 + \frac{k_2}{s} \tag{28}$$

The productivity of PLL is characterized by the properties of the estimated phase value $\hat{\Theta}(t)$. The Laplace transform at the output is given by expression:

$$\hat{\Theta}(s) = \frac{k_0k_pF(s)}{s + k_0k_pF(s)}\Theta(s) \tag{29}$$

where $\Theta(s) = \Delta\Theta/s$ is used for offset phase; $\Theta(s) = \Delta\omega/s^2$ is used for frequency offset; $F(s)$ is the transfer function of the loop filter [16].

For example, the transfer function for a proportional integration filter has the form:

$$H_a(s) = \frac{k_0k_pk_1s + k_0k_pk_2}{s^2 + k_0k_pk_1s + k_0k_pk_2} \tag{30}$$

and it can be expressed in the form:

$$H_a(s) = \frac{2\xi\omega_ns + \omega_n^2}{s^2 + 2\xi\omega_ns + \omega_n^2} \tag{31}$$

where ξ is called the damping factor, and ω_n is called the natural frequency.

$$\xi = \frac{k_1}{2}\sqrt{\frac{k_0k_p}{k_2}} \tag{32}$$

$$\omega_n = \sqrt{k_0k_pk_2} \tag{33}$$

The PLL acts as a low-pass filter when operating as a linear system. The damping factor controls the magnitude of the frequency response at $\omega = \omega_n$. For insufficient damping systems ($\xi < 1$) the frequency response has a peak at $\omega = \omega_n$. The amplitude of this peak increases with the decrease of ξ. In fact, when $\xi \to 0$ then the peak rises to the momentum and the loop feedback is sinusoidal. For systems with excessive damping when $\omega = \omega_n$ the peak disappears and the feedback loop is more similar to the traditional low pass filter. The presence of a spectral peak in $H_a(j\omega)$ creates the attenuating sinusoidal component in the PLL feedback in time. For overloaded systems ($\xi > 1$) transient characteristic does not change relatively to the ideal response, but gradually increases. So, there is a classic tradeoff for second-order systems: systems with insufficient damping have a quick installation time, but detect overshoot and oscillations. Overloaded systems have more installation time, but they don't have oscillations.

2.8 Constellation Diagram Error Research

Consider the mathematical signal model at the input of the receiver. Let the transmitter generates digital information using m signals $s(t) = \{s_m(t)\}$, where m is the number of points in the signal constellation of digital modulation [22]. For modulation QPSK $m = 4$.

$$s_1(t) = \sqrt{\frac{2E}{T}}\cos(\omega_0 t + \pi/4),\ s_2(t) = \sqrt{\frac{2E}{T}}\cos(\omega_0 t + 3\pi/4)$$

$$s_3(t) = \sqrt{\frac{2E}{T}}\cos(\omega_0 t + 5\pi/4),\ s_4(t) = \sqrt{\frac{2E}{T}}\cos(\omega_0 t + 7\pi/4)$$

Each signal is transmitted over a symbolic interval T by period of time $0 \le t \le T$. The channel distorts the signal using additive white Gaussian noise (AWGN). Thus, the signal at the input of the receiver at the interval $0 \le t \le T$ can be written by the expression:

$$r(t) = s_m(t) + n(t) \tag{34}$$

where $n(t)$ means the AWGN implementation with two-sided power spectral density $\Phi_m(f) = N_0/2$.

Suppose that a signal $r(t)$ arrives to the input of the matched filter behind the sampling device. The signal at the output of the matched filter can be represented by the filter transfer function $H(f)$ and the spectral density of the input signal $S(f)$:

$$S_{mf}(t) = \int_{-\infty}^{\infty} H(f)S(f)e^{j2\pi ft}df \tag{35}$$

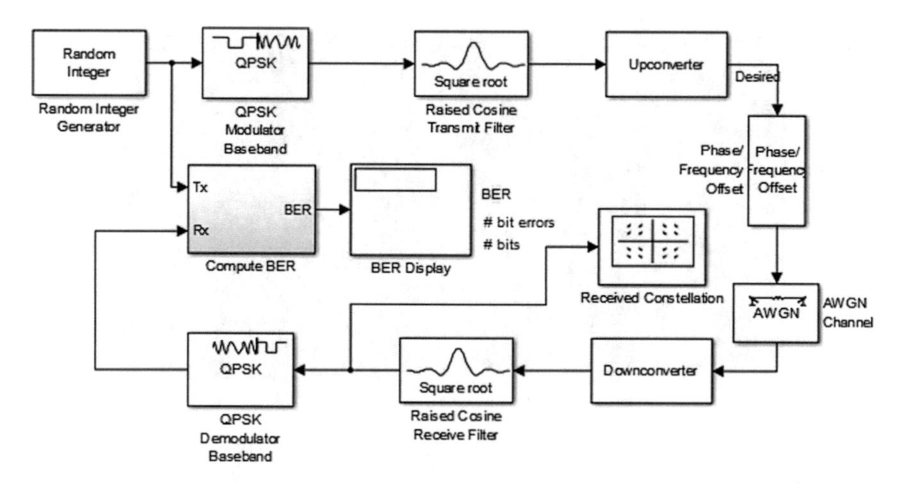

Fig. 18 A Simulink model of digital communication system with QPSK modulation

The variance of noise at the output σ_0^2 is determined by the expression:

$$\sigma_0^2 = \frac{N_0}{2} \int_{-\infty}^{\infty} \left| H(f)^2 \right| df \tag{36}$$

The receiver consists of two parts: a signal demodulator and a detector. Demodulator's function is to convert a signal *r*(*t*) into a vector $r = \{r_1, r_2, r_3, r_4\}$. Detector's function is to decide which of the possible signals *m* has been transmitted. Consider the Simulink model of the digital communication system with QPSK modulation presented in Fig. 18.

Random Integer Generator is a digital signal source. Therefore, according to the model there are: low-frequency QPSK modulation, matched filtering and transfer of the signal spectrum to the high-frequency range in the Up-converter (conversion) block. The Phase/Frequency Offset block applies phase and frequency offset to a complex high frequency signal [21]. If the input signal is *u*(*t*), then the output signal is:

$$y(t) = u(t) \cdot \left(\cos\left(2\pi \int_0^t f(\tau) d\tau + \varphi(t) \right) + j \sin\left(2\pi \int_0^t f(\tau) d\tau + \varphi(t) \right) \right) \tag{37}$$

where $f(t)$ is the frequency offset; φ(*t*) is the phase shift.

On the receiver side, the signal spectrum is transferred to the zero frequency range by the Down-converter (conversion) block, as well as matched filtering and low-frequency QPSK demodulation are performed.

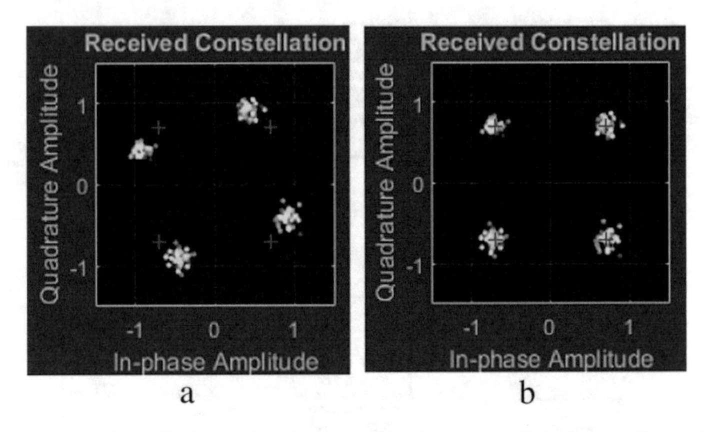

Fig. 19 A constellation of QPSK modulation caused by the phase shift, is the *a* and by the frequency shift, is the *b*

In this case, the constellation of QPSK modulation is shown in Fig. 19 a. If the Offset Frequency parameter is set to 2 Hz and the phase offset is 0, then the angles of the points in the constellation change linearly with time. This leads to a radial displacement of points on the scatter chart, it is shown in Fig. 19 b.

If the phase offset can be set to 20°, then each point in the constellation will be rotated at 20° counterclockwise. The bit error ratio is 0 when the SNR is 20 dB.

The constellation is rotating counterclockwise at 2 Hz. The bit error ratio is 0.5 when the SNR is 20 dB. The bit error ratio is 0.002 in the presence of a symbol synchronization circuit and frequency offset at 2 Hz. This means improving the efficiency at 250 times or at 24 dB.

3 Conclusions

A new radar system for digital signal processing before detection is proposed. These are the guidelines for selecting an intermediate frequency for signal processing. The features of signal processing in the case of echo-signal selection by the features of the correlation properties of their complex bypass are described.

The article is devoted to the study of the influence of the communication channel on the dependence of the number of errors on the SNR. It also explores the possibilities of reducing the effect of signal propagation conditions in a communication channel by using symbolic synchronization, which is based on a PLL.

The Simulink model of a digital communication system with a symbol synchronization circuit based on the PLL research is carried out in the paper. The mathematical model of the PLL and the discrete signals at the outputs of the model structural units are analyzed.

The study was conducted according to the dependence of number of bit errors on the SNR for two types of error synchronization detector. It is concluded that the error synchronization detector of early-late time of 1.5 dB is better than Gardner's detector with small SNR relations and has a more simple circuit implementation. A study of the energy efficiency of coherent digital communication system with QPSK at the phase offset in the propagation medium has been carried out. Increasing the phase shift from 0 to 40° reduces the energy efficiency by 3 dB at low SNR relations. Energy efficiency of non-coherent digital communication system with QPSK modulation at the phase shift of 30° in the propagation medium is reduced by 10 dB. Adding the symbol synchronization circuit compensates for the rotation of the signal constellation. The increase of the phase shift in the propagation medium to 45° for coherent communication system leads to a reduction in energy efficiency by 2 dB.

The presence of frequency offset greatly affects the energy efficiency of the communication system. Energy efficiency of a digital communication system is reduced by 10 dB when the frequency offset is 0.1 Hz and the absence of symbol synchronization. The symbol synchronization circuit increases the energy efficiency of 7 dB at a frequency offset of 0.1 Hz. Energy efficiency of coherent communication is increasing when the value of the frequency offset expands. The efficiency of coherent digital communication increases by 24 dB at frequency offset of 2 Hz.

References

1. Karpova L, Boiko J, Eromenko O (2019) Synthesis of ambiguity functions for complex radar signal processing. In: 2019 IEEE international scientific-practical conference problems of infocommunications, science and technology (PIC S&T). IEEE, Kyiv, Ukraine, pp 1–6
2. Lee J, Bang H (2018) Radial basis function network-based available measurement classification of interferometric radar altimeter for terrain-aided navigation. IET Radar Sonar Navig 12(9):920–930
3. Daim TJ, Lee RMA (2019) A weighted least squares consideration for IR-UWB radar based device-free object positioning estimation for indoor environment. Indonesian J Electr Eng Comput Sci 15(2):894–901
4. Nieh J (2018) Romero RA (2018) Comparison of ambiguity function of eigenwave form to wideband and pulsed radar waveforms: a comprehensive tutorial. J Eng 4:203–221
5. Boiko J, Karpova L, Eromenko O, Havrylko Y (2020) Evaluation of phase-frequency instability when processing complex radar signals. Int J Electr Comput Eng 10(4):4226–4236
6. Kaydok U (2018) Baseband signal modelling of chaff echoes for coherent pulsed radars. In: 2018 IEEE radar conference (RadarConf18), USA, 23–27 April 2018
7. Schvartzman D, Curtis CD (2019) Signal processing and radar characteristics (SPARC) simulator: a flexible dual-polarization weather-radar signal simulation framework based on preexisting radar-variable data. IEEE J Sel Top Appl Earth Observations Remote Sens 12(1):135–150
8. Xu-Dong H et al (2017) Research on multi-target resolution process with the same beam of monopulse radar. In: 2017 IEEE 17th international conference on communication technology (ICCT), China, 27–30 Oct 2017
9. Zeng R et al (2017) Joint estimation of frequency offset and doppler shift in high mobility environments based on orthogonal angle domain subspace projection. IEEE Trans Veh Technol 67(3):2254–2266

10. Zhang L (2016) A high resolution imaging method for sweep frequency continuous wave radar. In: 2016 CIE international conference on radar (RADAR), China, 10–13 Oct 2016
11. Rai A, Kumar VN (2016) Wideband acquisition technique for QPSK demodulator. In: 2016 IEEE international conference on recent trends in electronics, information & communication technology (RTEICT), India, 20–21 May 2016
12. Zhang L, He Z (2011) A modified timing synchronization algorithm for QPSK in digital receiver. In: 2011 2nd international conference on artificial intelligence, management science and electronic commerce (AIMSEC), China, 8–10 Aug 2011
13. Huang S et al (2018) Low-noise fractional-N PLL With a high-precision phase control in the phase synchronization of multichips. IEEE Microw Wirel Compon Lett 28(8):702–704
14. Chen Q, Li M (2013) Modified gardner algorithm for bit synchronization in high-order QAM system. In: 2013 international conference on computational problem-solving (ICCP), China, 26–28 Oct 2013
15. Brandonisio F, Kennedy MP (2014) Noise-shaping all-digital phase-locked loops. Springer, Switzerland
16. Stephens DR (2002) Phase-locked loops for wireless communications. Springer, Switzerland
17. Boiko J, Tolubko V, Barabash O, Eromenko O, Havrylko Y (2019) Signal processing with frequency and phase shift keying modulation in telecommunications. Telkomnika (Telecommun Comput Electron Control) 17(4):2025–2038
18. Bindima T, Elias E (2018) Design and implementation of low complexity 2-D variable digital FIR filters using single-parameter-tunable 2-D farrow structure. IEEE Trans Circ Syst I Regul Pap 62(2):618–627
19. Eghbali A, Johansson H (2011) Complexity reduction in low-delay farrow-structure-based variable fractional delay FIR filters utilizing linear-phase subfilters. In: 2011 20th European conference on circuit theory and design (ECCTD), Sweden, 29–31 Aug 2011
20. Ting-An C et al (2017) Super resolution using trilateral filter regression interpolation. In: 2017 IEEE 2nd international conference on signal and image processing (ICSIP), Singapore, 4–6 Aug 2017
21. Boiko J, Eromenko O, Kovtun I, Petrashchuk S (2019) Quality assessment of synchronization devices in telecommunication. In: 2019 IEEE 39th international conference on electronics and nanotechnology (ELNANO). IEEE, Kyiv, Ukraine, pp 694–699
22. Wadhwa A, Madhow U (2016) Near-coherent QPSK Performance with coarse phase quantization: a feedback-based architecture for joint phase/frequency synchronization and demodulation. IEEE Trans Signal Process 64(17):4432–4443
23. Cheng CC et al (2015) enhanced spatial modulation with multiple signal constellations. IEEE Trans Commun 63(6):2237–2248
24. Chen W, Huang M, Lou X (2018) Sparse FIR filter design based on interpolation technique. In: 2018 IEEE 23rd international conference on digital signal processing (DSP), China, 19–21 Nov 2018
25. Whelan KM et al (2007) A two-dimensional extension of the mueller and müller timing error detector. IEEE Signal Process Lett 14(7):457–460
26. Shaikh F, Joseph B (2017) Simulation of synchronous reference frame PLL for grid synchronization using Simulink. In: 2017 international conference on advances in computing, communication and control (ICAC3), India, 1–2 Dec 2017
27. Boiko J, Pyatin I, Eromenko O, Barabash O (2020) Methodology for assessing synchronization conditions in telecommunication devices. Adv Sci Technol Eng Syst J 5(2):320–327
28. Ageyev D, Al-Anssari A (2014) Optimization model for multi-time period LTE network planning. In: Proceedings of the 2014 first international scientific-practical conference problems of infocommunications science and technology (PIC S&T‘2014). IEEE, Kharkov, Ukraine, pp 29–30
29. Hu Z, Buriachok V, Bogachuk I, Sokolov V, Ageyev D (2021) Development and operation analysis of spectrum monitoring subsystem 2.4–2.5 GHz range. In: Radivilova T, Ageyev D, Kryvinska N (eds) Data-centric business and applications. Lecture notes on data engineering and communications technologies, vol 48. Springer, Cham

30. Loshakov V, Moskalets M, Ageyev D, Drif A, Sielivanov K (2021) Adaptive space-time and polarisation-time signal processing in mobile communication systems of next generations. In: Radivilova T, Ageyev D, Kryvinska N (eds) Data-centric business and applications. Lecture notes on data engineering and communications technologies, vol 48. Springer, Cham
31. Kryvinska N, Zinterhof P, van Thanh D (2007) An analytical approach to the efficient real-time events/services handling in converged network environment. In: Enokido T, Barolli L, Takizawa M (eds) Network-based information systems. NBiS 2007. Lecture notes in computer science, vol 4658. Springer, Berlin, Heidelberg
32. Boiko J, Pyatin I, Eromenko O, Stepanov M (2020) Method of the adaptive decoding of self-orthogonal codes in telecommunication. Indonesian J Electr Eng Comput Sci 19(3):1287–1296
33. Kryvinska N, Zinterhof P, Thanh D van (2007) New-Emerging service-support model for converged multi-service network and its practical validation. In First international conference on complex, intelligent and software intensive systems (CISIS'07). IEEE, pp. 100–110. https://doi.org/10.1109/CISIS.2007.40
34. Kushnir NY, Horley PP, Grygoryshyn AN (2005) Dynamic chaos in phase synchronization devices. In: 2005 15th international crimean conference microwave & telecommunication technology, Sevastopol, Crimea, 2005, vol 1, pp 346–347

Quality Assessment of Measuring the Coordinates of Airborne Objects with a Secondary Surveillance Radar

Valerii Semenets, Iryna Svyd, Ivan Obod, Oleksandr Maltsev, and Mariya Tkach

Abstract Based on a brief review of the place and role of Secondary Surveillance Radar (SSR) in the information support of airspace control and air traffic control systems, it is shown that the principle of constructing aircraft responders and SSRs as a whole predetermined the low quality of information support for the systems under consideration under the influence of intrasystem and deliberate interference. A brief description of the tasks solved by the considered information tools is given, as well as quality of information support integral indicator, the quality of which can be the probability of information support, which is defined as the product of the probability of detecting the airborne object of the requester, the probability of correctly receiving on-board information and the probability of combining flight and coordinate information. The effect of deliberate and unintended (impulse and fluctuation) interference on the quality of the assessment of measuring the range and azimuth of an air object by the considered information tool is evaluated. Based on the assessment of the influence of destabilizing factors, it is shown that in order to obtain higher accuracy in the range and azimuth measuring in the SSR, it is necessary to ensure a responder availability coefficient close to unity and high probabilities of detecting single pulses of response signals. It is shown that improving the quality of SSR information support can be achieved by searching for methods to reduce the influence of intentional and unintentional interference on the aircraft responder readiness coefficient, which is possible by changing the principle of service of request signals.

Keywords Secondary surveillance radar · Primary surveillance radar · Request signal · Aircraft responder · Air object

V. Semenets · I. Svyd (✉) · I. Obod · O. Maltsev · M. Tkach
Kharkiv National University of Radio Electronics, Nauky Ave. 14, Kharkiv 61166, Ukraine
e-mail: iryna.svyd@nure.ua

D. Ageyev et al. (eds.), *Data-Centric Business and Applications*, Lecture Notes on Data Engineering and Communications Technologies 69,
https://doi.org/10.1007/978-3-030-71892-3_5

1 The Place and Role of Secondary Surveillance Radar in the Information Support of the Airspace Control System

1.1 Tasks Solved by Secondary Surveillance Radar

The main sources of information about the traffic situation in the airspace and air traffic control system [1, 2] are primary surveillance radar (PSR) [3–7] and SSR [8–12].

SSR refers to independent cooperative systems and solves the following information problems [13–15]:

- determination of the air object (AO) coordinates;
- receiving flight information from the AO board;
- transfer on board the AO information necessary for the control and management of flights and guidance of the AO;
- dispatch identification AO;
- radar identification of AO state affiliation.

To solve these problems, the SSR has the following modes (A, B, C, D, S) and 1, 2, 3, 4 and 5 [16–20].

SSR belongs to the class of open asynchronous addressless systems by request channel, which allows the interested party unauthorized use of AO in their interests.

For request and response, SSR uses signals with primitive time-interval coding [21–25], which potentially cannot provide both energy secrecy and noise immunity of the considered information systems. All this also determines the low noise immunity of the SSR [26–31].

The principle of SSR construction predetermined the formation of an asynchronous SSR network, which caused, firstly, the lack of synchronization in time of the emission of the request signals of individual requesters and, secondly, the absence of any time synchronization between different SSRs.

AO SSR, based on the principle of servicing request signals, refers to queuing systems (QS) servicing the first correctly received request signal.

According to the construction principle, AOs belong to single-channel queuing systems with failures. The essence of such systems is that when servicing the AO request signal, the SSR is locked for a certain time, which is called the paralysis time [31]. The amount of paralysis time depends on the mode of operation of the SSR. The presence of AO paralysis time decreases the throughput of AO and SSR as a whole [13, 31]. As a characteristic of SSR throughput, can be used the probability of AO detecting SSR P_s, by the requester, which is defined as the probability of receiving a certain number of responses to requests from the considered requester. Throughput capacity is characterized by AO availability factor P_0, which is defined as the probability that it will emit response signals to the request signals of this SSR requester.

Since AOs serve any correctly accepted request (even simulated by an interested party), they belong to open queuing systems. The presence of only one channel for servicing request signals and the paralysis of AO during servicing of request signals makes it possible to attribute AO to single-channel open queuing systems with failures.

The operation of the SSR in the frequency range 1030/1090 MHz leads to significant intra-system interference [32–35] which significantly affects the quality of information support for the airspace control system.

The construction of an SSR based on the principle of an open single-channel queuing system with refuses and the use of primitive codes as request and response signals poses a problem of information security of these information systems [36–45].

Based on the above, we can conclude that the SSR is a request-response data transmission system. Indeed, it contains a channel for transmitting request signals and a channel for transmitting response signals. For coding information in these systems, primitive interval-time and positional codes are used, which significantly reduces the information capacity of data transmission channels, which are considered and, as a result, reduce the quality of information support for consumers.

1.2 Secondary Surveillance Radar User Information Quality Metrics

To solve the information problems described above, the SSR must detect the AO by the response signals, measure the coordinates of the AO, receive and decode the flight data from the board of the AO, form an information packet and give it to consumers. In general, the information package contains:

$$\vec{W}_p,\ \vec{C}_p^{-1},\ PI,\ "friend - foe", \tag{1}$$

where $\vec{W}_p$ coordinates of AO, $\vec{C}_p^{-1}$—is the accuracy matrix of the measured AO coordinates, *PI* AO flight data.

To solve the formulated problem, the detector processes incoming in accordance with some algorithm. The algorithm for detecting AO SSR [27, 46, 47] comes down to testing the hypothesis of the absence of AO against the alternative hypothesis of its presence, that is, the formation of a likelihood ratio and comparing this relationship with some predetermined number, which is selected based on the admissible probability false detection. The decision to detect an object with quality indicators goes to the AO coordinate meter. The estimation of the coordinates of the instantaneous position of the AO is done simultaneously with the detection of the AO. The task of the AO coordinate meter is to, based on the analysis of the obtained sequence of zeros and ones, evaluate the coordinates of AO in an optimal way.

The optimal algorithm for measuring the coordinates of AO is synthesized, as a rule, by the criterion of maximum likelihood. The type of likelihood function depends on the statistical characteristics of the signals and interference, the shape of the antenna system radiation pattern, and also on the method of scanning the SSR antenna during the measurement process.

Thus, when generating an AO detection signal from the output of the AO SSR coordinate meter, an estimate is given of the coordinate measurement vector, which are characterized by a correlation matrix of errors.

The integral indicator of the information support quality may be the probability of information support, which is determined by the total probability of compiling the information package.

In the problem under consideration, based on the expression (1) for SSR, the indicator of the quality of information support may be the probability, which can be of the following form

$$P_{\text{inf}} = D_1 P_{okp} P_{poe}, \tag{2}$$

where

$$D_1 = f(D_0, F_0, C, P_0) = f(q_0, F_0, C, P_0), \tag{3}$$

where $z_0(C)$ is analog (digital) AO signal detection threshold.

For primary data processing, partial indicators of the quality of information support may be the probability of correct detection of AO. When comparing and merging the data that is necessary for the automatic compilation of an AO form, the criterion is the quality of the measurement of coordinate data, because of the likelihood of actions that include:

- the probability of losing the correct flight data;
- probability of distortion of flight data;
- probability of combining coordinate and flight data SSR.

Briefly consider the given probabilities.

When processing flight data by a circuit according to criterion k/m there is a probability of loss of correct flight data in the processing device

$$P_{vtr} = 1 - P_{p.d}^{k}, \tag{4}$$

where $P_{p.d}$ the probability of issuing flight data from the SSR output in the first m information responses.

When using flight data confirmation processing devices using the criterion k/m, the probability of flight data distortion will be:

$$P_{isk.p.d} = \sum_{i=k}^{m} C_m^i P_{isk}^i (1 - P_{isk})^{m-i}, \quad (5)$$

where P_{isk} the probability of issuing SSR false flight data.

SSR flight data may arrive with some delay relative to the coordinate data. Then the number of discrete arrival of flight data:

$$N_d^{,} = N_d + T(KD)/r_d, \quad (6)$$

where N_d number of discrete inputs of coordinate data; $T(KD)$ is the delay for the SSR corresponding to the coordinate data code; r_d the price of the discrete range.

In practice, the probability of combining coordinate and flight data in the SSR will be:

$$P_{okr} = \left(1 - P_{vtr.p.d}\right)\left(1 - P_{isk.p.d}\right), \quad (7)$$

where $P\begin{Bmatrix} +N_0^{,} \\ -N_0^{,} \end{Bmatrix}$ is the conditional probability of arrival of flight data in the strobe from $+N_0^{,}$ to $-N_0^{,}$ relative to the coordinate data of AO.

The algorithm for combining data in the processing device is designed so that the single marks of the processing channels of the data of the monitoring systems are combined if the azimuthal angle between the centers of the packets does not exceed $\Delta\beta$, and their ranges Δr.

Provided that the deviation of the packet centers is independent and obeying the normal distribution, the probability of combining the packets can be determined from the following relation:

$$P_{poe} = \frac{1}{4}\left[1 + \Phi\left(\frac{\Delta\beta}{\sqrt{2}\sqrt{\sigma_{\beta 1}^2 + \sigma_{\beta 2}^2}}\right)\right]\left[1 + \Phi\left(\frac{\Delta r}{\sqrt{2}\sqrt{\sigma_{r1}^2 + \sigma_{r2}^2}}\right)\right], \quad (8)$$

where $\sigma_{\beta 1}, \sigma_{\beta 2}, \sigma_{r1}, \sigma_{r2}$ are the standard deviations of the azimuths (ranges) of the packet centers.

The justifications presented above show that the quality of information support for consumers of the SSR airspace use monitoring system is largely determined by the quality of measuring the location of an airborne object by the specified surveillance system. It should be noted that the accuracy of the AO location is determined by both the accuracy of the estimation of response signal parameters and the algorithm for estimating the coordinates of airborne objects.

2 Range Quality Assessment with Secondary Surveillance Radars

The slant range to AO in the SSR is measured by measuring the delay time of the response signals. The slant range to AO is determined by the delay time t_d of the response signals relative to the time of the emission of the request signals. Obviously, this time can be determined from the following relation

$$t_d = \frac{2r_{AO}}{c} + t_0, \tag{9}$$

where t_0 delay of the signal in the responder caused by the processing of the request signal; r_{AO} is distance to AO.

Since the value of t_0 is independent of the distance and it is the same for all responders (defined by the standard for all operating modes), it can be taken into account when measuring the slant range. The methodology for taking this quantity into account substantially depends on the degree of interaction of primary and secondary observation systems.

If we assume that the period of quantization pulses τ_q and taking into account that we are considering automatic measurement of coordinates AO, the last expression for estimating the distance to AO can be written as

$$\hat{r}_{AO} = \frac{c \cdot \tau_q \cdot (N_d - N_0)}{2}, \tag{10}$$

where N_d the number of the range ring in which the AO detection was carried out $N_0 =]t_0/\tau_q[$.

With enough accuracy for practice, we can assume that the position of the AO during the reception of the entire packet of signals does not change. Therefore, the range measured over all bursts of the burst can be averaged:

$$\hat{r}_{AO} = \sum_{i=1}^{N} \hat{r}_{AOi}, \tag{11}$$

and the variance of the range estimate is

$$\sigma_r^2 = \sigma_{ri}^2 / N, \tag{12}$$

where σ_{ri}^2 is the range variance estimates by a single measuring.

Considering that the processing of SSR signals is carried out both in the aircraft responder and in the ground requester, and also taking into account the specifics of the use of the request signals and response signals, the total variance of the range estimate for a single sample in the SSR can be written as

$$\sigma_{ri}^2 = \sigma_{1\,res}^2 + \sigma_{t\,res}^2 + \sigma_{0\,res}^2 + \sigma_{1\,req}^2 + \sigma_{t\,req}^2 + \sigma_{0\,req}^2, \tag{13}$$

where σ_1^2 is the variance of the time position of a single input pulse in the responder and requester, σ_t^2 is the variance of the instrumental error of the binding of RS (signals response) to the clock pulses of the responder (requester), σ_0^2 is the variance of the time position of the output decoded pulse, that is, taking into account the processing of RS and response signals in the decoders of the responder and requester, respectively.

Briefly discuss the components of the range measurement error variance in the SSR.

The potential measurement error of the range $\sqrt{\sigma_1^2}$, both for the responder and the requester, is determined by the width of the spectrum of the used request signal (response signal) and the signal/noise ratio.

The delay time, both in the aircraft responder and in the requester, is counted using a quantization pulse generator. Moreover, in the responder, the positions of the quantization pulses are not synchronized with the request signal (and in the responder, from the response signals) and, in this case, the variance of the instrumental error of the delay time (provided that the errors are independent values that are uniformly distributed over the interval $-\tau_q/2 \ldots + \tau_q/2$) can be written as $\sigma_t^2 = \tau_q^2/6$.

As shown above, when aircraft responder (AR) works in the field of many SSRs whose coverage areas overlap, its availability coefficient is always less than one. In the response channel, due to the influence of interference, high-frequency suppression of individual pulses of the received encoded signals is also carried out. Therefore, it is of interest to study the effect of interference in the request channel and the SSR response channel on the accuracy of the range estimate.

As follows from (13), to calculate the total error of the range measurement, it is necessary to calculate the variance of the estimate of the temporal position of the decoded pulses. Consider this.

2.1 Analysis of the Effect of Interference on the Time Position of Decoded Signals in the SSR

Let's study the estimation of the time position of decoded request (response) signals taking into account the high-frequency suppression of individual encoded signal pulses and the responder's availability factor.

Suppose that the action of interference in the radio channel is carried out independently for each pulse, and for this radio channel the probability of high-frequency pulse suppression is uniquely determined.

Consider the allocation of single request signals and response signals with fractional processing logic. This logic is used in the SSR and is a common case compared to processing with integer logic.

The integral and differential distribution function of the time position of decoded signals with fractional processing logics in decoders can be determined from the following relation

$$F_{k/n}(x) = 1 - [1 - F_1^k(x)]^{C_n^k},\ W_{k/n}(x) = kC_n^k W_1(x) F_1^{k-1}(x)[1 - F_1^k(x)]^{C_n^k - 1}, \quad (14)$$

where $F_1(x)$, $W_1(x)$ are the integral and differential probability distribution functions of the time position of the input pulses.

For the normal law of the probability distribution of the time position of the input pulses $F_1(x)$ and $W_1(x)$ can be written as

$$F_1(x) = [1/\sigma_1\sqrt{2\pi}] \int\limits_{-\infty}^{x} \exp[-y^2/2\sigma_1^2]\,dy,\ W_1(x) = [1/\sigma_1\sqrt{2\pi}]\exp(-x^2/2\sigma_1^2). \quad (15)$$

The mathematical expectation and variance of the time position of the decoded signals can be determined using the following expressions

$$M(x) = \int\limits_{-\infty}^{\infty} xW(x)\,dx,\ \sigma^2(x) = \int\limits_{-\infty}^{\infty} x^2 W(x)\,dx - M^2(x). \quad (16)$$

Calculations using expressions (15), taking into account (16), are presented in Fig. 1. The mathematical expectation and variance of the time position of the output pulses were: for 2/3 logic–0.128 and 0.351; 3/4–0.092 and 0.236; 3/5–0.259 and 0.156, respectively.

When receiving response signals at the requester, depending on where the inter-period signal processing circuit is switched on, two processing circuits can be used. Consider the effect of different signal extraction schemes on the time positions of decoded response signals.

The probability of suppressing the pulse at the output of the inter-period processing device at time x can be defined as

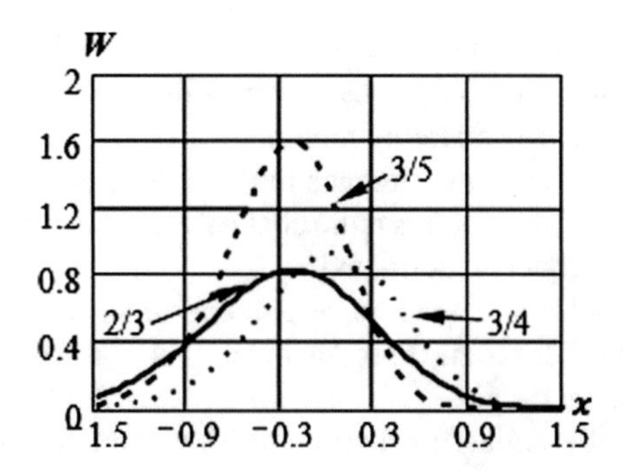

Fig. 1 Dependence $W = f(x, k/n)$

$$F(x) = \sum_{i=0}^{m-k} C_m^i P_0^{m-i} (1 - P_0)^i \sum_{l=0}^{m-k-i} C_{m-i}^l P_1^{m-l-i} F_1^{m-l-i}(x)[1 - P_1 F_1(x)]^l. \quad (17)$$

The probability of the appearance of a pulse at time x at the output of the decoder with logic n/n can be written as

$$F(x) = \sum_{i=0}^{m-k} C_m^i P_0^{m-i} (1 - P_0)^i \{ \sum_{l=0}^{m-k-i} C_{m-i}^l P_1^{m-l-i} F_1^{m-l-i}(x)[1 - P_1 F_1(x)]^l \}^n. \quad (18)$$

The probability distribution density of the time position of the initial pulses is found in the usual way

$$W_m(x) = dF_m(x)/dx. \quad (19)$$

After calculating, we get

$$\begin{aligned} W_m(x) &= nW_1(x) \\ &\times \sum_{i=0}^{m-k} C_m^i P_0^{m-i} (1 - P_0)^i \{ \sum_{l=0}^{m-k-i} C_{m-i}^l P_1^{m-l-i} F_1^{m-l-i}(x)[1 - P_1 F_1(x)]^l \}^{n-1} \\ &\times \sum_{l=0}^{m-k-i} C_{m-i}^l P_1^{m-l-i} F_1^{m-l-i}(x)[1 - P_1 F_1(x)]^{l-1} [m - l - i - (m - i) P_1 F_1(x). \end{aligned} \quad (20)$$

When decoding, followed by inter-period processing of the input signals, the probability of the appearance of pulses at time x at the output of the decoder with logic n/n can be written as

$$F_d = P_0 P_1^n F_1^n(x), \quad (21)$$

and the probability of an impulse at the time x at the output of the inter-period processing device with logic k/n

$$F_d(x) = \sum_{i=0}^{m-k} C_m^i [P_0 P_1^n F_1^n(x)]^{m-i} [1 - P_0 P_1^n F_1^n(x)]^i. \quad (22)$$

In this case, the probability distribution density of the time position of the output pulses can be written as

$$W_d(x) = nW_1(x) \times \sum_{i=0}^{m-k} C_m^i [P_0^{m-i} P_1^{n(m-i)} F_1^{n(m-i)-1}(x)][1 - P_0 P_1^n F_1^n(x)]^{i-1} \cdot [m - i - mP_0 \cdot P_1^n \cdot F_1^n(x)]. \tag{23}$$

Expressions (20) and (23) are given for the general case when P_0 and P_1 are variables.

When $P_0 = 1$ there is a case in which the influence of the availability factor of the responder is not taken into account; when $P_1 = 1$ the second separate case in which the influence of interference from the response channel is not taken into account; when $P_0 = P_1 = 1$ the third separate case (only the influence of the distribution of the time position of the input pulses is taken into account).

Let's determine the probabilistic characteristics of the time position of the output pulses for both decoding methods under the normal law of the probability distribution of the time position of the input pulses.

For this, the values of $F_1(x)$ and $W_1(x)$ were substituted into expressions (20) and (23), and then the probability densities of the time position of the output pulses were calculated.

In Figs. 2 and 3 show probability density plots for processing logics 2/3 and 3/5 in an inter-period processing device and different values of the aircraft responder availability factor and the probability of detecting single pulses of the response signal. From the presented dependencies it is seen that a decrease in the availability factor of the responder and the probability of detecting single pulses leads to an increase in the delay of the output pulses and an increase in the variance. Moreover, for the decoding method with subsequent inter-period processing of the response signals, the increase in both the delay time and variance of the estimate of the delay time is more significant than for the processing circuit with the previous inter-period processing.

For both methods of processing response signals, an increase in the significance of the time-interval code used increases the delay.

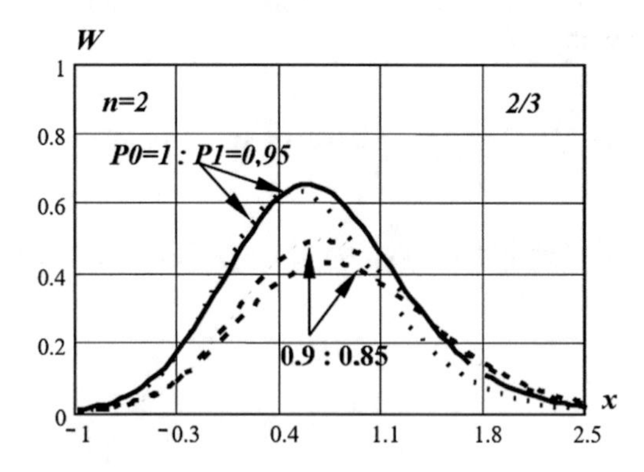

Fig. 2 Dependence $W = f(x, n, P_0, P_1, 2/3)$

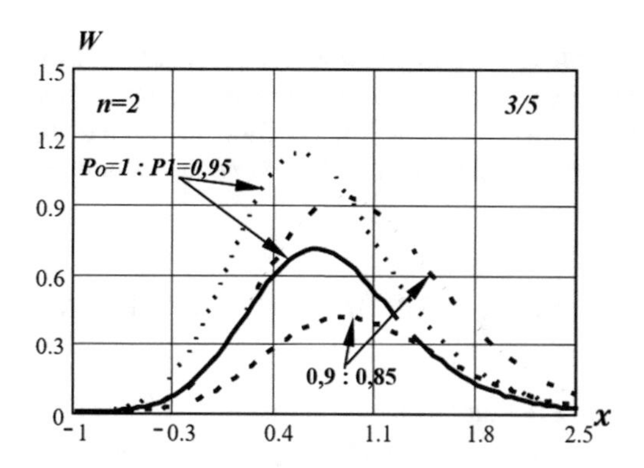

Fig. 3 Dependence $W = f(x, n, P_0, P_1, 3/5)$

As follows from the presented dependences, in order to obtain higher accuracy of range measurement in SSR, it is necessary to strive to ensure a responder availability factor close to unity and high probabilities of single pulse detection.

3 Azimuth Measurement with Secondary Surveillance Radars

SSRs perform azimuthal direction finding of AO by analyzing the envelope of the packet of n-pulse response signals [48–52]. Under the influence of intrasystem and intentional correlated interference in the request channel and the response channel, part of the response signals is lost with probability $1 - P_0(1 - P_p)$, which leads to a decrease in the length of the packet of received response signals [53, 54].

Consider the algorithms that are optimal by the maximum likelihood criterion for measuring the azimuth of the SSR packet of binary-quantized n-pulse response signals present in the observation with probability $1 - P_0(1 - P_p)$ and obtain an estimate of the potential accuracy of measuring the azimuth of AO SSR based on the Cramer-Rao inequality.

For digital processing of SSR signals, the initial ones when measuring the azimuth are binary-quantized values of the envelope of the sum of the signal observations assigned to one range ring

$$\vec{Z} = \begin{vmatrix} z_{11} & z_{12} & \cdot & z_{1N} \\ z_{21} & z_{22} & \cdot & z_{2N} \\ \cdot & \cdot & \cdot & \cdot \\ z_{n1} & z_{n2} & \cdot & z_{nN} \end{vmatrix}, \tag{24}$$

where $\vec{y}_k = |z_{1k}, z_{2k}, \ldots, z_{nN}|^T$ is a sample of binary-quantized observations obtained in the k-th time interval for receiving response signals.

Since the time of correlation, fluctuation interference with a uniform power spectral density in the receiver passband (affects the SSR radio channel) does not exceed the interval between pulses, the likelihood function of the k-th response signal obtained directly from the conditional probability of combinations of zeros and ones can be written as

$$L(\vec{y}_k) = \prod_{i=1}^{n} P_{11}^{z_{ik}}(k) P_{10}^{1-z_{ik}}(k), \quad k = 1, .., N. \tag{25}$$

Taking into account the fact that with probability $1 - P_0(1 - P_p)$, the possible loss of response signals in each interval of signal reception, the likelihood function for the observation matrix $\vec{Z}$ can be represented as follows

$$L(\vec{Z}) = \prod_{k=1}^{N} [P_0(1 - P_p) \prod_{i=1}^{n} P_{11}^{z_{ik}}(k) P_{10}^{1-z_{ik}}(k) + [1 - P_0(1 - P_p)] \prod_{i=1}^{n} P_{01}^{z_{ik}}(1 - P_{01}^{1-z_{ik}})]. \tag{26}$$

We denote by $s_k = \sum_{i=1}^{n} z_{ik}$ the number of units obtained in the k-th response signal. In this case, vector $\vec{s} = (s_1, s_2, \ldots, s_k)$ is a component part of the matrix $\vec{Z}$. Then (26) can be written as

$$Q(\vec{s}) = \prod_{k=1}^{N} [P_o(1 - P_p) P_{11}^{s_k}(k)\, P_{10}^{n-s_k}(k) + [1 - P_o(1 - P_p)] P_{01}^{s_k}(1 - P_{01})^{n-s_k}]. \tag{27}$$

This circumstance allows us to reduce the volume of the initial data from $\vec{Z}$ to $\vec{s}$ without losing information about the estimated azimuth of AO. Further we will use $Q(\vec{s})$ as a likelihood function.

To compose the likelihood equation, we differentiate the logarithm of the likelihood function $Q(\vec{s})$ by the estimated parameter:

$$d \ln Q/d\alpha = \sum_{k=1}^{N} \frac{(d P_{11}(k)/d\alpha)[P_{11}^{s_k - 1}(k) P_{10}^{n-s_k-1}(k)(s_k - n P_{11}(k))]}{[P_o(1 - P_p) P_{11}^{s_k}(k)\, \mathrm{P}_{10}^{\mathrm{n - s_k}}(k) + [1 - P_o(1 - P_p)] P_{01}^{s_k}(1 - P_{01})^{n-s_k}}. \tag{28}$$

In obtaining (28), the obvious relation was used:

$$d P_{11}(k)/d\alpha = -d P_{10}(k)/d\alpha, \quad k = 1, 2, \ldots, N. \tag{29}$$

The likelihood expression for azimuth estimation $d \ln Q/d\alpha$ is obtained in the following form:

$$\sum_{k=1}^{N} \eta(k, s_k) = 0, \tag{30}$$

where weights are:

$$\eta(k, s_k) = \frac{(dP_{11}(k)/d\alpha)[s_k - nP_{11}(k)]}{P_{11}(k)P_{10}(k)[P_0P_p + (1 - P_0P_p)(P_{01}(k)/P_{11}(k))^{s_k}[(1 - P_{01})/P_{10}(k)]^{n-s_k}]}. \tag{31}$$

In accordance with expression (30), the maximum likelihood estimate is obtained from the condition that the sum of the weighting coefficients be equal to zero, and depends on the number of the response signal in the packet and the number of pulses detected in this code s_k.

The complication of the circuit obtained meter in comparison with the schemes of known weight meters, is as follows:

- the use of a circuit for counting the number of pulses in the response signal s_k;
- storing in memory not a vector, but a matrix of weight coefficients, which allows you to take into account the availability factor of the responder and the probability of suppressed response signals.

In a particular case, there is no interference in the response channel and the re-sponse coefficient of the responder, which is one, the weighting coefficients of expression (31) are respectively equal to:

$$\eta(k, s_k) = \frac{dP_{11}(k)}{d\alpha}\frac{s_k - nP_{11}(k)}{P_{11}(k)P_{10}(k)}, \quad k = 1, 2, \ldots, N, \tag{32}$$

and the likelihood ratio expression for azimuth estimation (30) can be respectively written as:

$$\sum_{k=1}^{N} s_k \frac{dP_{11}(k)}{d\alpha} \cdot [P_{11}(k)P_{10}(k)]^{-1} = n \sum_{k=1}^{N} \frac{dP_{11}(k)}{d\alpha}\frac{1}{P_{10}(k)}. \tag{33}$$

For a symmetrical radiation pattern, the right-hand side of expression (33) is zero, and in other cases it can be calculated in advance because it is independent of observations.

To measure the azimuth at $P_0 = P_p = 1$, it suffices to calculate the left side of Eq. (33), linear with respect to s_k. The weighting coefficients at s_k are dependent on the number of response signals k and are independent of the value s_k, which allows the weight vector to be stored in the permanent memory instead of the matrix of weighting coefficients.

The variance of the error due to the potential accuracy of measuring the parameters of the signals against the background of interference can be estimated using the Cramer-Rao inequality, according to which the variance of the error in estimating the parameter α, from N independent observations $r_1, r_2, \ldots, r_N$ is not less than

$$\sigma_m^2 = \left[\sum_{k=1}^{N} E(d \ln W(r_k/d\alpha)^2\right]^{-1}, \tag{34}$$

where E is the mathematical expectation of a random variable (in brackets), with a distribution density $W(r_k)$, that corresponds to k—the observation in the packet.

Using the sufficiency of statistics $\vec{s}$, we obtain the expression of potential variance (34) in terms of distribution $W(\vec{s})$:

$$\sigma_m^2 = \left[\sum_{k=1}^{N} \sum_{s_k=0}^{n} E(dW(s_k/d\alpha)^2/W(s_k)\right]^{-1}, \tag{35}$$

where $W(s_k)$ is the distribution function of sufficient statistics s_k.

In binary quantization of a response signal packet with a relative threshold z_0 the distribution of random variables s_k is a probabilistic mixture of binomial distributions with parameters $P_{11}(k)$ and P_{01}:

$$W(s_k) = C_n^{s_k}\left[P_0(1-P_p)P_{11}^{s_k}(k)P_{10}^{n-s_k}(k) + [1-P_0(1-P_p)]P_{01}^{s_k}(1-P_{01})^{n-s_k}\right]. \tag{36}$$

Differentiating (36) with respect to α and making the necessary simplifications, we find

$$\frac{dW(s_k)}{d\alpha} = \frac{dP_{11}(k)P_0P_pC_n^{s_k}[s_k - nP_{11}(k)]P_{11}^{s_k-1}(k)P_{10}^{n-s_k-1}(k)}{d\alpha}. \tag{37}$$

Substituting the functions (36), (37) into the expression (35), replacing the summation s_k with l, we obtain the expression for the minimum variance of the measurement error of the SSR azimuth behind the packet of binary-quantized pulses with the responder availability factor P_0 and the probability of suppressing response signals P_p:

$$\sigma_m^2 = \left[\sum_{k=1}^{N} (P_0P_p)^2\left(\frac{dP_{11}(k)}{d\alpha}\right)^2 \times \sum_{l=0}^{n} C_n^l \frac{[P_{11}^{l-1}(k)P_{10}^{n-l-1}(k)(l-nP_{11}(k))]^2}{P_0P_pP_{11}^l(k)P_{10}^{n-l}(k) + (1-P_0P_p)P_{01}^l(1-P_{01})n-l}\right]^{-1}. \tag{38}$$

The above expression generalizes the well-known expression for n–pulse response signals and the possible, with probability $1 - P_0(1 - P_p)$, absence of separate response signals caused by the responder availability factor and chaotic impulse noise in the SSR channel. The expression for variance (38) is an estimate of the potential accuracy of the azimuth measurement algorithm that is optimal according to the maximum likelihood criterion based on expressions (30), (31).

When deriving the optimal SSR azimuth measurement algorithms and expressing the potential accuracy of azimuth measurement using a package of binary-quantized response signals, no specific expressions were used anywhere for the error probabilities of the first and other kind when detecting individual pulses of response signals. The conclusion was drawn only on the assumption of independence of errors for each of the pulses. Therefore, the results obtained are valid not only for fluctuation interference with a power spectrum uniform in the passband, but also for interference of an arbitrary form that affect the SSR response channel, if only the distortions of individual pulses of the response signals remain independent of each other. In particular, this condition is fulfilled for chaotic impulse noise.

Thus, to calculate the matrix of meter coefficients and the accuracy of azimuth estimation for a given type of interference, which have the indicated properties, it is enough to calculate the error probabilities P_{01} and P_{10}, and substitute them into expression (31), etc.

Optimal SSR azimuth measurement algorithms obtained under the assumption that P_0 and P_p are known in advance. The coefficients of the azimuth measurement algorithm (31) depend on these values. If these parameters are not known in advance or change (which most often happens) depending on the operating conditions of a particular SSR, then, according to the principle of overcoming a priori uncertainty, the algorithm for measuring the azimuthal position of AO is supplemented with an unknown coefficient estimation unit.

The dispersion of the azimuth measurement error by the optimal algorithm for unknown values will always be greater than if there is complete information about the parameters.

We'll make a specific calculation of the variance of the azimuthal position estimate for a Gaussian approximation of the SSR antenna radiation pattern.

$$g(\alpha_k) = \exp[-(\alpha_k - \alpha)/\beta_o]^2, \tag{39}$$

where β_0 the width of the antenna pattern of the SSR; α_k is values of the azimuthal angle at the time point of the k-th response signal.

The angle offset of the antenna between adjacent pulses of the response signals is neglected.

Passing to the conventional units $g(\alpha_k)$ we will present in the following form:

$$g(\alpha_k) = g(k) = \exp(-k/\varphi)^2, \tag{40}$$

where $\alpha_k - \alpha = k\Delta\beta$; $\varphi_0 = \varphi\Delta\beta$; $\Delta\beta$ angular distance between probing.

The value of φ is the width of the antenna pattern at the same level as φ_0, expressed in discrete.

Differentiating the expression for $P_{11}(k)$ with respect to α, we obtain

$$\frac{dP_{11}(k)}{d\alpha} = z_o q \exp[-(z_o^2 + q_k^2)/2] I_1(z_o q_k) dg(k)/d\alpha$$
$$= 2z_o q \exp[-(z_o^2 + q_k^2)/2] I_1(z_o q_k) g(k) k/\phi\Delta\beta. \tag{41}$$

Substituting expression (41) into (38), taking into account the symmetry of $g(\alpha_k)$ and relating the positions of AO to the center of the packet, we obtain:

$$\frac{\sigma}{\Delta\beta} = \frac{\dfrac{\phi^2 \exp(z_o^2/2)}{2\sqrt{2} z_o q_o P_o P_p}}{\left\{ \sum_{k=1}^{(N-1)/2} g^2(k) \exp(-q_k^2) k^2 I_1(z_o q_k) \sum_{i=0}^{n} \frac{\{P_{11}^{l-1}(k) P_{10}^{n-i-1}(k)(i - nP_{11}(k))]^2}{P_o P_p P_{11}^i(k) P_{10}^{n-i}(k) + (1 - P_o P_p) P_{01}^i (1 - P_{01})^{n-i}} \right\}^{1/2}}. \tag{42}$$

Expression (42) allows us to calculate the potential relative root-mean-square error of the AO SSR azimuth measurement by the optimal digital algorithm (30) for independent probabilities of suppressing response signals in the SSR channel.

The calculation results by expression (42) are presented in Fig. 4 (for integer processing logics) and Fig. 5 (for fractional processing logics) with different responder availability coefficients and the probability of suppressing response signals, which is 0.97.

As follows from Figs. 4 and 5, a decrease in the availability factor leads to a decrease in the accuracy of measuring the azimuth of the AO SSR. Comparative analysis of Figs. 4 and 5 shows that the use of fractional response signal processing logics

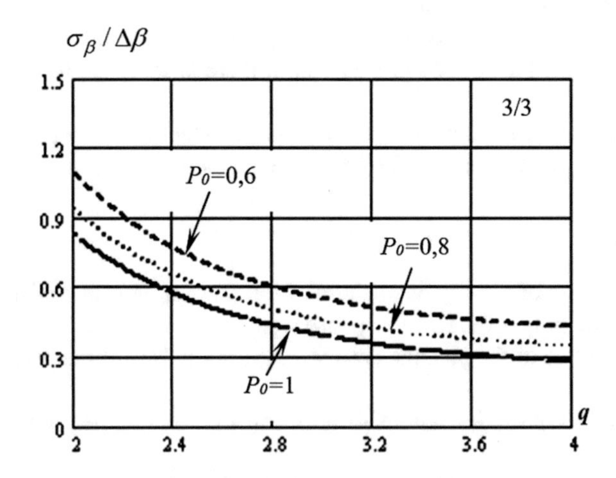

Fig. 4 Estimation of azimuth measurement at $k/m = 3/3$

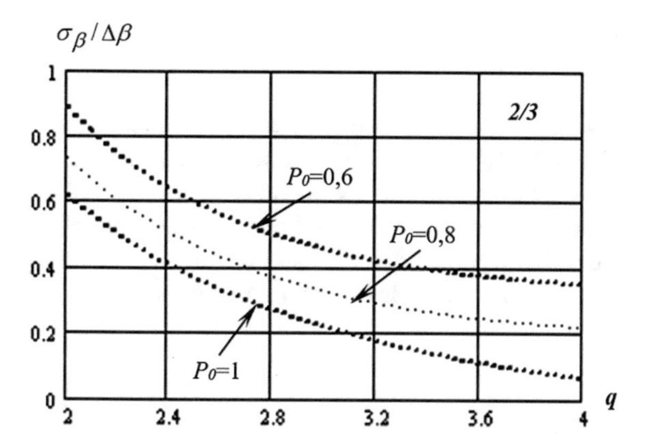

Fig. 5 Estimation of azimuth measurement at $k/m = 2/3$

can improve the accuracy of the AO SSR azimuth estimation and reduce the dependence of the azimuth measurement accuracy on the aircraft responder availability factor.

In Fig. 6 shows the dependences of the potential standard error of the azimuth AO measurement as a function of the AR availability factor. As shown in Fig. 6, the potential accuracy of measuring the SSR azimuth when using the fractional logic of processing the response signals is not worse, and when $P_0 > 0, 4$ the use of the fractional logic of processing the response signals is more accepted. I would like to note that using SSR with $P_0 < 0, 4$ loses its meaning.

As can be seen from the given dependences, it is possible to improve the quality of measurement of the azimuth of an air object by changing the principle of servicing request signals [17, 53, 54].

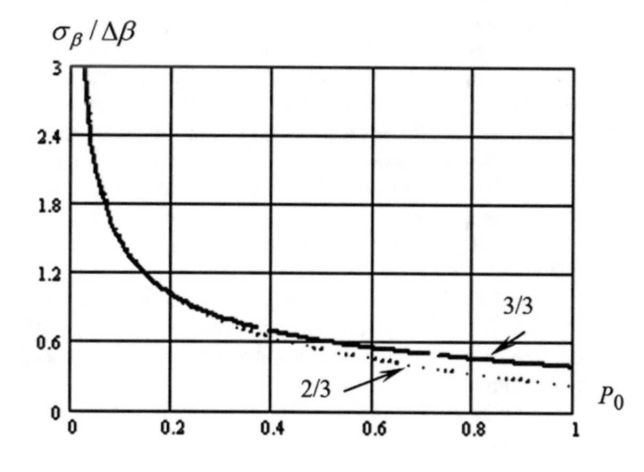

Fig. 6 Assessment of the effect of the aircraft responder availability factor on azimuth accuracy

4 Conclusions

Based on the results of the analysis and mathematical modeling, it has been shown that improving the quality of measuring the azimuth of an airborne object, and, consequently, the quality of information support with the considered information tool, can be achieved by searching for methods to reduce the effect of intentional and unintentional interference on the readiness factor of an aircraft responder. This is possible by changing the principle of serving request signals.

References

1. Stevens B, Lewis F, Johnson E (2016) Aircraft control and simulation: dynamics, controls design, and autonomous. Wiley
2. Lute C, Wieserman W (2011) ASR-11 radar performance assessment over a wind turbine farm. In: 2011 IEEE RadarCon (RADAR). https://doi.org/10.1109/radar.2011.5960533
3. Skolnik M (2008) Radar handbook, 3rd edn. McGraw-Hill, New York
4. Farina A, Studer F (1993) Digital processing of radar information. Radio i svyaz, Moscow
5. Lynn P (2013) Radar systems. Springer, New York
6. Skolnik M (2020) Introduction to radar systems, 3rd edn. McGraw-Hill Education, Boston
7. Richards M (2005) Fundamentals of radar signal processing. McGraw-Hill Professional, New York
8. Stevens M (1988) Secondary surveillance radar. Artech House, Norwood
9. Kim E, Sivits K (2015) Blended secondary surveillance radar solutions to improve air traffic surveillance. Aerosp Sci Technol (45)
10. Gao J, Zou J, Guo N (2020) A Secondary surveillance radar data analysis technique based on geometrical method. In: Liang Q, Liu X, Na Z, Wang W, Mu J, Zhang B (eds) Communications, signal processing, and systems. CSPS 2018. Lecture Notes in Electrical Engineering, vol 517. Springer, Singapore
11. Svabenik P, Zeman D, Balada R, Fedra Z (2011) Separation of secondary surveillance radar signals. In: 2011 34th International conference on telecommunications and signal processing (TSP). https://doi.org/10.1109/tsp.2011.6043683
12. Barott W, Johnson M, Scott K (2014) Passive radar for terminal area surveillance: performance feasibility study. In: 2014 IEEE/AIAA 33rd digital avionics systems conference (DASC). https://doi.org/10.1109/dasc.2014.6979455
13. Jackson D (2016) Ensuring honest behaviour in cooperative surveillance systems (CDT technical paper series). The centre for doctoral training in cyber security
14. Obod I, Svyd I, Maltsev O, Maistrenko G, Zubkov O, Zavolodko G (2019) Bandwidth assessment of cooperative surveillance systems. In: 2019 3rd International conference on advanced information and communications technologies (AICT). https://doi.org/10.1109/aiact.2019.8847742
15. Obod I, Svyd I, Maltsev O, Vorgul O, Maistrenko G, and Zavolodko G (2018) Optimization of data transfer in cooperative surveillance systems. In: 2018 International scientific-practical conference problems of info communications. Sci Technol (PIC S&T). https://doi.org/10.1109/infocommst.2018.8632134
16. Malyarenko A (2007) Secondary radar systems for air traffic control and state recognition. HUVS, Kharkiv
17. Svyd I, Obod I, Maltsev O, Shtykh I, Maistrenko G, Zavolodko G (2019) Comparative quality analysis of the air objects detection by the secondary surveillance radar. In: 2019 IEEE 39th International conference on electronics and nanotechnology (ELNANO). https://doi.org/10.1109/elnano.2019.8783539

18. NATO (2016) STANAG 4193 PT III. Technical characteristics of IFF MK XA and MK XII Interrogators and transponders. Part III: Installed system characteristics, 23 May 2016
19. Huan L, Feng Z, Bai L, Jian W (2015) One joint demodulation and despreading algorithm for MOD5. Open Autom Control Syst J 7(1). https://doi.org/10.2174/1874444301507010386
20. Guo Y, Yang J, Guan C (2013) A Mode 5 signal detection method based on phase and amplitude correlation. In: 2013 Ninth international conference on natural computation (ICNC). https://doi.org/10.1109/icnc.2013.6818164
21. Sirotkin S, Kon'kov A (2014) Methods of continuous processing of information from frequency sensors. Electr Data Process Facil Syst (3)
22. Tsikin I, Poklonskaya E (2017) Secondary surveillance radar signals processing at the remote analysis station. St. Petersburg State Polytechnical Univ J Comput Sci Telecommun Control Syst 10(2). https://doi.org/10.18721/jcstcs.10205
23. Svyd I, Obod I, Maltsev O, Okachova T, Zavolodko G (2019) Optimal request signals detection in cooperative surveillance systems. In: 2019 IEEE 2nd Ukraine conference on electrical and computer engineering (UKRCON). https://doi.org/10.1109/ukrcon.2019.8879840
24. Lenshin A, Lebedev V (2020) On problem of expert evaluation of air object identification system quality. Telecommunications, Nauka i tehnologii, Moscow (2)
25. Svyd I, Obod I, Maltsev O, Shtykh I, Zavolodko G, Maistrenko G (2019) Model and method for request signals processing of secondary surveillance radar. In: 2019 IEEE 15th international conference on the experience of designing and application of CAD systems (CADSM). https://doi.org/10.1109/cadsm.2019.8779347
26. Zhironkin S, Bliznyuk S, Kuchin A (2019) Jamming resistance of the inbound channel of an identification system with broadband signals and error control codes in the conditions of pulse noise and intra-system jamming. J Siberian Federal Univ Eng Technol. https://doi.org/10.17516/1999-494x-0166
27. Obod I, Svyd I, Maltsev O, Bakumenko B (2020) Spatial methods for increasing the bandwidth of a mobile information network. In: 2020 IEEE 15th international conference on advanced trends in radioelectronics, telecommunications and computer engineering (TCSET). https://doi.org/10.1109/tcset49122.2020.235388
28. Obod I, Svyd I, Maltsev O, Bakumenko B (2020) Comparative analysis of noise immunity systems identification friend or foe. In: 2020 IEEE 40th international conference on electronics and nanotechnology (ELNANO). https://doi.org/10.1109/elnano50318.2020.9088856
29. Piracci E, Galati G, Petrochilos N, Fiori F (2009) 1090 MHz channel capacity improvement in the air traffic control context. Int J Microwave Wirel Technol 1(3). https://doi.org/10.1017/s1759078709000191
30. Galati G, Piracci E, Petrochilos N, Fiori F (2008) 1090 MHz channel capacity improvement in the Air traffic control context. In: 2008 Tyrrhenian international workshop on digital communications—enhanced surveillance of aircraft and vehicles. https://doi.org/10.1109/tiwdc.2008.4649030
31. Honda J, Otsuyama T (2018) Statistical analysis of 1090 MHz signals measured during a flight experiment, 2018 international symposium on antennas and propagation (ISAP). Busan, Korea (South)
32. EUROCONTROL (2006) CASCADE programme: 1090 MHz capacity study–final report, edition 2.7, July 2006
33. Pollack J, Ranganatha P (2018) Aviation navigation systems security: ADS-B, GPS, IFF. In: International conference on security & management, SAM'18, Las Vegas, Nevada, USA
34. Li W, Kamal P (2011) Integrated aviation security for defense-in-depth of next generation air transportation system. In: 2011 IEEE international conference on technologies for homeland security (HST). https://doi.org/10.1109/ths.2011.6107860
35. Svyd I, Obod I, Maltsev O, Strelnytskyi O, Zubkov O, Zavolodko G (2019) Method of increasing the identification friend or foe systems information security. In: 2019 3rd international conference on advanced information and communications technologies (AICT). IEEE. https://doi.org/10.1109/aiact.2019.8847853

36. Anderson R (2008) Security engineering: a guide to building dependable distributed systems, 2nd edn. Wiley, Indianapolis
37. Skaves P (2011) Information for cyber security issues related to aircraft systems. In: 2011 IEEE/AIAA 30th digital avionics systems conference. https://doi.org/10.1109/dasc.2011.6095968
38. De Cerchio R, Riley C (2011) Aircraft systems cyber security. In: 2011 IEEE/AIAA 30th digital avionics systems conference. https://doi.org/10.1109/dasc.2011.6095969
39. Svyd I, Obod I, Maltsev O, Zavolodko G, Maistrenko G, Saikivska L (2019) Method of enhancing information security of requesting cooperative surveillance systems. In: 2019 IEEE international scientific-practical conference problems of infocommunications, science and technology (PIC S&T). https://doi.org/10.1109/dasc.2011.6095969
40. El-Badawy E, EL-Masry W, Mokhtar M, Hafez A (2010) A secured chaos encrypted mode-S aircraft identification friend or foe (IFF) system. In: 2010 4th International conference on signal processing and communication systems. https://doi.org/10.1109/icspcs.2010.5709756
41. Petrov A, Mikhalev V (2019) Bit-error rate in a digital data transmitting channel at chaotic impulse noise with random radio-pulse duration action. Syst Control Commun Secur 3. https://doi.org/10.24411/2410-9916-2019-10303
42. Bernhart S, Leitgeb E (2018) Evaluations of low-cost decoding methods for 1090 MHz SSR signals. In: 2018 International conference on broadband communications for next generation networks and multimedia applications (CoBCom). https://doi.org/10.1109/cobcom.2018.8443986
43. Svyd I, Obod I, Maltsev O, Shtykh I, Zavolodko G (2019) Model and method for detecting request signals in identification friend or foe systems. In: 2019 IEEE 15th international conference on the experience of designing and application of CAD systems (CADSM). https://doi.org/10.1109/cadsm.2019.8779322
44. Galati G, Studer F (1990) Maximum likelihood azimuth estimation applied to SSR/IFF systems. IEEE Trans Aerosp Electron Syst 26(1). https://doi.org/10.1109/7.53411
45. Svyd I, Obod I, Maltsev O, Maistrenko G, Zavolodko G, Pavlova D (2019) Fusion of airspace surveillance systems data. In: 2019 3rd International conference on advanced information and communications technologies (AICT). https://doi.org/10.1109/aiact.2019.8847916
46. Pavlova D, Zavolodko G, Obod I, Svyd I, Maltsev O, Saikivska L (2019) Optimizing data processing in information networks of airspace surveillance systems. In: 2019 10th international conference on dependable systems, services and technologies (DESSERT). https://doi.org/10.1109/dessert.2019.8770022
47. Lebedev V, Lenshin A, Tikhomirov N (2015) Effective suppression of the radar systems with active response codes jamming. The bulletin of Voronezh Institute of the Ministry of Internal Affairs of Russia (4)
48. IEEE Standard (2012) IEEE standard for distributed interactive simulation, application protocols. In: IEEE Std 1278.1-2012 (Revision of IEEE Std 1278.1-1995). https://doi.org/10.1109/ieeestd.2012.6387564
49. Svyd I, Obod I, Maltsev O, Vorgul O, Zavolodko G, Goriushkina A (2018) Noise immunity of data transfer channels in cooperative observation systems: comparative analysis. In: 2018 international scientific-practical conference problems of infocommunications, science and technology (PIC S&T). IEEE. https://doi.org/10.1109/infocommst.2018.8632019
50. Svyd I, Obod I, Maltsev O, Tkachova T, Zavolodko G (2019) Improving noise immunity in identification friend or foe systems. In: 2019 IEEE 2nd Ukraine conference on electrical and computer engineering (UKRCON). IEEE. https://doi.org/10.1109/ukrcon.2019.8879812
51. Hubacek P, Vesely J (2016) Probabilistic code extractor for low SNR SIF/IFF mode A, C respond. In: 2016 17th International radar symposium (IRS). https://doi.org/10.1109/irs.2016.7497367
52. Obod I, Svyd I, Maltsev O, Vorgul O, Maistrenko G, Zavolodko G (2020) Optimization of the quality of information support for consumers of cooperative surveillance systems. In: Radivilova T, Ageyev D, Kryvinska N (eds) Data-centric business and applications. Lecture notes on data engineering and communications technologies, vol 48. Springer, Cham. https://doi.org/10.1007/978-3-030-43070-2_8

53. Obod I, Svyd I, Maltsev O, Zavolodko G, Pavlova D, Maistrenko G (2021) Fusion the coordinate data of airborne objects in the networks of surveillance radar observation systems. In: Radivilova T, Ageyev D, Kryvinska N (eds) Data-centric business and applications. Lecture Notes on Data Engineering and Communications Technologies, vol 48. Springer, Cham. https://doi.org/10.1007/978-3-030-43070-2_31
54. Leonardi, M. and Gerardi, F. (2020) Aircraft Mode S Transponder Fingerprinting for Intrusion Detection. Aerospace, 7(3), https://doi.org/10.3390/aerospace7030030

Pulse and Multifrequency Van der Pol Generators Based on Transistor Structures with Negative Differential Resistance for Infocommunication System Facilities

Andriy Semenov, Olena Semenova, Oleksandr Osadchuk, Iaroslav Osadchuk, Kostyantyn Koval, Serhii Baraban, and Mariia Baraban

Abstract The paper proposes new circuits of Van der Pol generators based on nonlinear and reactive properties of transistor structures with negative differential resistance, which operate in relaxation and quasi-periodic modes when parameters of their self-oscillating systems are controlled in a wide range. New analytical relations were obtained, which described parameters of sawtooth-shaped and rectangular voltage pulses of the relaxation-type Van der Pol generators, built on bipolar and field-effect transistor structures with negative differential resistance. A new multifrequency Van der Pol generator of quasi-periodic electrical oscillation based on a field-effect transistor structure with negative differential resistance was developed and its mathematical model was proposed. There were obtained analytical ratios for calculating the stationary oscillation amplitude in single-frequency and multifrequency modes, the critical detuning between single-frequency and multifrequency modes, as well as lower and upper cutoff frequencies of the operating frequency band. Phase portraits, time diagrams and amplitude spectra of electrical oscillations of the Van der Pol generators in oscillatory, relaxation and quasi-periodic modes were studied. Influence of parameters of the Van der Pol generators' self-oscillating on generated oscillation dynamics was estimated.

A. Semenov (✉) · O. Semenova · O. Osadchuk · I. Osadchuk · K. Koval · S. Baraban · M. Baraban
Vinnytsia National Technical University, Khmelnytske shose 95, Vinnytsia, Ukraine
e-mail: semenov.a.o@vntu.edu.ua

O. Semenova
e-mail: semenova.o.o@vntu.edu.ua

K. Koval
e-mail: kkoval@vntu.edu.ua

S. Baraban
e-mail: serg@politex.org.ua

D. Ageyev et al. (eds.), *Data-Centric Business and Applications*, Lecture Notes on Data Engineering and Communications Technologies 69,
https://doi.org/10.1007/978-3-030-71892-3_6

Keywords Van der Pol generator · Transistor structure · Negative differential resistance · Phase portrait · Electrical oscillations · Amplitude spectrum · Relaxation mode · Quasi-periodic mode

1 Introduction

Nowadays, telecommunication and electronic systems [1–20] are widely used in various aspects of society's life [6, 11, 14–16, 20]. A key issue in operation of such systems is providing high quality of service [1, 5, 13, 18] and traffic management and control [2, 3, 7, 9, 19]. Moreover, modern mobile and telecommunication networks involve high-tech devices for information transmission, which require new and sophisticated circuits of signal generators [4, 8, 10, 12, 17].

The process of converting different types of energy into the energy of electrical oscillation is called a generation of electrical oscillations. An autonomous source operating in self-excitation mode is called an electrical oscillation generator. According to a type of sources of converted electrical energy, electrical oscillation generators are divided into two main classes [21, 22]:

- if primary electrical oscillations are converted into oscillations of required frequency and shape, generators with external excitation and parametric generators are utilized;
- if the energy of DC voltage sources is converted, reference generators autogenerators are utilized.

Electrical oscillation generators as functional units are a part of many various devices and communication systems. According to the form of generated signals without modulation, such generators are divided into two main groups: harmonic and pulse oscillations [23].

The electrical oscillation generator is a closed system with positive feedback that contains a power supply, a frequency selective system (resonator), which determines the frequency of generated oscillations, active element covered by a positive feedback circuit, which compensates for energy losses in a frequency selective system, a nonlinear element that limits the generated oscillation amplitude [24–26].

A promising approach to the construction of electric oscillation generators is to use devices with negative differential resistance to compensate for energy losses in passive tuning circuits and oscillating system of the generator, which will improve energy characteristics of these generators [27–31].

Harmonic oscillators using the effect of negative differential resistance have been studied in [31–36]. Among sources of non-sinusoidal signals, the dominant position is occupied by oscillators of rectangular oscillations. Multivibrators, which in turn are functional units of more complex communication devices, are most often used to obtain voltage or current pulses of rectangular shape [33, 37–40].

One of the promising areas of the pulse technology development is application of new devices and materials that allow designing pulse generators, which employ previously unknown effects and provide fundamentally new opportunities to control parameters of output pulses [41].

Pulse generators based on transistor structures with negative differential resistance (NDR) use a combination of a feedback two-pole and a nonlinear active two-pole, which yields negative differential resistance under certain conditions [42–44]. If the negative resistance is greater than the positive resistance of the oscillating circuit, then by including such resistance in a circuit, one can compensate for losses and thus create in the circuit oscillations undamped in time.

Self-oscillating systems based on devices with negative differential resistance can be complete or degenerate [45, 46]. Self-oscillating systems are divided into complete and degenerated ones depending on completeness of the set of reactive elements in a oscillating circuit of a generator. Degenerated self-oscillating systems are generally generators of the relaxation type. However, depending on a reactive component value of the impedance of a transistor structure with negative differential active resistance, the generated oscillations can be harmonic. The main disadvantage of such generators is the low output power, which is reduced due to filtering extra components of the output voltage spectrum.

The aim of this study is to develop and examine pulse and multifrequency Van der Pol generators based on transistor structures with negative differential resistance for application in facilities of infocommunication systems.

2 Relaxation Type Generators on Transistor Structures with Negative Differential Resistance by the Method of Van der Pol Oscillator

The relaxation mode of Van der Pol generators' operation is used to construct a wide class of pulse signal generators [47]. Standard types of pulse signals are electrical oscillations of sawtooth, triangular and rectangular shapes [48]. The relaxation mode of the generator, built by the method of Van der Pol oscillator differs by a rapid increase of losses in the oscillating circuit due to the inductive component [47, 48]. Therefore, an equivalent circuit of the relaxation type generator on the transistor structure with negative differential resistance (NDR) built by the method of Van der Pol oscillator will look like this (Fig. 1).

When constructing pulse generators of sawtooth, triangular and linear AC voltage by the method of Van der Pol oscillator (Fig. 1), transistor structures with NDR with S-type I–V curves are used as an active element [49]. When constructing pulse generators of rectangular voltage, transistor structures with NDR with N-type I–V curves are used as an active element [49].

Let's construct a linearly alternating voltage generator (Fig. 1). Figure 2 shows a combined static I–V curve of the transistor structure with NDR with the operating point trajectory the during one oscillation period.

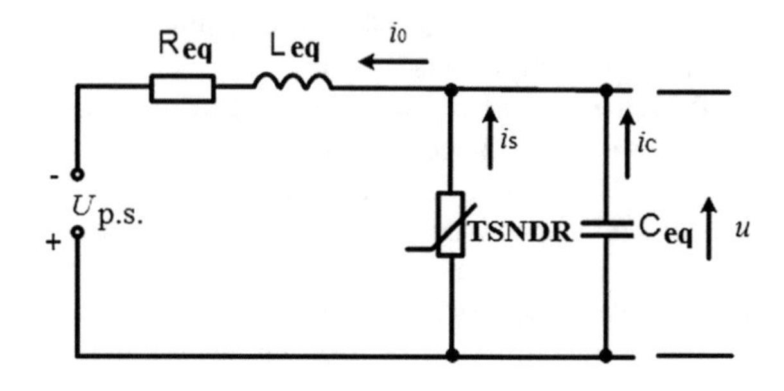

Fig. 1 Equivalent circuit of the Van der Pol generator based on the transistor structure with NDR (TSNDR) in relaxation mode

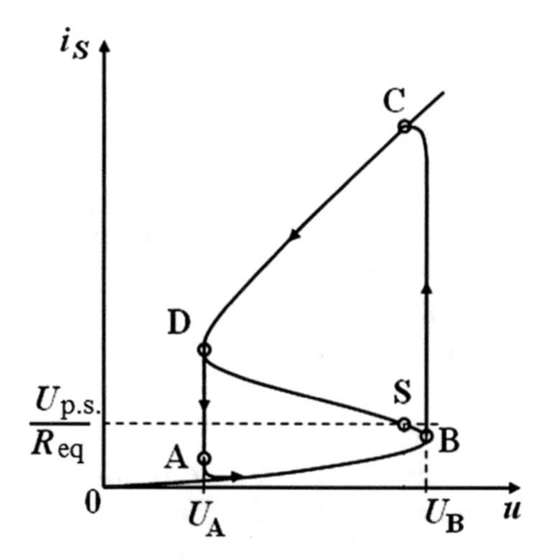

Fig. 2 Combined static I–V curve of the transistor structure with NDR with the operating point trajectory during one oscillation period

The equivalent circuit of the Van der Pol pulse generator of linearly alternating voltage in the region of slow motions is shown in Fig. 3.

From Kirchhoff's first law, dynamic processes in the electrical circuit are described by the equation

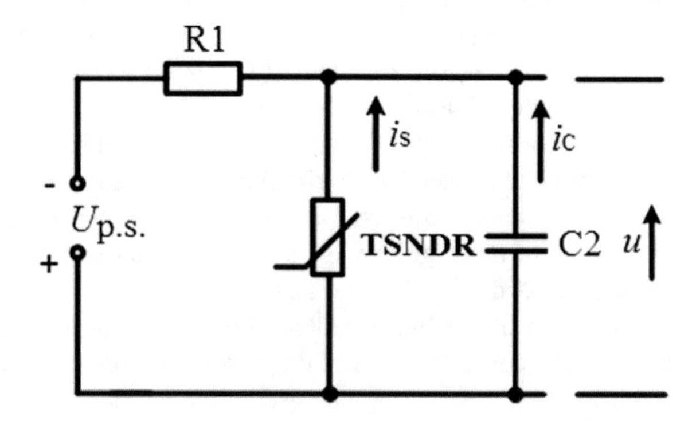

Fig. 3 Equivalent circuit of the linear-alternating voltage generator in the region of slow motions

$$C_{eq}\frac{du}{dt}+i_S=\frac{U_{p.s.}}{R_{eq}}, \tag{1}$$

or

$$C_{eq}R(i_S)\frac{di_S}{dt}+i_S=\frac{U_{p.s.}}{R_{eq}}, \tag{2}$$

where $R(i_S)=\frac{du}{di_S}$ is the differential resistance of the generator's active element.

Forming the section of linearly alternating voltage (forward time) corresponds to the operating point movement along the lower branch of the static I–V curve of the bipolar transistor structure (section AB in Fig. 2). The time of forming the linear voltage section is the solution of the linear differential Eq. (2) and, according to Fig. 2, is

$$t=C_2\int\limits_{I_A}^{i_S}\frac{R(i_S)}{I_0-i_S}di_S, \tag{3}$$

where I_A is the ordinate of point A on the static I–V curve (Fig. 2).

To integrate the right part of Eq. (3), the lower part of the static I–V curve of the transistor structure with NDR must be approximated. The subintegral function in Eq. (3) is approximated by the following power function [50]

$$i_S(u)=I_B-A(U_B-u)^{1/n}, \tag{4}$$

where A is the approximation coefficient; n is the approximating equation degree, n = 2–5 [50].

From (4) the differential resistance of the lower section of the static I–V characteristics for the transistor structure with NDR is

$$R(i_S)=\frac{du}{di_S}=\frac{n}{A^n}[I_B-i_S(u)]^{n-1}, \tag{5}$$

If the I–V curve is static (Fig. 2), it is sufficient to apply a second-degree approximation in (4) ($n=2$). In this case, the forward time of the linear AC voltage generator is

$$t=\frac{2C}{\sqrt{A}}\left[i_S-I_A-\left(I_B-\frac{U_{p.s.}}{R_1}\right)\ln\left|\frac{\frac{U_{p.s.}}{R_1}-i_S}{\frac{U_{p.s.}}{R_1}-I_A}\right|\right], \tag{6}$$

where i_S is described by Eq. (4).

The authors calculated the differential Eq. (2) considering approximation of the I–V curve of the transistor structure with NDR (4) by applying the standard function *odesolt ()* of the software package MathCad 11.0. Figure 4 shows the combined static I–V curve of the transistor structure with NDR with the operating point trajectory during one period. The generated current pulses are shown in Fig. 5.

The electric circuit of the linearly alternating voltage Van der Pol generator constructed by the authors is given in Fig. 6a. The linear alternating voltage generator by the method of Van der Pol oscillator operates as follows [51]. When the power

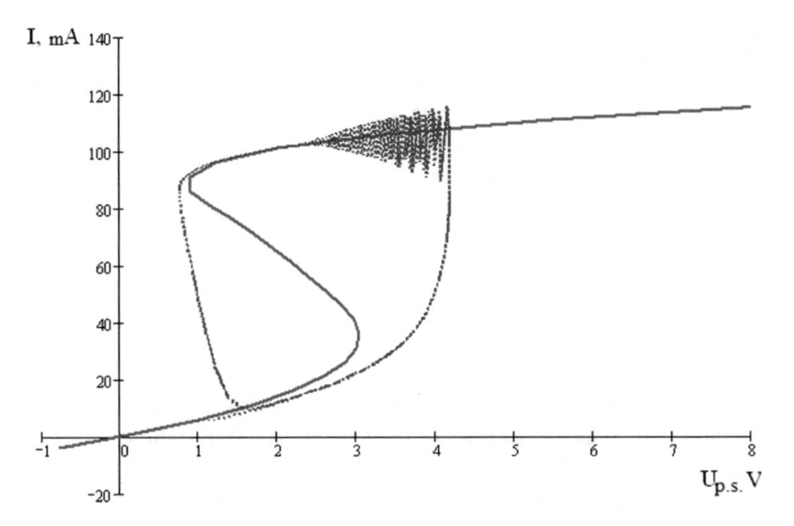

Fig. 4 Combined approximate I–V curve of the transistor structure with NDR with the operating point trajectory during one period

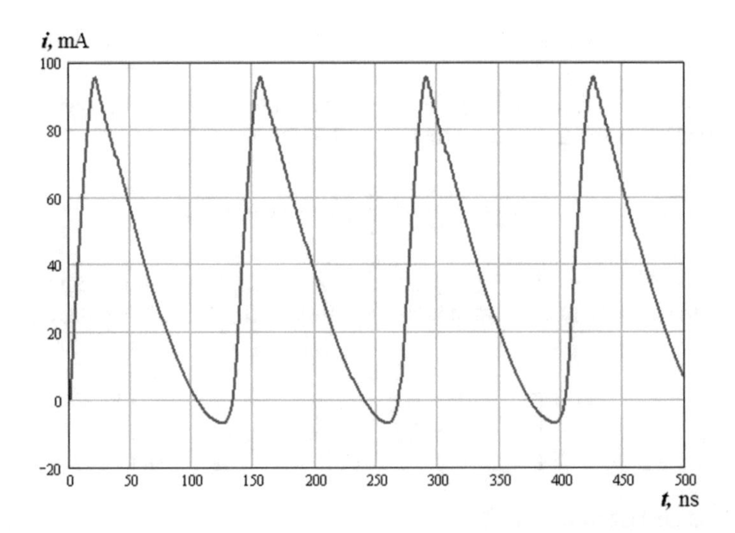

Fig. 5 Time dependence for the current

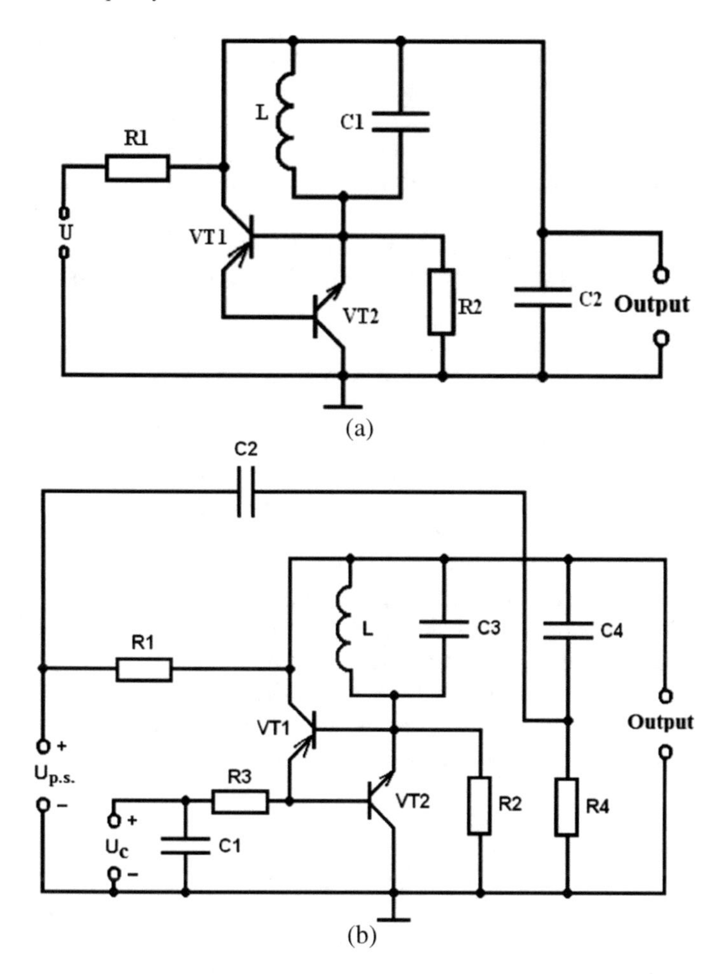

Fig. 6 Electrical circuits of the linearly alternating voltage generator based on the bipolar transistor structure with NDR: **a** without negative feedback, **b** with global negative voltage feedback

supply voltage U increases to a value providing appearance of the negative differential resistance on collector-collector electrodes of bipolar transistors VT1 and VT2, relaxation oscillations appear due to cyclic processes of charging and discharging of capacitor C2. The oscillating circuit of elements LC1 compensates for the impedance reactive component the transistor structure at frequency of the generated pulses.

Resistor R1 limits the supply current value of the bipolar transistor structure with NDR and together with resistor R2 forms a voltage divider. Capacitor C2 forms with the inductor a series oscillating circuit for transforming the generator's output resistance to match it with the load. The reactive component of the impedance at the collector-collector electrodes of bipolar transistors, which has a capacitive nature, depends on the supply voltage value. Resistor R2 forms a negative voltage feedback, which increases the frequency stability for the generated pulses. The change in the supply voltage value causes a change in the repetition frequency of linearly variable pulses of the generated voltage, as well as a change in its rate.

The global negative voltage feedback (NF) in the basic linearly alternating voltage generator circuit (Fig. 6a) provides reduction of the return time and increase of the forward voltage linearity. The circuit of such generator is shown in Fig. 6b. The generated pulses of linearly alternating voltage have a sawtooth shape. Changing parameters of the generated pulses of linearly alternating voltage can be performed by applying the control voltage Uc (Fig. 6b), that improves the energy characteristics of the generator.

Obtained by the authors results of experimental studies of the linearly alternating voltage generator (Fig. 6a) on two discrete bipolar transistors VT1 UKT3101 and VT2 UKT3102 are given in [51]. Nominal values of the circuit passive elements are R1 = 1.5 kΩ, R2 = 8.2 kΩ, L = 240 μH, C1 = 1.8 nF, C2 = 4.7 nF. It was experimentally found out that when the supply voltage changed within 5–9 V, the current consumption changed within 27–81 mA, and the generated pulse amplitude changed within 1.2–10 V at a load of 50 Ω when changing the period of linearly alternating voltage pulses within 0.13–1.67 μs.

When constructing rectangular voltage pulse generators according to the method of Van der Pol oscillator, it is necessary to ensure the required duration of the leading and trailing fronts and to obtain the required pulse or pause shape [52, 53]. Given that the negative differential resistance of the transistor structure with NDR in relaxation mode compensates for the equivalent loss resistance Req, for the equivalent circuit (Fig. 1) according to Kirchhoff's laws we can write the following system of differential equations [51]

$$\begin{cases} L_{eq}\frac{di_0}{dt} = U_{p.s} - u, \\ i_0 = C_{eq}\frac{du}{dt} + i_N, \end{cases} \tag{7}$$

where i_N is the current of the transistor structure with NDR with Λ-type I–V characteristic.

We reduced the system of differential Eqs. (7) in [51] to one differential equation of the second order against to the generated voltage

$$\frac{d^2u}{dt^2} + \frac{G(u)}{C_{eq}}\frac{du}{dt} + \omega_0\big(u - U_{p.s.}\big) = 0, \tag{8}$$

where $G(u) = di_T\big/du$ is the differential conductivity of the transistor structure with NDR; $\omega_0 = 1/\sqrt{L_{eq}C_{eq}}$ is the angular frequency of rectangular voltage pulses.

The equation of the Van der Pol relaxator at the limit cycle establishing [54–56] was numerically solved by the 4-th order Runge-Kutta method in MathCad 15.0 software that yielded the time dependence of the generated rectangular voltage pulses in normalized time shown in Fig. 7.

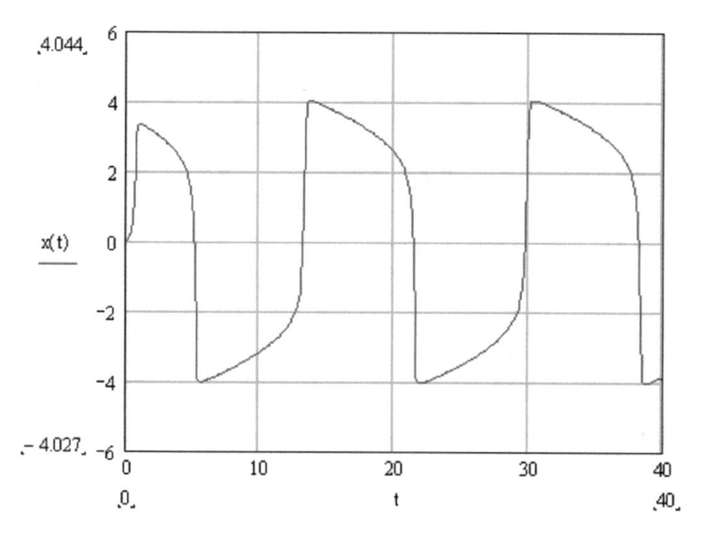

Fig. 7 Time dependence of the pulse voltage for the Van der Pol relaxer (time on the abscissa is in microseconds, voltage on the ordinate is in volts)

Constructing the rectangular voltage pulse generators by the method of Van der Pol oscillator provides the duration of the fronts or a given pulse shape and requires separation of the pulse generation process in the region of fast and slow motions (Fig. 8). For the slow motion region, capacitance C_{eq} of the transistor structure with NDR can be neglected. Then, an equivalent circuit of the Van der Pol relaxer will look like (Fig. 9a) [57, 58]. For the region of fast motions, the current is considered to be $i_0 = I_0 = const$, and therefore the equivalent circuit of the Van der Pol relaxer, has the form shown in Fig. 9b [57, 58].

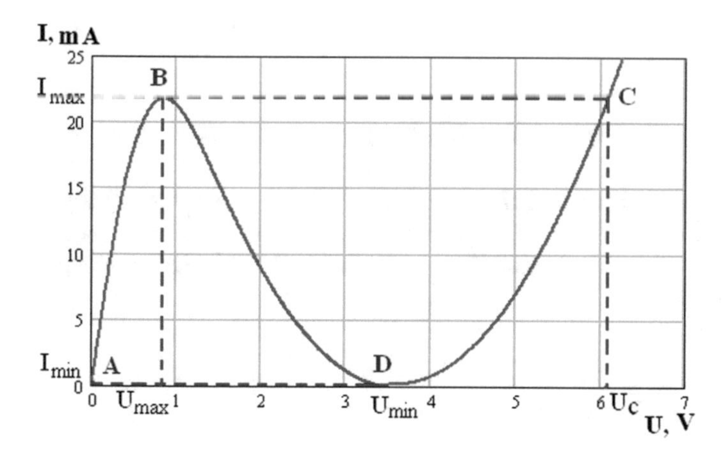

Fig. 8 Static I–V curve of athetransistor structure with NDR N-type with coordinates of the operating point trajectory

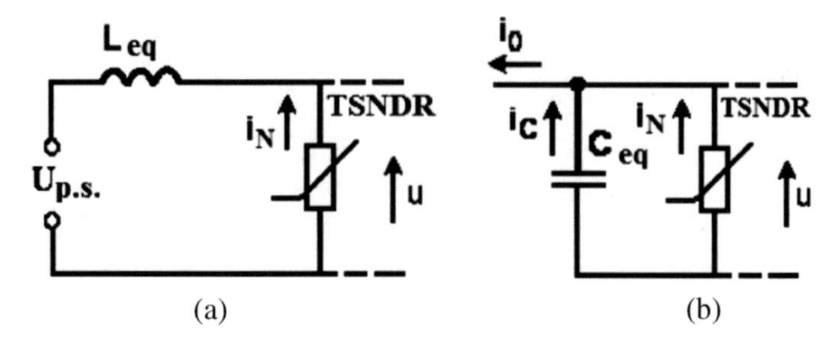

Fig. 9 Equivalent circuits of the Van der Pol generator of rectangular voltage pulses for regions of: **a** slow motions, **b** fast motions

The oscillation dynamics in the region of slow motions for the equivalent circuit (Fig. 9a) is described by the differential equation [51]

$$L_{eq}G(u)\frac{du}{dt} + u - U_{p.s.} = 0. \tag{9}$$

The solution of this differential equation is [51]

$$t = L_{eq}\int\limits_{U_C}^{u}\frac{G(u)}{U_{p.s.} - u}du, \tag{10}$$

where the lower bound of integration U_C is the abscissa of point C (Fig. 8).

Then the duration of pulse or pause equals [51]

$$t = L_{eq}\int\limits_{U_C}^{u}\frac{G(u)}{U_{p.s.} - u}du = \left.\frac{L_{eq}i(u)}{U_{p.s.} - u}\right|_{U_c}^{u} - L_{eq}\int\limits_{U_C}^{u}\frac{i(u)}{\left(U_{p.s.} - u\right)^2}du, \tag{11}$$

where $i(u)$ is the approximation function for the I–V curve of the transistor structure with NDR.

Equation (11) includes an integral that cannot be defined explicitly. Therefore, numerical methods should be applyed to determine the duration of pulse or pause.

To determine the pulse duration and pause, Eq. (11) is integrated along trajectories AB and CD

$$\tau_i = \frac{L_{eq}i(U_C)}{U_{p.s.} - U_C} - \frac{L_{eq}i(U_D)}{U_{p.s.} - U_D} - L_{eq}\int\limits_{U_D}^{U_C}\frac{i(u)}{\left(U_{p.s.} - u\right)^2}du, \tag{12}$$

$$\tau_n = \frac{L_{eq} i(U_B)}{U_{p.s.} - U_B} - \frac{L_{eq} i(U_A)}{U_{p.s.} - U_A} - L_{eq} \int\limits_{U_A}^{U_B} \frac{i(u)}{\left(U_{p.s.} - u\right)^2} du. \tag{13}$$

The dynamics of oscillations in the region of fast motions (Fig. 9b) is described by the differential equation:

$$C_{eq} \frac{du}{dt} + i(u) = I_0. \tag{14}$$

Equations of the duration of the leading and trailing fronts of the generated rectangular voltage pulses are solutions of Eq. (14)

$$t_f^{01} = C_{eq} \int\limits_{U_B}^{U_C} \frac{du}{I_0 - i(u)}; \tag{15}$$

$$t_f^{10} = C_{eq} \int\limits_{U_D}^{U_A} \frac{du}{I_0 - i(u)}. \tag{16}$$

The results of the study of dynamic processes in Van der Pol generators on the transistor structure with NDR in relaxation mode shows that using an inductor as electrical energy storage deforms the shape of the vertex and fronts of generated rectangular voltage pulses [51, 59]. Their shape can be improved by using a transistor inductor [53]. The authors have synthesized the electrical circuit of the Van der Pol generator of rectangular voltage pulses on the field-effect transistor structure with NDR (Fig. 10).

In [51], a Van der Pol generator of rectangular voltage pulses on discrete transistors KP103 (VT1, VT3), KP303 (VT4) and KT363 (VT2) was experimentally investigated. Nominal values of of passive elements are capacitor C1 (1 μF), resistors R1 (51 kΩ), R2 (5.6 kΩ) and R3 (200 Ω). The frequency of rectangular voltage pulses is controlled by changing f the resistor R1 resistance in a wide range. Figure 11 shows an oscillogram of the generated voltage rectangular pulses. It was experimentally established that the shape of generated voltage rectangular is affected by the supply voltage $U_{p.s.}$. This is explained by the fact that during the operating point movement along the limit cycle of the phase trajectory during one period of oscillations, it twice passes through the static I–V curve of the transistor structure with NDR, which determines the relationship between the shape of rectangular voltage pulses, its period and duration, with the shape and duration of the fronts. Therefore, the Van der Pols method of constructing rectangular voltage pulse generators based on the transistor structure with NDR allows its forming according to a required shape and duration of pulses and fronts.

In [51] a static Λ-type I–V curve was approximated to establish the relationship between the shape of voltage pulse oscillations $u(t)$ and parameters of the field-effect

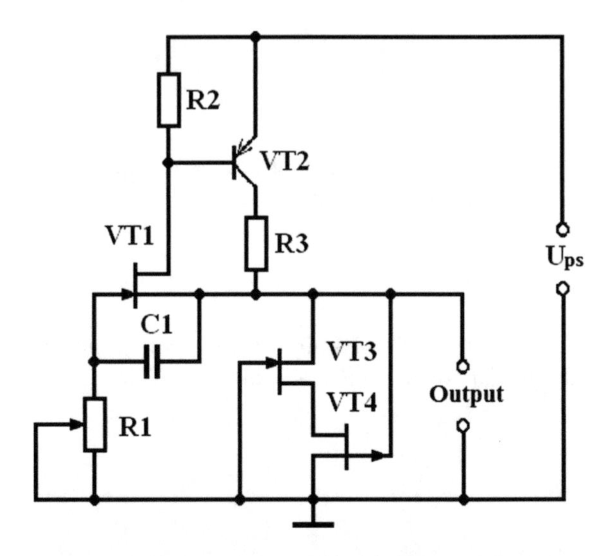

Fig. 10 Electrical circuit of the rectangular voltage pulses Van der Pol generator on the field-effect transistor structure with NDR

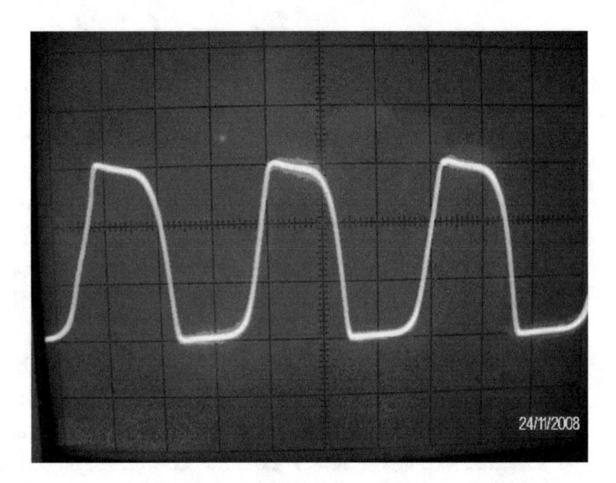

Fig. 11 Oscillograms of rectangular voltage pulses generated by the generator

transistor structure with NDR of the active element of the Van der Pol generator proposed in [60]. In case of choosing field-effect transistors with the same cut-off voltage values, but with different parameters of drain currents, we will have [51, 60]

$$I_{D3} = I_{D03}\left(1 - \frac{U(1-\Delta)}{2U_0}\right)^2 (thM)^{-1} th\left[M\frac{U(1+\Delta)/2U_0}{1 - U(1-\Delta)/2U_0}\right], \tag{17}$$

$$I_{D4} = I_{D04}\left(1 - \frac{U(1-\Delta)}{2U_0}\right)^2 (thM)^{-1} th\left[M\frac{U(1+\Delta)/2U_0}{1 - U(1-\Delta)/2U_0}\right], \tag{18}$$

where I_{D03}, I_{D04} are drain curents at $U_{GD} = 0$ and $U_{SD} = U_0$ respectively for field-effect transistors VT3 and VT4; the parameter Δ is defined as [60]

$$\Delta = (U_{SD3} - U_{SD4})/U_0. \tag{19}$$

The current equation of the field-effect transistor structure with NDR obtained by applying field-effect transistors with drain currents $I_{D03} > I_{D04}$, which differ by 1.5–2.0 times, with accuracy of the order Δ^2 is [51]

$$I = \frac{2I_{D03}}{1 + I_{D03}/I_{D04}}\left(1 - \frac{U}{2U_0}\right)^n (thM)^{-1} th\left[M\frac{U/2U_0}{(1 - U/2U_0)^{n-1}}\right], \tag{20}$$

where n equals 1.8–2.0 for low-power field-effect transistors.

The equation of negative differential conductivity of a transistor structure with NDR with different field-effect transistors, obtained by the authors in [51], has is

$$G(U) = \frac{2I_{D03}}{1 + I_{D03}/I_{D04}}(thM)^{-1}\left\{n\left(1 - \frac{U}{2U_0}\right)^{n-1} th\left[M\frac{U/2U_0}{(1 - U/2U_0)^{n-1}}\right]\right.$$
$$\left. + \frac{M}{2U_0}\left[1 + (n-1)\left(1 - \frac{U}{2U_0}\right)\right]\left(1 - \frac{U}{2U_0}\right)ch^{-2}\left[M\frac{U/2U_0}{(1 - U/2U_0)^{n-1}}\right]\right\}. \tag{21}$$

Having applyed the method of Van der Pol oscillator, the authors have constructed new circuits of generators on the transistor structure with NDR, operating in oscillatory and relaxation modes, which are a part of radio measuring transducers of physical quantities [61–65].

3 A Multifrequency Generator of Quasi-Periodic Oscillations on the Field-Effect Transistor Structure with Negative Differential Resistance by the Method of Van der Pol Oscillator

In modern electronics, analysis and synthesis of generators of multifrequency signals with required spectral characteristic and noise-like signals is a relevant task [66, 67]. A separate group of oscillators of special shape-signals are chaotic generators [68–70]. A group between them are generators of quasi-periodic oscillations, on their phase plane a strange non-chaotic attractor appearing [71–73]. Classically, the Van der Pol multifrequency generators consist of interconnected linear and nonlinear oscillating circuits, as well as an amplifier with a nonlinear characteristic that provides self-excitation of the self-oscillating system and limits the generated stationary oscillation amplitude [74–77]. The barrier capacitance of the p-n junction is used as a nonlinear reactive element of the oscillating circuit in the self-oscillating system [78].

The authors in [51] have proposed a method for constructing a multifrequency Van der Pol generator appluing the capacitive effect of a field-effect transistor structure with NDR. Oscillation dynamics of the generator in multifrequency mode were analized by the authors considering the following simplifications: (1) the equivalent circuit of the multifrequency generator is the circuit in Fig. 1 with a current source connected in parallel with the nonlinear resistance and the parallel connection of L_{eq} and R_{eq}; (2) the amplitude and frequency of harmonic oscillations of the current source correspond to the amplitude and frequency of the single-frequency stationary mode of the generator; 3) the occurrence of other harmonic components of the multifrequency mode is due to both nonlinear properties of the negative resistance and nonlinear properties of the capacitive component of the impedance of the transistor structure with NDR [78].

The current $i_{ex}(u, t)$ takes into account how the generator's active element affects the circuit, compensating for the losses in in. The system of differential equations, which describes the dynamics of oscillations of the inductance current and voltage on the capacitor in real time is [51]:

$$\begin{cases} \frac{di}{dt} = \frac{1}{L}u; \\ \frac{du}{dt} = \frac{1}{C}\left[i_{ex} - i - \frac{u}{R_{eq}}\right]. \end{cases} \quad (22)$$

The nominal frequency of the multifrequency Van der Pol generator on the transistor structure with NDR in the single-frequency mode is close to the resonant frequency of the generator's oscillating system:

$$\omega_{nom} = \omega_{ex} = \omega_0, \quad (23)$$

where ω_{ex} is the oscillation frequency of the current source.

In normalized time

$$\tau = \omega_0 t, \omega_0 = \frac{1}{\sqrt{L_{eq} C_{eq}}}. \quad (24)$$

Taking into account the secondary parameters of the oscillating circuit

$$\rho = \omega_0 L_{eq} = \frac{1}{\omega_0 C_{eq}} = \sqrt{\frac{L_{eq}}{C_{eq}}}, \quad (25)$$

$$Q = \frac{\rho}{R} = \omega_0 C_{eq} R, \quad (26)$$

under condition (23) the system of differential Eqs. (22) is reduced to a second-order differential equation [51]:

$$\frac{d^2 i}{d\tau^2} + 1 = i_{ex} - \frac{1}{Q}\frac{di}{d\tau} + \nu i, \tag{27}$$

where ν is the relative detune of the frequency relative to the resonant frequency of the generator's oscillating circuit [51]:

$$\nu = \frac{\omega_{nom}^2 - \omega_0^2}{\omega_0^2} \approx \frac{2(\omega_{nom} - \omega_0)}{\omega_0}. \tag{28}$$

The solution of the differential Eq. (27) can be given as [51]:

$$i = I_m \sin(\tau + \varphi) = I_m \sin \psi; \tag{29}$$

$$\frac{di}{d\tau} = I_m \cos(t + \varphi) = I_m \cos \psi. \tag{30}$$

Differential equations for establishing the amplitude I_m and phase φ of the generated oscillations are [51]

$$\frac{dI_m}{d\tau} = \frac{1}{2} I_{1C} - \frac{I_m}{2Q}; \tag{31}$$

$$\frac{d\varphi}{d\tau} = \frac{1}{2}\frac{I_{1S}}{I_m} - \frac{1}{2}\nu, \tag{32}$$

where I_{1C} and I_{1S} are cosine and sinusoidal components of the first harmonic of the decomposition function $i_{ex}(u, t)$ in the Fourier series.

Using a power approximation for the multifrequency Van der Pol generator on the transistor structure with NDR yields

$$i_T(u) = \left(I_S + gU_S - hU_S^3\right) - \left(g - 3hU_S^2\right) \cdot (u - e) - 3hU_S(u - e)^2 + h(u - e)^3, \tag{33}$$

where g, h are the approximation coefficients, U_S, I_S are the coordinates of the middle of the I–V curve descending section (initial setting of the operating point on the I–V curve descending section of the) and

$$e(t) = E_m \cos \omega_{nom} t \tag{34}$$

is the external equivalent effect on the oscillating circuit of the generator.

The amplitude of stationary oscillations of the generator in single-frequency mode is determined by the following expression [51]

$$U_{ST} = \frac{2}{\sqrt{3}}\sqrt{\frac{g - 3hU_S^2}{h} + \frac{1}{hQ\rho\cos\varphi_\beta}}. \tag{35}$$

The amplitude of stationary oscillations for the generator voltage in the multifrequency oscillation mode when deviating from the main frequency is determined by the following expression [51]:

$$U_m = -\frac{\rho E_m\left[3hU_S^2 - g + \frac{3}{4}hU_{ST}^2\right]}{\nu}\sin\varphi, \tag{36}$$

where U_{ST} is the amplitude of stationary oscillations (35).

The Van der Pol multi-frequency generator can operate in single-frequency, dual-frequency, three-frequency and multi-frequency modes. The operation modes of are changed electrically by changing the operating point position on the descending I–V curve of the transistor structure with NDR (active component of the impedance of the transistor structure with NDR) and on the volt-farad characteristic of the transistor structure with NDR (reactive component of the impedance of the transistor structure with NDR). Starting from the two-frequency mode, the beat of quasi-harmonic signals close in frequency can be observed. A critical detune in transition from single-frequency to multi-frequency mode is determined from the ratio (36)

$$\nu_{\kappa p} = \frac{E_m\rho\left[3hU_S^2 - g + \frac{3}{4}hU_{CT}^2\right]}{U_{CT}}. \tag{37}$$

The lower and upper cutoff frequencies of the operating band are respectively equal to

$$\omega_H = \omega_0\left(1 - \frac{E_m}{2QU_{CT}}\right), \tag{38}$$

$$\omega_B = \omega_0\left(1 + \frac{E_m}{2QU_{CT}}\right). \tag{39}$$

When the oscillating circuit of the multifrequency generator has a high quality factor

$$E_m \approx U_{CT} \text{ that is why } \omega_B \approx \omega_H \approx \omega_0. \tag{40}$$

The electrical circuit of the multifrequency Van der Pol generator synthesized by the authors is shown in Fig. 12. The reactive component of the impedance of the transistor structure with NDR at the drain-drain electrodes of field-effect transistors VT1 and VT2 and inductance L1 forms the nonlinear oscillating circuit. Those of elements L1 and C3 forms the linear oscillating circuit. At position of the minimum value of the variable resistor R2 slider, the equivalent capacitance of the field-effect

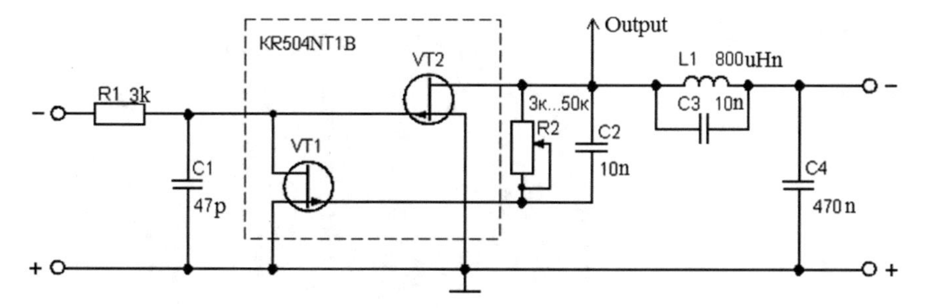

Fig. 12 Electrical circuit of the multi-frequency Van der Pol generator on the field-effect transistor transistor structure with NDR [51]

transistor structure with NDR is equal to the capacitor C3 capacitance, and therefore the natural resonant frequencies of linear and nonlinear oscillating circuits are the same, which leads to quasiharmonic oscillation. The change in the value of resistance R2 causes a change in the equivalent capacitance of the field-effect transistor structure with NDR, which leads to a change in the resonant frequency of the nonlinear circuit. The deviation from the resonant frequency in the linear circuit is insignificant, and therefore there is a beat of two quasi-harmonic oscillations in the generator [51].

A subsequent change in the value of resistance R2 changes the quasi-periodic operation mode of the Van der Pol generator from two-frequency to multi-frequency one [51]. The authors have investigated experimentally the operating modes, the results are shown in Figs. 13, 14, 15, 16, and 17 [51]. This allows to employ the multi-frequency Van der Pol generator to increase the noise immunity of radiocommunication systems [79, 80], as well as to transmit discrete information with spread spectrum using quasi-periodic and multi-frequency signals [81, 82].

From (22)–(39) the mathematical model of the multifrequency Van der Pol generator based on the field-effect transistor structure with NDR in normalized time has the form

$$\begin{cases} \frac{dx_1}{d\tau} = x_2, \\ \frac{dx_2}{d\tau} = \mu\left(1 - bx_1 - gx_1^2\right)x_2 - x_1 + \sin\Omega t + \sum_{k=2}^{n} U_k\left[\cos(k\Omega t) + \cos\left(\frac{\Omega t}{k}\right)\right], \end{cases} \tag{41}$$

where k is a harmonic number with amplitude U_k, Ω is the excitation frequency normalized to the average frequency of the operating range ω_0 [83, 84]. Results of mathematical modeling obtained by the authors in [51] under conditions $\Omega = 1, 2$, $\mu = 0, 1$, $b = 0, 5$, $q = 0, 1$ and different n and U_k, are given in Figs. 18, 19, 20 and 21. In Figs. 18, 19, 20 and 21 the voltage of the generated oscillations is given in volts, time and frequency are normalized relative to the frequency of stationary oscillations of the single-frequency mode ω_0.

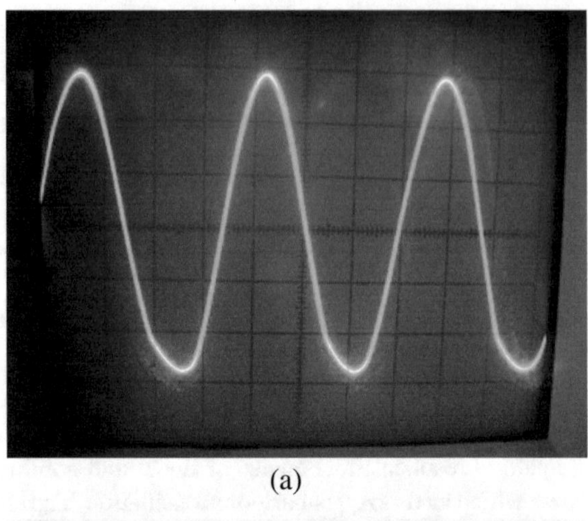

(a)

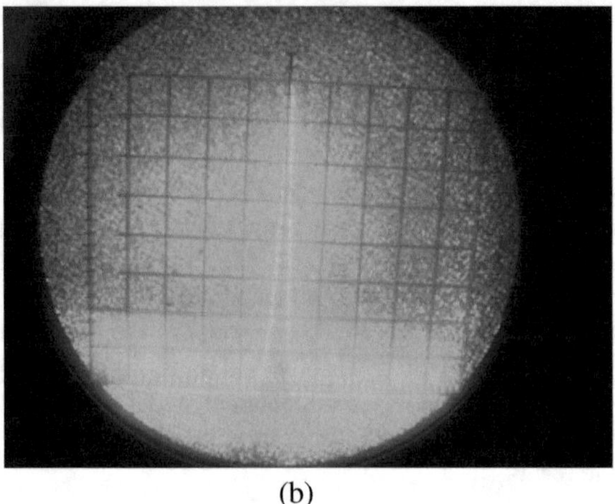

(b)

Fig. 13 Oscillogram (**a**) and spectrum (**b**) of generated oscillations in the single-frequency mode

4 Conclusion

In this section, signal generators with regular dynamics of relaxation and quasi-periodic types were constructed by the method of Van der Pol oscillator on transistor structures with NDR for facilities of infocommunication systems. The electric circuits of the constructed generators were presented, the principles of their operation were described, and dynamic processes in them were studied. The following theoretical and practical results were obtained:

1. There were developed Van der Pol pulse signal generators based on nonlinear and reactive properties of transistor structures with NDR, operating in relaxation and quasi-periodic modes with electrical control of the parameters of their self-oscillating systems in a wide range.

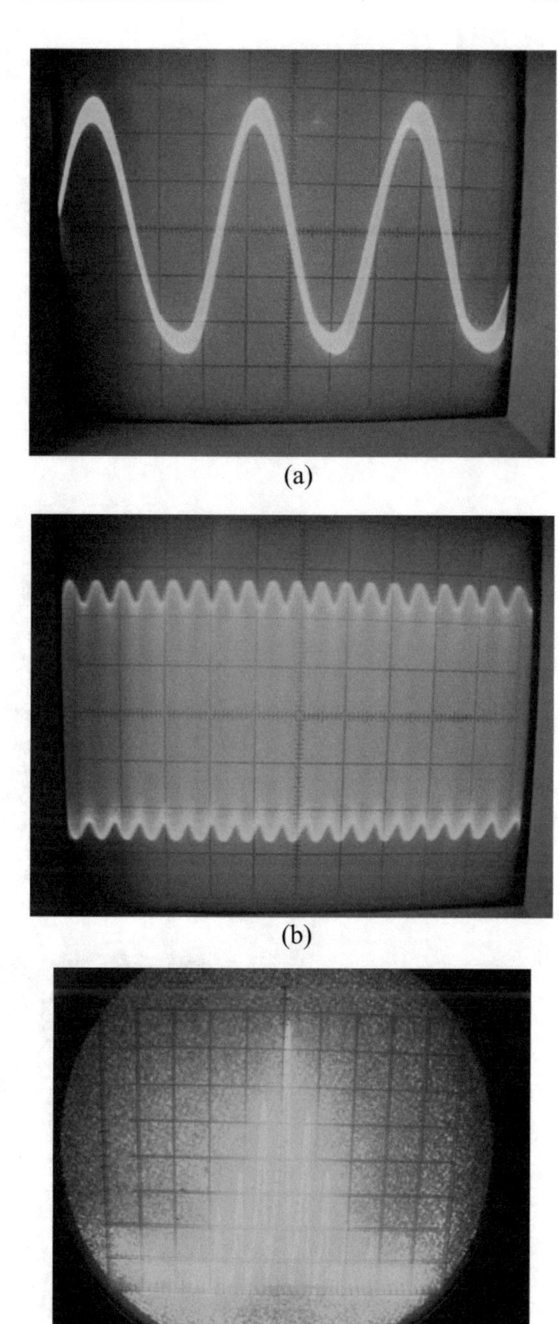

Fig. 14 Oscillograms (**a**) and (**b**) and spectrum (**c**) for the 5-frequency mode

Fig. 15 Oscillograms (**a**) and (**b**) and spectrum (**c**) for the 9-frequency mode

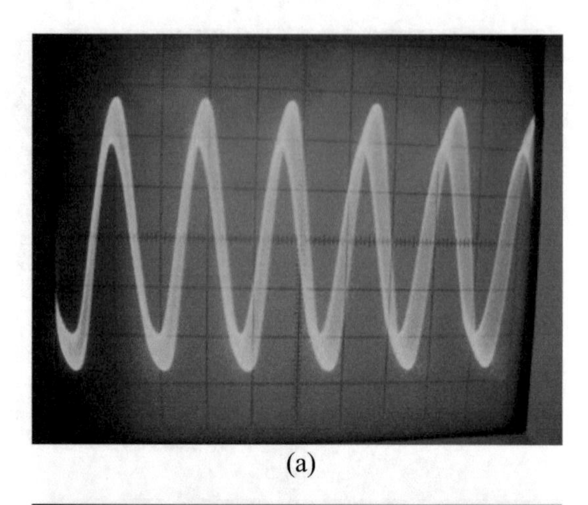

(a)

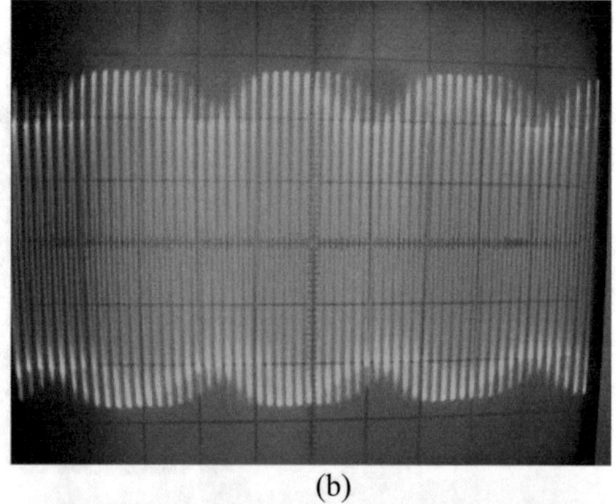

(b)

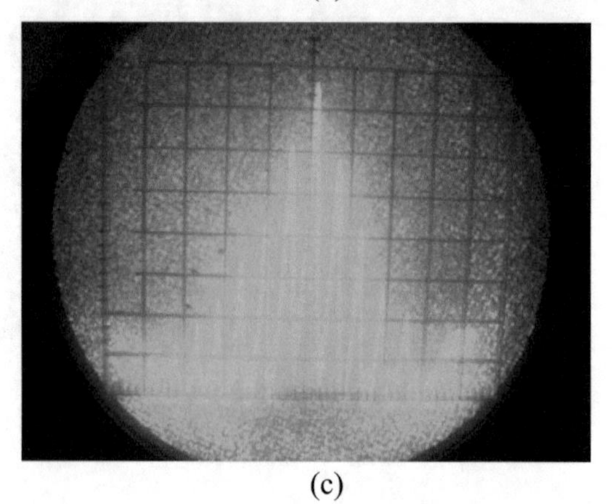

(c)

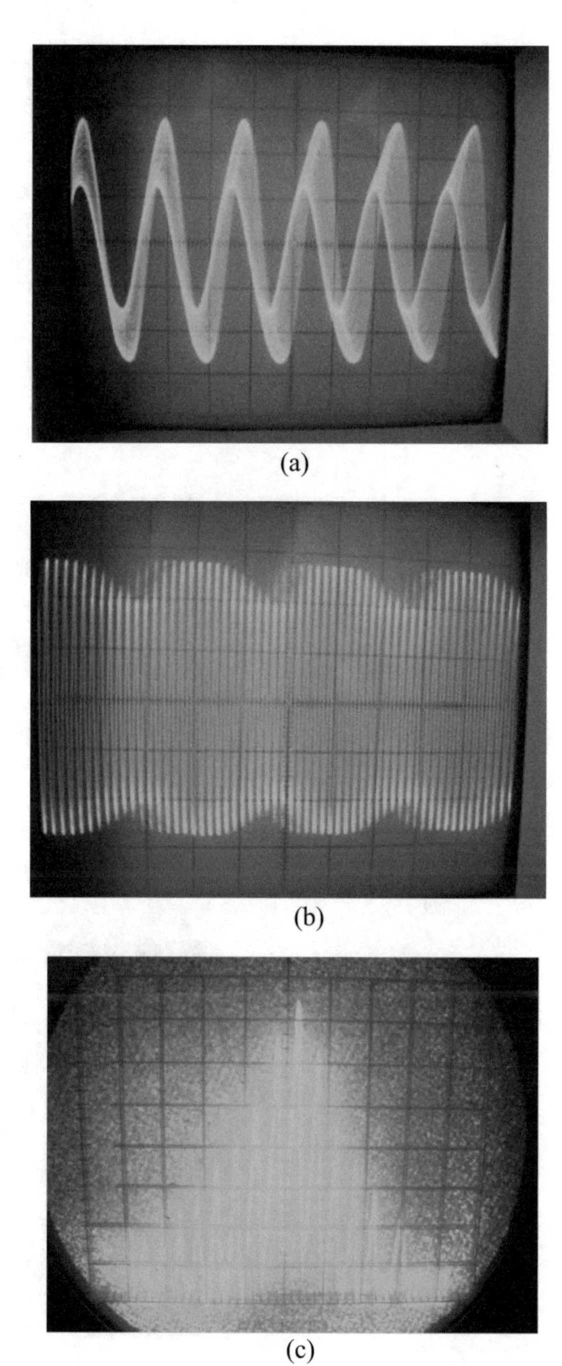

Fig. 16 Oscillograms (**a**) and (**b**) and spectrum (**c**) for the 12-frequency mode

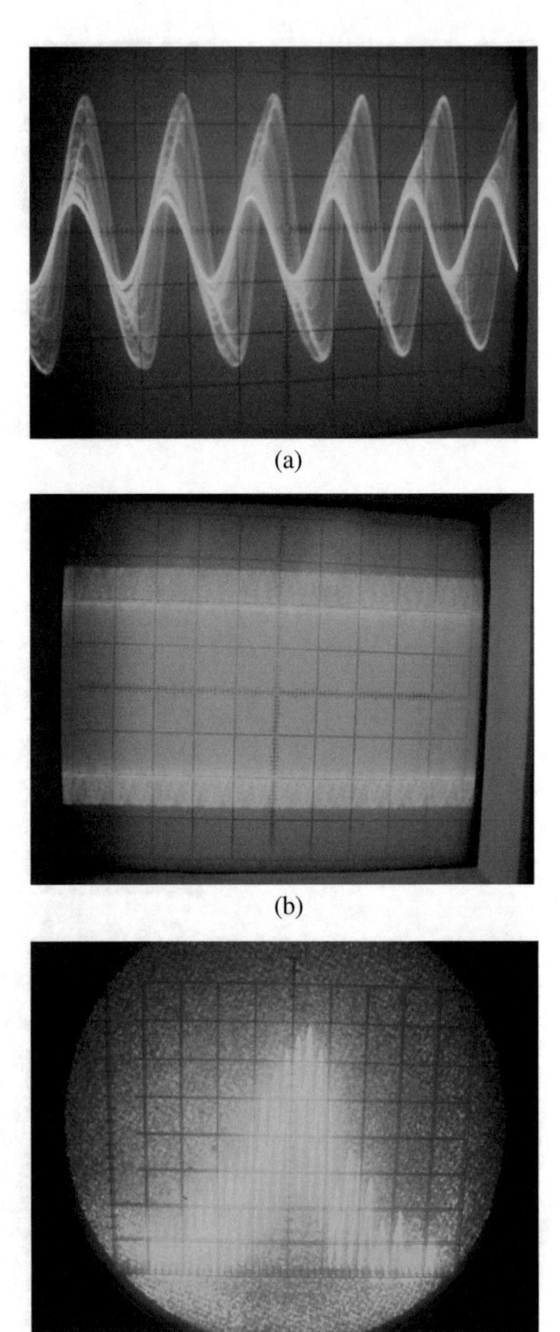

Fig. 17 Oscillograms (**a**) and (**b**) and spectrum (**c**) for the multifrequency mode of the generator operation

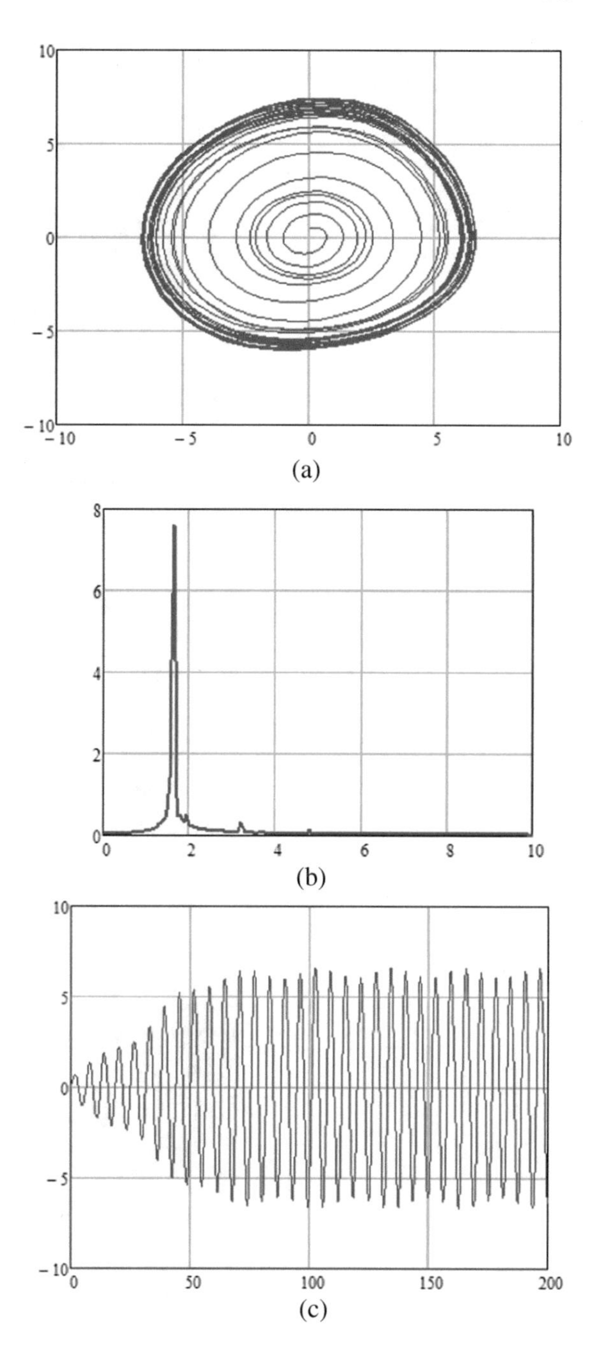

Fig. 18 Phase portrait (**a**) amplitude spectrum (**b**) and time diagrams (**c**) of the generated voltage fluctuations in the single-frequency mode of the generator operation

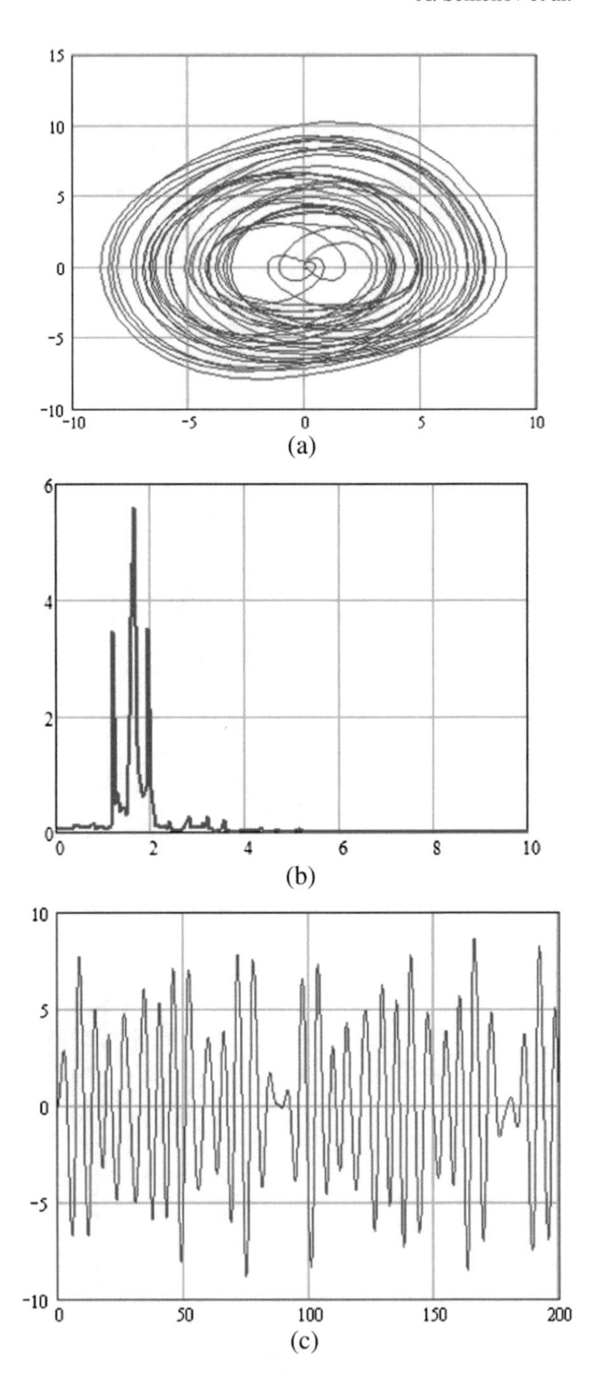

Fig. 19 Phase portrait (**a**) amplitude spectrum (**b**) and time diagrams (**c**) of the generated voltage fluctuations in the three-frequency mode of the generator operation

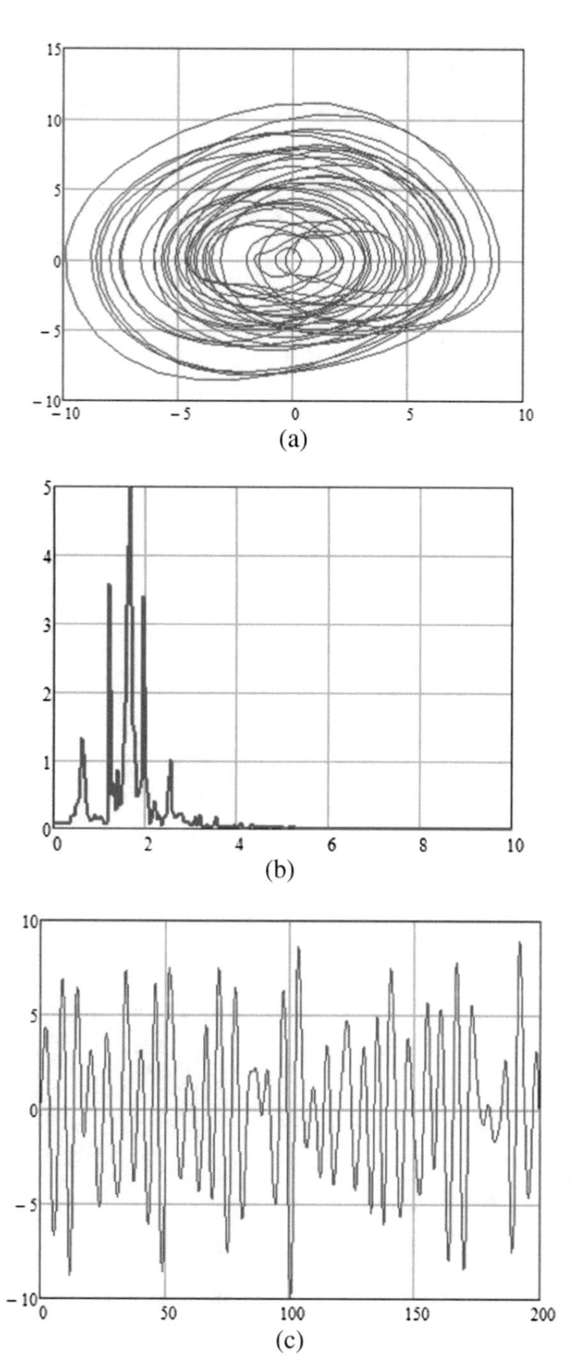

Fig. 20 Phase portrait (**a**) amplitude spectrum (**b**) and time diagrams (**c**) of the generated voltage fluctuations in the five-frequency mode of the generator operation

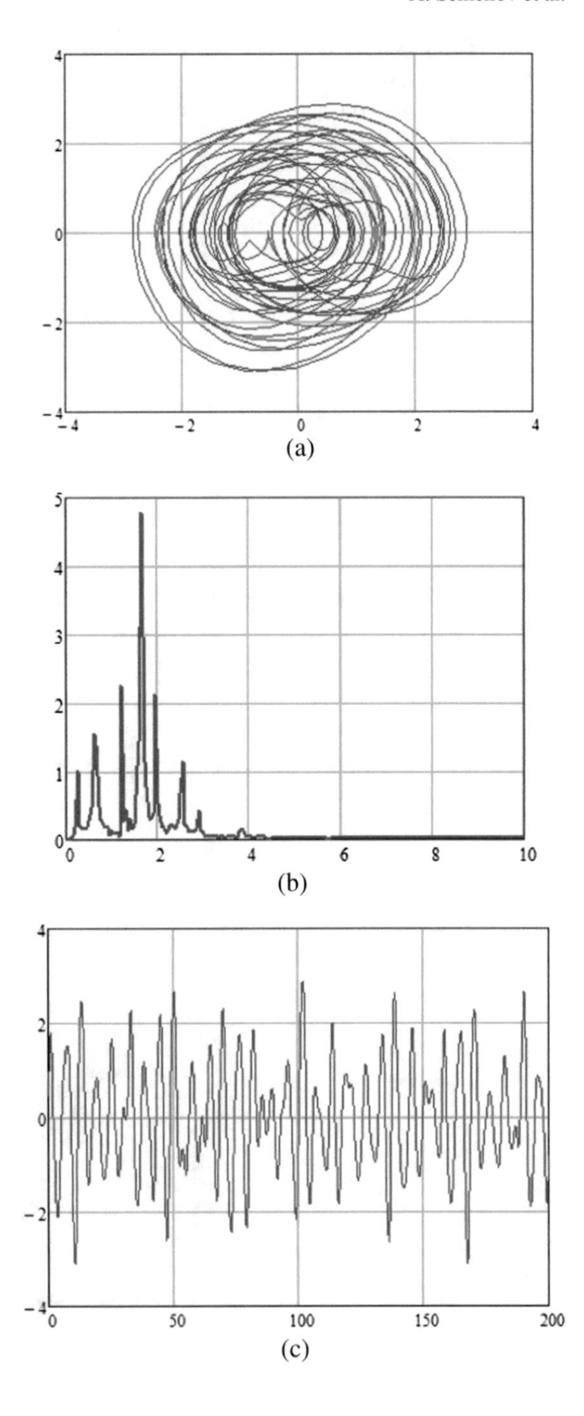

Fig. 21 Phase portrait (**a**) amplitude spectrum (**b**) and time diagrams (**c**) of the generated voltage fluctuations in the seven-frequency mode of the generator operation

2. New circuits of Van der Pol pulse signal generators based on nonlinear and reactive properties of transistor structures with NDR were constructed. Theoretical and experimental studies of these generators were carried out. Mathematical models of the Van der Pol generators with complex dynamics of generated oscillations were developed. Results of numerical modeling and experimental investigation are given. Convergence of the obtained results confirms adequacy of the developed mathematical models.
3. New analytical ratios were obtained, which described parameters of sawtooth and rectangular voltage pulses in the Van der Pol generators on bipolar and field-effect transistor structures with NDR. Phase portraits, time diagrams and amplitude spectra of electric oscillations of the Van der Pol generators in relaxation and quasi-periodic modes were obtained.
4. The ultifrequency Van der Pol generator of quasi-periodic electric oscillations based on the field-effect transistor structure with NDR was constructed and its mathematical model was proposed. New analytical rattios were obtained to calculate of the stationary oscillation amplitude in single-frequency and multi-frequency modes, the critical frequency detuning in transition from single-frequency to multi-frequency mode, as well as the lower and upper operating frequency limits.

References

1. Daradkeh YI, Kirichenko L, Radivilova T (2018) Development of QoS methods in the information networks with fractal traffic. Int J Electron Telecommun 64(1):27–32. https://doi.org/10.24425/118142
2. Kirichenko L, Radivilova T (2017) Analyzes of the distributed system load with multifractal input data flows. In: 2017 14th International conference the experience of designing and application of CAD systems in microelectronics (CADSM), Lviv, 21–25 Feb 2017, pp 260-264. https://doi.org/10.1109/cadsm.2017.7916130
3. Ivanisenko I, Kirichenko L, Radivilova T (2015) Investigation of self-similar properties of additive data traffic. In 2015 Xth International scientific and technical conference "computer sciences and information technologies" (CSIT), Lviv, 14–17 Sept 2015, pp 169–171. https://doi.org/10.1109/stc-csit.2015.7325459
4. Ageyev D, Bondarenko O, Alfroukh W, Radivilova T (2018) Provision security in SDN/NFV. In: 2018 14th International conference on advanced trends in radioelecrtronics, telecommunications and computer engineering (TCSET), Slavske, 20–24 Feb 2018, pp 506–509. https://doi.org/10.1109/tcset.2018.8336252
5. Radivilova T, Kirichenko L, Ageyev D, Bulakh V (2020) The methods to improve quality of service by accounting secure parameters. Adv Intell Syst Comput 938:346–355. https://doi.org/10.1007/978-3-030-16621-2_32
6. Kirichenko L, Radivilova T, Zinkevich I (2017) Forecasting weakly correlated time series in tasks of electronic commerce. In: 2017 12th International scientific and technical conference on computer sciences and information technologies (CSIT), Lviv, 5–8 Sept 2017, pp 309–312. https://doi.org/10.1109/stc-csit.2017.8098793

7. Kirichenko L, Radivilova T, Bulakh V (2018) Classification of fractal time series using recurrence plots. In: 2018 International scientific-practical conference problems of infocommunications. Science and technology (PIC S&T), Kharkiv, Ukraine, 9–12 Oct 2018, pp 719–724. https://doi.org/10.1109/infocommst.2018.8632010
8. Al-Dulaimi A, Al-Dulaimi M, Ageyev D (2016) Realization of resource blocks allocation in LTE downlink in the form of nonlinear optimization. In: 2016 13th International conference on modern problems of radio engineering, telecommunications and computer science (TCSET), Lviv, 23–26 Feb 2016, pp 646–648. https://doi.org/10.1109/tcset.2016.7452140
9. Ageyev D, Yarkin D, Nameer Q (2014) Traffic aggregation and EPS network planning problem. In: 2014 First international scientific-practical conference problems of infocommunications science and Technology, Kharkov, 14–17 Oct 2014, pp 107-108, https://doi.org/10.1109/infocommst.2014.6992316
10. Ageyev D (2010) NGN network planning according to criterion of provider's maximum profit. In: 2010 International conference on modern problems of radio engineering, telecommunications and computer science (TCSET), Lviv-Slavske, Ukraine, 23–27 Feb 2010, pp 256–256
11. Kryvinska N, Poniszewska-Maranda A, Gregus M (2018) An approach towards service system building for road traffic signs detection and recognition. Procedia Comput Sci 141:64–71. https://doi.org/10.1016/j.procs.2018.10.150
12. Tkachenko R, Izonin I, Kryvinska N, Dronyuk I, Zub K (2020) An approach towards increasing prediction accuracy for the recovery of missing iot data based on the grnn-sgtm ensemble. Sensors 20(9):2625. https://doi.org/10.3390/s20092625
13. Beshley M, Kryvinska N, Seliuchenko M, Beshley H, Shakshuki EM, Yasar A-U-H (2020) End-to-End QoS "smart queue" management algorithms and traffic prioritization mechanisms for narrow-band internet of things services in 4G/5G networks. Sensors 20(8):2324. https://doi.org/10.3390/s20082324
14. Kryvinska N, Kaczor S, Strauss C (2020) Enterprises' servitization in the first decade-retrospective analysis of back-end and front-end challenges. Appl Sci 10(8):2957. https://doi.org/10.3390/APP10082957
15. Kryvinska N, Bickel L (2020) Scenario-based analysis of IT enterprises servitization as a part of digital transformation of modern economy. Appl Sci 10(3):1076. https://doi.org/10.3390/app10031076
16. Larisa G, Mariia S, Anna Z (2015) The method of resources allocation for processing requests in online charging system. In: The experience of designing and application of CAD systems in microelectronics, Lviv, 24–27 Feb 2015, pp 211–213. https://doi.org/10.1109/cadsm.2015.7230838
17. Skulysh M, Globa L, Siemens E (2020) Resource sharing challenge for micro operator pattern in 5G SDN / NFV network. In: Siemens E, Mylnikov L (Eds) Proceedings of 8th international conference on applied innovation in IT, ICAIIT 2020, Koethen, Germany, 10 Mar 2020, vol 8, no 1, pp 21–28
18. Globa L, Skulysh M, Romanov O, Nesterenko M (2019) Quality control for mobile communication management services in hybrid environment. Lect Notes Electr Eng 560:76–100. https://doi.org/10.1007/978-3-030-16770-7_4
19. Skulysh M, Sulima S (2015) Management of multiple stage queuing systems. In: The experience of designing and application of CAD systems in microelectronics, Lviv, 24–27 Feb 2015, pp 431–433. https://doi.org/10.1109/cadsm.2015.7230895
20. Larysa G, Mariia S, Tetiana P, Reverchuk A (2014) Managing of incoming stream applications in online charging system. In: 2014 X International symposium on telecommunications (BIHTEL), Sarajevo, 27–29 Oct 2014, pp 1–6. https://doi.org/10.1109/bihtel.2014.6987632
21. Grebennikov A (2007) RF and microwave transistor oscillator design. Wiley, New York
22. Grebennikov A (2011) RF and microwave transmitter design. Wiley, Hoboken, New Jersey
23. Rybin YK (2014) Measuring signal generators. Springer, Dordrecht
24. Kazimierczuk MK, Krizhanovski VG, Rassokhina JV et al (2005) Class-E MOSFET tuned power oscillator design procedure. IEEE Trans Circuit Syst I Regul Pap 52(6):1138–1147. https://doi.org/10.1109/TCSI.2005.849127

25. Kazimierczuk MK, Krizhanovski VG, Rassokhina JV et al (2006) Injection-locked class-E oscillator. IEEE Trans Circuits Syst I Regul Pap 53(6):1214–1222. https://doi.org/10.1109/TCSI.2006.875176
26. Krizhanovski VG, Chernov DV, Kazimierczuk MK (2007) Low-voltage electronic ballast based on class E oscillator. IEEE Trans Power Electron 22(3):863–870. https://doi.org/10.1109/TPEL.2007.896512
27. Ulansky VV, Ben Suleiman SF, Elsherif HM (2016) Optimization of NDR VCOs for microwave applications. In: Proceedings of the 2016 IEEE 36th international conference on electronics and nanotechnology (ELNANO), Kiev, Ukraine, pp 353–357. https://doi.org/10.1109/elnano.2016.7493083
28. Ulansky VV, Elsherif HM (2015) A voltage-controlled oscillator based on negative inductance converter. In: Proceedings of the 2015 IEEE 35th international conference on electronics and nanotechnology (ELNANO), Kiev, pp 508–511. https://doi.org/10.1109/elnano.2015.7146939
29. Ulansky VV, Ben Suleiman SF (2013) Negative differential resistance based voltage-controlled oscillator for VHF band. In: Proceedings of the 2013 IEEE XXXIII international scientific conference electronics and nanotechnology (ELNANO), Kiev, Ukraine, pp 80–84. https://doi.org/10.1109/elnano.2013.6552016
30. Agarwal H et al (2018) Engineering negative differential resistance in NCFETs for analog applications. IEEE Trans Electron Devices 65(5):2033–2039. https://doi.org/10.1109/TED.2018.2817238
31. Gibson GA (2018) Designing negative differential resistance devices based on self-heating. Adv Funct Mater 28:1704175. https://doi.org/10.1002/adfm.201704175
32. Ulansky V (2011) Low phase-noise HEMT microwave voltage-controlled oscillator. In: Proceedings of the 2011 microwaves, radar and remote sensing symposium, Kiev, Ukraine, pp 55–58. https://doi.org/10.1109/mrrs.2011.6053600
33. Ulansky VV, Elsherif HM (2013) Optimization of LC voltage-controlled oscillators in 90-nm CMOS technology for 3G transceivers. In: Proceedings of the 2013 IEEE XXXIII international scientific conference electronics and nanotechnology (ELNANO), Kiev, Ukraine, pp 85–89. https://doi.org/10.1109/elnano.2013.6552040
34. Ulansky VV, Machalin IA, Elsherif H (2017) UHF voltage-controlled oscillator with inductive feedback. In: Proceedings of the 2017 IEEE 37th International conference on electronics and nanotechnology (ELNANO), Kiev, Ukraine, pp 475–479. https://doi.org/10.1109/elnano.2017.7939799
35. Ulansky VV, Kolesnik AA, Elsherif HM (2014) A new high-performance OPA based VCO for microwave applications. In: Proceedings of the 2014 IEEE microwaves, radar and remote sensing symposium (MRRS), Kiev, Ukraine, pp 50–53. https://doi.org/10.1109/mrrs.2014.6956663
36. Ulansky VV, Ben Suleiman SF (2013) A low phase-noise GaAs FET/BJT voltage-controlled oscillator for microwave applications. In: Proceedings of the 2013 international kharkov symposium on physics and engineering of microwaves, millimeter and submillimeter waves, Kharkiv, Ukraine, pp 407–414. https://doi.org/10.1109/msmw.2013.6622100
37. Akarvardar K et al (2006) Four-Gate transistor voltage-controlled negative differential resistance device and related circuit applications. In: 2006 IEEE international SOI conferencee proceedings, niagara falls, NY, 2–5 Oct 2006, pp 71–72. https://doi.org/10.1109/soi.2006.284438
38. Kobashi K, Hayakawa R, Chikyow T et al (2017) Negative differential resistance transistor with organic p-n heterojunction. Adv Electron Mater 3:1700106. https://doi.org/10.1002/aelm.201700106
39. Cojan N, Cracan A, Cojan R (2010) A balanced differential CMOS oscillator with simulated inductor and negative resistance. In: Proceedings of the 2010 IEEE region 8 international conference on computational technologies in electrical and electronics engineering (SIBIRCON), Listvyanka, 11–15 July 2010, pp 796–799. https://doi.org/10.1109/sibircon.2010.5555009
40. Krizhanovski VG, Chernov DV, Andrei G (2018) Low-voltage class E/F3 high frequency oscillator. In: Proceedings of the 2018 14th international conference on advanced trends in

radioelecrtronics, telecommunications and computer engineering (TCSET), Slavske, Ukraine, 20–24 Feb 2018, pp 607–611. https://doi.org/10.1109/tcset.2018.8336275
41. Rudiakova A, Krizhanovski V (2006) Advanced design techniques for RF power amplifiers. Springer, Dordrecht. https://doi.org/10.1007/1-4020-4639-1
42. Chua LO, Yu J, Yu Y (1985) Bipolar—JFET—MOSFET negative resistance devices. IEEE Trans Circ Syst 32(1):46–61. https://doi.org/10.1109/TCS.1985.1085599
43. Chu C-L et al (2012) Investigation of voltage-dependent drift region resistance on high-voltage drain-extended MOSFETs' I-V characteristics. Electron Lett 48(2):110–111. https://doi.org/10.1049/el.2011.3286
44. Kumar U (2002) A complication of negative resistance circuits generated by two novel algorithms. Act Passive Electron Compon 25(3):211–214. https://doi.org/10.1080/08827510213495
45. Pitchaimuthu ME, Kathamuthu T (2020) Dynamics of the forced negative conductance series LCR circuit. Int J Circuit Theory Appl 48(2):214–230. https://doi.org/10.1002/cta.2729
46. Jelbart S, Wechselberger M (2020) Two-stroke relaxation oscillators. Nonlinearity 33(5):2364–2408. https://doi.org/10.1088/1361-6544/ab6a77
47. Semenov A (2016) The Van der Pol's mathematical model of the voltage-controlled oscillator based on a transistor structure with negative resistance. In: Proceedings of the 2016 13th international conference on modern problems of radio engineering, telecommunications and computer science (TCSET), Lviv, 23–26 Feb 2016, pp 100–104. https://doi.org/10.1109/tcset.2016.7451982
48. Semenov A, Osadchuk O, Semenova O et al (2019) A deterministic chaos ring oscillator based on a MOS transistor structure with negative differential resistance. In: Proceedings of the 2016 2019 IEEE international scientific-practical conference problems of infocommunications, science and technology (PIC S&T), Kyiv, Ukraine, 8–11 Oct 2019, pp 709–714. https://doi.org/10.1109/picst47496.2019.9061330
49. Ulansky V, Raza A, Oun H (2019) Electronic circuit with controllable negative differential resistance and its applications. Electronics 8(4):409. https://doi.org/10.3390/electronics8040409
50. Dmitriev AS (ed) (2012) Generatsiya haosa [Generation of chaos]. Tehnosfera, Moscow, Russia (In Russian)
51. Osadchuk VS, Osadchuk OV, Semenov AO (2011) Funktsionalni vuzly radiovymiriuvalnykh pryladiv na osnovi reaktyvnykh vlastyvostei tranzystornykh struktur z vidiemnym oporom [Functional units of radio measuring devices based on the reactive properties of transistor structures with negative resistance]. VNTU, Vinnytsia (In Ukrainian). ISBN 978-966-641-405-5
52. Osadchuk AV, Semenov AA, Baraban SV et al (2013) Noncontact infrared thermometer based on a self-oscillating lambda type system for measuring human body's temperature. In: Proceedings of the 2013 23rd international crimean conference "microwave & telecommunication technology", Sevastopol, Ukraine, 8–14 Sept 2013, pp 1069–1070
53. Semenov AO, Baraban SV, Osadchuk OV et al (2020) Microelectronic pyroelectric measuring transducers. In: Tiginyanu I, Sontea V, Railean S (eds) 4th International conference on nanotechnologies and biomedical engineering, ICNBME 2019, Chisinau, Moldova, 18–21 Sept 2019. IFMBE Proceedings, vol 77, pp 393–397. https://doi.org/10.1007/978-3-030-31866-6_72
54. Chen H, Tang Y (2020) An oscillator with two discontinuous lines and Van der Pol damping. Bulletin des Sci Mathematiques 161: https://doi.org/10.1016/j.bulsci.2020.102867
55. Kengne KL et al (2020) Dynamics, control and symmetry breaking aspects of a modified Van der Pol-Duffing oscillator, and its analog circuit implementation. Analog Integr Circ Sig Process 103(1):73–93. https://doi.org/10.1007/s10470-020-01601-4
56. Shabunin AV (2020) Strange waves in the ensemble of Van der pol oscillators. Izvestiya Vysshikh Uchebnykh Zavedeniy. Prikladnaya Nelineynaya Dinamika 28(2):186–200. https://doi.org/10.18500/0869-6632-2020-28-2-186-200
57. Semenov A (2016) Mathematical simulation of the chaotic oscillator based on a field-effect transistor structure with negative resistance. In: Proceedings of the 2016 IEEE 36th international

conference on electronics and nanotechnology (ELNANO), Kiev, Ukraine, 19–21 Apr 2016, pp 52–56. https://doi.org/10.1109/elnano.2016.7493008
58. Semenov A (2016) Reviewing the mathemetical models and electrical circuits of deterministic chaos transistor oscillators. In: Proceedings of the 2016 international siberian conference on control and communications (SIBCON), Moscow, 12–14 May 2016, pp 1–6. 39. https://doi.org/10.1109/sibcon.2016.7491758
59. Osadchuk OV et al (2017) Optical transducers with frequency output. Proc SPIE 10445:104451X. https://doi.org/10.1117/12.2280892
60. Molotkov VI, Potapov EI (1991) Efficiency of current-voltage characteristics of low-power FET's and λ-diodes and amplitude estimation of the λ-diode self-oscillator. Izvestiya VUZ: Radioelektronika 34(11):108–110
61. Semenov A, Baraban S, Semenova O et al (2019) Statistical express control of the peak values of the differential-thermal analysis of solid materials. Solid State Phenom 291:28–41. https://doi.org/10.4028/www.scientific.net/SSP.291.28
62. Khutornenko S, Osadchuk O, Osadchuk I et al (2017) Mathematical model of piezoelectric oscillating system with electrodes of variable nonlinear and constant linear air gap. Telecommun Radio Eng (English translation of Elektrosvyaz and Radiotekhnika) 76(18):1639–1648. https://doi.org/10.1615/TelecomRadEng.v76.i18.50
63. Osadchuk OV et al (2017) Frequency pressure transducer with a sensitivity of mem capacitor on the basis of transistor structure with negative resistance. Proc SPIE 10445:1044559. https://doi.org/10.1117/12.2280958
64. Osadchuk AV et al (2019) Numerical method for processing frequency measuring signals from microelectronic sensors based on transistor structures with negative differential resistance. Proc SPIE 11176:111765Y. https://doi.org/10.1117/12.2536942
65. Naumenko VV, Totsky AV, Pidchenko SK et al (2017) Multi frequency synthezier of bispectral triplet-signal designed for digital communication system. Telecommun Radio Eng (English translation of Elektrosvyaz and Radiotekhnika) 76(2):147–155. https://doi.org/10.1615/TelecomRadEng.v76.i2.50
66. Pidchenko S, Taranchuk A, Totsky A (2017) Multi-frequency quartz oscillating systems using digital compensation of frequency instability caused by variations of temperature and vibrations. Telecommun Radio Eng (English translation of Elektrosvyaz and Radiotekhnika) 76(13):1193–1200. https://doi.org/10.1615/TelecomRadEng.v76.i13.70
67. Taranchuk AA, Pidchenko SK, Khoptinskiy RP (2015) Dynamics of temperature-frequency processes in multifrequency crystal oscillators with digital compensations of resonator performance instability. Radioelectronics Commun Syst 58(6):250–257. https://doi.org/10.3103/S0735272715060023
68. Kushnir NY, Horlsy PP, Grygoryshyn AN (2005) Dynamic chaos in phase syncronization devices. In: Proceedings of the 2005 15th international crimean conference microwave and telecommunication technology, CriMiCo'2005—conference proceedings, vol 1, pp 346–347. https://doi.org/10.1109/crmico.2005.1564936
69. Galiuk SD, Kushnir MY, Politanskyi RL (2011) Communication with use of symbolic dynamics of chaotic systems. In: Proceedings of the 2011 21st international crimean conference "microwave & telecommunication technology", sevastopol, pp 423–424
70. Kushnir M, Ivaniuk P, Vovchuk D, Galiuk S (2015) Information security of the chaotic communication systems. In: Proceedings of the 8th international conference on chaotic modeling and simulation, CHAOS 2015, Henri Poincare Institute, Paris, France, 26–29 May 2015, pp 441–452
71. Lukin K, Zemlyaniy O (2019) Chaos-based spectral keying technique for design of radar-communication systems. In: Proceedings of the 2019 signal processing symposium (SPSympo), Krakow, Poland, 17–19 Sept 2019, pp 51–56. https://doi.org/10.1109/sps.2019.8881982
72. Lukin KA, Zemlyaniy OV (2016) Digital generation of wideband chaotic signal with the comb-shaped spectrum for communication systems based on spectral manipulation. Radioelectronics Commun Syst 59(9):417–422. https://doi.org/10.3103/S0735272716090053

73. Zemlyaniy OV (2016) Keying of the broadband chaotic signal spectrum for data transmission. Telecommun Radio Eng (English translation of Elektrosvyaz and Radiotekhnika) 75(5):401–411. https://doi.org/10.1615/telecomradeng.v75.i5.20
74. Lukin KA, Kulyk V, Zemlyaniy OV (2009) Chaos generators for noise radar. Underst Complex Syst 2009:433–437. https://doi.org/10.1007/978-3-540-85632-0_40
75. Zemlyanyi OV (2007) Experimental study into a radio frequency band chaos oscillator. Telecommun Radio Eng (English translation of Elektrosvyaz and Radiotekhnika) 66(12):1067–1077. https://doi.org/10.1615/TelecomRadEng.v66.i12.30
76. Zemlyaniy OV (2006) Experimental Investigation of chaotic waveform generator for ultra wide band noise radar. In: Proceedings of the 2006 international radar symposium, Krakow, Poland, 24–26 May 2006, pp 1–4. https://doi.org/10.1109/irs.2006.4338091
77. Zemlyaniy OV, Lukin KA (2003) Correlation-Spectral properties of chaos in the nonlinear dynamical system with delayed feedback and asymmetric nonlinear map. Telecommun Radio Eng (English translation of Elektrosvyaz and Radiotekhnika) 60(7–9):137–149. https://doi.org/10.1615/TelecomRadEng.v60.i789.180
78. Tarnovskii NG, Osadchuk VS, Osadchuk AV (2000) Modeling of the Gate Junction in GaAs MESFETs. Russian Microelectron 29(4):279–283. https://doi.org/10.1007/BF02773276
79. Rudyk AV, Semenov AO, Kryvinska N, Semenova OO, Kvasnikov VP, Safonyk AP (2020) Strapdown inertial navigation systems for positioning mobile robots-MEMS gyroscopes random errors analysis using allan variance method. Sensors 20(17). https://doi.org/10.3390/s20174841
80. Pidchenko S, Taranchuk A, Totsky A et al (2019) Construction principles of technically invariant quartz generators based on the double-loop pulse phase locking systems. Telecommun Radio Eng (English translation of Elektrosvyaz and Radiotekhnika) 78(13):1167–1177. https://doi.org/10.1615/TelecomRadEng.v78.i13.30
81. Semenov A, Osadchuk O, Semenova O, Baraban S, Voznyak O, Rudyk A, Koval K (2021) Research of dynamic processes in the deterministic chaos oscillator based on the colpitts scheme and optimization of its self-oscillatory system parameters. Lect Notes Data Eng Commun Technols 48:181–205. https://doi.org/10.1007/978-3-030-43070-2_10
82. Maksimov NA, Panas AI (2017) A solid-state microwave-range self-oscillating chaotic system with a simplified structure. Tech Phys Lett 43(2):180–182. https://doi.org/10.1134/S1063785017020080
83. Semenov A, Baraban S, Baraban M, Zhahlovska O, Tsyrulnyk S, Rudyk A (2020) Development and research of models and processes of formation in silicon plates p-n junctions and hidden layers under the influence of ultrasonic vibrations and mechanical stresses. Key Eng Mater 844:155–167. https://doi.org/10.4028/www.scientific.net/KEM.844.155
84. Semenov AO, Voznyak OM, Osadchuk OV, Baraban SV, Semenova OO, Rudyk AV, Klimek J, Orazalieva S (2019) Development of a non-standard system of microwave quadripoles parameters. Proceed SPIE 11176:111765N. https://doi.org/10.1117/12.2536704

The Method of Redistributing Traffic in Mobile Network

Oleksander Romanov, Mykola Nesterenko, and Volodymyr Mankivskyi

Abstract Cellular network generates a lot of signaling data. A large part of signaling data is generated to handle the mobility of subscribers and contains location information that can be used to fundamentally change our understanding of mobility principle. However, location data available from standard interfaces in cellular networks is very an important research question is how this data can be processed in order to efficiently use it for traffic state estimation and traffic planning. The design of the mobile operator's network is carried out by the method of frequency spatial planning. It is believed that the solution to this problem provides the required indicators of electromagnetic compatibility of network elements, and as a result, performance of the network. Ideally, these findings should be replicated in a study where uniformity of traffic over network elements is relegated to the background. Results provide a basis for affects both throughput and quality of service. In this paper, it is proposed to use the sector analysis method for optimizing the load distribution between base stations when predicting the coverage areas of base stations, in addition to using the frequency-spatial planning method, when forecasting service areas of base stations. The technology of cellular systems is changing at such a speed that 4G networks have not yet had time to fully deploy, as 5G is already being introduced. The fourth generation is characterized by LTE-advanced technology, which implies an intelligent network with self-training and partial adjustment of its parameters. The distribution functions of the radio resource of the cellular communication network of this standard lie at the base stations. However, clear control algorithms for such networks have not yet been developed. As part of situationally adaptive planning of radio resources in radio communication systems, a method is proposed for determining the optimal coverage areas of base stations depending on the distribution of subscribers according to billing data. To this end, in addition to the statistics for base stations for servicing the load, enrich it with billing system data.

O. Romanov · M. Nesterenko · V. Mankivskyi (✉)
National Technical University of Ukraine "Igor Sikorsky Kyiv Polytechnic Institute", Peremoga Ave. 37, 03056 Kiev, Ukraine

O. Romanov
e-mail: a_i_romanov@ukr.net

D. Ageyev et al. (eds.), *Data-Centric Business and Applications*, Lecture Notes on Data Engineering and Communications Technologies 69,
https://doi.org/10.1007/978-3-030-71892-3_7

Keywords Poison flow · Planning · Loads · Markov's chain · Traffic loading

1 Introduction

In the End of XXI century Ukraine has active stage of deployment 4G and 5G technologies on mobile networks. The present findings confirm due to errors in forecasting development trends. The situation is aggravated when 2G and 3G networks are served for all voice traffic for a long time. In addition, the Operators make profit, both financial and commercial, from the use of the existing network segment. Hence, it is necessary to use techniques to improve the performance of existing networks.

According to [1], hybrid telecommunication service is a service that includes components of cloud and telecommunication services. The work of the telecommunications network is inextricably linked with computer systems. A mobile network consists of a local area network, a radio access network and a provider's core network. The emergence of cloud computing has expanded the possibilities of servicing telecommunications systems.

The volume of services and the volume of transmitted traffic is gently expanding [2–6], which provides a route to sustained increase in the resources required to assist a large number of applications types in operator system.

Deployment process of a mobile operator network begins with radio frequency planning, which also determines the electrical compatibility of network objects and the quality of services provided. To change the number of subscribers in a cell, the size of the incoming load requires an improvement in network structure. Mobile network operation differentiates between two methods: service area forecasting and location range coverage analysis.

Broadly translated our findings indicate that work of planning environments (RPLS ONEGA, Enterprise Asset) which are used mobile network for developing prior radio coverage proposal. The methodology for calculating radio coverage was carried out according to the technical characteristics of the BS. These characteristics can include: technical, geographical, BS coordinates, transmitter power, type, suspension height, elevation angles, azimuths of a sector antenna, allocation of a frequency resource, etc.

Next-generation mobile networks (not 5G), especially with the advent of the UMTS technology stack, the number of calls has decreased due to the increase in data transfer. Mobile network operators focus on optimizing the radio coverage of the data network. The main reason is that the voice load of the subscriber does not increase. This is achieved through an empirical approach to determining the optimal parameters of base stations. 3GPP standards are only recommendations for adjustment (lower quality limit), on the one hand, and on the other hand, customer satisfaction with the quality of communication (upper quality limit) [1]. The recommendation can be to pay attention to those operators who have their own optimal setup standard.

The number of complaints about radio coverage using this approach has increased. As a result, a significant proportion of such complaints are received for the provision

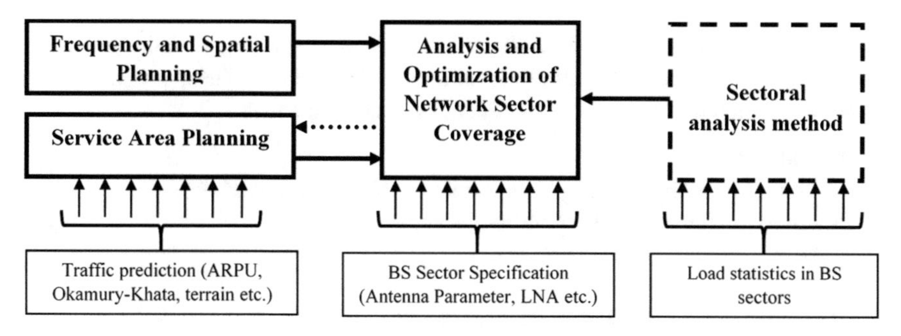

Fig. 1 Common GSM and CDMA network process planning

of voice services. Inefficient radio scheduling (from the point of view of the final subscriber) leads to an increase in the number of errors, and as a result, deterioration of the quality of communication during voice transmission. In terms of EMC, interference level and hysteresis, at the same time, the radio planning was excellent.

In this paper, it is proposed to supplement the existing approaches to planning with the data provided, taking into account the data from the Billing System. Therefore, to improve network efficiency and service quality, it is necessary to perform the initial calculation of the load associated with the BS sectors, and use them to assess the effectiveness of radio planning, as shown in general [7]. In this paper, suggest that the planning process be described as in the model (Fig. 1).

The traditional model of access to the radio resources of a mobile communication network is based on the conditional positioning of the subscriber in the network at the time of connecting the BS with the greatest radiation signal damage [8].

In such a communication system, there is no forecast of load distribution created by subscribers. The operator's network operates around the clock in the peak power mode of the emitters [8, 9] and receivers in anticipation of an abnormal phenomenon in the form of a sharp change in the concentration of subscribers inside the served cell. Moreover, in the event of a peak load increase as a result of a large number of calls to the base stations of the telecom operator in a single cell, blocking and denial of service occur.

The control task in the operator's network is the uniform distribution of all radio resources, taking into account anomalous phenomena. The proposed conceptual method (Fig. 1) of analysis of subscriber access to radio resources and traffic distribution in cellular communication systems differs from the known ones by the introduction of additional functions for billing data analysis, clustering and channel resource management, which makes it possible to increase the efficiency of cellular communication systems taking into account the requirements for QoS. The method allows forecasting in the considered mobile communication systems.

Thus, the method of optimal traffic Distribution Based on the Sectoral Method of Loading Network Element to serve the variable load of billing systems operator is an urgent task.

The reason for the increase in the load on the tariff system is that the calculation of payment for services requires multi-stage procedures. These procedures are necessary to determine their value, which, as a result, leads to:

- low call service efficiency;
- worsening of both flexibility and efficiency of tariff approaches;
- failure to guarantee due to quality of service.

2 Organization of Telecommunications in Mobile Network

Ensuring satisfactory quality of radio frequency coverage and ensuring a sufficient level of BS signal in the consumer area has always been the main strategy in GSM networks. The power of the radio resource provided to the Subscriber (virtual channel) in UMTS networks remains an advantage.

The network is adjusted based on the data of the use of the channel resource already during operation. However, in [10, 11] the analysis does not take into account the nature of the data, such as a direct call or a manual call, does not take into account the characteristics of the subscriber profile. This data is hardly used in network planning, although it is constantly accumulated in the billing system and is used to forecast the provision of services.

Mathematical models for calculation and forecasting At the planning stage, the network is used to determine such indicators as coverage and load of subscribers. Okumura-Hata (signal strength calculation model) is one of the most commonly used models. Therefore, according to [12], when using this method, corrections for manual motion in terms of mutual electromagnetic influence were taken into account. The proposed approach in this article proposes to take into account the cause of manual (hand-held) traffic as a result of suboptimal planning of BS sectors.

The most common practice in the CIS countries is that Operators use UMTS and GSM networks differently. Transfer of trust to networks LTE operators in the CIS countries in the near future do not pay much attention to the use of two networks simultaneously.

Therefore, it can be assumed that most of the network processes are performed in data centers, and the network is only a means of delivering informational messages [2, 8]. In the conditions of the distribution of software-managed routers network structure is shown in Fig. 2.

In Fig. 2 shows how a mobile subscriber communicates in a combined GSM/CDMA network, which converts the input radio signal into an optical one, and then the signal is sent to the core network, which is also located in the data center. Next, according to the technology, the flow is sent to the core of the operator for further processing.

The further direction of data channels is determined mainly by the service: the flow comes to the home network; the stream must be sent outside the operator's local network, it is routed to the margin router and then to the external network. This is an example of a next generation network.

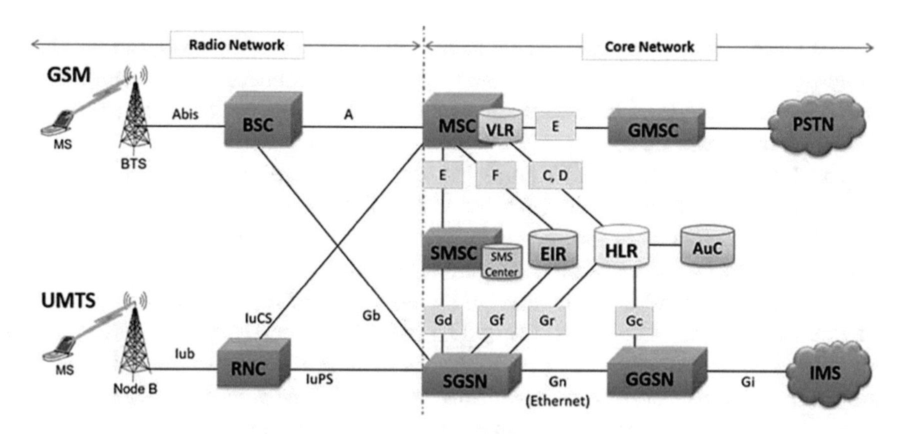

Fig. 2 GSM and UMTS architecture mobile operator

These problems cause the fact that in such conditions, operators face a number of problems. At the same time, Operators of developed countries were able to prevent this by switching to "pure" UMTS and, accordingly, to LTE.

The main focus of the article is the relative position of the UMTS and GSM cells. The main task of such placement is taking into account their mutual influence on the load distribution. Such are the scenarios for areas with moderate density. At the same time, improving UMTS zones will also improve the quality of communication in hybrid networks. The main criterion for improving the quality of communication is, first of all, reducing the number of connection failures. Optimal load distribution in the cells is the main step towards this. In [11] the optimal coverage can be characterized visually, the paper proposes to estimate the optimal coverage based on the indicators of the Payment System, which is an integral part of the Operator's network. The network load estimation system will help to assess the accuracy of the method.

Analysis of the generalized method showed that the process of service load operation was greatly influenced by many different factors. Given these factors, we were able to take advantage of the ability to adapt to changes in the load services sector. Such design of coverage areas (based on the forecast of the level of handover) will be able to improve the efficiency of load distribution in the network.

3 Method of Sector Analysis for Distributing Load in Net

The controller in the data center manages all subsystems for mobile communications. Interactions between controllers in subsystems for control only occur in the center of the date center. Subscriber search, search for physical elements involved in the transfer process, and transfer of instructions the are functions of managing service processes.

The recording device is connected to the base station-troller located in the data center. Depending on the protocols, the system-troller connects to the central data center, sending the final hardware solution to physical devices to initiate the data transfer processes (Fig. 3).

The design of the network and the customer of the highest quality control takes place locally. Traditionally, systems for that LTE network perform a set of tasks according to the standards and standards. The document seeks to separate subsystem management tasks and tasks related to the data transfer system directly to that LTE network. The structure is an increase in the efficiency of the systems compared to traditional network generations. More than half of the functions of the subsystem are related not to that of the service, but to the management of that communication system. Service type control occurs in small areas of NodeB, SGSN, PCRF.

In Fig. 4 shows the interactive games in the system systems and the sion-sion of operations management and data transmission functions. In fact, each ar-line in this

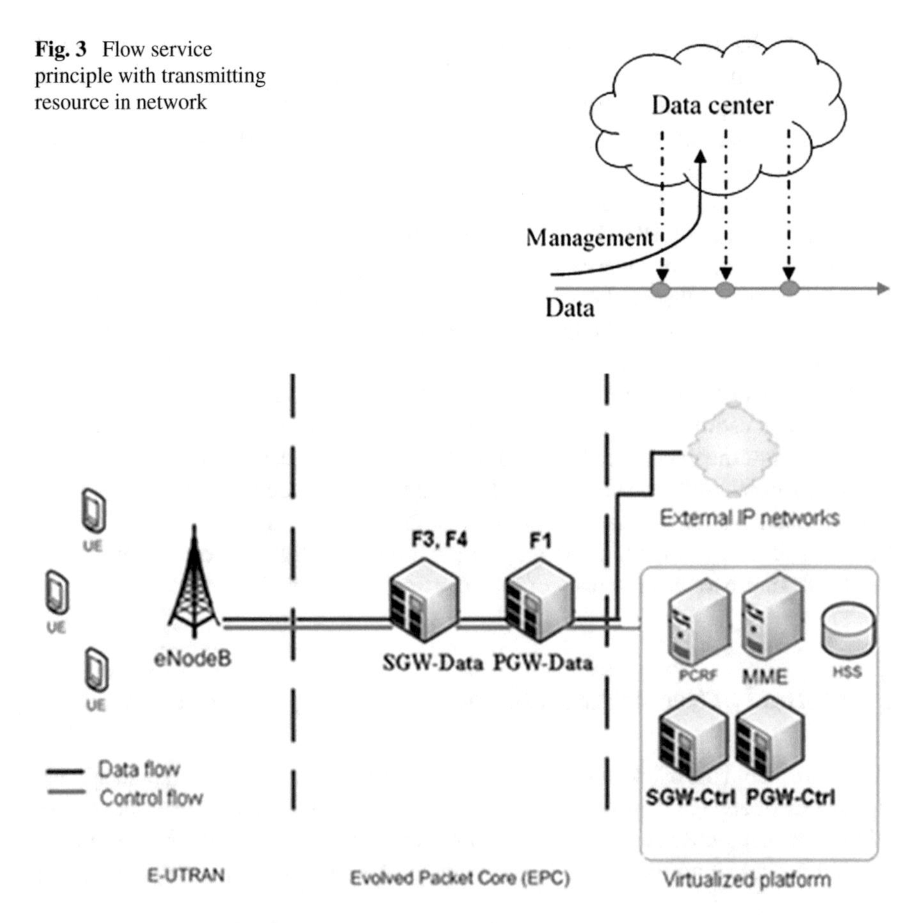

Fig. 3 Flow service principle with transmitting resource in network

Fig. 4 Distribution of loading in core between physical net elements

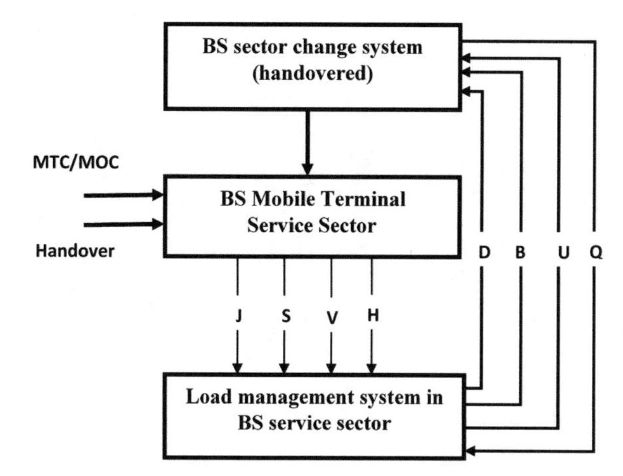

Fig. 5 Method analysis for GSM and UMTS net

schema is a request for protection on this physical node. The number of requests is the strength of the load to a given service node.

Performance analysis systems for load management in mobile systems allow us to see four components of this method of evaluation the optimal traffic Distribution (Fig. 5):

- Sector BS service mobile terminal.
- Sector BS changes system.
- The load management system in the BS sector.
- Factors affecting them.

The control object is the load in the BS service sector and is a set of functionally related and interacting with each other and subscriber's elements.

Operation system has elements (objects of control) and containing controls, information has exchanged, characterized by variables for different data types.

The quality of service in mobile network is understood as its property to provide processing of incoming applications with the required probabilistic—temporal characteristics. The quality of service for applications is mainly determined by the methods of servicing applications (Fig. 6), as well as algorithms that implement these methods.

A. **The following data can be transferred from the control elements to the control bodies:**

- Distribution of control the load, the cell are transmitted: CRO, TO, and PT parameters, which are transmitted on the BCCH channel in each cell ($h_1, h_2, \ldots, h_i \in H$).
- Parameter C1 for the serving cell is less than zero ($C1 < 0$) for 5 s when the loss on the radio interface is too large and the MS needs to change the cell ($j_1, j_2, \ldots, j_i \in J$).

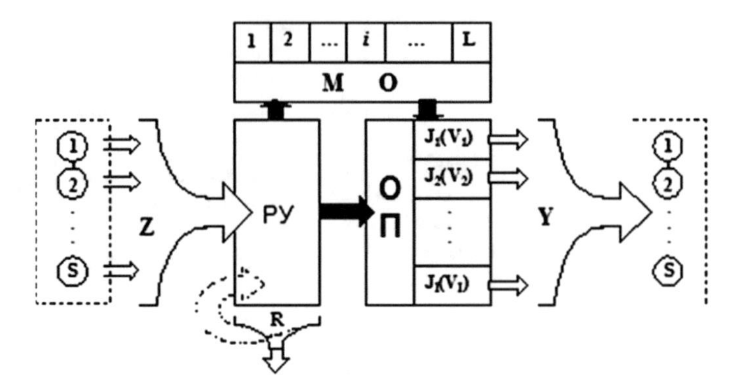

Fig. 6 Application process for incoming requests

- Parameter C2 for nearby cells exceeds same value for serving cell for 5 s ($q_1, q_2, \ldots, q_i \in Q$).

B. **The following instructions can be transmitted from control to control objects:**

- Cell changes within LA (Location Area) ($u_1, u_2, \ldots, u_i \in U$)—Information that a new cell belongs to the same LA, on BCCH channel of neighboring cells.
- Cell changes within one MSC ($s1, s2, \ldots, si \in S$)—MS when a cell changes, it detects LAI change based on the analysis of information transmitted over the BCCH channel.
- Update location when Subscriber entering the service area of a new MSC/VLR ($b_1, b_2, \ldots, b_i \in B$)—Location Update (LU) is performed when MS moves to a new LA.
- Use of necessary backup channels for handover ($v_1, v_2, \ldots, v_i \in V$)—Holding a call while in transit using the network segments sector;
- Measurement system control system, radio interface indicator for message delivery ($d_1, d_2, \ldots, d_i \in D$)—Measurements of BCCH signal levels at serving and neighboring cell frequencies.

4 Mathematical Interpretation of the Traffic Distribution Method

When designing a mobile network, it is necessary to take into account a wide variety of parameters—from load to cell size, which depends on the technical characteristics equipment of the base station and the switching center of mobile services. You should also consider the Quality of Service (QoS) requirements that must be met for services supported by the network. One of the approaches to designing networks is the decomposition principle: the network is divided into blocks, after which an analysis and calculation of the necessary characteristics are performed for a separate block.

We will consider a mobile network whose service area is covered with cells of regular hexagonal shape. Suppose that all cells in a network are identical in size, number of radio channels, and call service requirements. Assume also that in at any given time, network subscribers located within the service area are distributed evenly across it. Under the assumptions made, a single JCSS cell can serve as a separate unit for analyzing the network functioning process.

We focus on the physical model of the call service process in a cell of a mobile network, which will allow us to build mathematical models of one or another degree of adequacy in the future, study them and obtain various probabilistic characteristics. A feature of wireless mobile communication systems is the mobility of the subscriber, which entails the need to transfer the current connection of the mobile subscriber from one cell to another without interruption of communication. Thus, in each cell of a mobile network, two types of calls arise—the so-called new calls, which were triggered by the initiation of a connection by a subscriber located on the territory of a given cell, and handover calls.

Before you think about the check mode, say the following call in the tension parameter in the BS load components: incoming—λ_MOC, outgoing—λ_MTC and hand—λ_hand.

Initiator of BS load there following types of calls (Fig. 7):

- Started and finished in the same cell of a GSM or UMTS network.
- Started in a GSM cell and finished in a cell of the UMTS network and vice versa.
- Started in one and finished in another cell of the UMTS network.
- A handover that arrives in a GSM network cell from neighboring cells of the same network and finished in the considered cell (lines 3).
- A handover that arrives in a GSM network cell from neighboring cells of the UMTS network and break of in the considered cell (line 5).

The main way for dynamic quality control is as follows: the number of orders in the service of connection requests (disconnection, recovery) is compared with the quality of customer service. The developed approach makes it possible to analyze

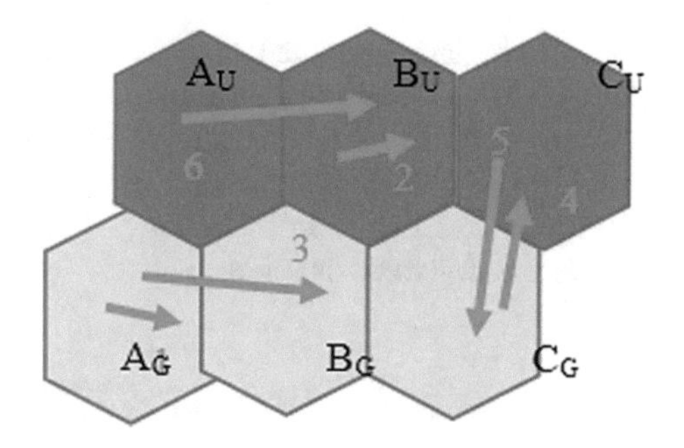

Fig. 7 Script deal with load in GSM and UMTS net

the following quantitative indicators of the effective operation of the system, such as: time delay in inviting the service flow in the node and the probability of losing requests to the service node.

Appropriate measures should be taken after determining the cause of the performance problem. If there is a problem during the transfer between service nodes, there is a problem, in which case only the system reconfiguration is an effective measure. Increasing the number of service resources is one way, provided that the problem is detected in the service node. If you find a decrease in the quality of service in an integral connected number of interfaces (which form a single core of the EPC network), it is recommended to limit the flow of programs.

4.1 Mathematical Model: Narrowing and Presumption

- The Mobile Originated Calls and Mobile Terminated Calls and Handover Calls flows are Poisson flow λ_{MOC}, λ_{MTC}, λ_{hand}. The sum up of all call is also Poisson flow in both TCH (Traffic Channel at GSM) and DTCH (Data Traffic Channel at CDMA) channels:

$$\lambda = \lambda_{hand} + \lambda_{MOC} + \lambda_{MTC} + \lambda_{hand\_DTCH} + \lambda_{ULCH} \tag{1}$$

- Some calls with the intensity (μ_1) has finished service within cell and with intensity (μ_2) goes to service in neighbor cell (Handover Calls). Simultaneously, the law distribution of duration service calls both within the cell and during handover has an exponential division with parameters μ_1 and μ_2. Besides, duration of a radio channel is determined by sum of these and has an exponential sharing.
- Service time length of any calls is described by exponential division (distribution) χ(t):

$$p_n = \lim_{t \to \infty} \{X(t) = n\}, \quad n \in X \tag{2}$$

- The serving load will be determined by tension of the serving flows (1), noted by Z. Besides, the finished load is the tension of finished requests (2), noted by Y.

Let's conduct a detailed analysis of the script. The call starts and ends in the same GSM cell, there is a transmission of input signal from UMTS cells, without restrictions on the reservation of radio channels for input transmission. For this script, we describe as a model of a system with failures, which already has a known description by Erlang's formula B in [10, 13]. The probability of failure (PR) in such a system is described as follows:

$$p_R = \frac{\frac{(\lambda * \tau)^n}{N!}}{\sum_{n=0}^{N} \left(\frac{(\lambda * \tau)^n}{N!} \right)} = \frac{\frac{Z^V}{V!}}{\sum_{n=0}^{V} \left(\frac{Z^n}{n!} \right)} \tag{3}$$

Both incoming and outgoing GSM MTC calls and for UMTS input, the probability of failure will be determined as (4).

Besides, process of servicing with events in the cell of base station, different stages of the life cycle can take place. One of the indicators of service quality, which can be excluded from the calculations, given the load on the cell, is the probability that during the time (t) all channels are not occupied:

$$p_{Y-Z,R=0} = 1/\left[\sum_{n=0}^{V}\left(\frac{Z^{V}}{V!}\right)\right] \tag{4}$$

The above indicator should be used when analyzing the input load in the sector of the base station. It should also be borne in mind that the base station may be intermediate throughout the route when transmitting data. It should be noted that the standard indicator of the base station is its loading at the longest time, an expression for estimating the probability of filling the specified number of channels (V):

$$p_{Y} = p_{not\_load} \cdot \left[\frac{Z^{V}}{V!}\right] \tag{5}$$

Using (3)–(5) outline the average number of busy channels:

$$\langle k \rangle = p_{Z-Y,R=0} \cdot \sum_{n=1}^{V}\left[\frac{Z^{V}}{(n-1)!}\right] \tag{6}$$

The proposed approach in the model allows to determine the overloaded and insufficiently loaded sections of the network. The essence of this approach is that moving from busy channels to the operating load in the cell of the base station. Then, in the process of managing the incoming request service, use the data to ensure a uniform download of network elements. It should be noted that it is necessary to identify the handover traffic that has passed to the neighboring base station, which is not taken into account in the standard planning, as shown in [8, 9].

The proposed script of load calls are formed in the sector CG (Fig. 4). In this module, the call handling process in the C_G cell is an N-channel (number of RC in the cell) fully accessible queuing system (Q_S). Such a system accepts two Poisson flows: 1—call flow (λ_{MTC}), 2—handover call from UMTS in GSM—intensity (λ_{hand}).

4.2 *Two Possible Work: Describes*

- Time line whit call goes in system occupies one channel, if channel has release places.

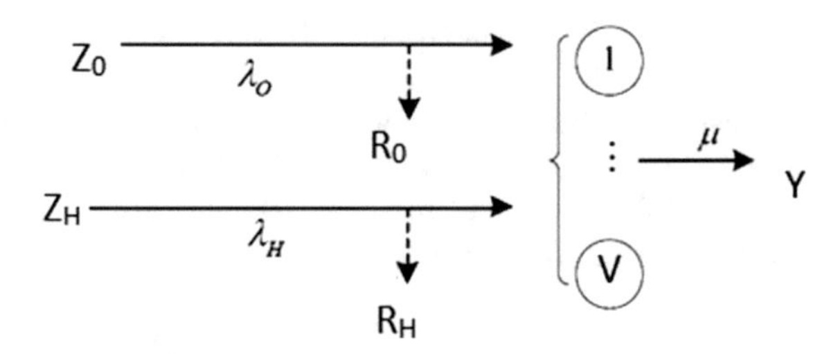

Fig. 8 Input loading in main script for GSM and UMTS net

- Time line whit call goes in system does not occupies any channel, if there are not release places in channel.

The main characteristics are the probability of losing the request p_{lost_1} (blocking the acceptance of a new call) and the probability of p_{lost_2} (blocking the handover). Using the above statements allows you to create a mathematical model of the cell.

A similar approach was used in [8] in the analysis of the GSM life cycle. In this paper, the main goal is given to the specifics of combined GSM and UMTS networks. Also, the work takes into account only the transfer of service between the BS, serving one controller BSC, which is a kind of limitation in the calculation.

Service life is an independent random variable that has an exponential distribution with a parameter μ (Fig. 8).

The status cell can be changed by receiving calls. The status cell can also be changed using the call termination service. Based on (1), the total call flow included in the base station has a Poisson flow character. The main feature of this flow, which will be used in the future, is the property of no consequences. Therefore, as shown in [9], the process of obtaining requests to the cell after time t does not depend on the operation of the system until time t.

From (2) the duration requests service that will go to cell after t does not depend on state system until time t. By the stationarity property of Poisson flow, the probability of call arriving at time from t to $t + \Delta t$ (small interval, in the limit—infinitely small) is $\Delta t/T$ ($\Delta t \ll T$). On the other hand, if the interval is comparable to T (or more), the formula will not work, because differential formula for infinite small interval.

According to [11], the duration of the call service depends only on the length Δt of the interval. This vortex is a property of the stationary Poisson flow. This property of the Poisson flow allows us to represent a flow that has both the property of a Markov flow—the property of the absence of aftereffects.

The Markov flow is determined by the graph of state transitions. These states are descriptions of various call life options in the base station sector. In such a chain, changes in states are determined by the probability of transition from one stage to another. To solve the problem, we used a Markov chain with n service states of calls in the cell, which is a property of the Poisson flow—stationarity.

Homogeneous Markov process with a discrete state is a process of simultaneous death and birth in a chain, and it has the form of Fig. 9. The state graph can be stretched in a chain with a finite number of states, in which jumping through two states is impossible.

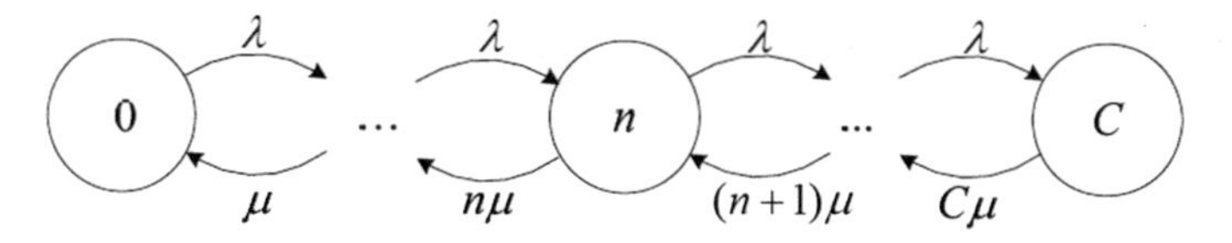

Fig. 9 Change for request describes by Markov chain

The creation of a mathematical model required determining the probability of the events described above. Express the conditions of change of states in the form of a Markov chain with each of its transitions. The following is a model of the solution in the form of a matrix of transitions, which is a reflection of Markov.

The sum of all events of the Poisson flow (1) on the interval of length Δt has a Poisson distribution with the parameter (λ * Δt). It follows that the probability of no calls during time Δt in the cell of the base station:

$$P(t) = e^{\frac{-\tau}{T}} = e^{-\lambda * \tau} \tag{7}$$

The average waiting time of the first call with the probability that before t there was no call, but in the interval (t + Δt) there was one call:

$$P(t) = e^{\frac{-\tau}{T}} * \frac{\Delta t}{T} \tag{8}$$

Multiplying (8) by Δt and taking the integral, we get the average waiting time for input happens:

$$\int_{0}^{\infty} t * e^{\frac{-\tau}{T}} * \frac{\Delta t}{T} = T \tag{9}$$

The main indicator is the average waiting time for the event (parameter T): incoming call request (MTC) or incoming handover event request. Subject to the replacement of the parameter by the coefficient λ, we obtain the probability of occurrence of the event (λ * Δt). The parameter T—characterizes the expectation of the event in the form of the time distance between events. As a result, the indicator Δt characterizes the intensity of the flow, which reflects the number of events completed per unit time [11].

We consider the main indicator as the average number of calls per unit interval which determines the Poisson distribution parameter. Starting from the unit interval (τ) to the intensity: for the interval there are λ events, with the probability that k-events will be completed for the interval:

$$P(k, \lambda) = \frac{(\lambda)^k}{k!} \cdot e^{-\lambda} \tag{10}$$

Using (10) we get probability during time Δt the cell will not receive a call, is equal to:

$$P(0, \lambda) = \frac{(\lambda)^0}{0!} \cdot e^{-\lambda} = \frac{1}{1} \cdot e^{-\lambda} = e^{-\lambda} \tag{11}$$

Probability that two or more calls or a handover call will have finished in cell during time line Δt can be described as simplest flow. This property characterizes fact that for a sufficiently little time period probability of occurrence two or more calls is a value of a small number o(τ).

Exponential distribution property determines probability that call will have not finished during Δt:

$$P(0, \mu) = \frac{(\mu)^0}{0!} \cdot e^{-\mu} = \frac{1}{1} \cdot e^{-\mu} = e^{-\mu} \tag{12}$$

After that probability that during timeline Δt cell services of n calls does not end, are equal:

$$P(0, \mu) = \left[\frac{(\mu)^0}{0!} \cdot e^{-\mu}\right]^n = \left[\frac{1}{1} \cdot e^{-\mu}\right]^n = e^{-n \cdot \mu} \tag{13}$$

Probability timeline Δt is served by one call is equals:

$$P(1, \mu) = \left[\frac{(\mu)^1}{1!} \cdot e^{-\mu}\right]^n = \left[\frac{\mu}{1} \cdot e^{-\mu}\right]^n = \mu \cdot e^{-n \cdot \mu} \tag{14}$$

All of the above probabilities are part of the life cycle of events in the mobile network. For example, at some point in the cell were served n calls: n = {1, V}, where V is the maximum number of data channels in the cell. This can be written as X(t + Δt) = n. Next time t + Δt = τ cell state can:

1. During timeline τ, new incoming call won't arrive in cell (10) and call won't have finished (13):

$$P_{n+0}(0, \lambda, \mu) = P(0, \lambda) \cdot P(0, \mu) = e^{-(n \cdot \mu + \lambda)} \tag{15}$$

2. Let's go to the next state +1: during timeline (t + Δt) one call arrives in cell (11) and other call does not finish (12):

$$P_{n+1}(0, \lambda, \mu) = P(1, \lambda) \cdot P(0, \mu) = \lambda \cdot e^{-(\mu + \lambda)} \tag{16}$$

3. Let's go to the next state −1: during timeline (t + Δt) one call serve in cell (14) and new calls does not receive (10):

$$P_{n-1}(1,\lambda,\mu) = P(0,\lambda)\cdot P(1,\mu) = \mu\cdot e^{-(n\cdot\mu+\lambda)} \tag{17}$$

4. Probability that during timeline Δt two or more calls for a sufficiently short period of time will arrive in cell, probability is value of small number o(τ).

As a result, we get that for time τ in the cell the load will increase under the condition due to a new call. On the other hand, there is a possibility of reducing the load due to call service. Simultaneous event of two types of calls will be considered as a sequential event in the time interval (t + Δt) at Δt → 0, expressed by the property of the normal flow of events [2]. As a result, expressions (15–17) describe the life cycle of the input load as a process of birth and death with probabilities of transition from one state to another n y (n + 1):

$$\begin{cases} P_{n+0}(0,\lambda,\mu) = e^{-(n\cdot\mu+\lambda)}, \\ P_{n+1}(0,\lambda,\mu) = \lambda\cdot e^{-(\mu+\lambda)}, \\ P_{n-1}(1,\lambda,\mu) = \mu\cdot e^{-(n\cdot\mu+\lambda)}, \\ P_{n+2}(2,\lambda,\mu) = \mathrm{o}(\tau) \end{cases} \tag{18}$$

5 Prediction Needed Channel Resource in Mobile Network: Ensuring Efficient Operation for Traffic Loading with Different Traffic Types

The main recommendations in this paper are based on a large amount of data from monitoring systems. Based on this data, it is necessary to develop configurations of service systems that will meet the requirements. Upon receipt of configuration data and statistics, a plan for the use of resources was developed, in which it is quite convenient to check the availability of network resources (Fig. 10).

Starting from the presented forecasting algorithm, we will expand the mathematical model for a homogeneous process.

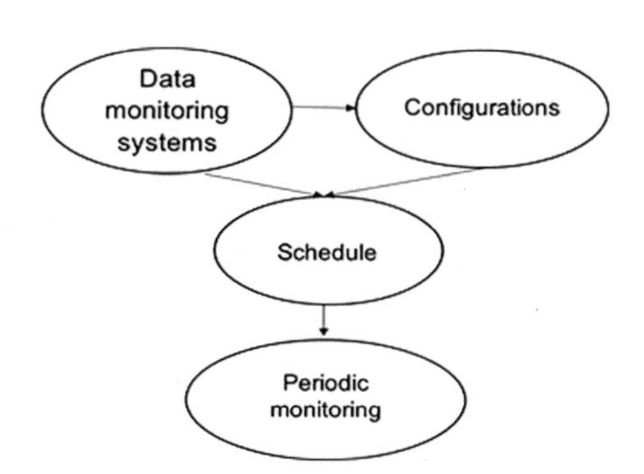

Fig. 10 Prediction algorithm for using channel resource

5.1 Mathematic Modelling of Homogeneous Process for Traffic Distribution in Mobile Network

Markov processes has specific feature is that their conditional functions satisfy differential equations of part derivative for parabolic type link to [13]. This system of Eqs. (18) will have solved either by the method of the Cauchy method, or Kolmogorov equations, the last will be preferred.

The system of Eqs. (19) for the homogeneous process of death and birth according to the drawing (Fig. 5) can be rewritten as a matrix of order n. This tridiagonal matrix describes a homogeneous system of linear algebraic equations with respect to the vector of limit probabilities of states, according to the source [2]. The main approach in solving such a system of equations is the use of local balance and marginal rationing:

$$P(1, \lambda) = 1 - P(0, \lambda) = 1 - e^{-\lambda} \tag{19}$$

In this manner, for n:

$$P_n = \frac{P_{k-1}}{k!} P_k \cdot \frac{P_k}{(k+1)!} P_{k+1} \cdots \frac{P_{k-n}}{(k+n)!} P_{k+n} = \frac{P^n}{n!} \cdot P_0 \tag{20}$$

Let's use normalization process (19), for P_0:

$$P_0 = \sum_{n=0}^{V} \frac{Z^n}{n!}, \quad n \in \{0, V\} \tag{21}$$

Let's describe probability of request loss in a cell, based on conditions that at time τ (receipt of the request) there are no release channels (V) in cell. Using expressions (20, 21), have got:

$$P_R = \frac{Z_0^V}{V!} \cdot \left[\sum_{n=0}^{V} \frac{Z^n}{n!} \right]^{-1} \tag{22}$$

Formula (22) is used provided that the probability of receiving two or more requests is a value of small order from o(τ). Let's move on to the next step: determine the probability of requesting a loss in the cell. We are talking about the probability of transmission from neighboring cells, the value of which depends on the type of service signal (GSM, UMTS):

$$P_{R\_handover} = \frac{Z_h^V}{V!} \cdot \left[\sum_{n=0}^{V} \frac{Z^n}{n!} \right]^{-1} \tag{23}$$

The presented paper considers the access strategy in the absence of redundancy. Under such conditions, radio channels in the sector should be served as internal calls (single BS), as well as manual calls and without backup radio channels. Under such conditions, the probabilities of denial of service for receiving calls and calls on the handover are equal.

5.2 Method of Forecasting Load at Short Time Intervals

The method of load prediction at short intervals is improved on the basis of the ARIMA method (autoregressive integrated moving average). This method can be described as a method of auto regression with sliding mathematical expectations. In contrast to the method presented in [2], you are offered a solution to the problem of finding the minimum slip interval. The use of this condition will satisfy the requirements, which will minimize the number of operations to fulfill the forecast. The presented solution is able to provide the optimal forecasting speed; the criterion of optimality remains ambiguous.

In this paper, we present a method, part of which are the following steps: forecasting interval and frequency of load forecasting. The forecast interval is calculated based on the statistics of the node. The second indicator is based on controlling the adequacy of resources in the node.

In the experimental part of the work the local part of the Mobile Operator of Kyiv was used. The topology of the BS sectors monitored is shown in Fig. 1. The experiment took place in a network on three sites GSM-1800 BS. These sectors of the BS were connected to one BSC controller with the highest percentage of utilization in the network. Conditionally depicted these areas in Fig. 11, each zone has an average diameter of up to 1 km.

For the experiment, densely populated area was chosen, where there are large transport interchanges with the presence of public, private and commercial vehicles, and municipal institutions, and business centers, and social institutions.

According to the steps described above, it is necessary to position the network subscribers. After analyzing the selected area according to existing city maps, we conclude that five-story buildings with a standard street width and building density predominate. Given that the GSM 900/1800 system is being considered, we select a positioning system based on the Okamura-Hata model of calculating the attenuation of the radio signal. The positioning accuracy is 200 m, the area of the study area will be 22 km^2.

At the second stage of the analysis of the existing network, it is necessary to identify the nodes of demand (to cluster the masses of subscribers received). For this, it is proposed to use algorithms clustering demand nodes.

At the third stage of the analysis of the existing network it is necessary to identify nodes with no demand, that is, concentrating areas with free channel resource. For this purpose, as well as in the previous point, the algorithm of clustering of nodes with a negative sign is used.

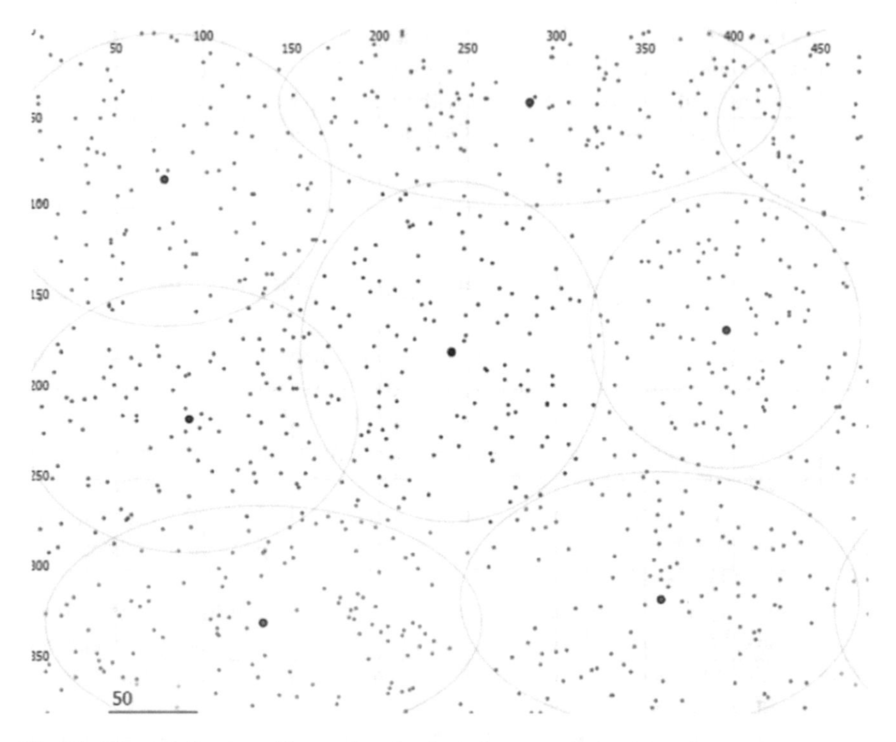

Fig. 11 The virtualization of forecasting the channel resource in local area base station

Algorithm of the method:

The operators organizing such network structures are faced with a number of problems associated with setting up the systems. At the stage of network planning, designers resort to variables mathematic models for predicting customers (subscribers) loading. But all these models find their place only at the stage of network planning. Time operation, section has one group of sector, but when starting operation, sectors was increased by another BS group. But the question of the effectiveness of this approach is to reduce the number of complaints. Therefore, in showed example, it is proposed to evaluate the feasibility of increasing the coverage area of this site.

When the network is already in operation, the analysis of the network is carried out by evaluating the statistical data on events in the system using various analytical software systems. The latter include: handover characteristics, dropped connections, unsuccessful attempts to establish connections, interference. But, for an optimal assessment of the system's operation, it is not enough to know the number of events in the system, you need to know the characteristics of the load that was served by this network section: during event, all-time beginning, quality indicator of the event, specific mapping for each cell in BS.

Data is generated in time the service, accumulation center of such data is the billing system of mobile network. This system acts not only as a system of mutual settlement with customers, but also as a system of aggregation and analysis for incoming data. To solve this problem, in this paper proposed use CDR (Call Data Records—specific format data saving on income and outcome events in mobile network).

The results were evaluated by comparing the planned part (it is laid down in the planning) with the projected load distribution on the network. In Figs. 12 and 13 you can see results of comparing obtain from archiving data set of mobile network during the week.

The Gray areas (Fig. 12) show areas with insufficient channels for re-sublease to service the load. The proposed method increased by +20% compared to the planned one (Fig. 13) in the area of border sectors. On the other hand, an error of about 10% was found, which is the reason for the lack of sub-release channels to service the load at the border areas. This result is explained by the fact that data from border sectors are absent in the preliminary calculation.

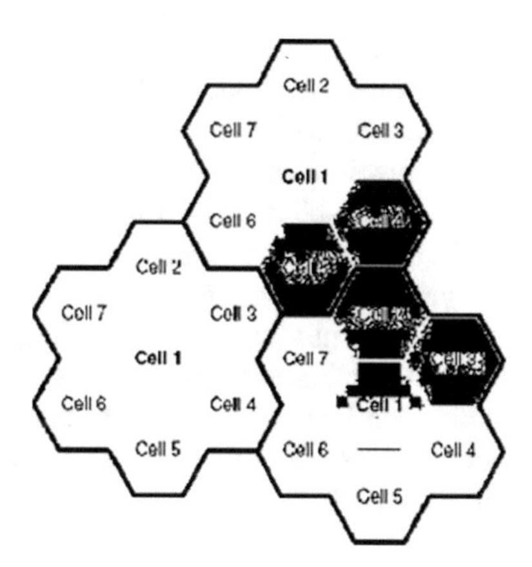

Fig. 12 The principle of forecasting channel resource

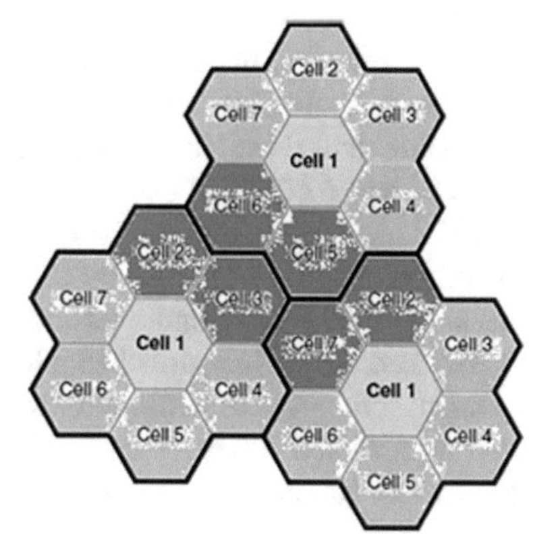

Fig. 13 The principle of existing channel resource

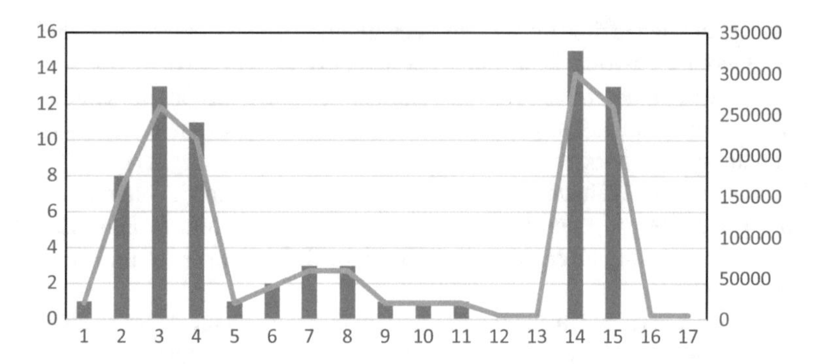

Fig. 14 The call data records for incoming requests

Let's pay attention once again in Fig. 13: the adjacent sectors which are a part of other controller of BSC are represented. As a result, the transfer led to an increase in the load that was carried out from the neighboring BS, so this load was absent in the model. To determine subscribers load, it is necessary to bind the number of received measurement results to subscriber traffic when the channel is occupied. It was mentioned above that with any access to the system by the mobile station, the measurement results are transmitted not only in traffic channels, but also in control channels, which must be taken into account.

At the beginning of this section there was only one group of sectors, when the BS groups were enlarged, at the end of the work additional sectors were received, which were put into operation. Because the effectiveness of the evaluation of the method was based on subscribers 'complaints, namely in reducing the number of subscribers' complaints. At the same time, in this example, the article proposes to assess the impact of increasing the coverage area with decreasing the number of complaints (Fig. 14).

If use the scale "percentage of accepted measurements" [12, 14], then we can only determine the effective service area, where 60–80% of the traffic is carried out, as well as the current service area. It also allows you to highlight the location of the MT accumulation, which is determined by the peculiarity of the terrain (buildings, travel routes, and other features of an urban or other type of relief).

When establishing a connection, the BSC uses the AGCH to inform the MS which SDCCH and SACCH it can use, and also provides TA information. SDCCH and SACCH are used to establish a connection. The busy time of the SDCCH channel for different types of access by the mobile station is different, and, consequently, the number of transmitted measurement results. In addition, the total load on the traffic channels and control channels also affects the occupation time of the SDCCH channel. If the mobile station fails to occupy the SDCCH channel, the connection is not disconnected, but is resumed. The BSC assigns a TSN channel, and the SDCCH is released. MS and BTS switch to the frequency of the TCH channel and the time slot allocated for this channel. If the subscriber answers, then the connection is established. During the conversation, the radio connection is controlled by the information transmitted and received by the MS via the SACCH.

Based on results, it can be accomplished that this sectoral approach provides a better solution in the implementation of such applications compared to conventional radio planning. In addition, this article explains the poor use of communication lines in conventional GSM networks. It can be seen that this method, which uses data from both billing and the radio part, is able to more efficiently distribute incoming traffic. With an increase in subscriber traffic, the shortest path from the original destination BS is heavily overloaded and leads to loss of transmission data.

It has been modeled and demonstrated that traffic accounting in terms of quality of service (billing) is able to handle incoming traffic more efficiently than traditional approaches to radio planning, by distributing data traffic across several BSs, which it is not able to achieve with conventional radio-planning. Performance in mobile network is greatly improved by implementing sectorial analyze in mobile network. To implement this technology on networks, a significant simplification of the network architecture is necessary.

6 Conclusion

In the network that is put into operation, it is necessary to apply a specific network analysis by evaluating statistical data. Such statistics include information about events in the system, both business and useful data. Similar characteristics are included in the analysis in accordance with the recommendations in [3]: successful connections, failed connections, as well as unsuccessful attempts to establish connections and others. First of all, you need to know the characteristics of the load that served this segment of the network: event duration, start time, event quality indicators, binding to a specific BS. These network characteristics are necessary for optimal evaluation of the system, in addition to the events in the system.

Modern technologies require modernization of its logical part (control, etc.) with minimal rearrangements of the physical component ("topology" of networks). Cellular networks are already deployed and are functioning quite successfully. Transformations leading to an improvement in the quality of service, an increase in the data transfer rate, an increase in the number of operator functions and the services offered to the final consumer should occur unnoticed by network subscribers. Such methods of computer modeling and prediction of various technological processes using radio waves are already used in various fields of science and practice [15]. The proposed method and software products allow the use of signals and commands available on the operator's network for an additional function of selecting demand nodes. The commissioning of such software will allow not only to improve the quality of service [5], but will also allow a more detailed study of traffic processes and radio networks [16].

The billing network of the Operator is a source of test data for customer service in the network. The billing system acts in this work not as a system of calculations, but as a system of aggregate source data for the method.

In the process of network planning, the Erlang B failure model was used to calculate the bandwidth of base station cells. The initial calculation of the network capacity is performed with a limitation of the amount of losses (P_lost) in the range from 0.01 to 0.05%. The main indicator is the probability of failure. This decision is based on the fact that the level of service—the quality of service, in terms of the probability of failure due to the deterioration of the signal-to-noise ratio in the radio channel [17]. In the test part of the network, this figure is set no worse than 5%.

It was demonstrated that the use of the load forecasting model provides recommendations that can lead to the processing of incoming traffic more efficiently up to 17%. The comparison is made in relation to traditional approaches to radio path planning. In the traditional approach, the identification of traffic "received" to the neighboring BS, falls out of consideration when planning.

Modern technologies require the modernization of its logical part with minimal rearrangements of the physical component. Cellular networks are already deployed and are functioning quite successfully. Transformations leading to an improvement in the quality of service, an increase in the data transfer rate, an increase in the number of operator functions and the services offered to the final consumer should occur unnoticed by network subscribers. These methods of modeling and predicting various technological processes using radio waves are already used in various fields of science and practice [18]. The proposed model allows the use of signals and commands available on the operator's network for the additional function of highlighting the optimal position of base stations, I highlight the most concentrated zones. The commissioning of such a model will allow not only to improve the quality of service (QoS and CoS) [19], but will also allow a more detailed study of subscriber data flows [20].

References

1. Romanov O, Mankivskyi V (2019) Optimal traffic distribution based on the sectoral model of loading network elements. In: IEEE international scientific-practical conference problems of infocommunications, science and technology (PIC S&T), pp 683–688
2. Gaydamak Y, Zaripova E, Samuylov K (2008) Cellular mobile call service models. Russian University of Friendship, Moscow Russia
3. Fuchs C, Aschenbruck N, Martini P, Wieneke M (2011) Indoor tracking for mission critical scenarios: a survey. Pervasive Mob Comput 7(1):1–15
4. Cuevas A, Moreno JI, Einsiedler H (2006) IMS service platform: a solution for next-generation network operators to be more than bit pipes. IEEE Commun Mag 75–81
5. Ho W-C, Tung L-P, Chang T-S, Feng K-T (2013) Enhanced component carrier selection and power allocation in LTE-advanced downlink systems. In: Wireless communications and networking conference (WCNC), IEEE, pp 574–579

6. Romanov O, Dong TT, Nesterenko M (2020) The possibilities for deployment eco-friendly indoor wireless networks based on LiFi technology. In: 8-th International conference on applied innovations in IT, (ICAIIT)
7. Skulysh M, Romanov O (2018) The structure of a mobile provider network with network functions virtualization. In: TCSET 2018: 14-th international conference on advanced trends in radioelectronics, telecommunications and computer engineering, 20–24 February 2018: conference proceedings. Lviv–Slavske, pp 1032–1034
8. Romanov O, Nesterenko M, Veres L (2017) IMS: model and calculation method of telecommunication network's capacity. In: Proceedings of the 2017 international conference on information and telecommunication technologies and radio electronics (UkrMiCo) 11–15 Sept 2017, Odessa, Ukraine. IEEE Conference Publications, pp 1–4
9. Popoola S, Oseni O (2014) Empirical path loss models for GSM network deployment in Makurdi, Nigeria. Int J Sci 3(6):85–94
10. Skulysh M, Klimovych O (2015) Approach to virtualization of evolved packet core network functions. In: The 13th international conference experience of designing and application of CAD systems in microelectronics (CADSM). IEEE, pp 193–195
11. Globa L, Skulysh M, Romanov O, Nesterenko M (2018) Quality control for mobile communication management services in hybrid environment. In: The international conference on information and telecommunication technologies and radio electronics. Springer, Cham, pp 76–100
12. Romanov O, Nesterenko M, Mankivskyi V (2016) Application of the regression model of the coefficient of use of channels for forming the plan of load distribution in the network. In: Bulletin of NTUU "KPI". Radio engineering series, radio apparatus construction, No 67, pp 34–42
13. Degollado-Rea A, Vidal-Beltrán S, López-Bonilla J, Thapa GB (2015) Okumura-Hata, walfish-ikegami and 3GPP propagation models in urban environments for UMTS networks. SciTech J Sci Technol 4(1):70–78
14. Tahcfulloh S, Riskayadi E (2015) Optimized suitable propagation model for GSM. Telkomnika Indonesian J Electr Eng 14(1):154–162
15. Ilchenko M, Uryvsky L, Moshynska A (2017) Developing of telecommunication strategies based on the scenarios of the information community. Cybern Syst Anal. 53(6):905–913
16. Skulysh M, Romanov O (2018) The structure of a mobile provider network with network functions virtualization. In: 14th International conference on advanced trends in radioelecrtronics, telecommunications and computer engineering (TCSET). IEEE, 1032–1034
17. Romanov O, Hordashnyk Y, Dong T (2017) Method for calculating the energy loss of a light signal in a telecommunication Li-Fi system. In: Proceedings of the 2017 international conference on information and telecommunication technologies and radio electronics (UkrMiCo), 11–15 Sept 2017, Odessa, Ukraine. IEEE Conference Publications
18. Daradkeh YI, Kirichenko L, Radivilova T (2018) Development of QoS methods in the information networks with fractal traffic. Int J Electron Telecommun 64(1):27–32. https://doi.org/10.24425/118142
19. Ageyev D et al (2019) Infocommunication networks design with self-similar traffic. In: IEEE 15th international conference on the experience of designing and application of CAD systems (CADSM). IEEE, pp 24–27. https://doi.org/10.1109/cadsm.2019.8779314
20. Kryvinska N (2004) Intelligent network analysis by closed queuing models. Telecommun Syst 27:85–98. https://doi.org/10.1023/B:TELS.0000032945.92937.8f
21. Skulysh MA, Romanov OI, Globa LS, Husyeva II (2019) Managing the process of servicing hybrid telecommunications services. Quality control and interaction procedure of service subsystems. In: Advances in intelligent systems and computing, vol 889, pp 244–256
22. Kurdecha VV, Zingaeva NA (2011) Optimal reconfigurable base stations (R-BS) architecture and requirements to R-BS. In: 21st international crimean conference "microwave and telecommunication technology", Sevastopol, pp 465–466
23. Moshynska A, Osypchuk S, Pieshkin A, Shmihel B (2018) The effect of the features of signal-code constructions forming on indicators of functionality and reliability of communication

systems based on the 802.11 N/AC standards. J Sci Europe 2(26):38–47. Praha, Czech Republic. (ISSN 3162-2364)
24. Ageyev D, Qasim N (2015) LTE EPS network with self-similar traffic modeling for performance analysis. In: Proceedings of the 2015 second international scientific-practical conference problems of infocommunications science and technology (PIC S&T). IEEE, Kharkov, Ukraine, pp 275–277. https://doi.org/10.1109/infocommst.2015.7357335
25. Kryvinska N (2008) An analytical approach for the modeling of real-time services over IP network. Math Comput Simul 79(4):980–990. https://doi.org/10.1016/j.matcom.2008.02.016
26. Barabash O, Lukova-Chuiko N, Sobchuk V, Musienko A (2018) Application of petri networks for support of functional stability of information systems. In: IEEE 1st international conference on system analysis and intelligent computing, SAIC 2018—Proceedings. IEEE, Kyiv, Ukraine, pp 1–4. https://doi.org/10.1109/SAIC.2018.8516747
27. Ghosh A, Ratasuk R, Mondal B, Mangalvedhe N, Thomas T (2010) LTE-advanced: next-generation wireless broadband technology. IEEE Wirel Commun 17(3):10–22
28. Romanov O, Fediushyna D, Dong T (2018) Model and method of Li-Fi network calculation with multipath light signals. In: International conference on information and telecommunication technologies and radio electronics (UkrMiCo), 10–14 Sept 2018

Development of System for Registration and Monitoring of UAVs Using 5G Cellular Networks

Roman Odarchenko, Yaroslav Horban, Oleksandr Volkov, Mykola Komar, and Dmytro Voloshenyuk

Abstract The uncontrolled distribution and usage of unmanned aerial vehicles (UAVs) in the world, in combination with the risks associated with aircraft, other property, people's lives, privacy, violation of secure territories, and overall security, require the creation of a UAV registration and monitoring system. This work aimed to analyze the current state of the UAV registry in Ukraine, to identify the main weak points, and to develop an architecture for the UAV registration and monitoring system. The following tasks were solved to achieve the goal: analysis of the current state of the UAV registry in Ukraine; identifying the main weaknesses of UAV registration and monitoring in Ukraine and determining the directions for their solutions; development of a UAV registration and monitoring method; development of a model for the registration and monitoring of UAVs using 5G cellular networks; development of software for UAVs' registration and monitoring. As a result of the work, a system was developed that allows registering UAVs when they are turned on and monitoring the current coordinates of the UAVs which are using this system.

1 Introduction

Analysis of the UAVs' usage dynamics. More than 150 years have passed since the first UAV was created in 1849, but only since the beginning of World War I, the development of UAVs actively began and until the beginning of the twenty-first century, most of the UAVs were created for military purposes. The main milestones in the history of UAV development are [1, 2]:

R. Odarchenko (✉) · Y. Horban
National Aviation University, Kyiv, Ukraine
e-mail: odarchenko.r.s@bundleslab.com

R. Odarchenko
Bundleslab KFT, Budapest, Hungary

O. Volkov · M. Komar · D. Voloshenyuk
International Research and Training Center for Information Technologies and Systems of the National Academy of Sciences of Ukraine, Kyiv, Ukraine

D. Ageyev et al. (eds.), *Data-Centric Business and Applications*, Lecture Notes on Data Engineering and Communications Technologies 69,
https://doi.org/10.1007/978-3-030-71892-3_8

- August 22, 1849—the first usage of unmanned aerostat;
- 1924—the first full-controlled flight;
- 1960—a US reconnaissance aircraft was shot down with the capture of a pilot, which stimulated the development of UAVs;
- 1990s—the rapid development of UAVs associated with the development of global positioning system (GPS);
- 1995—"G.A. MQ Predator"—the world's first military "hunter-killer" aircraft without a human on board was created;
- 2014—"X 47B"—first unmanned (remotely piloted) aircraft to launch and land on the aircraft carrier.

The dominant leader in the UAV market for the last years is the Chinese company DJI, accounting for more than 70% of the world market and more than 80% of the Ukrainian market [3]. According to recent data, the UAV market in Ukraine reached a value of $1 million per month. While in the world, the value will reach $185 billion in 2019 [4].

Figure 1 shows that the Total UAVs' market cost is increasing by about 15% every year. Analysts say that at this rate it will reach 2.8 trillion in 2030. USD.

In terms of UAVs' diversity, Ukraine ranks 27th in the world, but most of the produced UAVs by Ukraine are rather outdated and inefficient. However, UAVs of domestic production make up only 7% of the total number of UAVs in Ukraine. Almost 90% of the total number are purchased in the USA, China, Israel, and Russia [6].

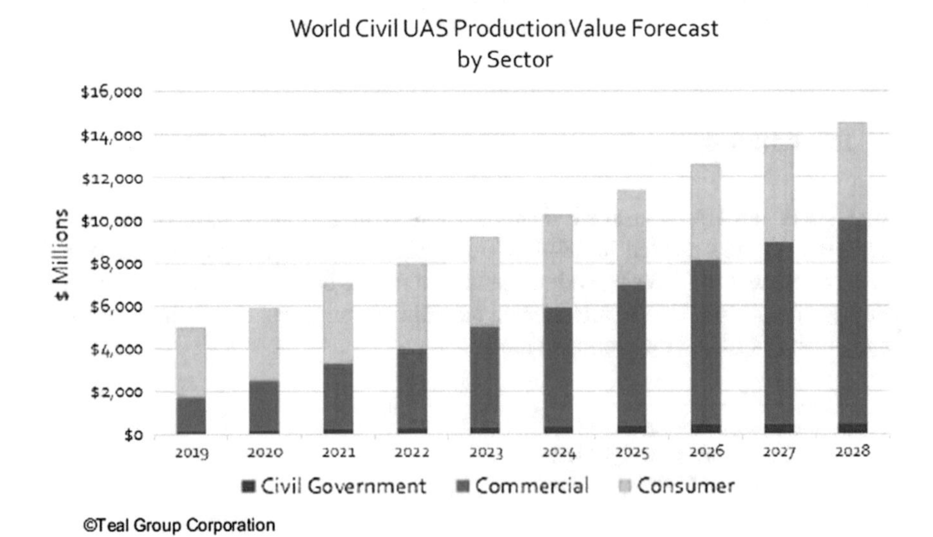

Fig. 1 UAVs world market [5]

In recent years, we have seen an impressive growth in the use of Unmanned Aerial Vehicles (UAV, also commonly referred as drones) for a wide range of applications, spanning from military to commercial domains. Several successful applications of drones already available in the market include surveillance, reconnaissance, remote sensing, search and rescue, aerial photography, crop surveys, on-demand emergency communications, traffic control, monitoring natural resources like oil or gas exploration etc.

With the growing appearance of unmanned aerial vehicles (UAVs) both in business and people's daily lives one of the most important tasks that need to be addressed is to ensure the safe use of UAVs. At the same time, the introduction of promising information technologies plays a significant role.

In Ukraine, the priorities in the UAVs' development strategy include the creation of a unified registry and the direct registration of every UAV. For each of them, the information about the owner, place, date, and time of registration. In addition, the complete monitoring of UAV movements is planned, which means that all UAVs can be tracked at any point in time. For citizens, this will guarantee safety, since today the airspace of Ukraine is relatively empty, but it will gradually fill up with the development of UAVs. In-air accidents that could lead to a crash and further casualties can be prevented with the M2M (machine-to-machine) communication technology [7]. The new opportunities for the developments in this technology, which includes communication between UAVs, are provided with the usage of 5G cellular networks.

One of the key issues when organizing UAV monitoring is communication between multiple objects. The task of transmitting information as fast as possible, i.e. with minimal delay and high data transfer speed, remains as well. In order to properly support radio access technologies, widespread coverage of modern cellular networks should be created in Ukraine. Mobile communication technologies should significantly improve the quality of the basic criteria required for UAV monitoring.

2 Background Works Analysis

The ICAO Statute is the ninth edition of the Convention on International Civil Aviation (also called the Chicago Convention), which includes amendments from 1948 to 2006 [8]. According to the definition, an aircraft is any machine that can derive support in the atmosphere from the reactions of the air other than the reactions of the air against the earth's surface [2]. An unmanned aerial vehicle is a pilotless aircraft, which is flown without a pilot-in-command on-board and is either remotely and fully controlled from another place (ground, another aircraft, space) or programmed and fully autonomous. An unmanned aircraft, piloted from a remote piloting point, is called RPA (remotely piloted aircraft) [9]. All unmanned aircraft, regardless of whether they are remotely piloted, fully autonomous, or combined, are subject to the provisions of Article 8 of the Chicago Convention relating to obtaining special authorization. According to Article 31, every aircraft engaged in international navigation shall be provided with a certificate of airworthiness issued or rendered valid by the state's authority in which it is registered [10]. The system consisting of UAVs and a control system on the surface of the earth is called RPAS (remotely piloted aircraft system), while it does not have to be self-sufficient. According to ICAO documents, of the components included in the RPAS, only the RPA is recorded in the aircraft register [11].

UAVs present a wide range of dangers for the civil aviation system [12]. These dangers must be identified and mitigated for safety risks [11, 42], as well as the introduction of airspace redesign, new equipment or procedures must be done. To achieve this, all countries, which are members of ICAO, must implement the relevant laws. Permission to use the airspace of European countries and the conditions for its use are granted by a joint civil-military air traffic management system based on an application for the use of airspace, except cases provided by the Air Codes. Regarding the Ukraine, currently, Ukraine as well does not have a registry created specifically for UAVs, that is, the use of UAVs is not defined by in the Air Code of Ukraine law [14]. That is why it is very necessary first to define the best rules for UAV registration and monitoring and afterwards development of novel architecture solutions for doing this.

Among various enabling technologies for UAV registration and monitoring, wireless communication is essential and has drawn significantly growing attention in recent years [15, 43]. Indeed, the standardization bodies are currently exploring possibilities for serving commercial UAVs with cellular networks. Industries are beginning to trial early prototypes of flying base stations or user equipment, while academia is in full swing researching mathematical and algorithmic solutions to address interesting new problems arising from flying nodes in cellular networks [16]. In [16] was provided a comprehensive survey of all of these developments promoting smooth integration of UAVs into cellular networks. Also, a large number of scientific works were devoted to the analysis of wireless technologies, new opportunities and challenges of these technologies if use them for drone communications [17–19], enhancements for LTE networks to support aerial vehicles [20, 21] and other open issues [22].

Thus, in order to optimize existing and build new 4G and 5G networks for drone communications to support remote registration and monitoring process, it is necessary to develop methods that will improve the performance of cellular communication networks so that they can meet a number of criteria in modern urban area: ensure the introduction of new mobile systems and support existing (saving of investments, that were made); meet the requirements of next-generation network architectures; have effective traffic management and quality assurance tools; provide convenient means for maintenance and operation. Therefore, to properly support new broadband radio access technologies in today's 5G cellular networks, the efficiency of data transmission should be improved while reducing the cost of delivering each megabyte of traffic and providing the quality of service (QoS) required by each type of traffic. In addition, new, more sophisticated network architectures for data transmission and management are emerging with the development of cellular networks. However, there are a number of unresolved tasks and issues that need to be addressed and resolved accordingly.

3 Research Objectives Statement

This chapter should analyze 5G network architectures for different UAV usage use cases. A gap analysis should be performed to evaluate which requirements are not met by other wireless technologies. After this will be developed the 5G network architecture for.

Thus, the **object** of research is the process of registration and monitoring of unmanned aerial vehicles using cellular networks. The **subject** of research is the methods, architecture design and technologies for recording and monitoring unmanned aerial vehicles using 5G cellular networks. The **goal** of the work is to ensure the registration and monitoring of unmanned aerial vehicles. The following **scientific tasks** were solved to achieve this goal:

The current state of UAVs' usage regulation in European countries and in Ukraine is analyzed.
5G network architecture for UAV registration and monitoring was developed.
A model for tracking UAV movements was developed.
Network-centric method for UAV registration and monitoring was developed.
Procedures for the registration and monitoring of UAVs were developed.

4 Research of Cellular Networks' Characteristics to Provide Registration and Continuous Monitoring of UAVs

To ensure the communication needs of the UAV different cellular networks can be used in separate or even parallel way (multilink solution [23, 24]) (Fig. 2). Cellular networks can act as a control and/or data transmission channel between the drone and the operator or between drones (UAV networks). In this case, the "range" of the control channel is limited only by the cellular network's area of coverage in the city

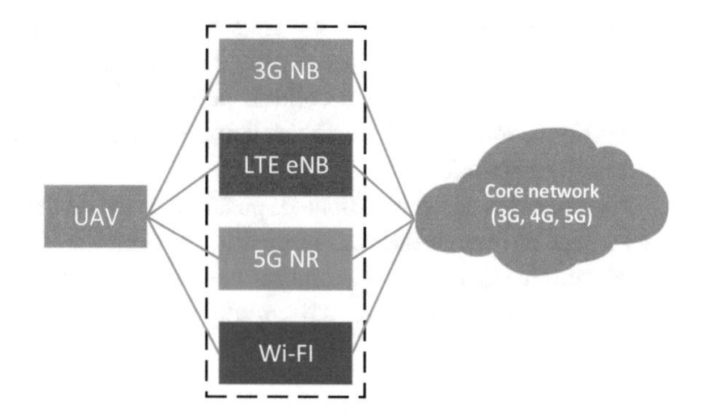

Fig. 2 Wireless networks for UAV communications

Table 1 Comparison of cellular communication networks

Technology	Data transfer speed
1G	2.4 kbit/s
2G	63 kbit/s
3G	144 kbit/s–2 Mbit/s
4G	100 Mbit/s–1 Gbit/s
5G	1 Gbit/s–35 Gbit/s

or country. After losing the network connection, the drone can return to the coverage area independently and choose a different route.

Moreover, the modern generation of cellular communication networks can provide most of the UAV needs [18]. Table 1 compares the data transfer rates of modern cellular communication networks generations.

5G technology is more suited to provide drone communications because it has low power consumption, which is especially important for long UAV flights. The advantages of using this technology include the high range of the radio signal, up to 30 km in the open field and up to 8 km in the city. The unique bandwidth of the radio signal, which provides stable communication in difficult to reach areas, should be noted. The usage of this technology is most appropriate in this case.

During 2020 the final 5G standards will be set. In August 2019, more than 1 million people used 5G networks in South Korea [25]. 5G is evolutionary not only in data transfer rate but also in other important requirements. One of the main advantages is the delay time reduction from 5 to 1 ms. Reducing the delay to such a short time increases the possibility to develop new projects, such as [26, 27].

Along with this, 5G networks will use Massive MIMO technology—communication system with separate transmit and receive antennas [28]. Their use will allow spatial and temporal processing of signals, more efficient use of the power emitted by the transmitter, and reducing the interferences' negative impact. As a result, the number of antennas will be increased to 8–16 compared to conventional communication systems using single-element antennas.

Furthermore, 5G networks will support D2D (device-to-device) technology [29]. Device-to-device technology allows devices that are close to each other to communicate directly without the involvement of a 5G network. Only signaling traffic will pass through the core of the cellular network. The advantage of this technology is the ability to transfer data in the unlicensed part of the spectrum, which will further offload the network.

All these characteristics make it possible to create new original projects and appliances. One of them is the communication system for UAVs. 5G will allow unmanned aerial vehicles to operate at a qualitatively new level. 5G will allow UAVs to exchange data almost instantly, which will greatly assist in resolving airspace issues.

5 Development of UAV Communication Architecture Using 5G Networks

Today, one of the most important problems is the shortage of the radio frequency spectrum of the wireless networks supported by UAVs. This problem is affected by the factors, such as a large number and increasing usage of portable mobile devices (e.g. smartphones and tablets), various wireless networks (Bluetooth, WiFi, LTE, and cellular networks) coexist in UAVs' operating spectrum ranges. This leads to extremely intense competition for the spectrum usage, and thus UAV communication systems face a spectrum deficit. Therefore, it is necessary to obtain a further range of access through the dynamic use of existing frequency bands for the drones' communication. So far, many researchers and members of standardization groups have presented a concept of using UAV communication systems to increase the spectral capabilities, which are called cognitive UAV communications [36]. This concept is a promising network architecture that provides communication between UAVs and ground mobile devices operating in the same frequency band. In this case, UAV communications can cause strong interference with existing ground devices, since UAVs are usually located on LOS with ground users. Several articles could be found in the literature that studied the UAV cognitive communication system [36–38]. Joint optimization of the UAV trajectory and power transfer was performed to achieve the maximum impulse throughput of cognitive UAV communication, while simultaneously limiting the obstacles placed on the primary receivers below the acceptable level [36]. Recently, a method for distributing the contracting spectrum between an unmanned network cell and a traditional terrestrial cellular network with various scenarios was introduced, i.e., dividing the spectrum of single-tier unmanned cells in a 3D network and distributing the spectrum between an unmanned network cell and a traditional 2D cellular network. Using stochastic geometry, they obtained explicit expressions for the probability of drone cells' coverage and achieved the optimal density of UAVs to maximize throughput.

One of the main role in cognitive UAV networks belongs to 5G cellular networks. They are expected to be the foundation of the next generation UAV networks.

The first 3GPP release of 5G technology (Release 15) [30], also known as 5G NR (New Radio), was completed in 2018. Was introduced 5G NR to enhance the user plane performance and efficiency using dual connectivity across the LTE and NR bands [31]. During the second-half of 2018, the standalone (SA) version of 5G was standardized, including the 5G core network (5GC), that enables deployments without any LTE infrastructure [32]. The 5G NR SA deployment can be also in combination with LTE but using only 5G NR for the user plane as in the early drop. The last drop of Release 15 specification was at the end of 2018, and it enables more architecture options for hybrid LTE and 5G NR deployments using the 5GC [31].

5G NR [33] should be used as a basis for UAV communications, and it has been acknowledged that diverse use cases from media, IoT, V2X and public safety should be considered when designing the architecture solution [31]. The subtask vision is to incorporate point-to-point, point-to-multipoint, multilink capabilities in

5G under one common framework and enabling dynamic use of different modes and network slices to maximize network and spectrum efficiency for more safe, reliable and cheaper UAV communications, registration and monitoring.

This section introduces the integrated space-air-to-ground network architecture in future 5G wireless communications (Fig. 3).

For example, densely deployed terrestrial networks in urban areas will be able to maintain access to high-speed data transmission, satellite communications systems can provide wide coverage and stable unhindered connection with the most remote and sparsely populated areas, while UAV communication can help existing cellular networks effectively manage very crowded areas. Currently, it is widely believed that an individually existing network cannot satisfy the need to process huge amounts of data and execute complex programs such as IoT (Internet of Things) applications. Therefore, there is a growing demand among scientific communities for the development of an integrated network architecture for space, air, and ground networks [34].

Such architecture will emphasize the need for communication between all devices that belong to the same category. In addition, the connection between satellite and UAVs is a key component in the formation of an integrated space-air-to-ground

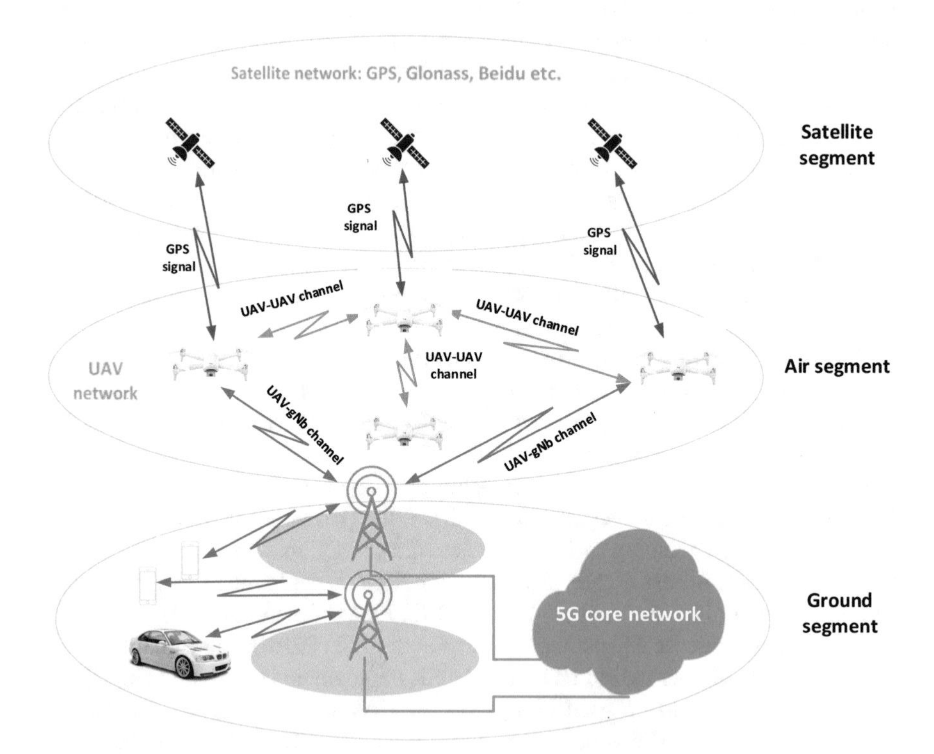

Fig. 3 Architecture of 5G networks

network. That is, for the regulation of airspace is necessary a high-quality data transmission [35].

For the architecture (Fig. 3) common IP packet format for UAV registration and monitoring system can be presented as follows:

IP header	UAV Serial number	Current coordinates of the UAV	Additional information

Each such packet has to consist of the information regarding the UAV identification number (ID) in the registry. It can be i.e. UAV serial number, which is given by the certified manufacturer in accordance with the state rules. Also, current coordinates of the UAV have to be transmitted in such packets. Additional information may carry special commands, requests etc. and the different types of additional payload have to be agreed on the layer of state regulatory bodies in the field of aviation.

Registration and monitoring system should build upon the 3GPP 5G network architecture [30]. This 5G network architecture should provide some key principles and concepts as follows [31]:

- Separate the User Plane (UP) functions from the Control Plane (CP) functions, allowing independent scalability, evolution and flexible deployments, for e.g. centralized location or distributed (remote) location of remote registration and monitoring system.
- Modularize the function design.
- Wherever applicable, define procedures (i.e. the set of interactions between network functions) as services, so that their re-use is possible.
- Enable each Network Function to interact with other NF directly if required.
- Support a unified authentication framework.
- Support concurrent access to local and centralized services. To support low latency services and access to local data networks, UP functions can be deployed close to the Access Network.
- The main design principles for the novel architecture are aligned with 3GPP direction.

The 5G system architecture for remote UAV registration and monitoring system is shown in Fig. 4. Figure 4 shows how UAVs data IP packets will be transferred using 5G network architecture.

In this architecture functionality as it exists is concentrated in "Remote UAV monitoring and registration server" (RUMRS). The RUMRS is placed between the external users or regulatory bodies and the DN so the RUMRS would contain specific functionality for the user plane:

- registration of the UAV in 5G network;
- real-time monitoring of the current coordinates of UAV;
- restricted area approach tracking;
- emergency messaging to the UAV users/owners and regulatory bodies.

Control of the registration and monitoring session is performed by the RUMRS.

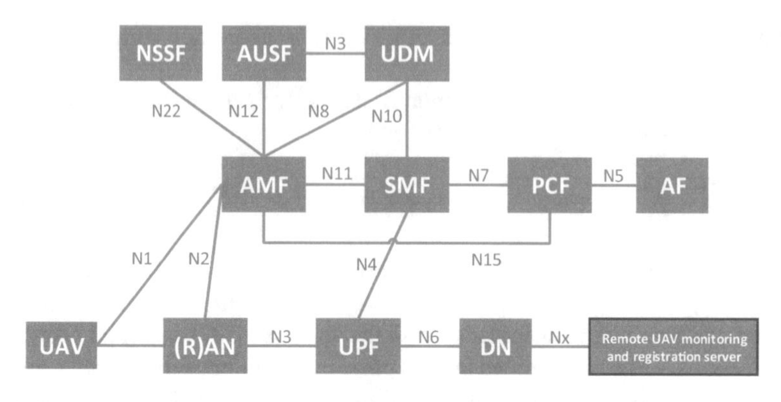

Fig. 4. 5G network architecture for UAV remote registration and monitoring

Session control (Start, Update and Stop) messages are sent (via the service-based interface) to the AMF through the SMF which distributes them (over the N2 reference point) to the (R)AN.

The user plane session is managed by the SMF and the N4 reference point.

6 Development of Network-Centric Method for UAV Registration and Monitoring

Countering the occurrence and elimination of the consequences of violations using UAVs includes:

- continuous computer monitoring of potentially dangerous locations and objects to determine the necessary measures to eliminate the consequences of each type of possible violations;
- the implementation of the necessary measures to prepare to deal with the consequences of possible violations;
- the implementation of a quick response to violations involving UAVs and the interaction of all used resources;
- organizing a possible set of parallel operational measures, dispatching, synchronizing and maneuvering resources in the dynamics of control [9].

In a broad sense, monitoring is understood as the systematic accumulation and processing of data on the state and dynamics of changes in the parameters of the object or process under consideration and the presentation of the results in a form convenient for the manager or expert [39, 45]. In dynamics, this is the collection and analysis of data on losses from violations. The monitoring system (Fig. 5) combines the monitoring tools of all levels and areas of management into a single whole.

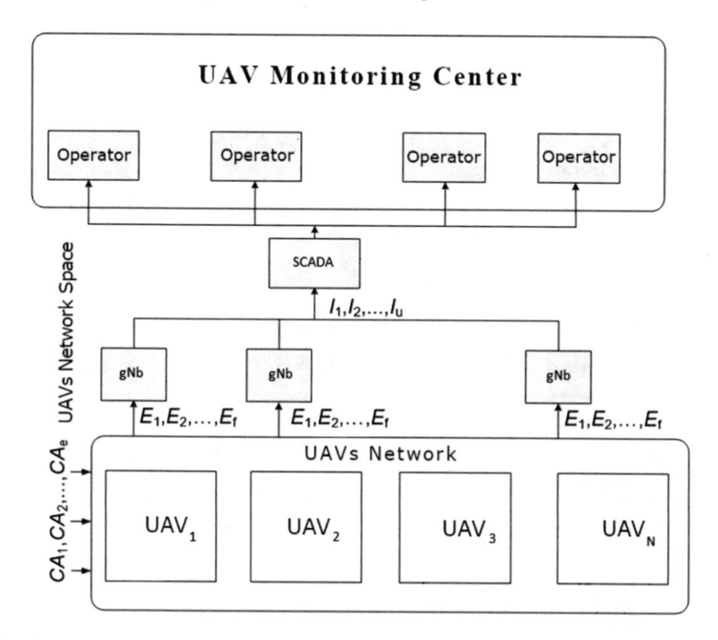

Fig. 5 Implementation of network-centric monitoring in the cellular network

The system must ensure delivery of all necessary information to the recipients in real-time or close to it as it is received and, very importantly, using the information for all levels and directions of management [46].

The following sequence of actions must be observed while operating such monitoring centers:

A certain security violation $E_1, \ldots, E_f$ occurs in the UAV network. It can be caused accidentally, or be a consequence, for example, of a certain type of terrorist attack$CA_1, \ldots, CA_e$.
Data on the security events are transmitted to specialized dedicated network sensors gNb. Each sensor can only be connected to a separate group of network devices.
Sensors compare the received information with the corresponding patterns, conditions, and flight zones. After that, the data on the occurred terrorist attacks $\mathrm{I}_1, \ldots, \mathrm{I}_u$ is transmitted to SCADA (Supervisory Control And Data Acquisition).
Based on a set of rules for responding to terrorist attacks, SCADA decides how to respond.
After the violation is eliminated, SCADA exchanges information with operators and the SCADA system between operators.

Based on the above concept, the network-centric monitoring and response to violations in the UAV network are functioning as shown in Fig. 6.

The first step of the algorithm is to collect data from the network about the occurring violations from the set $E_1, \ldots, E_f$, caused by $CA_1, \ldots, CA_e$—both intentional

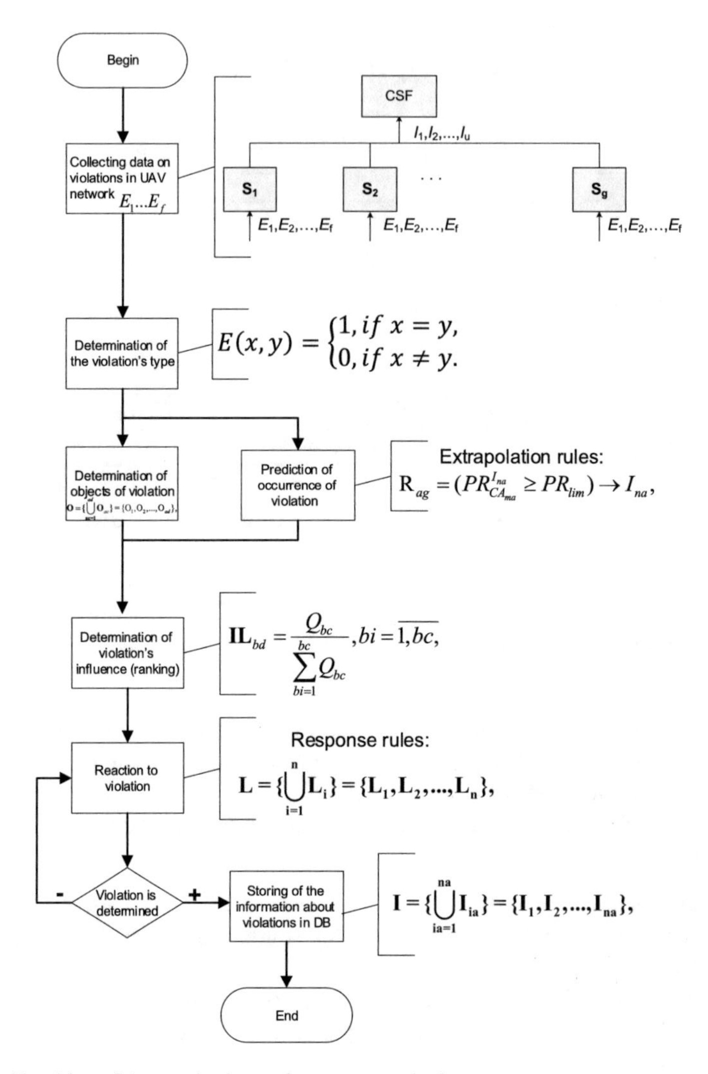

Fig. 6 Algorithm of the monitoring and response method

and unintentional actions—transmitted to the gNb sensors. Further, these violations $I_{1,}\ldots, I_u$ are identified by a specific set of their parameters through comparison with the corresponding patterns. This determines the type of known violation, or is identified as "not yet known."

After this, the identification of objects $O = \left\{\bigcup_{ac=1}^{ad} O_{ac}\right\} = \{O_1, O_2, \ldots, O_{ad}\}, (ac = \overline{1, ad})$ that are exposed to the violation takes place in parallel, and a set of extrapolation rules $R = \left\{\bigcup_{ab=1}^{ag} R_{ab}\right\} =$

$\{R_1, R_2, \ldots, R_{ag}\}$, $\left(ab = \overline{1, ag}\right)$ is formed that allows assessing the potential impact of the violation.

Determining the degree of violations' influence is carried out according to the formula:

$$IL_{bd} = \frac{Q_{bc}}{\sum_{bi=1}^{bc} Q_{bc}}, bi = \overline{1, bc},$$

where the assessment of comparative significance can be calculated by the formula:

$$Q_{bc} = \sum_{bi=1}^{bc} a_{bi}, x_{bibj}, bi = \overline{1, bc},$$

where x_{bibj} is the value of the bi-th criterion of the bj-th type of violation; a_{bi} is the "weight" of the bi-th criterion.

Violations response occurs according to the rules from the set $R = \left\{\bigcup_{i=1}^{n} R_i\right\} = \{R_1, R_2, \ldots, R_n\}$, followed by checking if the consequences of the violation have been eliminated. After a successful response, corresponding data on it is saved and a set $I = \left\{\bigcup_{ia=1}^{na} I_{ia}\right\} = \{I_1, I_2, \ldots, I_{na}\}$ is generated describing those violations. Data is transmitted and processed in the 5G radio interface secure unloading method and the reservation of network resources and load balancing methods. The requirements of the local and country laws must be taken into account.

Basing on the proposed algorithm it is possible to provide an implementation of the UAV monitoring and registration center to the network architecture (Fig. 7).

In Fig. 7 the multi-access edge computing (MEC) was used as an additional functionality to provide closer access to the remote registration and monitoring services. MEC paradigm aims to explore the potential that could be achieved through the convergence of diverse fields such as communication and information technology (IT) [31, 42]. Such a convergence would help to the development of new applications for remote monitoring and registration enabled by the provisioning of cloud computing at the edge of the fixed and/or wireless access network. The overall mobile edge computing framework presented in [40], adapted to RUMRS is as shown in Fig. 7. MEC could play a key role in hosting the low-latency UAV monitoring applications which could then be delivered to the UAVs using mobile access networks. The caching of frequently fetched content at the edge can enable MNOs to significantly optimize the transport network load, thereby minimizing deployment costs [31, 47].

One possible application of MEC from a RUMRS perspective, is as shown in Fig. 8, where the remote UAV monitoring application and related high-capacity, low-latency content is hosted. The low-latency constraints for mentioned above traffic requires the content to be hosted closer to the access network, with possible dynamic update.

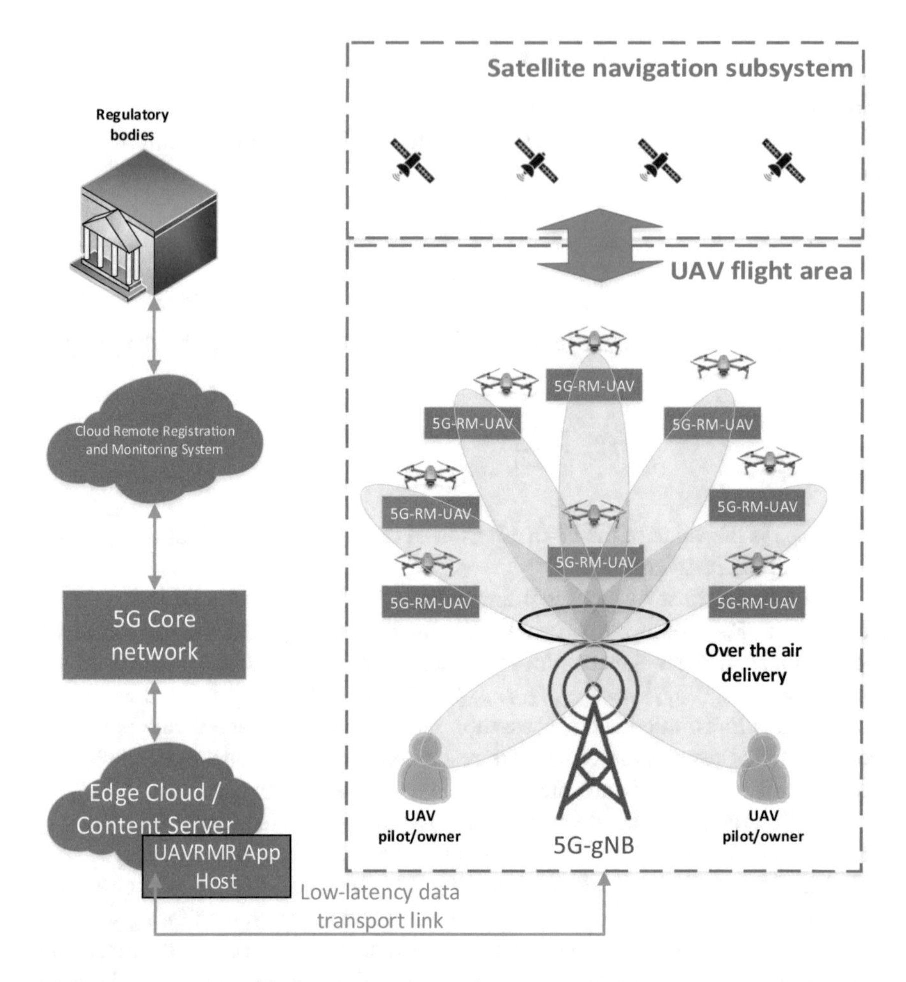

Fig. 7 Implementation of the UAV monitoring and registration center to the network architecture

7 UAVs' On-Board Registration and Monitoring System Architecture

After analyzing modern registration models of unmanned aerial vehicles in the world [3, 5, 41], there were made several conclusions. Thus, in many developed countries: China, Japan, Austria, Russia, UAVs' regulations are too strict and do not provide sufficient opportunity for the development of this area in the future.

From our point of view, the United States is an excellent example of the right policy with UAVs, which stimulates interest and development.

There are a lot of differences for different states on the moment of registration itself. Registration should be carried out directly with the purchase, that is, full control over the UAV market in the country. It is better, because everything is simple since

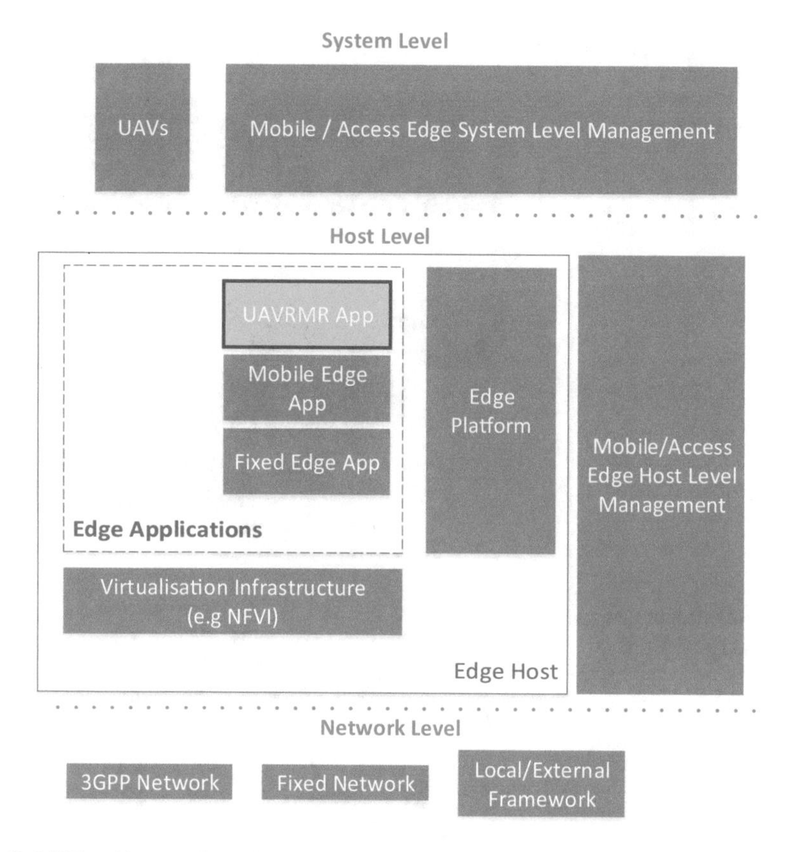

Fig. 8 MEC architecture for RUMRS

when buying we can use it (so far no one sees it). However, when the registration takes place at the time of purchase, it is possible to establish the identity of the pilot. Along with this, another important issue of the time allotted for UAV registration disappears. The greater convenience and effectiveness of this system (Fig. 9a) compared to the existing one (Fig. 9b). This greatly facilitates the process of purchasing a UAV. Along with this, it will reduce the number of violations due to late registration.

The UAV's registration and monitoring system usually is a small device that is attached to the UAV and connected directly to the battery, and it is polling from it (Fig. 10). Variations in which polling comes from batteries are possible.

At the first launch of the UAV, the registration and monitoring system is also started. Immediately after the first power-up, the first signal is sent to the server. This signal contains a regular UAV number and after the transmission is received by the server the number is linked to the device in the registry. The date and place of the registration are also determined at the same time. Thus, in the state register, information about the name and passport number of the owner is added to the UAV's number and model.

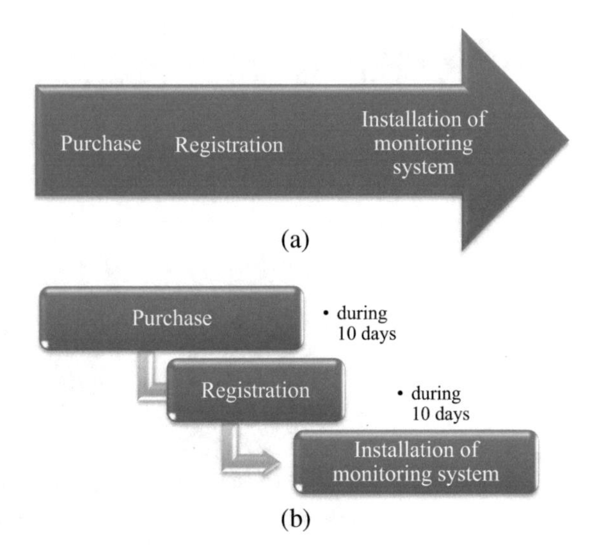

Fig. 9 UAV registration and monitoring process **a**—roposed procedures, **b**—existing procedures

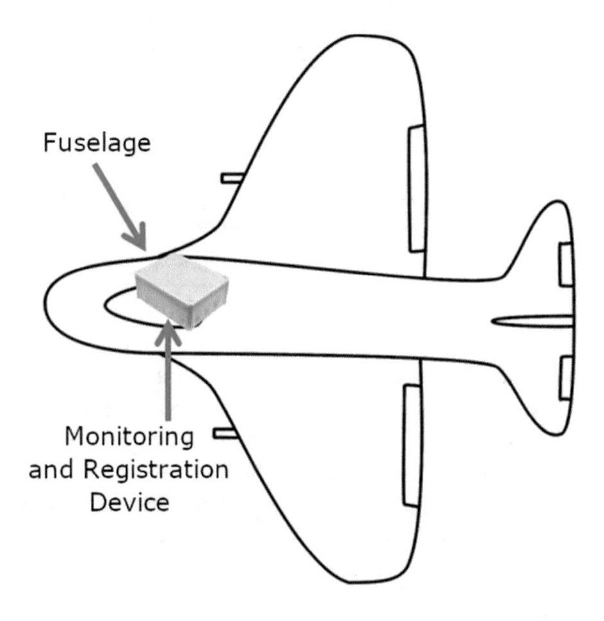

Fig. 10 Registration and monitoring system in UAV

Physically, this concept can be implemented in the form of a Center for Monitoring and Responding to Violations (data processing center), which will receive data on the state of security, violations, etc. directly from various network nodes (Fig. 11).

Figure 12 shows the stages of signal transmission and processing according to the UAV monitoring model (Fig. 9). The UAV sends a signal that reaches the radio tower, where it immediately forwarded to the local center, and from that to the UAV

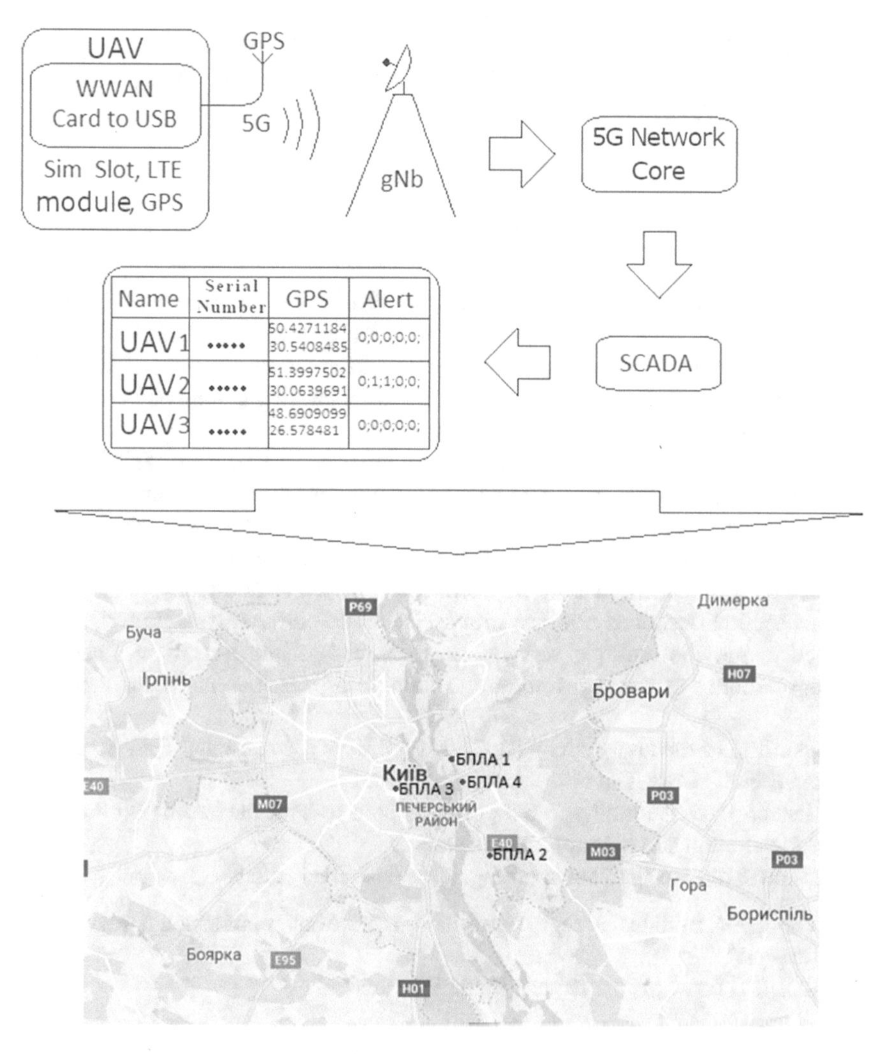

Fig. 11 Proposed model for UAVs' monitoring

Fig. 12 Signal path

monitoring center. The received signals are analyzed, reformatted, and stored in RUMRS (Fig. 7).

The UAV's monitoring and registration system sends a signal with a frequency of 5 s. This interval is enough to track UAVs during the flight. In case of possible law violations, the UAV sends a normal signal, which is defined as a violation directly in the monitoring center, after which the signal is separately stored in the database. After the identification of violation, the following steps depend on the regulatory legal acts of the country/area in which the violation occurred. Possible options: penalization of the owner, UAV removal, calling the law enforcement officers.

8 Procedures for the Registration and Monitoring of UAVs

To register a UAV at purchase, a buyer must confirm its identity. After familiarization with the terms of UAVs' use in the airspace of a certain state, the drone passes the first stage of registration. The second stage occurs during the first turn-on of the drone. If a person disconnects the UAV registration and monitoring system on his own, this will cause various penalties, depending on the laws.

Since the signal is sent with a frequency of once per 5 s, so during a full flight, the monitoring and registration system spends 1000 times less charge than the average battery charge, that is, the monitoring and registration system reduces the flight time by 0.1%, which actually can be neglected. However, it would be more effective to have a monitoring and registration system built into the UAV directly.

This section describes how the proposed mobile core network (Fig. 4) can be enabled to provide multiple services in developed architecture. It was achieved through defining of the registration and monitoring sessions and their main call flows.

According to the above architecture (Fig. 4), it is assumed that UAV is already registered in 5G Core network.

After the UAV is coming to 5G coverage area in-built device should decide to create registration and monitoring session.

Session setup process shown on Fig. 13 is described below.

1. The UAV initiates a registration and monitoring connection (Registration Request).
2. RUMRS forwards the request to the SMF and AMF.

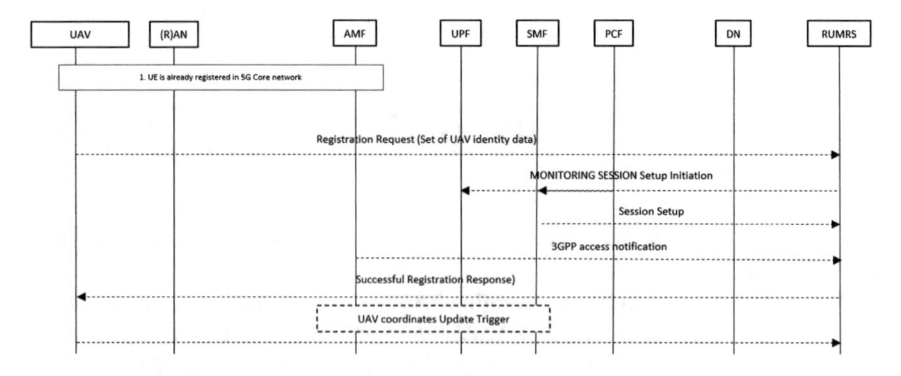

Fig. 13 Registration and monitoring session setup process

3. AMF forwards the request to the RAN (3GPP connection establishment).
4. AMF sends 3GPP access notification to the RUMRS.
5. RUMRS sends Session Setup Response to the UAV.
6. UAV sends UAV coordinates Update Trigger to the RUMRS.

9 Conclusions

The work is dedicated to solving the problem of monitoring and registration of UAVs. The regulation documents on unmanned aerial vehicles were examined and the conclusion was made that in these documents the problem of registration and tracking is at the initial stages of formation. This problem must be solved, since every year the use of UAVs is growing, i.e., drones are becoming increasingly popular among the population of the entire world. If the problem is not resolved, such aircraft can be used for criminal purposes and harm people.

To solve the problem of UAV administration, cellular networks were studied and compared. As a result, a modern 5G cellular network was chosen to develop a communication system. This network has a high data transfer rate with ultra-low latency and is the most progressive among all existing alternatives. This chapter describes in detail the main advantages and disadvantages of this technology.

A new UAV registration model was also proposed. This model can provide the highest level of safety when using UAVs.

This chapter examined the call flows within the Core Network that support the identified mechanisms. This level of evaluation and detail may be used as a solid basis for further 3GPP design and standardization in the direction of 5G usage for drones.

Acknowledgements This work was supported in part by the European Commission under the 5G-TOURS: SmarT mObility, media and e-health for toURists and citizens (H2020-ICT-2018-2020 call, grant number 856950). The views expressed in this contribution are those of the authors and do not necessarily represent the project.

References

1. Nalty BC (ed) (1997) Winged shield, winged sword: a history of the USAF. Air Force History and Museums Program, United States Air Force, Washington, DC
2. Fahlstrom P, Gleason T, Introduction to UAV systems, 4th edn, p 287
3. 3 reasons why China is the global drones leader. https://www.weforum.org/agenda/2018/09/china-drones-technology-leader/
4. The Drone Market 2019–2024: 5 things you need to know. https://www.droneii.com/the-drone-market-2019-2024-5-things-you-need-to-know
5. The drone market. https://www.commercialuavnews.com/infrastructure/teal-group-analyzes-and-predicts-drone-market
6. What Are Drones? https://drones.cnas.org/reports/what-are-drones/

7. Anis M, Gadallah Y, Elhennawy H (2016)Machine-to-machine communications over the internet of things: private wi-fi access prospects. In: 2016 Wireless days (WD), Toulouse, pp 1–3
8. ICAO 10019 Remote Manned Aviation Systems Manual: https://aeronet.aero/biblioteka/2016_07_06_10019_rukovodstvo_po_distantsionno_pilotiruemym_aviatsionnym_sistemam_dpas
9. "Acronyms and Glossary of Key Terms," Appendix E, Unmanned Aircraft Systems (UAS) at Airports: a primer. National Academies of Sciences, Engineering, and Medicine, 2015. The National Academies Press, Washington, DC, pp 70–81
10. The Legal Framework for RPAS/UAS Suitability of the Chicago Convention and its Annexes. https://www.icao.int/Meetings/RPAS/RPASSymposiumPresentation/Day%202%20Workshop%2010%20Legal%20Christopher%20Petras%20-%20The%20Legal%20Framework%20for%20RPAS_UAS.pdf
11. Manual on Remotely Piloted Aircraft Systems (RPAS) (2015) ICAO Doc 10019 1st edn
12. ICAO Cir 328, Unmanned Aircraft Systems (UAS)
13. Zeng Y, Wu Q, Zhang R (2019) Accessing from the sky: a tutorial on UAV communications for 5G and beyond. Proc IEEE 107(12):2327–2375
14. Flight space organization https://bpla.ua/praktichni-novini-vikoristannja-bpla/organizacija-povitrjanogo-prostoru/
15. Zeng Y, Wu Q, Zhang R (2019) Accessing from the sky: a tutorial on UAV communications for 5G and beyond. Proc. IEEE 107(12):2327–2375
16. Fotouhi A, Qiang H, Ding M, Hassan M, Giordano LG, Garcia-Rodriguez A, Yuan J (2019) Survey on UAV cellular communications: Practical aspects, standardization advancements, regulation, and security challenges. IEEE Commun Surv Tutor pp 1–28
17. Mozaffari M, Saad W, Bennis M, Nam YH, Debbah M (2018) A tutorial on UAVs for wireless networks: applications. challenges, and open problems. ArXiv e-prints
18. Hayat S, Yanmaz E, Muzaffar R (2016) Survey on unmanned aerial vehicle networks for civil applications: A communications viewpoint. IEEE Commun Surv Tutor 18(4):2624–2661
19. Zeng Y, Zhang R, Lim TJ (2016) Wireless communications with unmanned aerial vehicles: Opportunities and challenges. IEEE Commun Mag 54(5):36–42
20. NTT DOCOMO Inc., Ericsson (2017) New SID on enhanced support for aerial vehicles. 3GPP RP-170779 RAN#75
21. 3GPP Technical Document RP 181644 (2018) Summary for WI enhanced LTE support for aerial vehicles
22. Gupta L, Jain R, Vaszkun G (2015) Survey of important issues in UAV communication networks. IEEE Commun Surv Tutor 18(2):1123–1152
23. Odarchenko R, Aguiar R, Altman B, Sulema Y (2018) Multilink approach for the content delivery in 5G networks. In: 5th International scientific-practical conference problems of infocommunications science and technology, PIC S and T 2018—Conference proceedings, pp 140–144
24. Odarchenko R, Gimenez J, Sulema Y, Altman B, Petersen S (2019) Multilink solution for 5G: efficiency experimental studies. In: 2019 3rd International conference on advanced information and communications technologies, AICT 2019—Proceedings 8847862, pp 336–339
25. South Korea reaches 2 million 5G subscribers in four months. https://www.rcrwireless.com/20190808/5g/south-korea-reaches-2-million-5g-subscribers-four-months
26. SmarT mObility, media and e-health for toURists and citizenS https://5g-ppp.eu/5g-tours/
27. 5G for Drone-based Vertical Applications. https://5g-ppp.eu/5gdrones/
28. Geraci G, Garcia-Rodriguez A, Galati Giordano L, Lopez-Perez D, Bjoernson E (2018) Supporting UAV cellular communications through massive MIMO. In: 2018 IEEE international conference on communications workshops (ICC workshops), pp 1–6
29. Asadi A, Wang A, Mancuso V (2014) A survey on device-to-device communication in cellular networks. IEEE Commun Surv Tutor 16(4):1801–1819
30. 3GPP TS 23.501, System architecture for the 5G system
31. 5G-Xcast WP4 D4.1, Mobile core network

32. 5G-Xcast WP4 D4.2, Converged core network
33. 3GPP TS 38.413, NG-RAN; NG application protocol (NGAP)
34. LoRaWAN—Innovative IoT network. https://www.satellite31.ru/?page_id=3528
35. Zhao J, Gao F, Ding G, Zhang T, Jia W, Nallanathan A (2018) Integrating communications and control for uav systems: opportunities and challenges. IEEE Access 6:67519–67527
36. Huang Y, Mei W, Xu J, Qiu L, Zhang R (2019) Cognitive UAV communication via joint maneuver and power control. IEEE Trans Commun 67(11):7872–7888
37. Li P, Jie Xu (2020) Fundamental rate limits of UAV-enabled multiple access channel with trajectory optimization. IEEE Trans Wirel Commun 19(1):458–474
38. Gao Y, Tang H, Li B, Yuan X (2020) Robust trajectory and power control for cognitive UAV secrecy communication. IEEE Access 8:49338–49352
39. Hernantes J, Gallardo G, Serrano N (2015) IT infrastructure-monitoring tools. IEEE Softw 32(4):88–93
40. ETSI GS MEC 003, Mobile edge computing (MEC); Framework and reference architecture
41. Unmanned Aerial Vehicle (UAV) Market, https://www.marketsandmarkets.com/Market-Reports/unmanned-aerial-vehicles-uav-market-662.html
42. Ageyev D et al (2018) Classification of existing virtualization methods used in telecommunication networks. In: Proceedings of the 2018 IEEE 9th international conference on dependable systems, services and technologies (DESSERT), pp 83–86
43. Barabash OV, Dakhno NB, Shevchenko HV, Majsak TV (2018) Dynamic models of decision support systems for controlling UAV by two-step variational-gradient method. In: 2017 IEEE 4th international conference on actual problems of unmanned aerial vehicles developments, APUAVD 2017—Proceedings, Kyiv, Ukraine, IEEE, pp 108–111. https://doi.org/https://doi.org/10.1109/APUAVD.2017.8308787
44. Barabash O, Dakhno N, Shevchenko H, Sobchuk V (2018) Integro-differential models of decision support systems for controlling unmanned aerial vehicles on the basis of modified gradient method. In: 2018 IEEE 5th international conference on methods and systems of navigation and motion control, MSNMC 2018—Proceedings, Kyiv, Ukraine, IEEE, pp 187–190. https://doi.org/https://doi.org/10.1109/MSNMC.2018.8576310
45. Radivilova T, Kirichenko L, Ageyev D, Bulakh V (2019) Classification methods of machine learning to detect DDoS attacks. In: 2019 10th IEEE international conference on intelligent data acquisition and advanced computing systems: technology and applications (IDAACS), Metz, France. IEEE, pp 207–210. https://doi.org/https://doi.org/10.1109/IDAACS.2019.8924406
46. Kryvinska N (2008) An analytical approach for the modeling of real-time services over IP network. Math Comput Simul 79(4):980–990. https://doi.org/10.1016/j.matcom.2008.02.016
47. Kryvinska N, Zinterhof P, van Thanh D (2007) New-emerging service-support model for converged multi-service network and its practical validation. In: First international conference on complex, intelligent and software intensive systems (CISIS'07). IEEE, pp 100–110. https://doi.org/10.1109/CISIS.2007.40

Complex Tools for Surge Process Analysis and Hardware Disturbance Protection

Yuriy Rudyk, Victor Kuts, Oleg Nazarovets, and Volodymyr Zdeb

Abstract By analyzing the national normative and technical requirements of the lightning protection systems it seeing that there is a demand for an extra layer of protection measures from a direct lightning strike or voltage induction by a remote one. Protective devices and charts of their settings need to be determined in order to achieve that. Information and communication systems as well as their components have protection centered around counteracting thunderstorms in the environment. Ukraine's standardization implements the recommended methods of calculation and protecting schemes from impulsive overvoltage. Authors has led a thorough analysis is made for the case of existing protectively-switching devices. A comparison of low-voltage electric networks surge protection devices in combination with park protection devices at defective contact connections was performed. The sensitivity limits for the proper operation of protective devices are determined as furthermore performed an analysis on signals indicating the reception of information on an occurrence of a defective contact connection. This, overall, displays the necessity of applying measurements means against impulse overvoltage in information and communication circuits. Those means control detection and are capable of fixing impulse irregularities from the usual form of network voltage. Constructing of measuring instruments with high metrological characteristics for impulse disturbances in information and communication lines it is expedient to use the principle of optimal distribution of measurement and control functions between elements of the measuring complex. Based on this it was complemented the application of control devices schemes. In turn, the risk of fires and device damage from lightning discharge will be greatly reduced. The measures are recommended to be implemented in the operation and electronic devices maintenance fields.

Keywords Surge protection · Safety equipment · Measurement · Overvoltage · Lightning · Technical requirement

Y. Rudyk (✉) · O. Nazarovets
Lviv State University of Life Safety, 79007 Lviv, Ukraine

V. Kuts · V. Zdeb
National University Lviv Polytechnic, 79013 Lviv, Ukraine

D. Ageyev et al. (eds.), *Data-Centric Business and Applications*, Lecture Notes on Data Engineering and Communications Technologies 69,
https://doi.org/10.1007/978-3-030-71892-3_9

1 Analysis of Standardized Requarements for Space and Construction Protection Zones

A stable electrical supply is necessary for interruptible industrial and domestic energy and process supply of electronic equipment [1–4] in regards to the voltage of a network. Proper operation of equipment depends on rapid changes in the shape of the industrial frequency voltage, seen as fast impulse distortions of the sinusoidal voltage of a network.

The electromagnetic impulse of lightning will cause damage to electronic devices as the electromagnetic effects of the lightning current are accompanied with transient wave processes and the effects of the radiates electromagnetic field. On top of that, the impulse surge will cause overvoltage and/or supercurrent in the conducting parts. It is a result of transitional wave process caused by electromagnetic impulse of the lightning.

The direct hazard of lightning is fire, mechanical destruction, injuries and deaths of humans and animals, as well as damage to electrical and electronic equipment. The effects of lightning may be explosions and the release of hazardous products—radioactive and toxic chemicals, as well as bacteria and viruses. Lightning flashes can be especially hazardous for electronic systems and computer equipment. In order to implement modern means and methods of calculating lightning protection in Ukraine, a national standard [5–7] has been developed that complies with IEC 62305-2010 Protection against lightning. Lightning currents can affect the facility, directly or indirectly, when lightning is discharged into a lightning protection system or structures in the immediate vicinity and trees. But most often there are cases of secondary action when lightning strikes remote facilitys (transmission lines, substations, etc.), connected by any communications with the protected facility or in discharges between clouds. They cause pulse currents of great value in the metal elements of structures and communications.

In Europe, a number of standards [8] have been developed and implemented on the basis of extensive scientific research by the International Electrotechnical Commission (IEC), which regulate methods and means of protecting buildings, structures and electrical equipment against lightning.

In addition, lightning discharges can be especially hazardous for electronic systems and computer equipment, even without direct strike. The requirements of international and national standards in this field are based on the "Zone concept of protection", the main principles of which are:

- the use in structures metal elements (reinforcement, frames, load-bearing elements, etc.) electrically interconnected and to the grounding system, as well as the creation of a shielding environment to reduce the effect of external electromagnetic influences inside the facility ("cell Faraday");
- the presence of a properly executed system of grounding and equipotentialization;
- the division of space into conditional protection zones and the use of special surge protection devices;

- compliance with the rules for the placement of protected electrical equipment and its conductors, with respect to other equipment and conductors capable of transmitting hazardous effects or causing potentialization.

Reinforced concrete structures of buildings that perform the function of a natural grounding and have an electrical connection with the equipotentialization system, shield the inside of the equipment quite well from electromagnetic actions, diverting a large amount of lightning current into the ground upon direct contact with the facility. However, even the residual value is hazardous to preserve the working condition of modern electronic devices.

Although shielding is the primary means of reducing electromagnetic interference, it may not always be possible to achieve the desired protection reliability, or to create such a design. To limit the transient overvoltage and to remove the impulse current, a device is designed to protect against impulse overvoltage (SPD). This device has at least one nonlinear element. Based on the risk assessment of a direct lightning strike or remote surges, it is necessary to choose the type of protective device and the scheme of their installation. The choice of type of protective device is considered in [7].

The selected surge protection device must withstand after part of the lightning current (Fig. 1), limit the overvoltage and cut off the accompanying currents after the main lightning pulses. For the leading parts of communal communications included in the facility at ground level, the part of the lightning current produced by them is

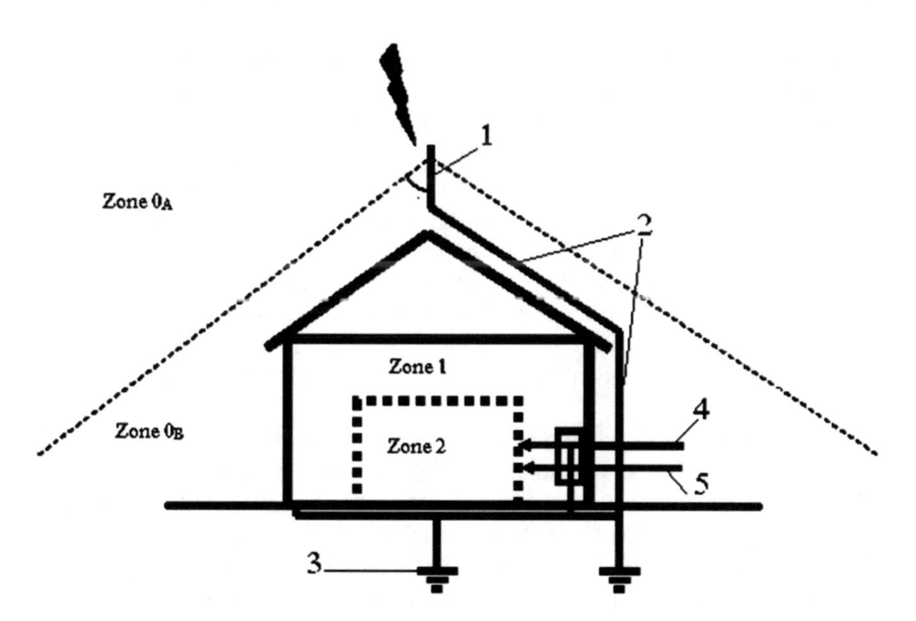

Fig. 1 Scheme of lightning protection zones in terms of direct and indirect lightning impact (1—air-termination system, 2—down conductor, 3—earth-termination system, 4—electrical and 5—non-electrical communications)

estimated. Therefore, it is necessary to improve the existing methods of protection of buildings and structures against lightning discharges.

IEC standard 62305-4 defines lightning protection zones in terms of direct and indirect lightning impact [7]. Protective equipment such as an external lightning protection system (LPS), shielding, equipotential connections of conductive parts and surge protection devices determine the lightning protection zones (Fig. 1). As the number of the protection zone increases, the effect of the electromagnetic field and the lightning current decreases.

Zone 0_A is the zone of the external environment of the facility, all points of which can be subjected to direct lightning strike and the influence of the resulting electromagnetic field.

Zone 0_B—zone of the external environment of the facility, the points of which do not undergo a direct lightning strike, as they are in a space protected by an external lightning protection system. However, a complete electromagnetic field operates in this zone.

Zone 1 is the inner zone of the facility, the points of which are not subjected to direct lightning strike. In this zone, the currents in all conductive parts are significantly smaller than in the zones 0A and 0B. The electromagnetic field is also reduced compared to the 0A and 0B zones due to the shielding properties of the building structures.

Other zones (2, etc.)—are installed to further reduce the current and/or weaken the electromagnetic field; the requirements for the zone parameters are determined according to the requirements for protection of different zones of the facility (increasing the number of the protective zone reduces the influence of the electromagnetic field and lightning current). Within the sections of individual zones, it is necessary to provide a protective consistent connection of all metal parts, with their periodic observation.

Methods of forming links at the interface between zone 0A, zone 0B and zone 1 are given in EN 62305-1 [5]. The energy distribution of electromagnetic fields inside an facility is affected by various elements of building structures: openings or cracks (e.g. windows, doors), sheet steel paneling (gutters, curtain rods), as well as the I/O cables for power, telecommunication, and other communications. External metallic communications shall enter the protection zone 1 at one point and connect their shielding or metal parts to the main earth bus at the interface between zones 0A–0B and zone 1.

Separation of a facility into conditional zones allows in practice to effectively organize the protection of networks up to 1000 V, as well as communication lines, data transmission, computer networks and other communications that enter into the facility, through the use of different types of SPD [9–15]. However, for the guaranteed protection of the facility from overvoltages arising from the flow of lightning currents to the grounding device or when the “surge” of the current wave from the supply network (in case of a remote lightning strike), the zone protection concept provides a three-stage scheme for the inclusion of protective devices, which is determined by the national standards of Ukraine today [16].

Power supply and computer networks have protective measures and devices in place [3]. Information about overvoltage induction parameters needs measurement. The installation of an internal lightning protection system is justified by taking into account the cost of electrical equipment operation failure, data loss, and consequences.

2 Protection Schemes and Lightning Current Parameters

2.1 Analysis of Current Spreading Calculation Method

For all levels of lightning protection (LPL) maximum and minimum (Table 1 ÷ 4 [5]) values of lightning current parameters and the probability (decreasing with increasing LPL) that these interrelated parameters will not exceed the natural values of lightning current parameters (Table 5, A.1–A.3 [5]). The fixed parameters of the lightning current are described by the values of the duration of the front T1 = 10 mks and the half-life T2 = 350 mks of the first pulse of the lightning current. The maximum values of the lightning current parameters are used to calculate the cross section of conductors; the thickness of the metal roof and the housing of the tanks that may have contact with lightning; the nominal discharge current of the SPD; the distance of the section to prevent hazardous sparking; determining the test parameters of the lightning protection system or its individual components.

The calculated values of lightning amplitudes and switching voltage impulses at the points of general purpose electric network connection for phase nominal voltages according to international standards [6] are given in Table 1.

When lightning enters the system of external lightning protection the distribution of currents between the metal elements of the building structure can be determined by calculating the resistance of grounding devices, pipelines, power supply, communication cables, etc. In cases where it is difficult to make a precise calculation, a so-called qualified assessment is carried out on the basis of the following considerations:

Table 1 The value of impulse amplitude voltage (kv) electric grid points

Location of connection points	Nominal voltage of the network, kV								
	0.38	6	10	35	110	220	330	500	750
Air line	10	100	125	325	800	1580	1890	2730	3570
	6/10	160 2000	190 2000	575 2000	1200 2000	2400 –	3000 –	3200 –	4800 –
Cable line	6	100	125	325	800	1580	–	–	–
	6	34	48	140	350	660	–	–	–
Power transformer	–	60	80	200	480	750	1050	1550	1950
	–	34	48	140	350	660	–	–	–

- 50% of total current Iimp = 200 kA (10/350)—IS1 = 100 kA (10/350) is discharged through the grounding device of the LPS;
- 50% of the total current Iimp = 200 kA (10/350)—IS2 = 100 kA (10/350) is evenly divided (about 17%) between the external inputs into the facility of the three main types of communications: telecommunication cables, metal pipelines and cables/wires for power supply 220/380 V.

It is a classical example of the current distribution of direct lightning strike at a facility (Fig. 2).

The value of the current passing through the separate input is denoted by I_i, while n is the number of inputs:

$$I_i = I_{s2}/n. \tag{1}$$

For evaluating of the current I_V in separate cores of the unshielded cable, the current in the cable is divided by the number of wires m: $I_V = I_i/m$.

According to the requirements of IEC standards, surge protection devices, depending on the installation location and the ability to transmit different impulse currents, are divided into the following classes: I, II, III (or A and B, C, D according to German standard DIN VDE 0675-6) [7]. The main requirements for SPD of different classes are shown in Table 2.

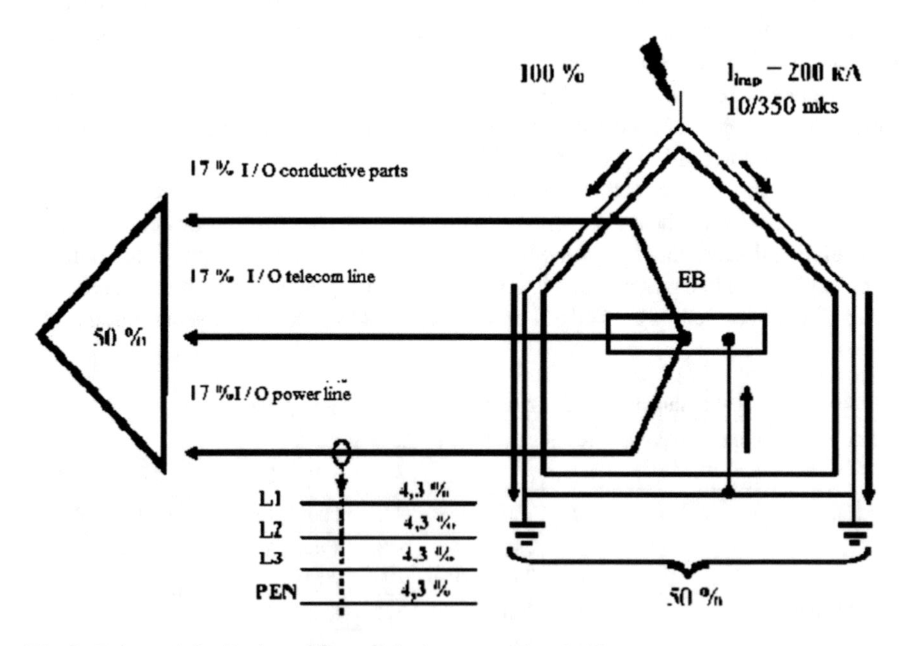

Fig. 2 Estimated distribution of direct lightning current in a facility

Table 2 Requirements for SPD's of different classes

Class SPD	Purpose of the device
I (A i B)	Designed to protect against direct lightning strikes in the building's LPS system (facility) or overhead power line (transmission line). They are installed at the entrance to the building in the main switchboard (EB) (category B) or on the power line (category A). Normalized by the impulse current with a front/half-pulse duration of 10/350 μs (Cat. B) and 8/20 μs (Cat. A)
II (C)	Designed to protect the facility's distribution network from interference or as a second level of lightning protection. Installed in switchboards. Normalized by a impulse current of 8/20 μs front/half-pulse duration
III (D)	Designed to protect consumers from residual voltage surges, protection against differential (asymmetric) overvoltage (e.g. between phase and zero operating conductor in the TN-S system), high-frequency interference filtering. Installed directly next to the consumer. They can have a wide variety of designs (in the form of sockets, mains plugs, and separate modules for mounting on a DIN rail). Normalized by impulse current with a front/half-pulse duration of 8/20 μs

The necessity of selecting the type of protective device and the scheme of their installation is based on the risk assessment of a direct lightning strike or remote voltage surges.

The need for lightning surge protection depends on:

- the intensity of lightning flashes at the location N (average annual number of lightning strikes per km^2 per year. In Europe, these statistics can be easily obtained by an automated lightning strike detection system. These systems consist of a large number of single control network sensors located all over Europe [17]. Information from sensors is sent to monitoring servers in real time and accessible via the Internet special password and get it using thunderstorm activity maps by region, but the resulting parameter will be very approximate.);
- risk assessments of the vulnerability of the electrical installation itself, for example, underground power supply systems are for obvious reasons considered less vulnerable than aerial ones;
- the cost of connected to the protected electrical installation equipment, as this may be an important criterion for complicating the protection scheme and vice versa.

When choosing protective devices on arresters or oxide-zinc varistors, it is necessary to pay attention to their parameters:

1. Rated operating voltage (Un). This is the rated voltage of the network for which the protective device is intended.
2. Maximum allowed operating voltage of the protective device (maximum operating voltage) (Uc). This is the highest value of AC voltage that can be applied for the entire life of the protective device.
3. Classification voltage (parameter for varistor surge suppressors). This is the value of the industrial frequency voltage that is applied to the varistor limiter to

obtain a valid classification current (the usual value of the classification current is assumed to be 1.0 mA).

4. Impulse current (Iimp). This current is determined by a peak of the test impulse of 10/350 microseconds and a charge Q. Applied for testing of Class I protective devices (category B).
5. Rated impulse discharge current. (In) This is the peak value of the test current pulse of 8/20 μs current passing through the protective device. The current of this value can be maintained by the safety device many times. Used for the testing of Class II SPD (Category C and D). The action of this impulse determines the protection level of the device. This parameter also coordinates the other characteristics of the SPD, as well as the norms and methods of its testing.
6. Maximum impulse discharge current (Imax). This is the peak value of the 8/20 μs test current pulse that the protective device may miss once and fail. Used for the testing of Class II SPD (Category C and D).
7. Accompanying current (If) (parameter for SPD based on arresters). This is the current that flows through the arrester after the end of the surge pulse and is supported by the current source itself, that is, the power system. In fact, the value of this current goes to the calculated short-circuit current (at the location of the arrester for this particular electrical installation). Therefore, to install in the circle "L-N; L-PE" gas-filled (and other) arresters with an If value of 300–400 A. Do not use as a result of prolonged operation of the accompanying current, they will be damaged and may cause a fire. For installation in this circuit it is necessary to use arresters with value If, which exceeds the calculated short-circuit current, that is, values from 2 to 3 kA and above.
8. Level of protection (Up). This is the maximum value of the voltage drop on the protective device when an impulsed discharge current flows through it. The parameter characterizes the device's ability to limit the surges that appear on its terminals. It is usually determined by the rated impulse discharge current (In).
9. Trigger time. For zinc oxide varistors, its value does not exceed 25 ns. For dischargers of various designs, the operating time can be in the range of 100 ns to several microseconds.

For contact clamps and SPD current parameters are evaluated in each case individually. The maximum pulse overvoltage at the boundary of each zone is coordinated with the permissible voltage of the internal system. The SPD at different zones is also coordinated by energy performance. Class I SPD based on the arrester have Up = 4 kV, on the basis of varistor even lower; Class II SPD have Up = 1.3–2.5 kV; Class III have Up = 0.8–1.5 kV.

It is necessary to choose the level of the SPD at a voltage lower than the maximum value proceeding from the requirements of insulation impulse resistance coordination in power plants and the stability of the equipment to damage. As a result, the effect on the equipment will always be below the permissible voltage. If the level of damage resistance is unknown, an indicative or test level should be used. Use a measuring instrument for determining and controlling the level of impulse overvoltage in this case [9]. It was developed proposed means for this purpose.

There are a number of other parameters (Fig. 3), which are also taken into account when choosing a SPD: leakage current (for varistors), maximum energy released on the varistor, fuse current (for protective devices with built-in fuses). To find the value of the current passing through the limiter of the first level of protection in the case of direct lightning strike into a building protected by an external BZS, it is recommended to proceed from the configuration of the grounding system and the equipotential bonding of the building, as well as the communications supplied to it (pipelines, power cables, communication cables). information and communication, etc.).

The most complex scheme of the protection system must be performed for facilities located on the open ground and which have highly placed structural elements. Such facilities include cottages in rural areas, high-pipe industrial facilities, antenna-mast communication facilities, etc., in which lightning is likely to strike, as well as facilitys having aerial power inputs [16, 18].

In the case when it is necessary to protect, for example, a building located in an urban settlement, the decision is somewhat simplified. In urban environments,

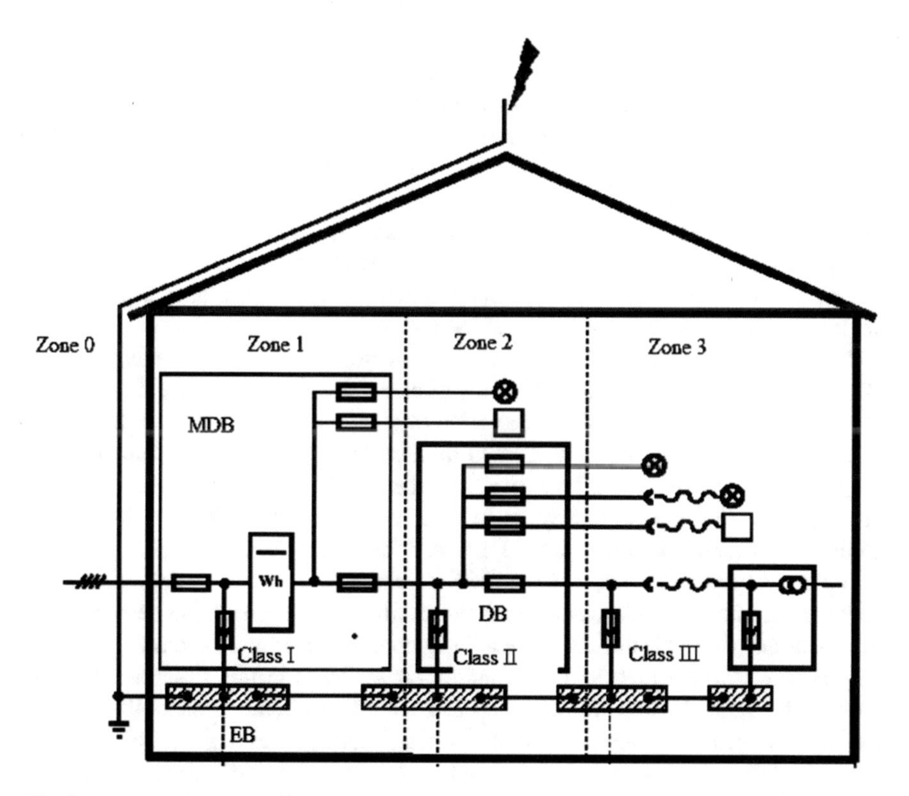

Fig. 3 Relationship between lightning protection zones, classes of protective devices and categories of equipment insulation resistance to impulse overvoltages (EB—main earth bus, MDB—main distribution board, DB—distribution board)

lightning strikes are most likely in industrial pipes, power lines, television towers, or some of the tallest buildings (especially if they have antenna-mast structures of cellular base stations).

2.2 *Balancing Characteristics at Hazardous Working Scenario for SPD*

The basic principle of the above viewed protection measures is the equalization of potentials between two conductors, one of which is usually a phase conductor and the other a zero working or zero protective conductor. However, in the event of failure of the limiter, a short circuit may occur between these conductors, which can cause failure of the electrical installation and even a fire.

The varistor overheating device (thermal protection), which is available in varistor limiters, usually works when the varistor ages when leakage currents increase, or when the actual discharge current through the limiter exceeds the maximum [7]. Given the short duration of the latter process, the varistor may not even malfunction, but it will still be disconnected from the protected circuit due to the release of a large amount of thermal energy. A somewhat different situation arises in case of constant overvoltage in the network above the maximum allowable operating voltage for the SPD. The consequence of this situation may be that the zero working conductor is burned off at entering the electrical installation. As is known, in this case, the load may be applied phase voltage of 380 V. The varistor opens and through it a long time flowing current. The value of this current is close to the short circuit current and can reach several hundred amperes. It is known from the practice that the heat protection device does not always work in such situations.

It should also be noted that arresters based on arresters do not have a thermal shut-off device in their composition. As a result of the described action, the protective device, as a rule, is destroyed by the action of a large amount of thermal energy. It is even possible for an arc to occur and the terminals of the device to the cabinet or DIN rail to melt when the body plastic is melted. Therefore, to protect electrical installations and limiters of all types from short circuits, it is necessary to provide additional protection in the form of high-speed fuses with the characteristic of gG or gL (classification according to VDE 0636 (Germany), which are installed in series with each limiter [15]. Designed to protect conductors and switching devices from overloads and short circuits and have a fairly complex internal structure.

It is worth noting that the use of automatic switches in this situation may not provide the desired result. Experience has shown that circuit breakers themselves can be damaged by a lightning current surge. This may result in welding of the contacts of the coupler to each other and it is probable that the machine will not malfunction at short circuit in load. The fuse completely eliminates this situation. In addition, the correct choice of the nominal virtually eliminates the likelihood of the fuse when passing through the protective device of the impulse current when struck by lightning.

It must also be understood that if the fuses are not installed, in the event of a short circuit in at least one of the SPD, the input automatic switch will be triggered and the power supply of the consumer will be interrupted until the fault is resolved. The use of fuses in the circuit of each SPD greatly reduces the likelihood of such a situation. When choosing their denominations, the manufacturer's recommendations for surge protection devices should be followed. The values of common and individual fuses are determined by the selectivity of their operation, as well as by the ability of the protective devices to withstand the calculated short-circuit currents for a particular electrical installation.

2.3 *Summarized Circumstances of Complex Measurement Tools Implementation*

Choosing the right type of protective devices and the scheme of their installation will supplement the measures of protection against direct lightning strike or voltage guidance by remote discharge, provided for [7].

As the first level of protection, it is recommended to install:

- when air power supply is introduced regardless of the presence of an external LPS, when lightning can directly penetrate the power lines directly near the facility—lightning arresters, capable of passing through themselves 10/350 μs impulse currents with an amplitude value of 50–100 kA and providing a level of protection (Up) less than 6 kV (SPD A);
- with underground power supply and in the presence of an external LPS, when there is a probability of discharge in the lightning arrester, it is possible to install varistor surge arresters capable of passing through themselves impulse currents of the form 10/350 μs with an amplitude value of 10–25 kA and also guarantee a level of protection 4 kV (SPD B);
- in the absence of an external LPS, it is recommended to install SPD A, since a direct lightning strike in this case usually results in dynamic action on the building structures, and can cause a fire by sparking and overlapping air gaps between the conductive elements of the facility.

Modules with a maximum impulse current of 20–40 kA of 8/20 μs form and a protection level (Up) of less than 2.5 kV (SPD C) are used as the second protection level.

Modules with a maximum impulse current of 6–10 kA of 8/20 μs form and a protection level of less than 1.5 kV are used as the third level of protection (SPD D). Combined devices may be used, including an additional noise-bandwidth filter in the range of 0.15–30 MHz.

Class I protective devices shall be installed on entry into the building (in an inboard panel or special box) after the introductory panel (at the boundary of zone 0 and zone 1). Class II protective devices are in the following switchgear (e.g., floor or other panels). It is advisable to place them in group. The location of this class of devices may be between zone 1 and zone 2. Class III protection may also be installed in distribution boards or directly adjacent to the consumer (protection zone 3). At a distance of more than 10–15 m from the point of installation of the SPD to the consumer, it is desirable to install an additional Class III device directly next to the protected equipment, in order to guarantee possible removal of the specified cable lengths.

3 Measurement and Yield of Quality Norms of Electrical Energy in General Purpose Grids

3.1 Working Parameters of Supply Voltage in Public Grids

Switching processes or emergency situations in the power supply system, as well as certain unpredictable processes on the side of electricity consumers will result in the appearance of individual impulses or a series of them. They can be caused by radiation of conductive electromagnetic interference from individual facilitys or from atmospheric processes, like lightning discharges.

According to the regulatory requirements and the works of leading specialists [3, 8–17], impulse distortions of the network voltage are characterized by the following basic parameters of the electric power: the amplitude and duration of a single impulse or a series of single-impulses or bipolar impulses of a diverse form.

Any voltage impulse can be represented in the form

$$u(t) = U_{mi} \cdot \varepsilon(t), \tag{2}$$

where U_{mi}—amplitude and $\varepsilon(t)$—is a normalized function that describes the shape of the investigated impulse. At the same time, Table 1 summarizes the most common types of fronts (recessions) of impulse signals.

Since the expression of the voltage signal distorted by impulse $f_U(t)$ can be regarded as the sum of the industrial frequency sinusoid functions and (2), then when realizing the means of measuring these parameters of electricity, a characteristic feature of the appearance and flow of such distortions is used. It consists in increasing the velocity of the temporal change of the distorted signal $s_U(t)$ in comparison with the sinusoid [16].

3.2 Design of Measuring Instrument

To construct measuring instruments with high metrological characteristics for dynamic parameters of impulse disturbances in power grids it is expedient to use the principle of optimal distribution of measurement and control functions between elements of the measuring structure.

The structure of the developed instrument for controlling impulse disturbances dynamic parameters in power grids, which contains two parts: analog AU and calculating CU units [18–20] in Fig. 4 shows.

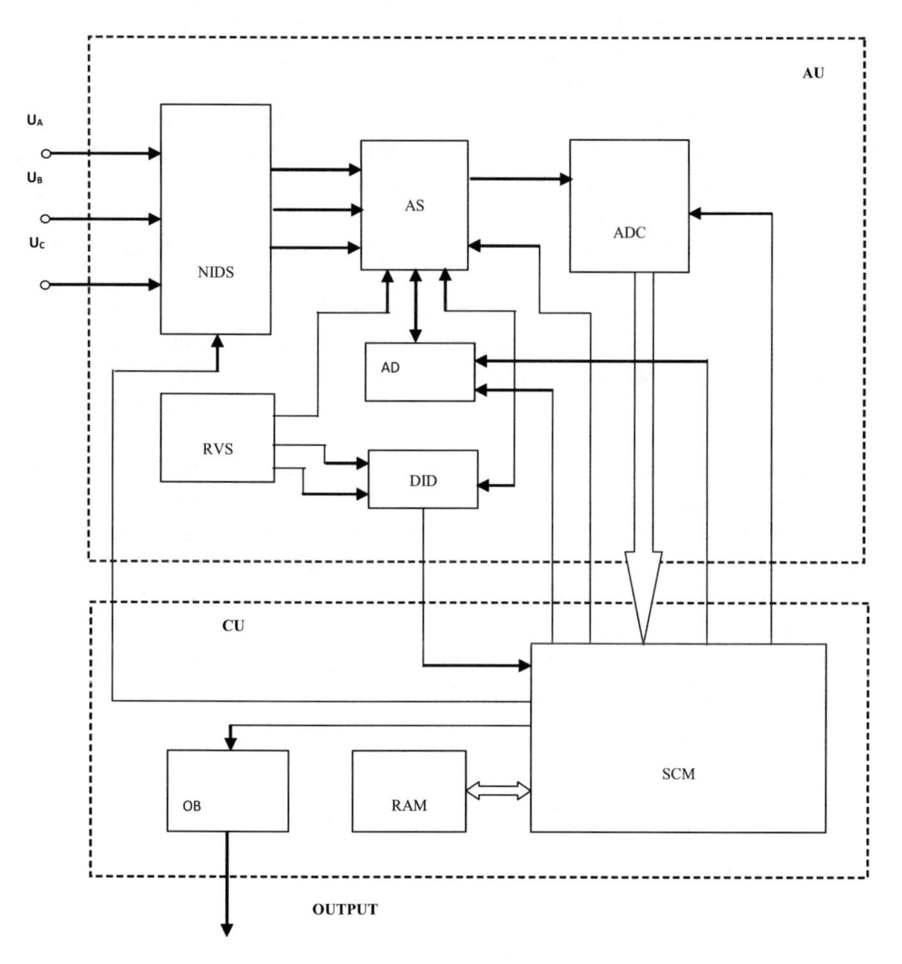

Fig. 4 Structure means of impulse disturbances dynamic parameters measurement in electrical networks

AU consists of: NIDS—normalizing input device with software control, AS—analog switch, AD—amplitude detector, DID—block of detecting impulse distortions $f_U(t)$, RVS—reference voltage sources and ADC—analog-digital conversion unit.

CU consists of: SCM—single-chip microcontroller, RAM—operative storage device and OB—output buffer.

In general, the controlled signals of three phase voltages U_A, U_B, U_C after scaling in the NIDS are alternately in time transmitted through the AS to the DID input, the main element of which is the differentiator.

At the moment of the voltage impulse distortion DID stimulates the start of the monitoring program in the SKM and measures this phenomenon of electric energy quality deterioration—the emergence of impulses imposed on the low-frequency voltage signal of the industrial network.

In this case, the location of the momentum or series of pulses in one or another half of the repetition period of the sinusoidal voltage is fixed, as well as the cyclic work of the ADC is initiated and the bipolar AD is activated.

The ADC is a circuit with an independently controlled range of input signals $f_U(t)$, which outputs a code signal of optimal bit, regardless of the presence or absence of impulse distortion.

ADCs receives arrays of instantaneous values of phase voltages that are written to the RAM using the SKM. At the same time, the SKM program searches for a time interval t_{ia} that contains the values U_{mi} of a single or each of the a-th detected impulses. In addition, measurements of impulse duration t_i or series $t_{i\Sigma}$ are performed.

A specific feature of the developed measurement device is to obtain, as a result of finding and controlling the impulse voltage of two arrays of codes—the digital instantaneous values $\{u_i(k)\}$ of the impulse part of the investigated function $f_U(t)$ and their corresponding time values $\{t_{iU}(k)\}$. Thanks to the analysis of these arrays, the start and end of the detected $\{l_i\}$ impulses that constitute distortion $u_i(t)$ are recorded.

For each l_i impulse, the start $(t_{ni})_{li}$ and end $\left(t_{зi}\right)_{li}$ times are fixed in the SCM, which are used to calculate the time parameters:

- duration of any impulse of series

$$(t_i)_{li} = (t_{\text{зi}})_{li} - (t_{\text{пi}})_{li}, \tag{3}$$

- the duration of the actual series

$$t_{i\Sigma} = \sum_{li=1}^{N_{\text{iм}}} (t_i)_{li}. \tag{4}$$

The proposed means for measuring control detects and fixes impulse deviations from sinusoidal form of network voltage.

After this, the following signal processing is possible: data accumulation, overvoltage settings fixation, the formation and transmission of a control signal to the protective devices and switching, etc.

The developed mathematical model of thermal balance [21] shows the possibility of applying the obtained parameters. This enables the calculation of the heating processes of internal electrical wiring copper conductors of different cross-sections, which are laid in different ways in rooms that are at a set temperature. We can determine when the degree that can cause the ignition of insulating materials and building structures that are in contact with the conductor is reached—the heating time. This allows to detect the minimal heating time values, as well as verify the parameters of the protection appliances in order to prevent overheating of the conductors by impulse surges.

4 Experimental Determination of the Characteristics of Defective Contact Compounds in Low-Voltage Line

Also characteristic of sections of low-voltage line (SLVL), there are signs of overvoltages such as arcing or heat from increasing the transient resistance of contact connections [22–24].

In recent decades, patent solutions have emerged in the area of SLVL disconnection devices with the appearance of defective contact connections [22] (hereinafter referred to as spark protection device), which allow to increase SLVL safety. Existing works describe in detail the principle of circuit design of spark protection devices based on the tracking of the appearance of high frequency harmonic components (output signal) in SLVL caused by the appearance of defective contact connections. At the same time, there is almost no research in the field of analysis of the output signals and determination of the sensitivity limits of spark protection devices.

In order to obtain information about the parameters that accompany the appearance of defective contact connections, an analysis of the output signals in SLVL to be protected was performed.

To increase the effectiveness of spark protection devices, the limits of the level of sensitivity required to activate them are defined.

To determine the characteristic features of defective contact compounds in SLVL an experimental stand was developed (Fig. 5) and such an experiment was conducted.

A digital oscilloscope "RECON-08MS" was used to obtain current oscillograms, with the shunt SU 10/50 with the matching device. Specialized software WinRec-MC version 2.8.3.0 was used for signal processing and MathCAD software for waveform construction.

In the absence of simulation of the sparks in the contact connection in the circuit (Fig. 5), the oscilloscope will be approximated to the standard sinusoid wave shape. The distortion of the sine waveform, that is, the quality of electricity as a whole [3], is caused, first of all, by the widespread use in modern power supply systems of

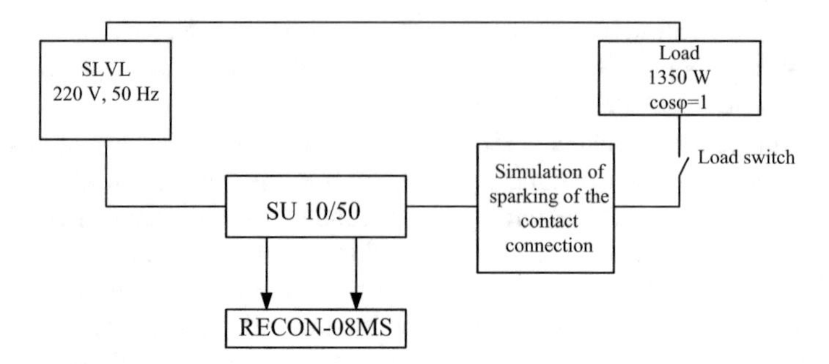

Fig. 5 Scheme of the experimental board for the analysis of defective contacts in SLVL

electric receivers, which clearly and significantly distort the form of mains voltage, such as electric welders, controlled thyristor electric drive, computer and other electronic machinery. A large part of these consumers generate higher harmonics into the network, significantly reducing the quality of SLVL electricity.

Spectral analysis of such a signal shows the presence of higher harmonics, multiples of 50–100, 150 and 300 Hz, in a working current with a frequency of 50 Hz due to load [25]. Spectral analysis was performed using the MCA3000 frequency analyzer. The waveform shown in Fig. 6, taken in the simulation of sparks in the contact connection in the scheme of Fig. 5. The waveform shows the periods of absence of the contact connection, as well as the periods of its recovery, accompanied by a

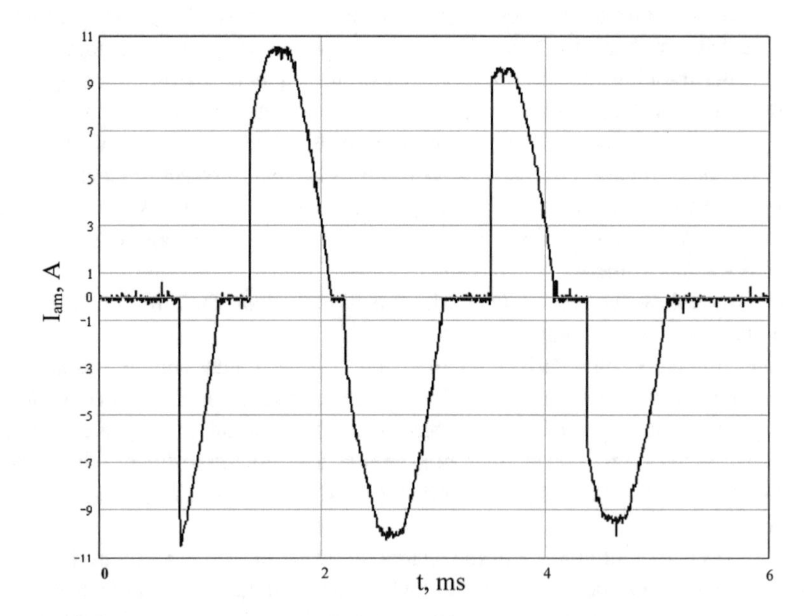

Fig. 6 Waveform chart of the current I_{am} single-phase SLVL voltage of 220 V, frequency 50 Hz with simulation of spark in the contact connection

sharp jump of current, in some places it exceeds the instantaneous values of the load current for a circle without disconnections—the process of arcing in the detachable connections of contacts, as well as the effect of spurious capacitive resistances of SLVL [26].

Spectral analysis of the waveform (Fig. 6) showed that the instability of the contact connection during the increase of transient resistances or the occurrence of arcing in such places leads to the appearance in the current due to loading of high frequency harmonic components with frequency from 500 Hz and above (Fig. 7).

The sensitivity limit of the spark protection device is determined by the required minimum range of change in the contact resistance of the contact connection (or the value of its derivative over time—the rate of change) at which the required current is triggered by the spark protection device.

The operating threshold is determined by the choice of elements of the schematic diagrams of spark protection device realization [22] and is ≈1.2 V.

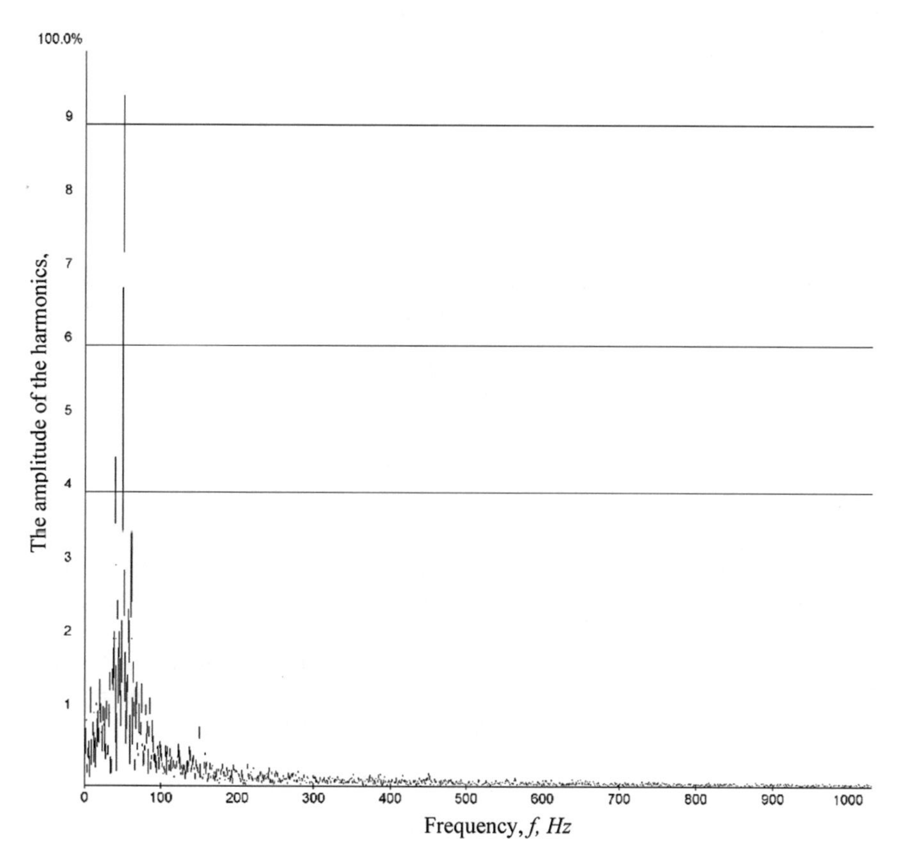

Fig. 7 Spectral analysis of sinusoid single-phase SLVL voltage of 220 V, frequency 50 Hz with simulation of spark in the contact connection

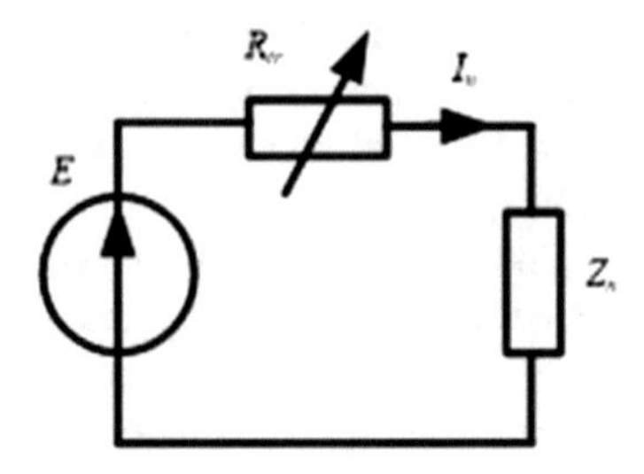

Fig. 8 Scheme of substitution of SLVL to be protected

The equivalent scheme of substitution of SLVL to be protected is represented in the following way (Fig. 8), where E = 220 V is the voltage of SLVL; Z_n—the set of all full load connected resistances (assume that cos φ = 1, and the values of the resistances are due to the nominal series of currents of possible loads in the range from 6 to 32 A); R_{tr} is the set of all transient resistances of the SLVL contact joints to be protected (R_{tr} is a variable value in the process of sparking in contact).

Determine the magnitude of the load current (I_n, A) for the replacement circuit (Fig. 8):

$$I_n = \frac{\mathrm{E}}{Z_n + R_{\mathrm{tr1}}} \tag{5}$$

where *E* is the electromotive force in the circle, Volt; R_{tr1}—the initial value of the transient resistance of the SLVL contact connection to be protected, to simplify the calculations, take $R_{tr1} = 0\ \Omega$; Z_n—full load impedance, Ohm.

$$I_n = \frac{E}{Z_n}. \tag{6}$$

$$I_n \pm I_{trspd} = \frac{E}{Z_n \mp R_{\mathrm{tr2}}}, \tag{7}$$

where I_{trspd} is the value of the tripping current of the spark protection device, A, (depends on the value of the time derivative R_{tr} (speed) and the range of change ΔR_{tr} from the value R_{tr1} to the value R_{tr2}).

Find ΔR_{tr} from Eqs. (6) and (7). For the sake of certainty, suppose that the value R_{tr1} increases to the value R_{tr2}. Therefore, ΔR_{tr} will have a plus sign and, as a consequence, the value of I_{trspd} will enter the minus sign in equation:

$$\Delta R_{\mathrm{tr}} = \frac{I_{trspd} \cdot Z_n^2}{E - I_{trspd} \cdot Z_n}. \tag{8}$$

After analyzing experimentally established dependence of the current of the actuation of the spark protection device [3] as a function of the frequency spectrum of the spark in contact (Fig. 8), seen that as the frequency of the spark spectrum increases, the required threshold value for the triggering of the I_{trspd} is reduced to tens of milliamperes with decreasing frequency. The I_{trspd} value rises to tens of amperes. Analyzing the Eq. (8) and the dependence (Fig. 9) get that the value ΔR_{tr} correlates with the frequency (f_{SP}) of the actuation current of the spark protection device and with the value Z_n. Using Eq. (8), we construct the dependences $\Delta R_{tr} = f(I_{trspd})$ shown on the Fig. 10 and $\Delta R_{tr} = f(f_{SP})$ (Fig. 11), for different load currents.

Thus, the output signal of obtained information on the occurrence of defective contact compounds in the SLVL subject to protection, was considered the appearance in the working current of the load of high frequency harmonic components with a frequency of 500 Hz and above, due to the non-stationarity of the characteristics of the contact connection.

Also, the qualitative state of the contact connection can be estimated by the presence in the spectrum (Fig. 9) of both low-frequency harmonic components due to mechanical instability of the contact connection (dynamic influence of current) and high-frequency harmonic components caused by the change of transient resistance or sparking in contact connections [23, 24].

The dependence analysis (Figs. 10 and 11) shows that the sensitivity of the spark protection device is determined by the speed and range of change of the contact resistance of the contact connection.

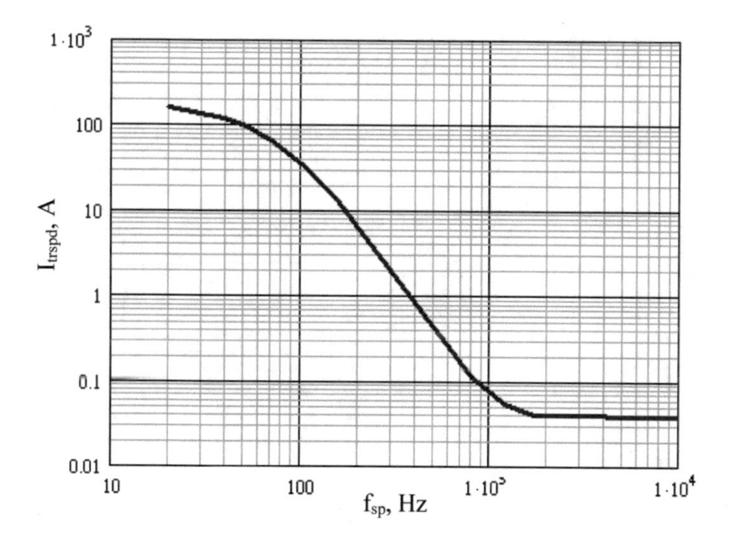

Fig. 9 Dependence of I_{trspd} on the frequency spectrum of the spark in the contact

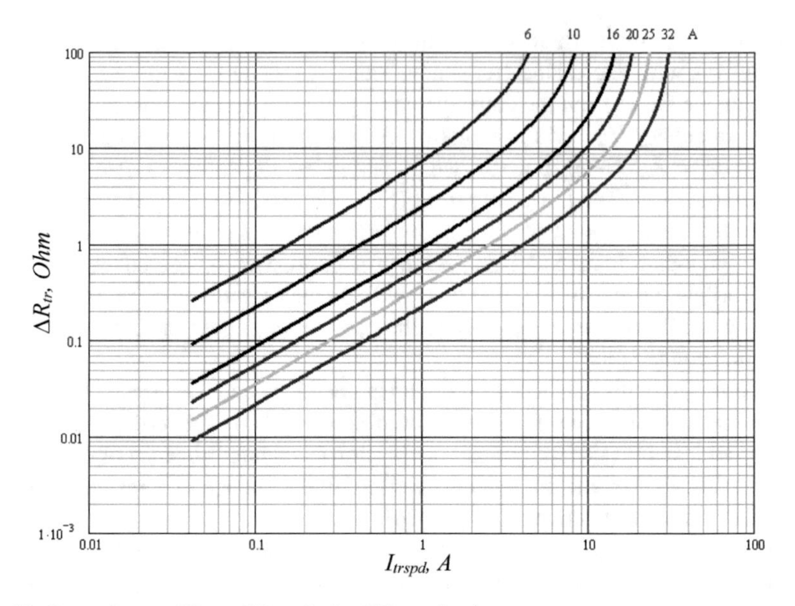

Fig. 10 Dependences $\Delta Rtr = f(Itrspd)$, for different load currents

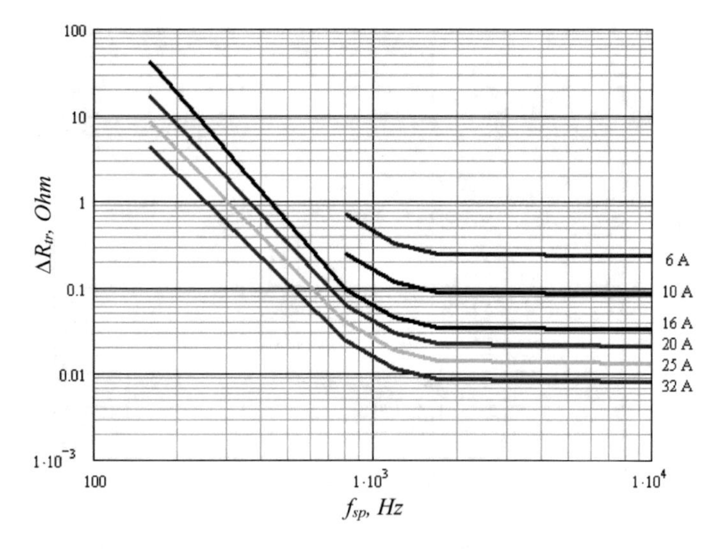

Fig. 11 Dependences $\Delta Rtr = f\ (fSP)$, for different load currents

5 Conclusion

The assessment of risk, the owner's responsibility, the influence of the control bodies affect the arrangement of the lightning protection system. Thus, making a decision on the use of protection measures in the procedures for assessing the risks of lightning is important. But if there is a desire to avoid unacceptable risk the decision to implement lightning protection systems can be made regardless of the results of risk assessment. The value of the measured influence factor—overvoltage in the network, can be used as one of the parameters in the formula for assessing the components of risk.

Choosing the type of protective devices and the circuit of their installation recommended by international standards of the EN [5] will supplement the protection against a direct lightning strike or voltage disturbances by the remote discharge provided for in Table I and [3]. The use of the suggested structure of the instrument for measurement and control of pulsed surges will enable conducting in the required volume of research of impulse distortions processes in power grids, identify and eliminate the electricity quality deterioration causes. The risk of fires and electronic device damage will be minimized as a result. You can expect an increase of reliability of power systems and protection of electronic and electrical equipment of different values.

The formation of arcing or heating of the sections of low-voltage line contact leads to a 50 Hz high frequency (from 500 Hz or higher) harmonic components in the working current.

It is established that in case of mechanical instability of the contact compound in the spectrum, low-frequency harmonic components predominate, and the appearance of transient resistances or spark instability in the contact connection leads to an advantage in the spectrum of high-frequency harmonic components with multiplicities up to 50 Hz. The necessary sensitivity limits of the device for spark protection of the order of three periods of a current of the network, which depends on the frequency and amplitude of change of the transient resistance of the contact connection, as well as the size of the inductive and capacitive components of the nominal load of sections of low-voltage line are revealed.

The conducted research resulted in the following: the duration of the distortion impulse or their series was determined, instantaneous (maximum) amplitude values of the impulse overvoltage were established. This gives us reason to propose implementing our solution in operation and maintenance fields of electronic and telecommunication equipment. The application of the proposed method and means in assessing the risk of economic loss from the damage to such equipment in particular [27]. We didn't carry out any direct introduction into production or operation as of yet. But measurement, parameter analysis, and study of the impact of thunderstorm activity on IT devices, telecom, and power lines were performed. A possibility of other non-security related applications is being considered. One of the implementations of this article is the improvement of risk management, as it is impossible without taking into account the state of the safe operation of technical devices. The same applies to critical infrastructure facilities.

References

1. Ageyev D, Al-Ansari A (2015) LTE RAN and services multi-period planning. In: 2015 second international scientific-practical conference problems of infocommunications science and technology (PIC S&T). IEEE, Kharkov, Ukraine, pp 272–274. https://doi.org/10.1109/infocommst.2015.7357334
2. Kisielewicz T, Battista Lo Piparo G, Mazzetti C (2016) Probability of surge protective device systems to reduce the risk of failure of apparatus due direct flashes to overhead low voltage lines. In: Proceedings of the 33th international conference on lightning protection (ICLP), Estoril, Portugal
3. Rudyk Y, Solonyi S (2010) The analyse of protecting schemes of electrical devices from impulsive overvoltage caused by thunderstorms and commutations. Fire Saf 17:20–25. https://journal.ldubgd.edu.ua/index.php/PB/article/view/959
4. Loshakov V, Moskalets M, Ageyev D, Drif A, Sielivanov K (2021) Adaptive space-time and polarisation-time signal processing in mobile communication systems of next generations. In: Radivilova T, Ageyev D, Kryvinska N (eds) Data-centric business and applications. lecture notes on data engineering and communications technologies, vol 48. Springer, Cham. https://doi.org/10.1007/978-3-030-43070-2_21
5. DSTU EN 62305-1:2012 Blyskavkozakhyst—Part 1: general principles [Effective from 2012-07-01]. K. Derzhstandart of Ukraine, 2012 (National Standard of Ukraine)
6. DSTU IEC 62305-2:2012 Blyskavkozakhyst—Part 2: Risk management [Effective from 2012-07-01]. K. Derzhstandart of Ukraine, 2012 (National Standard of Ukraine)
7. DSTU EN 62305-4:2012 Blyskavkozakhyst—Part 4: Electrical and electronic systems within structures [Effective from 2012-07-01]. K. Derzhstandart of Ukraine, 2012 (National Standard of Ukraine)
8. Maier V, Pavel SG, Beleiu HG, Fărcaş V (2016) Additional indicators for short time overvoltages and voltage impulses. In: 2016 international conference and exposition on electrical and power engineering (EPE), Iasi, pp 827–831. https://doi.org/10.1109/ICEPE.2016.7781452
9. Kryvinska N, Zinterhof P, van Thanh D (2007) New-emerging service-support model for converged multi-service network and its practical validation. In: First internationalconference on complex, intelligent and software intensive systems (CISIS'07). IEEE, pp 100–110. https://doi.org/10.1109/cisis.2007.40
10. Ageyev D et al (2018) Method of self-similar load balancing in network intrusion detection system. In: 2018 28th international conference radioelektronika (RADIOELEKTRONIKA). IEEE, pp 1–4. https://doi.org/10.1109/radioelek.2018.8376406
11. Shishlakov V, Vataeva E, Reshetnikova N, Shishlakov D, Solenaya O (2020) Synthesis of nonlinear impulse systems at "Proceedings of 14th international conference on electromechanics and robotics "Zavalishin's Readings"". Springer Science and Business Media LLC, pp 469–476
12. Lin H, Wang C, Lei W (2016) The design of overvoltage testing system for communication equipment in smart grid. In: 2016 IEEE PES Asia-Pacific power and energy engineering conference (APPEEC), Xi'an, pp 1812–1816. https://doi.org/10.1109/appeec.2016.7779802
13. Lennerhag O, Lundquist J, Engelbrecht C, Karmokar T, Bollen M (2019) An improved statistical method for calculating lightning overvoltages in HVDC over-head line/cable systems. Energies 12(16):3121
14. Kryvinska N, Zinterhof P, van Thanh D (2007) An analytical approach to the efficient real-time events/services handling in converged network environment. In: Enokido T, Barolli L, Takizawa M (eds) Network-based information systems. NBiS 2007. Lecture notes in computer science, vol 4658. Springer, Berlin, Heidelberg
15. Stanković KD, Perazić LS (2019) Determination of gas-filled surge arresters life-time. IEEE Trans Plasma Sci 47(1):935–943. https://doi.org/10.1109/TPS.2018.2874938

16. Kryvinska N, Zinterhof P, van Thanh D (2007) New-emerging service-support model for converged multi-service network and its practical validation. In: First international conference on complex, intelligent and software intensive systems (CISIS'07). IEEE, pp 100–110. https://doi.org/10.1109/cisis.2007.40
17. Rudyk Y, Shunkin V (2019) Compliance with lightning location systems indicators for risk assessment, loss prevention, investigation of causes. Fire Saf 35:54–62. https://doi.org/10.32447/20786662.35.2019.09
18. Barabash O, Shevchenko G, Dakhno N, Neshcheret O, Musienko A (2017) Information technology of targeting: optimization of decision making process in a competitive environment. Int J Intell Syst Appl 9(12):1–9. https://doi.org/10.5815/ijisa.2017.12.01
19. Vanko VM (2005) Sposib vyyavlennya ta vymiryuvannya parametriv yakosti impul'snoyi napruhy elektrychnoyi merezhi. [A method for detecting and measuring the quality parameters of the pulse voltage of the electric network]. Vidbir ta obrobka informatsiyi, (23), pp 69–74 (In Ukrainian)
20. Vanko VM (2005) Patent of Ukraine №72638. Tsyfrovyy vol'tmetr zminnoyi napruhy elektromerezhi [Digital active currence network voltmeter] Bulletin of inventions (3) (In Ukrainian)
21. Hudym V, Yurkiv B, Nazarovets O (2018) Mathematical modeling of heating processes in conductors of internal electrical networks in housing and public buildings. Fire Saf 26:59–64. Retrieved from https://journal.ldubgd.edu.ua/index.php/PB/article/view/273
22. Soloniy SV, Kovalev OP, Belousenko IV, Ershov MS, Demchenko GV, Vasin OO Pat. of the Useful model 59344 Ukraine, IPC (2011.01) N02H 3/16. Device protection against sparking in electrical networks of objects related to human life. Owner Donetsk National Technical University. № u201012991; declared 01.11.10; publ. 10.05.11, Bull. № 9 (In Ukrainian)
23. Solyonyj S, Rudyk Y, Demchenko G, Bennis Usef A (2011) The analysis of signals for prevention of a low-voltage electric network isolation ignition. Fire Saf 19:149–155. Retrieved from https://journal.ldubgd.edu.ua/index.php/PB/article/view/951
24. Wang S et al (2018) Analysis of lightning-induced overvoltage waveform parameters. In: 2018 34th international conference on lightning protection (ICLP), Rzeszow, pp 1–5. https://doi.org/10.1109/iclp.2018.8503345
25. Lappe R, Fisher F (1986) Izmerenia v energeticheskoy electronice [Measurements in power electronics]. Moskow (In Russian)
26. Nosov VV (1987) Promyshlennye pomekhi i obespechenie nadezhnosti funktsionirovaniya sistem upravleniya tekhnologicheskim protsessami [Industrial interference and ensuring the reliability of the operation of process control systems]. Izmereniya, kontrol, avtomatizatsiya, (2), pp 61–71 (In Russian)
27. Barabash O, Lukova-Chuiko N, Sobchuk V, Musienko A (2018). Application of petri networks for support of functional stability of information systems. In: Proceedings of 2018 IEEE 1st international conference on system analysis and intelligent computing, SAIC 2018. IEEE, Kyiv, Ukraine, pp 1–4. https://doi.org/10.1109/SAIC.2018.8516747

Development of Evaluation Templates for the Protection System of Wireless Sensor Network

Olexander Belej, Tamara Lohutova, and Liubov Halkiv

Abstract The main goal of the study is to formalize ontologies based on templates using software tools. With its help, the protection system in sensor wireless networks is analyzed and the viability of the proposed approach is noted. The implementation of the model of multi-agent systems is described and the simulation results for the protection system are analyzed. Also, we describe the implementation of the proposed estimation algorithms for the coefficient of deviation of the request to the database and the implementation results for sensor wireless networks. Particular attention is paid to defense agents, who calculate the coefficient of deviation in the work of the ward component and identify a possible attack. Adaptive algorithms for the operation of protection agents are proposed for estimating the coefficients of deviations of requests to the database based on statistics and using a neural network.

Keywords Protection system · Database · Algorithm · Wireless sensor network · Defense agent · Neural network · Ontology

1 Introduction

Modern society cannot imagine its existence without a multitude of information systems, including complex automation systems for controlling robotic systems. And, of course, without a wireless sensor network (WSN), which provides the collection and processing of large volumes of valuable information.

O. Belej (✉)
Lviv Polytechnic National University, 5 Mytropolyt Andrei Str., Building 4, Room 324, Lviv 79000, Ukraine
e-mail: Oleksandr.I.Belei@lpnu.ua

T. Lohutova
State Higher Education Institution "Pryazovskyi State Technical University", Str. Universytets'ka 7, Mariupol 87555, Ukraine
e-mail: Logutova_t_g@pstu.edu

L. Halkiv
Lviv Polytechnic National University, 12, Stepan Bandera str, 79013, Lviv 79000, Ukraine
e-mail: Lubov.I.Halkiv@lpnu.ua

D. Ageyev et al. (eds.), *Data-Centric Business and Applications*, Lecture Notes on Data Engineering and Communications Technologies 69,
https://doi.org/10.1007/978-3-030-71892-3_10

WSN quality directly affects the performance of enterprises and organizations, and one of the WSN quality indicators is their security. Providing protection is complicated by the fact that it is impossible to formally describe and predict the actions of an attacker. The constant development and increasing complexity of WSN is another factor affecting their security. The problem of creating templates for assessing the current security of WSN, taking into account not only known but also new vulnerabilities and threats, becomes very important. Such security patterns could be applied both in the development of new systems and for assessing the security of already functioning systems. The problem of obtaining such standards is poorly formalized and its solution requires a detailed analysis of many parameters of the systems.

Today, one of the most promising areas for research in the field of artificial intelligence (AI). At the moment, research is actively being conducted in the field of AI, the results of which are being successfully implemented in various spheres of human activity. Therefore, a promising area of research seems to be the use of methods used in AI systems to solve security control problems in WSN.

Based on the designed wireless sensor nodes, the wireless sensor network is deployed to gather, process, and transmit strain gauge signals and monitor results under different static test loads [1]. When an emergency is detected, the network coordinator alarms an operator through the GSM/GPRS or Ethernet network, and may autonomously control the source of gas emission through the wireless actuator [2].

Authors [3] also show that measurement accuracy obtained from wired systems cannot be obtained with the present wireless technology, and we do not recommend their use at present for fire monitoring and mitigation. The objective of the advanced volume sensor task was to develop an affordable detection system that could identify shipboard damage control conditions and provide real-time threat [4].

The performance evaluation of this real tests-bed scenario demonstrates the feasibility of the platform designed and confirms the simulation and analytical results [5]. Authors [6] first derive sufficient access-point density conditions that ensure that the data loss rates are statistically guaranteed to be below a given threshold.

A new technique for finding nodes' importance in the network is designed for better manipulation of nodes [7]. Authors [8] derive theoretic bounds on the expected hitting time between two consecutive visits of a mobile node to access points for both the square and hexagonal access point deployment structures.

The system is based on a server that can request and collect data from several nodes and store them in a database (DB) [9]. The paper [10] identifies several possible causes of latency in IWSNs and can be used as a basis for deploying Internet Protocol (IP) based IWSNs requiring IoT connectivity.

The paper [11] presents the current state of the art research in sensor-based M2M communication networks for agriculture monitoring, several existing agricultural monitoring systems and compare them on different design factors, the technical framework of some recent deployment of agriculture monitoring systems in developing countries, and identify their design challenges and major design and implementation differences of these monitoring systems in developed and developing countries. The NS2 simulation tool was conducted to validate the performance of the detection mechanism, the experimental results show that the IDM can improve the detecting rate and lower the false detecting rate in appropriately selected detection mechanism parameters, and every node-link can obtain a relatively fair throughput to improve the performance of overall networks when following the mechanism of proposed [12].

In earlier research [13] we considered an algorithm of constructing an electronic-digital message signature based on encryption using elliptic curves. This article discusses all the key tools provided by the protocol of transport layer security to protect information [14].

The paper [15] presents an analysis of the main mechanisms of decryption of SSL/TLS traffic. In paper [16] the problem of intrusion detection by classification of the network traffic realizations was considered.

Other authors evaluate and compute expected waiting time and time in the system and present numerical results of our calculations and provide corresponding curves for them [17]. A modeling method developed in paper [18] will be used for the fast configuration and testing of new converged network applications and services.

In the paper [19] Application of the variational-gradient method to the control tasks will allow expanding the range of tasks under consideration. The article [20] concludes that it is advisable to use such a model when building scanners to search for illegal bugs and transmitters.

There are various languages for the formal definition of ontologies. Their main problem is to provide an opportunity to describe both the data itself and the metadata. Based on metadata when interpreting data, machines can more fully take into account the state of things in the real world. This simplifies machine data processing and machine learning. Recently, languages based on Web standards have become widespread. They are used for exchanging data over a network and most of them are based on the XML format. But unlike XML, semantics are present in languages for describing ontologies.

The main problem in developing a view for analyzing and obtaining a static template of wireless sensor networks (WSN) is the lack of the required number of circuits. To facilitate their creation and distribution to describe the corresponding ontology, we will use one of the languages based on Web standards. These languages have powerful expressiveness and are familiar to most IT professionals. Also, these languages are designed to describe resources, and WSN in our view is a set of resources—components.

2 Development of Ontology for Assessing the Security of Wireless Sensor Networks

The description of the ontology using OWL is based on the statement of many facts-triplets: (“object”, “property”, “subject”). Such facts indicate that the specified object has the selected property, and the value of the property is the subject. Various notations are used to describe triplets.

For the subsequent addition of individuals belonging to classes, properties are needed. In the OWL language, there are two types of properties: characteristic properties (DatatypeProperty) and communication properties (ObjectProperty). The former associate classes with data of certain types, and the latter associate classes with each other. Properties may be subject to restrictions in the form of a domain and ranges. Domains define classes whose objects may possess the specified properties. And ranges specify classes whose objects can be values of the specified properties. Add the following properties-characteristics: "Data Characteristic" (DataCharacteristic) with the "Data" domain and a range of positive integers, "Impact" with the "Vulnerability" domain and a range of positive integers. Clarify the "Data Characteristic" property with three sub-properties to describe the data that requires accessibility (DataAvailability), confidentiality (DataConfidentiality) and integrity (DataIntegrity); and "Damage" by sub-properties to determine the impact on these data characteristics: damage to accessibility (ImpactAvailability), confidentiality (ImpactConfidentiality) and/or integrity (ImpactIntegrity).

To link classes, we will use the following properties:

- "Implements" and the inverse property "Realizes" for the connection of prototypes and their implementations.
- The transitive property "Contains" and Contains "to associate elements with children. As well as sub-properties "Contains a prototype" and "Contains an implementation" with drop-down domains, ranges, and inverse properties.
- "Data transfer" and the inverse of the property "Receive data from" to indicate the exchange of data between elements.
- "Directed at" to indicate the purpose of the threat.
- "Has a vulnerability" for vulnerability and implementation relationships.

The inverse of property A to property B means that if A (x, y) is true, then B (y, x) is also valid. Transitivity assumes that if A (x, y) and A (y, z) are true, then A (x, z) is also true. Domains and ranges of added properties can be seen in Fig. 1.

The owl: ObjectProperty element introduces a new connection property, owl: ObjectProperty introduces a characteristic property, rdfs: domain and rdfs: range specifies the domains and ranges for them, respectively. In the study, the "Contains" property is introduced with the domain and range the data element inverse to "Contained in", the property "Data Characteristic" and the child "Characteristic of accessibility".

To clarify properties in the ontology, restrictions are imposed on them, such as a domain and a range. These are global characteristics of properties. There are other, local characteristics and restrictions on cardinality or a range within a subclass. Enterprise Architect does not allow you to set several necessary properties of an ontology; therefore, the ontology was further developed at Protégé. Protégé not only allows you to introduce several additional restrictions but also takes them into account when describing individuals. Also, SPARQL is built into Protégé, with which you can make ontology queries.

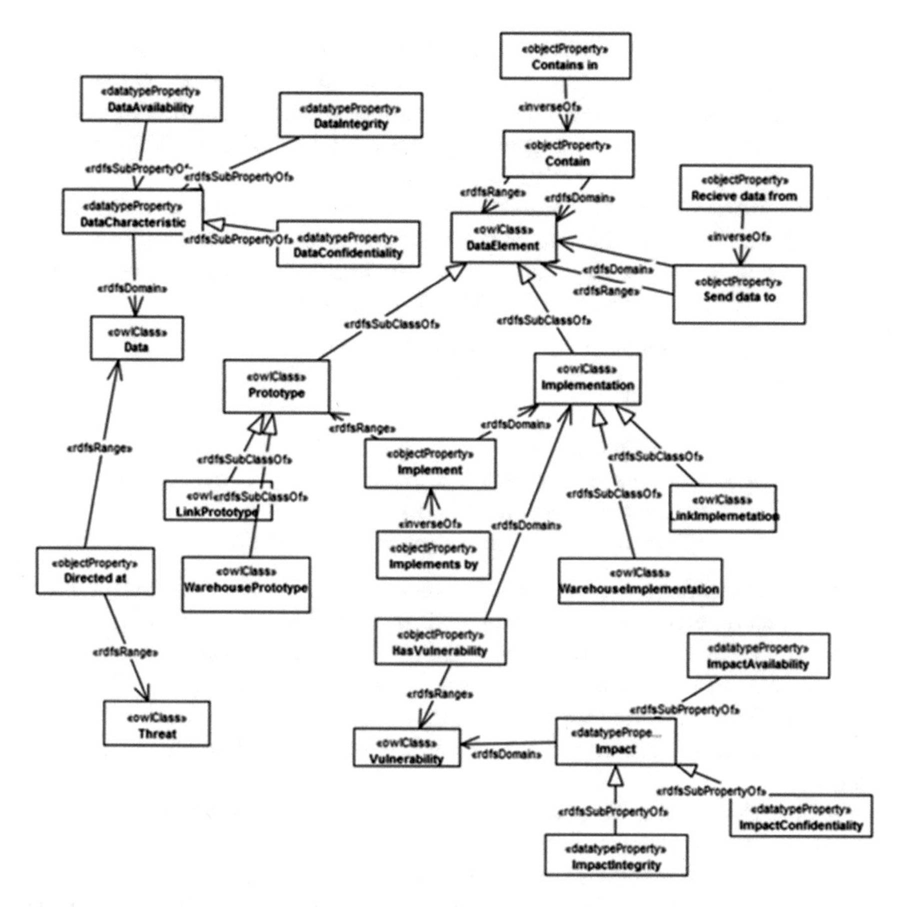

Fig. 1 Description of ontology classes and their properties in Erwin

One limitation that can be used to describe equivalent classes is the limitation of cardinality or the number of communication elements. To clarify the class of vulnerability as a vulnerability that damages data availability, the AvailabilityVulnerability class was created, which is equivalent to a class that has at least one ImpactAvailability property. Similarly, the ConfidentialityVulnerability and IntegrityVulnerability classes were defined. It is worth noting the difference between using a subclass and an equivalent class. In the first case, the necessary condition for belonging to the specified class is satisfied, and in the second, also sufficiency. That is, if for an individual the equivalence condition for a certain class is fulfilled, then it belongs to this class. In our case, if an individual vulnerability is detrimental to accessibility, then he also belongs to the class of “accessibility vulnerability”.

Similarly, data can be refined in three classes: data for which values of confidentiality, integrity, and accessibility are set, fall into vulnerable classes by the corresponding property: "Vulnerable by privacy property" (ConfidentialityVulnerableData), "Vulnerable by integrity property" (IntegrityVulnerableData), "Vulnerable by Privacy Property" (AvailabilityVulnerableData).

Before that, we defined classes according to restrictions on properties-characteristics, however, this can be done according to properties-relations. In an ontology, this is used to separate threats into threats to privacy, integrity, and availability. The access threat class is described as follows:

<owl:Class rdf:about="&sec;AvailabilityThreat">
<owl:equivalentClass>
<owl:Restriction>
<owl:onProperty rdf:resource="&sec;Directed_at"/>
<owl:someValuesFrom rdf:resource="&sec;AvailabilityVulnerableData"/>
</owl:Restriction></owl:equivalentClass>
<rdfs:subClassOf rdf:resource="&sec;Threat"/> </owl:Class>

The owl: equivalent class element begins the definition of an equivalent class, owl: Restriction enforces a restriction, owl: property indicates which property is affected by the restriction, owl: some values from requires at least one connection to AvailabilityVulnerableData, that is, data whose availability may be affected by the threat. Thus, a threat based on its purpose can fall into one or more subclasses of threats. After refinement of the ontology in Protégé, it took the following form (Fig. 2).

On the tabs of classes, properties-characteristics and properties-relations (Fig. 3), you can see certain elements of the ontology.

Finally, we add complex property relationships that will be chains or superpositions of other properties. These connections will embody the following conclusions: if the threat is directed to data, then during its implementation a prototype containing this data will be attacked. That is, the threat will be directed at the prototype. However, since the prototype itself is not more than an element of the circuit, its implementation will undoubtedly be attacked. It turns out that the threat will be aimed at implementation. If we now take into account the "Contains prototype" property, which defines the elements that specify the prototype, it turns out that the threat is directed not only at the prototype but also at its "parents". Also, the "Contains implementation" property assumes that if the threat is aimed at implementation, then it is also aimed at specifying implementations. To implement a threat, vulnerability is required, therefore if the threat is aimed at implementation, and the implementation has a vulnerability, then the threat can exploit this vulnerability.

If the threat is directed at the prototype, the condition is described in one chain, since its description includes the transitive property "Contained in". And the prototype, which has data properties set, will be both "Data" and "Prototype", which will allow us to organize a transitive connection and find additional prototypes that are targeted for the threat.

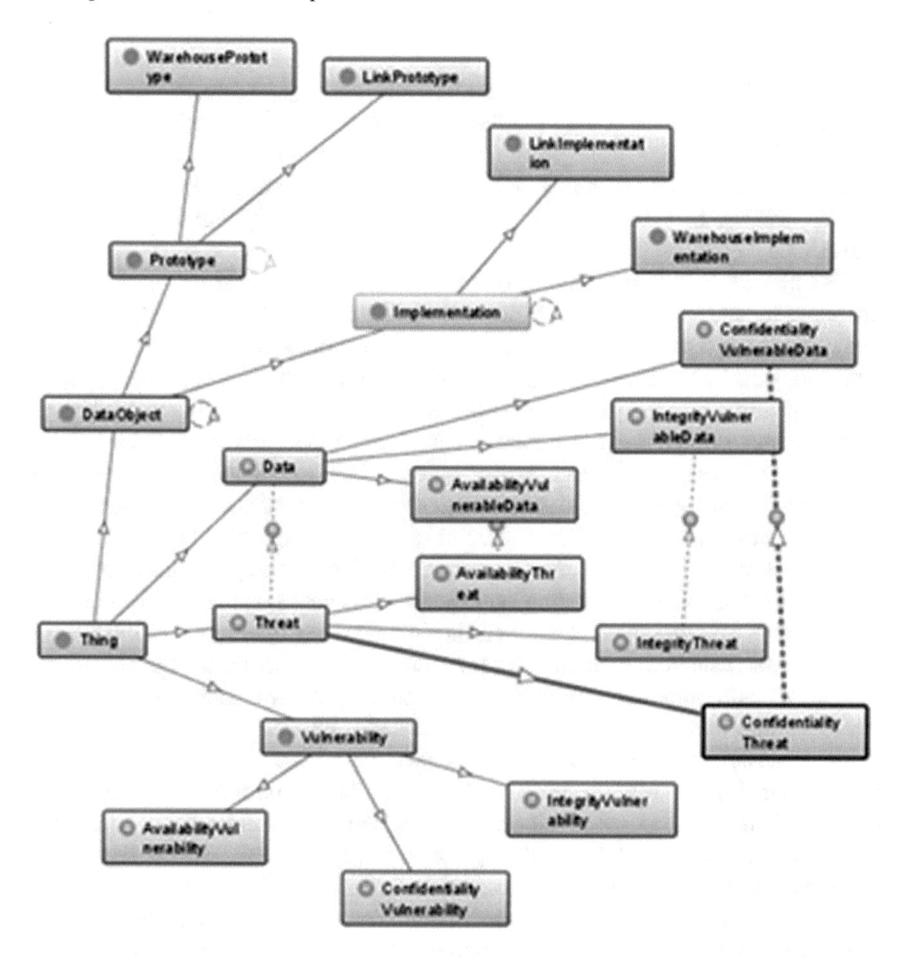

Fig. 2 Diagram of ontology classes and their relationships in Protégé

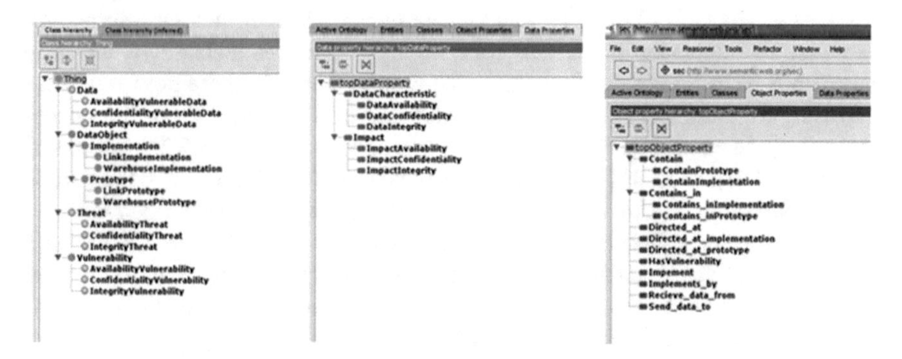

Fig. 3 Description of classes, property characteristics, and relationship properties in Protégé

Another factor to consider is data transfer. If a circuit element sends data, then the element that receives it and its parent will contain it. Therefore, they will also be at risk. These relationships are described as follows: Contain o Recieve_data_from –> Contain and Recieve_data_from o Contain –> Contain.

The introduction of property-link chains allows Protégé to automatically display them. Consider how the ontology will work in our study. For this we define individuals.

The first of them is intended to provide information support for most business processes related to the training of specialists in the higher education system. The SZ system supports the centralized management of directories, the search for any information objects by any criteria, the flexible mode of differentiation of powers between users depending on their job responsibilities and administrative affiliation, supports the maintenance of an ordered archive, an archive of reporting forms, and storage of the history of changes of all information objects. The system is a real movement of programmers through projects. Any changes in the properties of programmers are made by putting order into effect. A flexible mechanism is provided for customizing order templates and routes for moving draft orders. The developed solution provides powerful functionality for building statistical reports based on the OLAP cube.

The WSN system covers the entire educational process in terms of curriculum development, calendar schedules of the educational process, theoretical education plans, and semester curricula. The module provides convenient tools for creating educational process schedules, curricula, including theoretical training plans, semester curricula, and individual learning plans. The entire educational process planning process is accompanied by operational control for compliance with state standards. Documents developed in the system can be exported to WORD or Excel formats. By analogy with the "protection system" (PS), several reports are provided, but the OLAP cube is not used to build them.

PS is implemented on the Microsoft Dynamics CRM platform. This platform consists of a "wrapper" over MS SQL Server, with the help of which the user interacts with the database to view and modify data. Interaction is carried out through a Web interface on top of HTTP. HTTP requests are handled by MS Internet Information Services. To implement the OLAP cube, Analysis Services technology, which is part of MS SQL Server, was selected. The reports themselves are prepared using Reporting Services technology, which is also part of MS SQL Server. The system operates on three servers that are running Windows Server 2008 R2. The first of them is used to process HTTP requests, the second supports DBMS, and the third is responsible for processing the OLAP cube. The most valuable information is stored in the DB and for a short time, it can be cached by the webserver. An OLAP cube contains anonymized statistical information. Schemes that correspond to such an architecture can be seen in Fig. 4. It turned out five schemes of the first tier and two schemes of the second tier. The dashed lines indicate the "Contains" relationship, and the small dashed lines indicate the "Implements" relationship.

The same technologies were used to implement the WSN system, except for the MS CRM Dynamics 2011 platform. Data processing is also organized through the Web interface, however, its implementation is completely "self-written". Therefore,

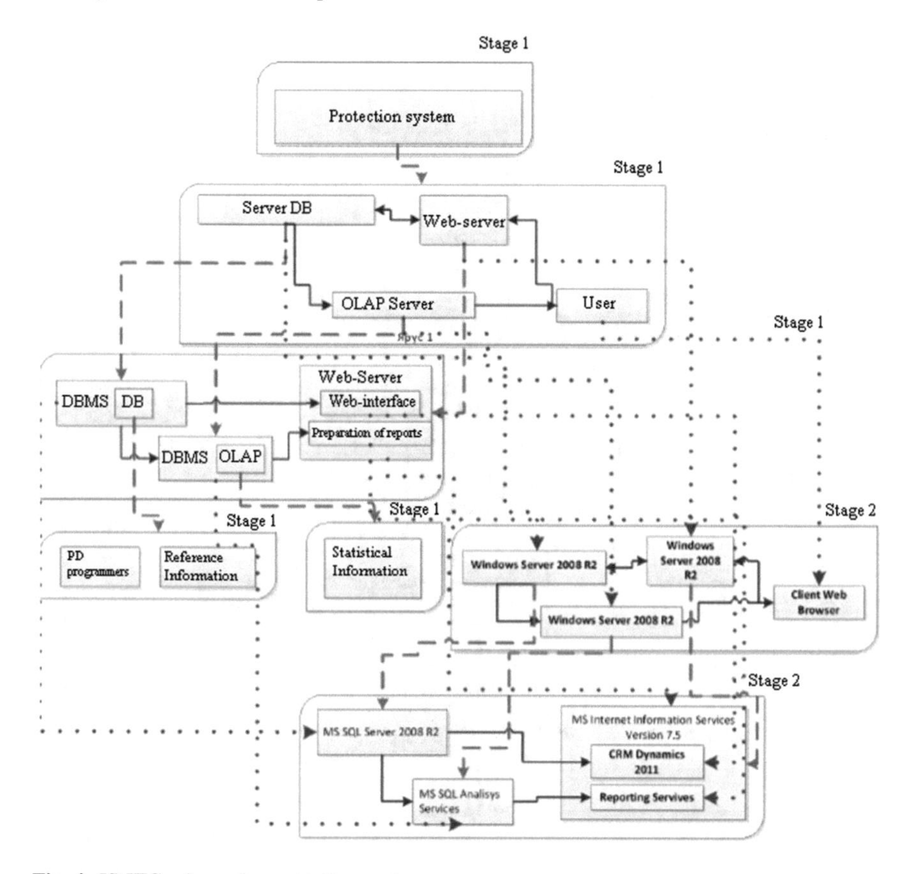

Fig. 4 IS-ITC scheme for protection system

the development of several additional services as required, such as a service for reading data from Active Directory and a service for downloading files to a server from a user's machine and vice versa, from a server to a user's machine. Silverlight 5 is used to create the web user interface along with WCF RIA Services. Support services are also implemented using WCF technology. The solution works on two servers: the first response to HTTP requests from users, the second runs MS SQL-Server. Windows Server 2008 R2 is responsible for the OS level. IS-ITC schemes for WSN are shown in Fig. 5.

To test ontology capabilities, schemas of the systems in question were introduced at Protégé. The "personal data" (PD) table was set to 100% for the data confidentiality property, the OLAP cube data was 100% integrity, the webserver was 100% available. The property set means the importance of maintaining the specified property as a percentage. Three potential security policy violations were examined. Firstly, it is a privacy risk for PD programmers. An attacker may try to capture this data and take advantage of it. Thus, the target for this threat was the PD table, which is present in both systems. Secondly, it can be considered relevant—a threat to the availability

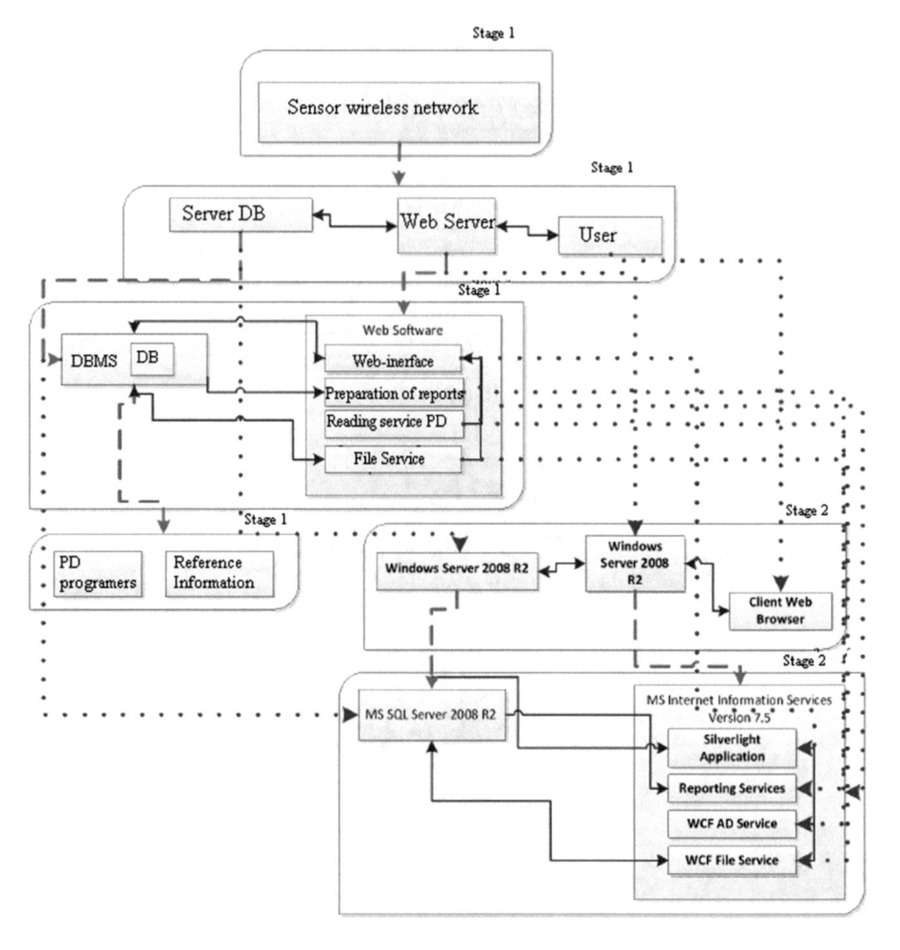

Fig. 5 IS-ITC diagram for WSN system

of a web server. If it is temporarily incapacitated, this will lead to certain losses. In case of loss of user trust and loss of relevance of information in DB. In this case, the "Aimed at" property was associated with a web server prototype. Thirdly, an attacker can violate the integrity of the data contained in the OLAP cube so that the reports generated based on them are incorrect.

After entering the initial data, the results of the work of the "Reason", which is part of Protégé, were analyzed. His work was possible since the OWL DL language was used to describe classes, their properties, and individuals.

The first thing that became noticeable was that the threats introduced were classified and divided into three classes: the first got into the privacy threat class, the second the accessibility threat, the third the integrity threat. Threats were divided based on the data classes they are targeted at.

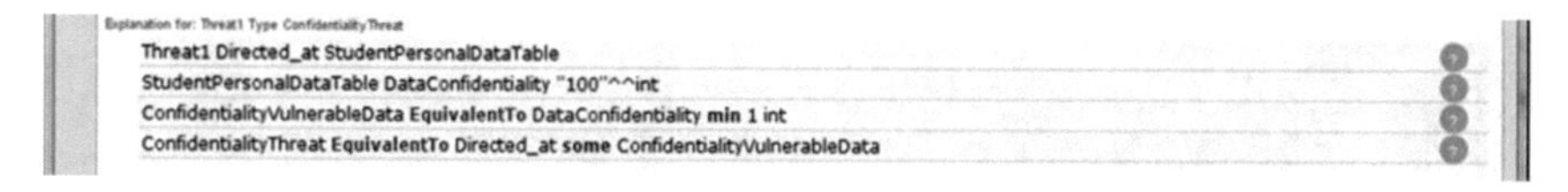

Fig. 6 Protégé explain window

When investigating the threat of access violation, the conclusion was constructed as follows (Fig. 6):

- the threat is directed to a table with PD programmers;
- the programmer's PD table has a property-characteristic—"Data privacy";
- the class "Vulnerable to Threats to Confidentiality" includes those individuals who have at least one property called "Data Privacy";
- "Threats to confidentiality" class includes those individuals who are "Directed at" individuals belonging to the classes "Vulnerable to privacy threats".

The second thing that the "Reasoning Person" was able to identify was potential vulnerabilities that could be involved in the implementation of the threat. In the study, three individuals were added for MS SQL Server and MS IIS, which each have the "Damage" sub-properties, and thus fall into the class of vulnerabilities. The list of calculated relations is presented in Fig. 7. For the first threat, the Reasoner identified all six vulnerabilities as potential to implement the threat, since sensitive data can appear in both DBMS and the web server cache. For the second, only three vulnerabilities can be exploited that are directly directed to the webserver. The "conclusion" of these connections looks more complicated (Fig. 8). It is based on the chains of connections defined above.

However, when calculating the described UseVulnerability relationship corresponding to potential vulnerabilities for realizing a threat, the Reason does not take into account the type of damage that will be done and the characteristic that should

Fig. 7 The results of the "Reason" Protégé

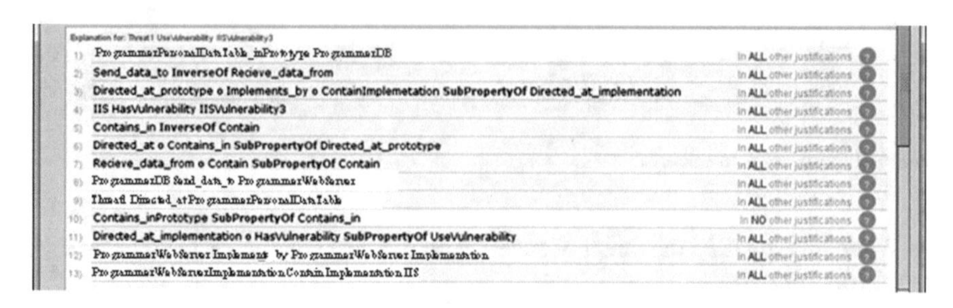

Fig. 8 Explanation of the work of the "Reasoner"

be protected. This feature can be taken into account by the Protégé system built-in query language for data presented in the RDF model—SPARQL. Using SPARQL, you can make a complex query on the information contained in the ontology. The following query will list the vulnerabilities that could be used to implement threats aimed at SQLAnalysisServices:

SELECT DISTINCT ?a ?b ?dataProp ?data value ?e ?f ?i ?j WHERE { ?a :Directed_at ?b.
?dataProp rdfs:subPropertyOf :DataCharacteristic.
?b ?dataProp ?dataValue.
?b :Contains_inPrototype ?c.*
?c :Implements_by ?d.
?d :ContainImplemetation ?e.
?e rdfs:label ?label. filter (str(?label) = "SQLAnalysisServices").
Optional {?e :HasVulnerability ?f.
?f :ImpactIntegrity ?i}. } order by ?a

The built-in interpreter of SPARQL queries in Protégé does not allow using the properties that were output by the "Reason". Therefore, the query partially duplicates the previously entered chains of property-relationships.

Now, if you add the second possible implementation of DBMS—MySQL, you can compare the two components of the implementation and choose a less vulnerable one. For such a comparison, five vulnerabilities for MySQL and three for MS SQL Server were added to the model. These are the vulnerabilities of the compared DBMS that were identified at the date of analysis. Information about vulnerabilities and the extent of damage to confidentiality, integrity, and availability is taken from OSVDB.

For the "Reasoner" to take MySQL into account as a possible implementation of DBMS, and to be able to find potential vulnerabilities, another "Contains" relationship was added. The "Reasoner" Protégé deduced all five vulnerabilities. Now compare the components under consideration for possible damage. The following query will display all possible values of the sub-properties of the "damage" property for all threats directed at presentation elements:

SELECT DISTINCT ?a ?b ?c ?d ?e ?f ?i ?j
WHERE { ?a :Directed_at ?b.

?b :Contains_inPrototype ?c.*
?c :Implements_by ?d.
?d :ContainImplemetation ?e.
Optional {?e :HasVulnerability ?f.
Optional { ?i rdfs:subPropertyOf :Impact.
?f ?i ?j }. }. } order by ?a

The results of this query are as follows (Fig. 9).

By adding filtering and aggregation of the results by the compared components, we obtain the results presented in Fig. 10. It can be seen that the probable damage when using MySQL as a DBMS is higher than when using MS SQL Server, therefore it is better to opt for Microsoft DBMS.

Since the potential damage is based on CVSS coefficients, comparing it with the value of characteristics that may be violated will answer the question: is it worth strengthening the protection or will the risk be accepted? Automate this process allows the ESSA system.

For further testing, we add individuals who will describe the WSN system. The scheme of all individuals in Protégé will take the following form (Fig. 11). You may notice that in the first tier scheme there are similar information storages (IS)—servers with DBs that contain PD and interact with the webserver. Signs of similarity are getting into the same classes, the content of the data of the same classes, the presence

a	b	c	d	e	f	i	j
Threat1	ProgrammarPersonalDataTable	ProgrammarDB	ProgrammarDBImplementation	SQLServer	SQLServerVulnerability3	ImpactConfidentiality	"20"^^<http://w
Threat1	ProgrammarPersonalDataTable	ProgrammarDB	ProgrammarDBImplementation	SQLServer	SQLServerVulnerability1	ImpactConfidentiality	"40"^^<http://w
Threat1	ProgrammarPersonalDataTable	ProgrammarDB	ProgrammarDBImplementation	SQLServer	SQLServerVulnerability2	ImpactConfidentiality	"30"^^<http://w
Threat1	ProgrammarPersonalDataTable	ProgrammarDB	ProgrammarDBImplementation	MySQL	MySQLVulnerability4	ImpactConfidentiality	"10"^^<http://w
Threat1	ProgrammarPersonalDataTable	ProgrammarDB	ProgrammarDBImplementation	MySQL	MySQLVulnerability5	ImpactConfidentiality	"30"^^<http://w
Threat1	ProgrammarPersonalDataTable	ProgrammarDB	ProgrammarDBImplementation	MySQL	MySQLVulnerability2	ImpactConfidentiality	"30"^^<http://w
Threat1	ProgrammarPersonalDataTable	ProgrammarDB	ProgrammarDBImplementation	MySQL	MySQLVulnerability3	ImpactConfidentiality	"40"^^<http://w
Threat1	ProgrammarPersonalDataTable	ProgrammarDB	ProgrammarDBImplementation	MySQL	MySQLVulnerability1	ImpactConfidentiality	"20"^^<http://w
Threat2	ProgrammarWebServer	ProgrammarWebServer	ProgrammarWebServerImplementatio	IIS	IISVulnerability1	ImpactConfidentiality	"20"^^<http://w
Threat2	ProgrammarWebServer	ProgrammarWebServer	ProgrammarWebServerImplementatio	IIS	IISVulnerability1	ImpactAvailability	"20"^^<http://w
Threat2	ProgrammarWebServer	ProgrammarWebServer	ProgrammarWebServerImplementatio	IIS	IISVulnerability2	ImpactConfidentiality	"10"^^<http://w
Threat2	ProgrammarWebServer	ProgrammarWebServer	ProgrammarWebServerImplementatio	IIS	IISVulnerability2	ImpactAvailability	"20"^^<http://w
Threat2	ProgrammarWebServer	ProgrammarWebServer	ProgrammarWebServerImplementatio	IIS	IISVulnerability3	ImpactAvailability	"10"^^<http://w
Threat3	ProgrammarStatisticalOLAPDat	ProgrammarOLAP	ProgrammarOLAPImplementation	SQLAnalisesServices			

Fig. 9 Probable damage DBMS-MySQL in WSN

component	count	sum
SQLServer	"3"^^<http://www.w3.org/2001/XMLSchema#integer>	"90"^^<http://www.w3.org/2001/XMLSchema#integer>
MySQL	"5"^^<http://www.w3.org/2001/XMLSchema#integer>	"130"^^<http://www.w3.org/2001/XMLSchema#integer>

Fig. 10 The amount of probable damage for the prototype DBMS

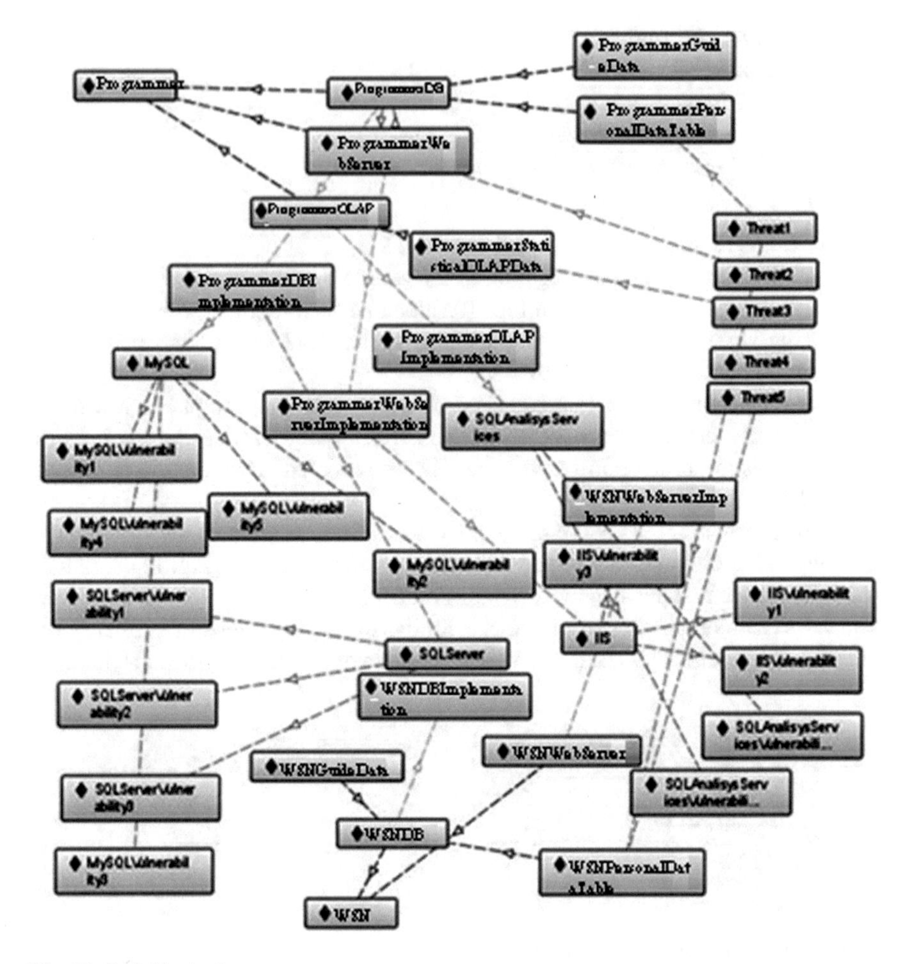

Fig. 11 Individuals chart

of the same link properties that describe the data transfer. Of course, there are not so many signs that suggested a similarity, but among them, there is both a WSN structure and simple properties. In this context, the identity of IS is not mentioned, but their similarity is only assumed. Based on the proposed assumption, it is concluded that when selecting components for implementation, it is necessary to pay attention to how they are presented in the second-tier implementation diagrams for similar ISs. In this case, there is one "similar" IS—this is IS from the PS scheme. Therefore, if there was a question about the components for implementing a DB server, then, first of all, it would be necessary to consider Windows Server and SQL Server—the software with which PS is implemented.

The developed ontology was used to compare various technology stacks: LAMP (Linux, Apache, MySQL, PHP), WAMP (Windows, Apache, MySQL, PHP), WIMP (Windows, IIS, MySQL, PHP), WISA (Windows, IIS, SQL Server, ASP.NET), WISP

(Windows, IIS, SQL Server, PHP). As of the date of the comparison, the minimum cost was provided by the WISA stack.

Thus, a static template is a prototype-implementation bundle. The properties of the prototype determine the similarity of implementation requirements and place in the structure of WSN. After that, the implementation of a similar component is taken as the basis for the implementation of a new component.

For the template to be not only already implemented and used by software, but also the least exploited by an attacker, we will connect information about successful attacks to the ontology. To do this, add the UseCount property for the Vulnerability domain. It will allow you to take advantage of the implicit connection of vulnerabilities and threat classes. The value of this property is taken from the logs on the functioning of software and reports. Several more threats were added to the test data, for which Protégé defined threat classes. Moreover, part of the threats did not fall into one class, but into several. For this, threats were directed to different data with different classes. In Fig. 12 shows the query by which the sum of the incidents for each of the threat classes is obtained.

Undoubtedly, this information is useful both for risk analysis and for the gradation of patterns. Implementations in which a large number of errors were found and used are not templates. They can only be called "anti-patterns." However, many companies will not want to share information about internal security incidents. A large number of attacks, even if not brought to a successful conclusion, are likely to scare the company's customers, and it will lose profit. On the other hand, global corporations have a large number of projects working on various software. In this case, problem data can be disclosed and successfully used by various departments of the same company.

The collection of information about security incidents and errors, in general, is built into much modern software. Microsoft solutions have a mechanism for automatically sending problem data to the server. This is a good data source for a developed view. Theoretically, the problem of falsification of this data is possible to violate the mechanism for constructing static templates. It is solved with the help of such

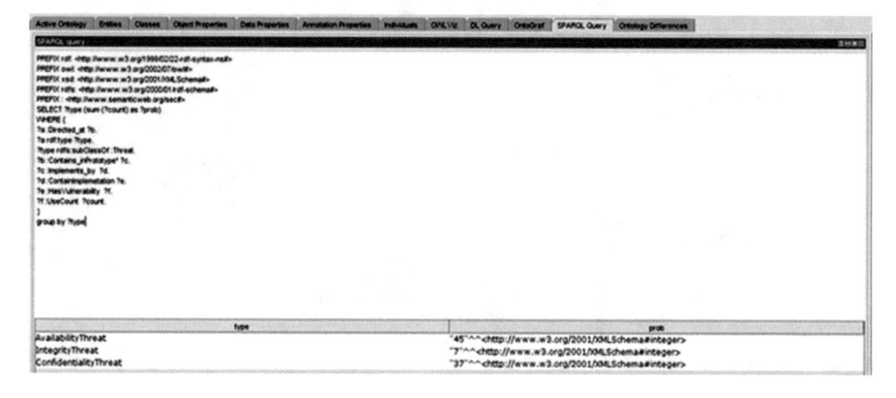

Fig. 12 Sum of incidents by threat class

a feature of the ontology description language as reformation or materialization. Revision allows you to make a statement of approval. Using helper classes, you can determine who made the statement and test its reliability.

The main problem and the reason that slows down and greatly complicates the further study, development, and use of static templates is the lack of a set of IS/ITC schemes that would allow testing the approach on a large set of "test data". To increase interest in the first tier schemes, it is proposed to add additional capabilities for describing the WSN according to the guidance documents. This is conveniently done by developing a new ontology for defining the class IS. Such an ontology will consist of the "Category" class and four possible values, the connection properties of WSN with the category, the property-characteristics "personal data volume" and from 4 classes, one of which will be assigned to IS. For each class, elementary equivalence conditions are described. Belonging to the class K4 is equivalent to the connection with the fourth category. The scheme of such an ontology is presented in Fig. 13. By connecting it to your data set, the user will be able to easily determine the class IS.

Also, the construction of schemes of the second tier can be automated. Since OWL is a language that is "understandable" to machines, the implementation will consist of polling software, obtaining values for properties, and arranging XML documents describing individuals. As for the definition and use of new functions, this is realized by introducing new classes and properties, and, possibly, developing their interpreter. The decision is time-consuming, but the benefits of its implementation can be great. Another possible area for further research is integration with other existing ontologies.

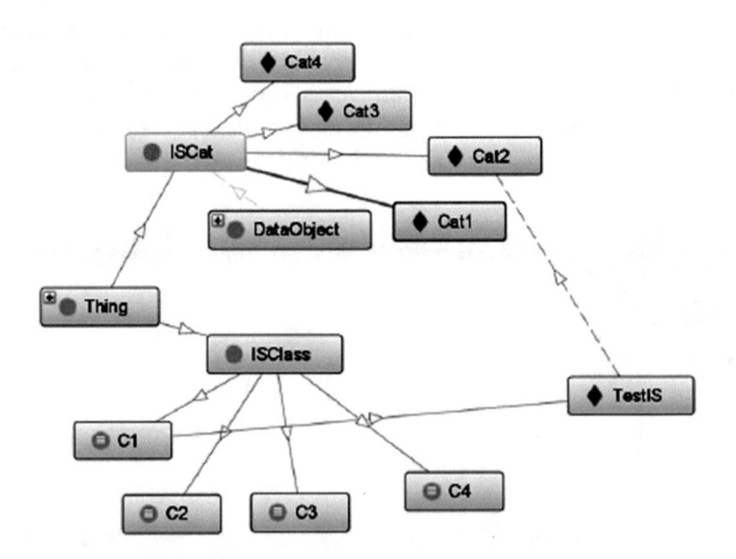

Fig. 13 Ontology for class definition IS

The above studies show the validity of the idea as a whole and reflect the possibilities of the proposed approach. However, it is noted that further development requires a larger number of schemes. The described solution will allow you to initially design WSNs with security in mind, which is an important part of development. However, system maintenance is not taken into account in any way. Initially laid down security must be able to maintain and expand. Therefore, the dynamic security template is further considered—a template that takes into account changes in the course of WSN operation.

3 Modeling of a Multi-agent Protection System for Wireless Sensor Networks

There are five classes of agents that are to be included in such a system to provide MAC-based security. These are Analysis Agents, Configuration Agents, Protection Agents (PAs), Counteraction Agents, and Learning Agents. The most interesting for the present work is the PA, as it is during their work that a dynamic security pattern of one component and WSN as a whole is obtained. Such agents bind to WSN components and estimate the deflection ratio (DC) of their operation. Each agent can use its algorithm to estimate the deviation coefficient (DC) and must be able to adapt it. The PA team, along with host security agents (HSA), detects a common host failure and provides this information to a software protection agent (SPA), which ultimately informs the system administrator of the problem.

Developing from scratch MAC protection is quite time-consuming and beyond the scope of this work. Also, some development requires some prototype that confirms the assumptions made. It is possible to obtain and evaluate the results of such a prototype using simulation. Therefore, we will try to show that the proposed concept is valid and does represent several advantages in protecting real systems based on simulation modeling. To do this, we first model the behavior of PA to estimate the deviation in the operation of models of real systems and then implement the algorithm previously considered for DC estimation of one of the possible components of WSN–DBMS.

We will use AnyLogic software for modeling. It is the tool of one of the leading companies in the field of tools and business applications of simulation modeling—The AnyLogic Company. AnyLogic supports various imitation modeling approaches, among which are system dynamics, “process” discrete-event, and agent modeling. The latter will be used in this section. AnyLogic supports the description of Java agent behavior as well as nested state diagrams that allow you to create agents of almost any complexity. Also, this tool allows you to make many parameters at the level of the user interface and to change the course of events during the “run” of the model. For the values of different variables, it is possible to build summary graphs that allow you to compare and analyze their values over some time.

We will use simplified variants of WSN agents and components to model the MAC. PA, counteraction agents, and agents modeling the behavior of the common user, the attacker, and the component—agent-user, agent-attacker, and agent-component, respectively, are involved in the analysis.

First, let's consider the logic behind the work of individual agents. The simplest part of the model is the user agents. Their behavior is quite primitive. They address components with requests that need to be processed, thereby mimicking the normal behavior of WSN. The return occurs with a certain probability, which is rendered in the parameter of each agent instance and is modeled by sending a message to the associated resources. All connections are specified at the model level. In the event of receiving a response to a request, the user agent enters the "Successful Response" intermediate state and in one step enters the "Waiting" state from which the request can be repeated. In the absence of a response from the component, the user enters the "No Response" state by timeout (5 steps) and then into the "Standby" state (1 step). The state diagram of this class of agents is shown in Fig. 14.

Arrows with a "timer" denote a transition after a specified period, arrows with an "envelope"—a transition when receiving messages from other agents, a rhombus—check that the condition is fulfilled, and arrows with labels—a transition when a certain condition is met. It is worth noting that the design of the agents consisted of both creating a state diagram and describing the actions for transitions in Java. Therefore, it is possible to move from one state to another by two different transitions, but with different actions.

The next part of the model is a team of attack agents, which consists of their malefactors. Since PA's require a certain period of training during which they accumulate statistics about the habitual state of components, attack agents begin their work with a delay. The magnitude of this (*WaitTime*) delay depends on the *MinWaitTime* parameter:

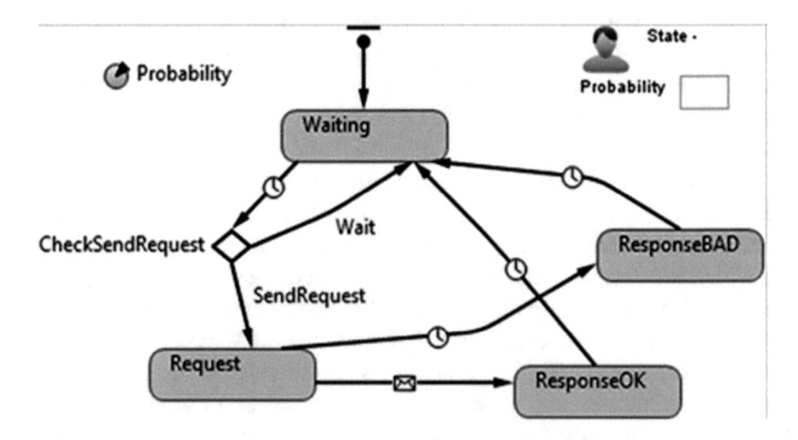

Fig. 14 User-agent status chart

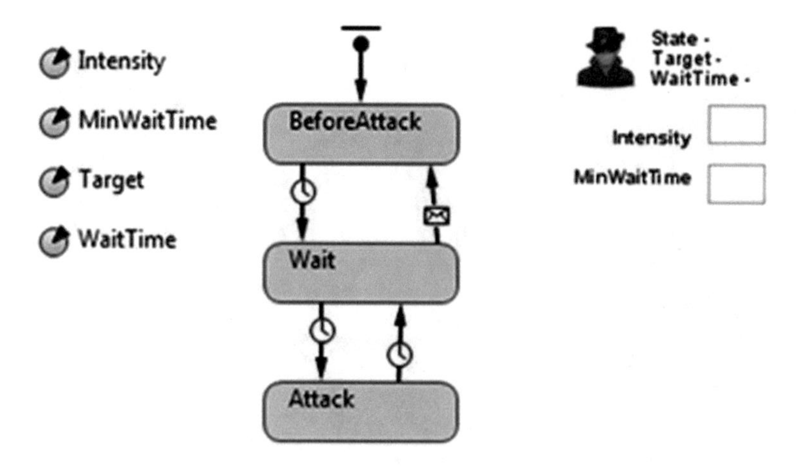

Fig. 15 Attacks gent status diagram

$$WaitTime = (1 + random()) * MinWaitTime \tag{1}$$

The function *random ()* gives a random value from 0 to 1. The MinWaitTime parameter is set at the model level for each of the attack agents individually. After the delay is over, the agent enters the Wait state, randomly selects the attacked component, and simulates the attack, sending a corresponding request. The logic of the probability of success or failure of an attack is rendered by the agent-component, since, regardless of the result, the very fact of the attack must exist. The second attack is carried out with a certain intensity. The logic of the work is simple and is presented in Fig. 15.

An agent component was designed to model arbitrary WSN components. This agent has a rather complicated mechanism of operation, the state diagram of which is presented in Fig. 16.

We assume that each agent component has two characteristics. The first characteristic is the current load on the component. It increases by any of the agents in the interaction and after a while returns to its default value of zero. Upon receipt of the message, if the load does not exceed one hundred percent, the request is processed. If the load is maximum, then the component goes into the state of "overloaded" and does not process the received request. After three steps of the model, the component load begins to decrease. Upon receiving a request from a user, the component processes this request in one step. If the current component needs a request to other components, it executes these requests and waits for a response. After processing the request, the component responds to the user with the message "ResponseOK".

The second characteristic is the degree of protection of the component from attack by the intruder. For brevity, let's call this characteristic a "barrier". The value of the barrier decreases when requested by an attacker. AttackProb is used to simulate the attack result. It sets the probability with which a successful attack occurs. Only in case of a successful attack, the barrier value decreases. If the attack agent performs

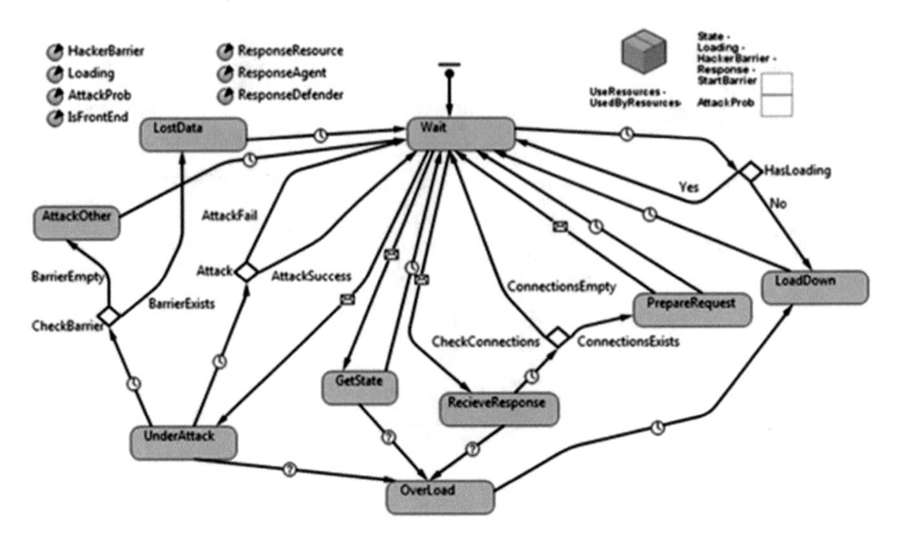

Fig. 16 State diagram of the agent component

a successful request-attack against a component that lacks a barrier, we will assume that this component is compromised. Compromise—the fact of access of the unauthorized person to the protected information. That is, the attacker grabs the component, receives valuable data, and can use it for its purposes. If a component with a zero barrier is repeatedly attacked, it sends an attack message to the connected components involved in request processing.

Briefly, the model can be described as user agents addressing the components and simulating normal WSN behavior. This changes the characteristic responsible for the component load. The attacker agent "waits" and starts attacking the components, reducing the barrier. When a barrier is lifted, attackers receive sensitive component data and use the captured component to attack the next. The developed components were added to the model and connections were established between them (Fig. 17). Since the definition of the relationships between agents and their parameters in the implementation were made in the parameters of the agent, it made it easy to model the various relationships between components, between users and components, as well as between attackers and components.

User agents, attack agents, and a graph that displays the load of the components under consideration were added to the model. During the test different numbers of users (from 3 to 5), different probabilities of request generation (from 0.1 to 0.9), different waiting time of the attacker (from 5 to 100), different intensity (from 5 to 10), and different the probability of success of the attack (from 0.1 to 0.9). Depending on the values selected, different results were obtained, which can be divided into two groups:

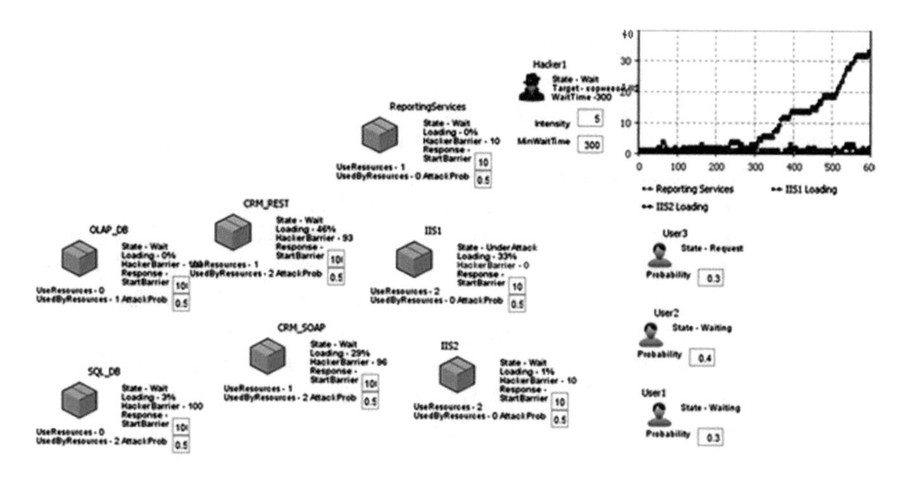

Fig. 17 PS modeling

1. User agents have overloaded the attacked components before the attacker began the attack. In this case, the attack of the attacker still led to success after some time, as part of his requests were processed.
2. User agents did not overload the attacked components and worked quietly. The attack agent after the attack increased the load of the attacked component depending on the intensity selected. As in the first case, regardless of whether the component overloaded or not, the attacker after some time "broke" the barrier.

In the real world, the architecture of the components, their capabilities, and the number are selected so that users can work properly with the system, that is, receive all their requests without delay. Thus, the first group of results can be considered incorrect in terms of the selection of the original parameters. The graph for the second group is presented in Fig. 17 (top right corner). It clearly shows that starting with the three hundredth step, and so much is expected by the attacker, the load on the component IIS1 increases. However, the load on a completely similar component of IIS2 does not change. Thus, in this model, the load is the characteristic that is suitable for obtaining DC. That is, the goal of security agents is to get high DC before an attacker can access sensitive data.

During the simulation, different PA algorithms were tested to calculate the DC load of the components. Attempts have been made to determine the average load and to consider the deviation from it. In this case, the average load was determined both for the entire period of work and for a small, recent period.

The deviation was calculated as a percentage of the average load. This approach proved to be ineffective. Another method that also did not produce positive results is the calculation of the root mean square load variance. Following failures in the analysis of load values, similar methods for determining DC were applied to the size of the discontinuities between the readings. In this case, there were a large number of false positives. Finally, a method was found that successfully determined a long-term increase in load and skipped the small changes likely to be in the habit. This approach is concluded in the analysis of the angle between the two lines. For this, PA removes N load readings. The angle is calculated for every N-th lift. Based on the readings, 3 points are obtained: (x1, y1), (x2, y2), and (x3, y3). Here, x1 is the time of the first shot, y1 is its value, x2 is the average time of N shots, y2 is the average value of N shots, x3 is the time of the last shot, y3 is the value. Time is understood as the conditional time in the AnyLogic model, which is a rational number. Other values were tested for the values of y1, y2, and y3—different averages of the readings. The values described above gave the largest number of true triggers. For the three points obtained, two lines are constructed, each of which is described by the point equation and the angular coefficient. The first line passes through the first and second points, the second through the second and third points. The angular coefficients compute the tangent of the angle between the lines, and the angle itself:

$$DC = \tan^{-1} \frac{k2 - k1}{1 + k1 * k2}, \tag{2}$$

where $k1 = \frac{y2-y1}{x2-x1}$, $k1 = \frac{y3-y2}{x3-x2}$.

This approach is somewhat "delayed" since when analyzing N values, the angle is determined for N/2 removal. However, it gives a minimal amount of false positives, which is a problem for other approaches. Adaptability is based on N recently taken values. The described approach was implemented in the final version of PA. An additional feature that required outside intervention was a load limit of 100. Regardless of the DC value, a counteracting agent is notified under such load. In Fig. 18 shows a PA state diagram and a code for DC analysis in Java, which is triggered when switching from CheckState to HandleState.

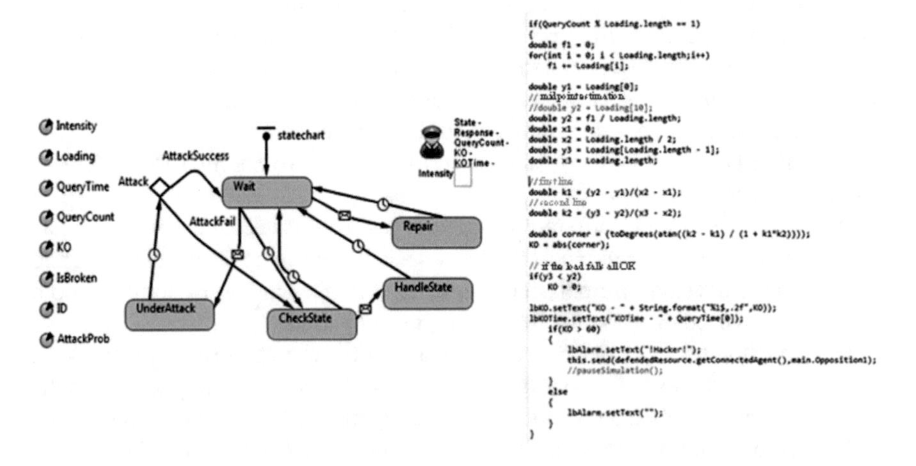

Fig. 18 PA state diagram and DC analysis

PA's algorithm also implies that it can be attacked like any component. With a successful attack, the PA fails and stops collecting readings of the protected component. Security agents monitor each other's work and, in the event of a breakdown, notify the counter-agent. These checks were implemented in Java.

Finally, a counter-agent was implemented and added to the model, which, when an attack was detected, was notified by the PA. Upon receipt of the alert, the Counteraction Agent detects the attack agent of the attack and translates it into a BeforeAttack state, that is, completes the attack. It is worth noting that the attacker after a certain time again chose a random target and began its attack. If the problem was a broken AZ, then it was transferred to the "Repair" state and soon returned to service. If the load on the component was 100, then the countermeasure agent "rebooted" the component and set the load to 0. Also, a SPA was added to the model, which collected values from the AZ, determined the maximum and average values, and displayed them on the conditional admin screen.

One of the models runs is shown in Fig. 19. There are two intruders who, with a delay of 600 and 700 steps, begin to attack the WSN. Their choice falls on IIS1 and IIS2 components, respectively. After the attack, the load on these components increases: the blue and red lines on the right graph reflect the corresponding changes. Defender2 and Defender3 security agents successfully identify suspicious activity for protected components. The left graph shows the jumps of the KO2 and KO3 functions, respectively. It is seen that the jump is lagging behind the beginning of the attack by the attacker. DC values rise only in 100–150 steps. This is due to the algorithm described above and the intensity with which the PA polls the components. The presence in the readings of not only the values of the load but also the time of measurement allows you to more accurately determine the period when the attack began. The plus is that PA notifies suspicious activity before the load reaches its

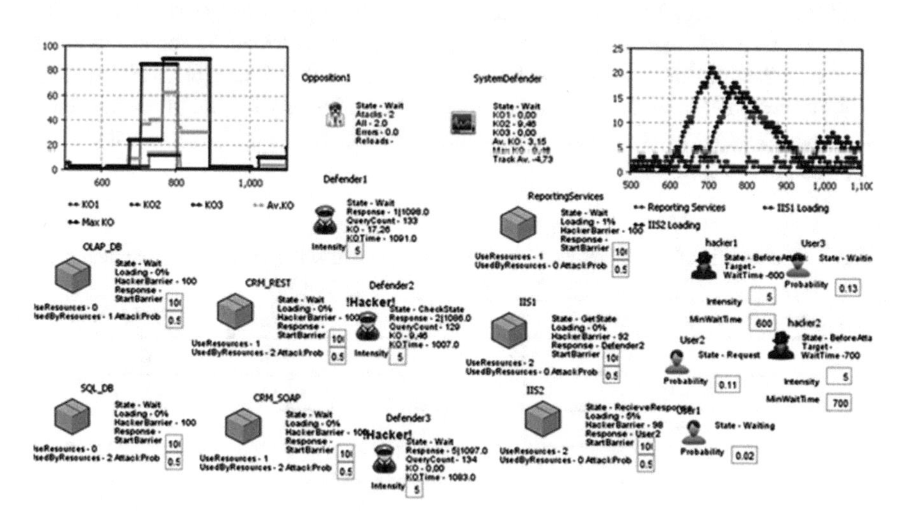

Fig. 19 Simulation of the interaction of 4 attack agents with the counteraction agent

maximum value. The potential result of such an increase is a refusal to serve a portion of the requests, which may be the attacker's ultimate goal.

However, there is also a downside, which is that the delay can be critical, i.e. the component and its data may be compromised before determining the PA attack. This can be seen from the fact that for the attacked components the barrier value is reduced. Determining the attack, the PA reported this to the Counteraction Agent (Opposition1) and he "zeroed in" on the targets of the attacker agents, thus completing their attack. In this case, the load on the component stopped rising and gradually began to return to its values before the attack. The corresponding changes are shown in the right graph—the values of the IIS1 Loading and IIS2 Loading functions are reduced. It is worth noting that when the load was reduced, PA also noted suspicious activity. These are the consequences of the work of the counteracting agent and the load drop. To eliminate false positives, the PA was modified to analyze load increases only. SPA coped with its problem and successfully received DC values from PA. The result of his work is shown in the left graph. This is a gray line that shows the average value (Av.KO) and a black line that displays the maximum DC (Max.KO). It is seen that a sequential attack of IIS1 and IIS2, which can be interpreted as an attack consisting of several steps, causes an increase in both average and maximum DC values. This is noted by the SPA (SystemDefender).

The simulation was performed for 1–6 attack agents. In this case, a threshold for DC of 60 was determined to be the most effective for the correct PA operation. SPA was used only as a monitor that collects common values, and all protection worked automatically without human intervention. In all cases, the PAs successfully identified the attackers, but they managed to do little damage. At the expense of what gradually the "barrier", separating the attackers from confidential data, disappeared. However, it is worth noting that the agents of defense and counteraction know nothing about the barrier and do not use it. In their work, they rely on component loading and can be considered agents that reflect attacks aimed at overloading the component. They coped with this type of attack well. This shows that PAs effectively identify DCs for the threat of breach of accessibility and ineffectively for the threat of breach of privacy. The number of times a component reboot per 100,000 model steps was required is shown in Table 1.

Table 1 The results of the modeling of the interaction of 6 attack agents with the counteraction agent

Malefactors	Total attacks	Warnings from PA	False	Reboots
1	125	133	8	0
2	315	326	11	1
3	507	509	2	2
4	707	698	–	6
5	886	843	–	9
6	1076	984	–	16

False triggers were defined as the difference between the total number of attacks and the number of PA warnings. And starting with 4 attackers, this approach began to give incorrect results, since on one detected attack the counter-attack agent could eliminate more than one attacker. Attack agents' goals could overlap.

The results show that a small number of attacks—no more than 2%—were successful. Most likely, this happened when the plans of the attackers coincided, that is, they simultaneously attacked one component or component and its defender. Additional tests were performed to confirm this hypothesis. They showed that to increase the likelihood of a successful attack up to 5% requires at least two attackers who will attack one target. PA's attack is unhelpful, as it quickly returns to service. In this case, the times of the attack should be correlated so that the PA several times in a row had time to find the first attacker and did not have time to detect the second and be so selected that the component did not have time to drop the load received from the attacks. Before the attack is successful, the PA will detect the attackers several times. In this case, if the counter-attack agent takes more serious protection measures, each time increasing MinWaitTime by 5%, then the success of the attack will be reduced to 2%. Thus, it can be concluded that a successful attack requires the cooperation of the attacker agents and the planning of the attack. For the attack not to be detected, it is necessary to "deceive" the PA. This is difficult because the attacker knows neither his existence nor the algorithms of his work.

Another possible and advantageous option for us to act as an attack agent is to give up our goal. One of the assumptions of the model is that the attacker is detected by the load on the component. But it can hide behind the average user and not cause a load change. Therefore, it is very important to choose the right set of characteristics for DC calculation. PAs were modified to determine DCs based on the change in barrier size, as well as a pheromone—messages left by PA intruders impersonating themselves. The results of these tests were similar to the previous ones and showed that the successful operation of the MAC depends largely on the DC used by each of the PAs.

When modeling a MAC, the processing time of a user request component was determined empirically based on the processing time of real responses. In the course of modeling the actions of intruders among the servers serving requests to the PS, the server was identified, which is the first to "fail" in case of attack. To confirm the possible problems, a DDoS attack on the frontend servers was conducted using the HULK software. The attack involved 3000 virtual users. The attack results coincided with the simulation results.

Figure 20 shows two graphs showing the load on the servers being attacked, the total number of requests, and the number of requests that they failed to process. As in the case of modeling, the first to fail was the reporting server. The results were explained by the fact that the report server has technical characteristics that are weaker than the application server, and is designed for the less active user activity. The coincidence of the simulation results and the actual attack confirms the robustness of the developed MAC model and the need for WSN protection based on dynamic security templates.

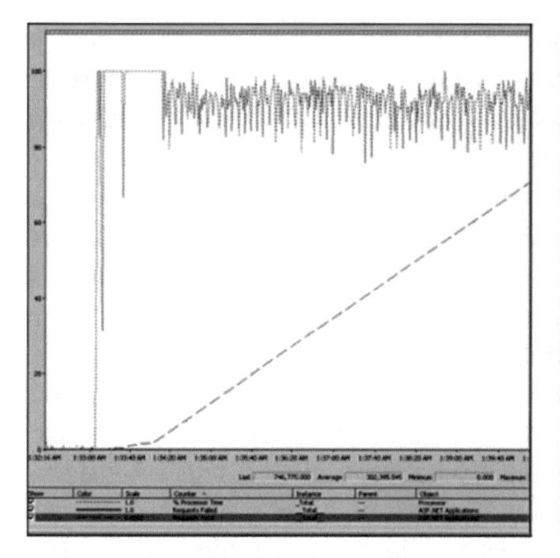
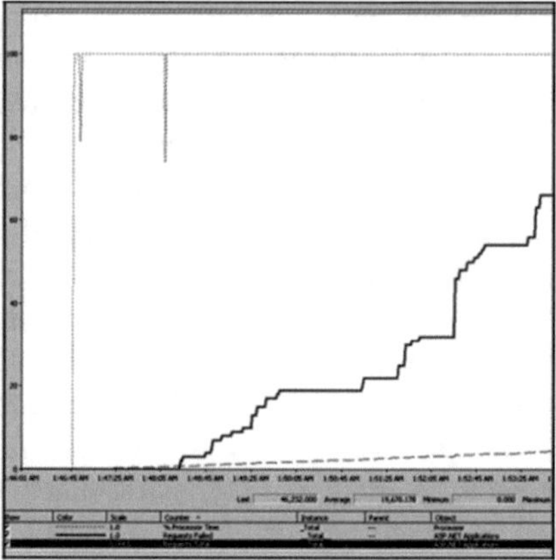

Fig. 20 Results of DDoS attack on the application server and report server

4 Result and Discussion

The model shows that the MAC can solve the problem—to detect deviations in the WSN. It can be concluded that the MAC-based protection does offer some advantages, given that each PA can correctly distinguish the characteristics for DC computation. The distribution of agents by WSN components reduces the load on the protected system. The architecture is flexible, as different PAs can protect one component from different threats, and is resistant to attacks, as PA failure is quickly detected and eliminated. The scalability of the solution is achieved through the ability of new agents to cover new WSN components. Implementing a multi-step attack on WSN is more difficult, as it must bypass several PAs, which in turn will mark an increase in DC, which will alert the system administrator via SPA. The attacker must either cheat the PA or "remove" it. To "remove" it is difficult, so the alleged actions of the attacker—cheating agent. We now implement and analyze the described DC estimation algorithms for PA DBMS against the threat of integrity violation.

As a malpractice action to which the PA must respond, we have selected the DB integrity breach threat to analyze SQL request rejection.

WSN database was used as the protected DB. And since this system uses DBMS SQL Server, the implementation of the estimation algorithm is closely linked to this DBMS and the version of SQL it supports. With this in mind, the problem was to determine the SQL SQL query for the DBMS SQL Server. A subset of the SQL query language—Data Manipulation Language (DML)—was considered a data management (manipulation) language.

This subset consists of four manipulation operators: SELECT (INSERT), UPDATE (DELETE), and DELETE (DELETE). You can break the integrity with one of three operators: INSERT, UPDATE, DELETE. To implement the proposed methods, it is necessary to collect statistics on the performance of these operations. Thus, in preparation, a mechanism for collecting information security required for

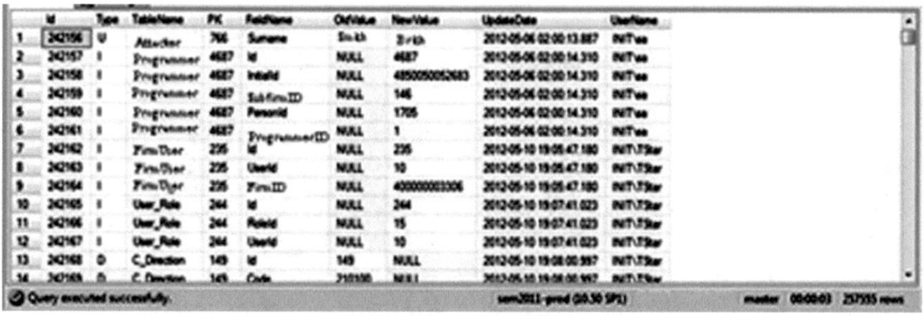

Fig. 21 Audit table

the agent was implemented. A trigger was used to collect data—an event handler for inserting, updating, and deleting data.

Dynamic SQL code made it possible to implement a stored procedure for mass creation of triggers on all tables of the protected DB. Besides, a procedure was developed to quickly disable all triggers.

The resulting solution allows you to process any data access and parse the request for components that will fall into a special table "Audit". To reduce the connectivity and volume growth of the primary DB, this table is placed in a separate DB on the same instance of SQL Server. The Audit table consists of the following fields: record identifier (Id); Type of operation; table name (TableName); variable record identifier (PK (primary key)); field name (FieldName); old value (OldValue); new value (NewValue); change date (UpdateDate); username (UserName).

Although not all fields participate in the DC calculation mechanism, they are presented in the audit table, as this may be useful for a detailed analysis of the incident. The volume of the table at the time of the experiments—260,000 records (Fig. 21).

Many heuristics have been proposed for DC estimation, each of which is implemented by a separate SQL Server scalar function. The first heuristic is the assumption that each user has a working window of time in which he interacts with the DB. On this basis, DC is calculated as:

$$DC_{time} = (P\max_{time} - P_{realtime}), \quad (3)$$

where $P\max_{time}$—the probability of the most frequent hour of requests; $P_{realtime}$—the probability of the current hour.

The developed SQL function looks like this:

```
CREATE FUNCTION [dbo].[KO1] (@userName nvarchar(200), @hour int)
RETURNS int AS BEGIN
declare @maxprobHour float, @probHour float, @allQueruesCount int;
— Total number of requests
```

```
SELECT @allQueruesCount = count(*) FROM [CM2011_log].[dbo].[Audit]
where UserName = @userName
Maximum probability
SELECT top1 @maxprobHour=round(cast(count(*)asfloat)/@allQueruesCount,
5)
FROM [CM2011_log].[dbo].[Audit]
where UserName = @userName
group by DATEPART(hh, UpdateDate)
order by round(cast(count(*) as float) / @allQueruesCount,5) desc
Probability of the hour
SELECT @probHour = round(cast(count(*) as float) / @allQueruesCount,5)
FROM [CM2011_log].[dbo].[Audit]
where UserName = @userName
and DATEPART(hh, UpdateDate) = @hour
The deviation coefficient
return (@maxprobHour - @probHour)*100
END
```

The second heuristic is the assumption that each user is for the most part associated with a restricted set of tables, and his access to tables not included in this set is a suspicious deviation. In this case, DC is calculated as:

$$DC_{table} = (P\max_{table} - P_{realtable}) * 100\%, \quad (4)$$

where $P\max_{table}$ is the probability of the most used table; $P_{realtable}$—the probability of accessing the table from the query.

The last assumption for DC computation is that for any query, the user uses an interface provided by the developer, and the attacker can bypass this interface. In the first case, the query is generated automatically, and in the second, the variable fields are entered manually. It is assumed that the deviation can be detected in the set of table fields in the query.

$$DC_{column} = (P\max_{column} - P_{reacolumn}) * 100\%, \quad (5)$$

where $P\max_{column}$ is the probability of the most "popular" set of fields; $P_{reacolumn}$—the probability of the current set of fields.

Since the number of all combinations of fields in the table can be high, in this case, it is better to determine the probabilities of the sets in advance. A stored procedure was implemented to calculate all column combinations in the executed queries. This procedure starts on a schedule and fills in an auxiliary table that contains the probabilities of all sets for each of the tables. The set of columns is stored in a row in which the ordered names of all columns are separated by a separator. At the time of development, the audit database contained 260 lines that described 60,000 requests. The WSN database consisted of 95 tables. Using the developed procedure,

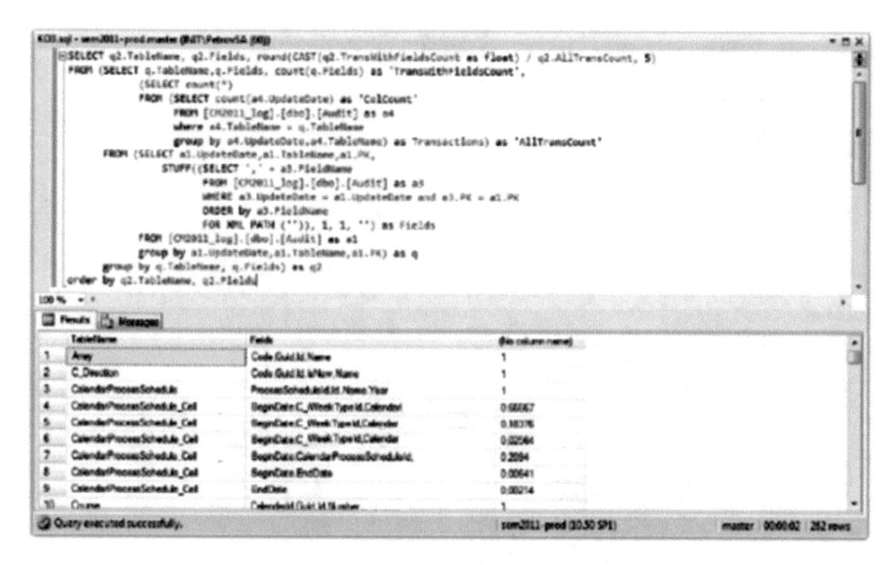

Fig. 22 Calculation of the probability of sets of columns in the query

the auxiliary table was filled with 262 variants of possible combinations of fields in the query and their probabilities (Fig. 22). Combinations that do not fall into it have zero probability. By analogy with the two previous methods of DC estimation, the third heuristic was implemented as a function.

As a result, three functions were obtained for DC estimation. The first two are used to calculate the DC of one entry from the audit table. The third one applies to the entire query, that is, to aggregate multiple records from the audit table. Recent queries have been evaluated during the testing phase.

In Fig. 23 shows the result of two queries. The first uses DC_{time} and DC_{table}, the second uses DC_{column}. It can be seen that there are both small DC values (0; 6) and higher values (30; 39).

It has also been found that the first two functions are more demanding on host resources than the third. The source of these requirements is a dynamic calculation of all probabilities, i. during DC assessment. The third function uses an auxiliary spreadsheet, which is filled daily on a schedule and runs faster during the evaluation phase. However, in general, the machine coped with the evaluation of DC requests.

To evaluate current DB requests and simulate PA operation in a trigger that parses the queries and forms an audit table, functionality has been added to run functions to evaluate DCs and display suspicious query information in Windows Event Viewer. In this case, the Event Viewer played the role of SPA, that is, the monitor that was receiving messages from the PA (Fig. 24). With the help of the xp_logevent operator, the ID of those queries whose DCs exceeded 50 for one of the heuristics was displayed in the log. If the value falls within the range of 50–75, then the message was a warning, and if above 75—an error.

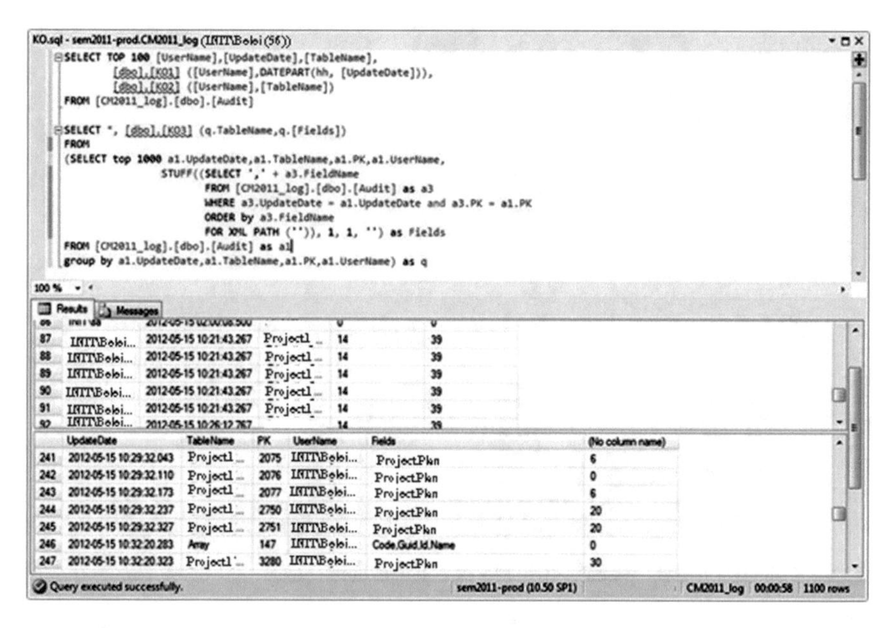

Fig. 23 Calculation DC_{time}, DC_{table} and DC_{column}

New View Number of events: 599

Number of events: 599

Level	Date and Time	Source
Warning	2/27/2014 5:18:06 PM	MSSQLSERVER
Warning	2/27/2014 5:18:05 PM	MSSQLSERVER
Error	2/27/2014 5:18:05 PM	MSSQLSERVER
Error	2/27/2014 5:18:03 PM	MSSQLSERVER
Error	2/27/2014 5:18:03 PM	MSSQLSERVER
Warning	2/27/2014 5:18:02 PM	MSSQLSERVER
Information	2/27/2014 5:18:00 PM	MSSQLSERVER
Information	2/27/2014 5:18:00 PM	MSSQLSERVER
Information	2/27/2014 5:18:00 PM	MSSQLSERVER
Information	2/27/2014 5:18:00 PM	MSSQLSERVER

Event 17063, MSSQLSERVER

General | Details

Error: 50001 Severity: 16 State: 1 KO - 89, AuditRowID - 267138

Fig. 24 Windows event viewer

Event Viewer was periodically analyzed to check for PA warnings or errors. The presence of a request identifier as part of the data from PA made it possible to delve deeper into the search for the reason for the rejection. Some of the alarms were false. To reduce them, the trigger threshold was increased. The DC agent's analysis of the assessment made it possible to identify two types of abuse, some of which actually led to a breach of DB integrity.

First, some developers manually corrected the values of the records. In doing so, they bypassed various validation tools that worked at the web application level. Validation tools checked the consistency of the fields. When setting a range using the minimum and maximum values, it is obvious that the minimum should be less than the maximum. Or required fields when creating new records. Having the fields filled in is a series and a number for the new passport line. Editing records directly did not contain such checks. These violations helped to identify a third way to calculate DC. When inserting values through the user interface, the query contained at least all required fields, and a direct query to the DB set the fields to be arbitrary.

The second that was discovered is suspicious requests from a dedicated account that was used for replication. It turned out that some of the developers were using this account to fix data that they lacked the rights to. Exceeding the threshold showed both DC-based on normal operating time and DC-based on a set of tables. The replication account worked at night and synchronized the values of a number of tables, and developers in the afternoon called other tables in the afternoon. An error in the account privilege for replication has been resolved.

The results obtained prove the applicability of the proposed heuristics for PA operation and DC estimation. The approach can be modified and applied to other PAs observing other types of WSN components.

To find the most appropriate neural network (NN) architecture and implement it, we will use the STATISTICA Neural Networks suite, which provides extensive neural network data analysis methods. It is part of the STATISTICA software product developed by StatSoft. STATISTICA Neural Networks has mechanisms that make it easy to sort through different neural network architectures, activation and error functions, and the ability to generate source code for one of the high-level languages.

The first thing to do to find the right NN architecture is to prepare a training dataset. For this purpose, a subsidiary table "TrainNS" was created, which was filled with data on the basis of the table "Audit". Each TrainNS table entry contained the query hour, its type, user name, target table, and set of columns that were present in the query. A total of about 60,000 records.

Since training on such a lot of raw data took quite a long time (about 30 min), a sub-sample consisting of 1400 records and seven users, with 200 requests for each user, was used to search for the NN architecture. The training subsample included 31 tables, 145 columns, and 8 h for user queries. Together with 3 neurons responsible for the type of operation, 187 input neurons were obtained. Since the number of potential users was seven, the output neurons were seven.

The whole set of raw data was broken into three parts. The first, the most voluminous part (70%) is the educational one. It was used to train NA. The second part (15%) is the test. Used to test network performance during training. And finally, the third (15%)—control, served for the final test of trained NN.

STATISTICA Automated Neural Networks was used to test different NN architectures. It was used to test a multilayer perceptron with one hidden layer and a network of a radial basis function type. The quadratic error function and cross-entropy were used as the error function. Both on the hidden layer and on the output, the following activation functions were tried: identical, logistic, hyperbolic, exponential and sine functions. The standard for the STATISTICA package was the Broiden-Fletcher-Goldfarb-Shannot (BFGS) algorithm.

Initially, a hidden layer with 20 neurons was used for the radial basis function type NN, and their value increased in steps of 5–120. As a result, none of the trained networks classified more than 20% of the records from the control sample. Similarly, the multilayer perceptron was tested, but the result was much better. Already, up to 85% of records were successfully classified for 4 neurons on the hidden layer. By increasing the number of neurons in the hidden layer, the percentage of positive results was increased to almost 93% (92.82%). In Fig. 25 presents the results obtained by NN in the classification of the training sample, test, and control.

For the five most efficient NN's in Fig. 26 summarizes the results of the classification. It is seen that for all trained NNs, the most difficult was to correctly classify the user to whom the penultimate, seventh column corresponds. For this user, the percentage of successful recognitions does not exceed 50%.

However, it is worth noting that in this case, NN works on the principle of "the winner takes everything". This means that the record corresponding to the request refers to NN to the user whose probability is maximum. We do not need to select a specific user, but just to get the probability of belonging to the request. In more detail, the decision-making mechanism used by the NN can be studied using a table of confidence levels on a case-by-case basis (Fig. 27).

Summary of active networks (Spreadsheet6)

Index	Net. name	Training perf.	Test perf.	Validation perf.	Training algorithm	Error function	Hidden activation	Output activation
6	MLP 187-6-7	91,82840	92,82297	92,82297	BFGS 52	Entropy	Tanh	Softmax
45	MLP 187-15-7	91,82840	92,82297	92,82297	BFGS 34	SOS	Tanh	Tanh
10	MLP 187-14-7	92,03269	92,34450	92,34450	BFGS 29	Entropy	Logistic	Softmax
38	MLP 187-14-7	92,03269	92,34450	92,34450	BFGS 55	SOS	Logistic	Tanh
31	MLP 187-12-7	91,52196	92,34450	92,34450	BFGS 19	Entropy	Identity	Softmax
4	MLP 187-9-7	91,21553	91,86603	92,34450	BFGS 36	SOS	Identity	Identity
3	MLP 187-9-7	92,03269	92,34450	91,86603	BFGS 37	Entropy	Tanh	Softmax
18	MLP 187-14-7	91,62411	92,34450	91,86603	BFGS 16	Entropy	Logistic	Softmax
19	MLP 187-17-7	91,21553	92,34450	91,86603	BFGS 11	Entropy	Tanh	Softmax
16	MLP 187-14-7	91,62411	91,86603	91,86603	BFGS 19	Entropy	Identity	Softmax
9	MLP 187-11-7	91,01124	91,86603	91,86603	BFGS 57	SOS	Identity	Identity
21	MLP 187-14-7	90,90909	91,86603	91,86603	BFGS 15	Entropy	Logistic	Softmax
27	MLP 187-7-7	92,33912	91,38756	91,86603	BFGS 112	SOS	Logistic	Identity
20	MLP 187-14-7	91,11338	92,82297	91,38756	BFGS 21	Entropy	Logistic	Softmax
22	MLP 187-17-7	91,41982	92,34450	91,38756	BFGS 17	Entropy	Identity	Softmax
51	MLP 187-6-7	91,21553	92,34450	91,38756	BFGS 53	SOS	Logistic	Identity
34	MLP 187-7-7	91,62411	91,86603	91,38756	BFGS 51	SOS	Logistic	Identity
5	MLP 187-8-7	91,01124	91,86603	91,38756	BFGS 102	SOS	Identity	Tanh
54	MLP 187-12-7	90,80695	91,86603	91,38756	BFGS 29	SOS	Identity	Identity
2	MLP 187-90-7	89,78550	91,86603	91,38756	BFGS 15	SOS	Identity	Identity
7	MLP 187-4-7	89,47906	91,86603	91,38756	BFGS 50	SOS	Identity	Identity
28	MLP 187-11-7	92,13483	91,38756	91,38756	BFGS 35	SOS	Tanh	Tanh

Fig. 25 Trained neural networks

		Var1-	Var1-	Var1-	Var1-	Var1	Var1- Ik	Var1 oll	Var1-All
6.MLP 187-6-7	Total	142,0000	138,0000	142,0000	147,0000	133,0000	135,0000	142,0000	979,0000
	Correct	142,0000	138,0000	141,0000	147,0000	133,0000	134,0000	64,0000	899,0000
	Incorrect	0,0000	0,0000	1,0000	0,0000	0,0000	1,0000	78,0000	80,0000
	Correct (%)	100,0000	100,0000	99,2958	100,0000	100,0000	99,2593	45,0704	91,8284
	Incorrect (%)	0,0000	0,0000	0,7042	0,0000	0,0000	0,7407	54,9296	8,1716
10.MLP 187-14-7	Total	142,0000	138,0000	142,0000	147,0000	133,0000	135,0000	142,0000	979,0000
	Correct	142,0000	138,0000	142,0000	147,0000	133,0000	133,0000	66,0000	901,0000
	Incorrect	0,0000	0,0000	0,0000	0,0000	0,0000	2,0000	76,0000	78,0000
	Correct (%)	100,0000	100,0000	100,0000	100,0000	100,0000	98,5185	46,4789	92,0327
	Incorrect (%)	0,0000	0,0000	0,0000	0,0000	0,0000	1,4815	53,5211	7,9673
31.MLP 187-12-7	Total	142,0000	138,0000	142,0000	147,0000	133,0000	135,0000	142,0000	979,0000
	Correct	141,0000	138,0000	141,0000	146,0000	133,0000	133,0000	64,0000	896,0000
	Incorrect	1,0000	0,0000	1,0000	1,0000	0,0000	2,0000	78,0000	83,0000
	Correct (%)	99,2958	100,0000	99,2958	99,3197	100,0000	98,5185	45,0704	91,5220
	Incorrect (%)	0,7042	0,0000	0,7042	0,6803	0,0000	1,4815	54,9296	8,4780
38.MLP 187-14-7	Total	142,0000	138,0000	142,0000	147,0000	133,0000	135,0000	142,0000	979,0000
	Correct	140,0000	138,0000	142,0000	147,0000	133,0000	135,0000	66,0000	901,0000
	Incorrect	2,0000	0,0000	0,0000	0,0000	0,0000	0,0000	76,0000	78,0000
	Correct (%)	98,5915	100,0000	100,0000	100,0000	100,0000	100,0000	46,4789	92,0327
	Incorrect (%)	1,4085	0,0000	0,0000	0,0000	0,0000	0,0000	53,5211	7,9673

Fig. 26 Query classification results

Var1 Target	Var1 - Output 6.MLP 187-6-7	Var1-l o 6.MLP 187-6-7	Var1- 6.MLP 187-6-7	Var1- k 6.MLP 187-6-7	Var1- 6.MLP 187-6-7	Var1- 6.MLP 187-6-7	Var1- Ik 6.MLP 187-6-7	Var1-l Ik 6.MLP 187-6-7
Pl aVV	P aGV	0,000196	0,000000	0,000000	0,000000	0,000081	0,620382	0,379340 F
Pl aVV	P aGV	0,000084	0,000000	0,000000	0,000000	0,000116	0,631523	0,368277 F
Pl aVV	P aGV	0,000084	0,000000	0,000000	0,000000	0,000116	0,631523	0,368277 F
Pl aVV	P aGV	0,000196	0,000000	0,000000	0,000000	0,000081	0,620382	0,379340 F
Pl aVV	P aGV	0,000084	0,000000	0,000000	0,000000	0,000116	0,631523	0,368277 F
Pl aVV	P aGV	0,000196	0,000000	0,000000	0,000000	0,000081	0,620382	0,379340 F
Pl aVV	P aGV	0,000084	0,000000	0,000000	0,000000	0,000116	0,631523	0,368277 F
Pl aVV	P aGV	0,000196	0,000000	0,000000	0,000000	0,000081	0,620382	0,379340 F
Pl aVV	P aGV	0,000196	0,000000	0,000000	0,000000	0,000081	0,620382	0,379340 F
Pl aVV	P aGV	0,000196	0,000000	0,000000	0,000000	0,000081	0,620382	0,379340 F
Pl aVV	P aGV	0,000084	0,000000	0,000000	0,000000	0,000116	0,631523	0,368277 F
Pl aVV	P aGV	0,000196	0,000000	0,000000	0,000000	0,000081	0,620382	0,379340 F
Pl aVV	P aGV	0,000196	0,000000	0,000000	0,000000	0,000081	0,620382	0,379340 F
Pl aVV	P aGV	0,000084	0,000000	0,000000	0,000000	0,000116	0,631523	0,368277 F
Pl aVV	P aGV	0,000196	0,000000	0,000000	0,000000	0,000081	0,620382	0,379340 F
Pl aVV	P aGV	0,000196	0,000000	0,000000	0,000000	0,000081	0,620382	0,379340 F
Pl aVV	P aGV	0,000084	0,000000	0,000000	0,000000	0,000116	0,631523	0,368277 F
Pl aVV	P aGV	0,000084	0,000000	0,000000	0,000000	0,000116	0,631523	0,368277 F

Fig. 27 A detailed analysis of the query classification process

After analyzing these records, it becomes clear that NN does not correctly refer the request to the user. However, the second most likely user is just the user to whom the request belongs. It can be said that NN also "suspects" him, that is, he considers as one of the variants of the solution. To calculate DC we use:

$$DC_{user} = (P\mathrm{max}_{user} - P_{realuser}) * 100\%, \tag{6}$$

where $P\mathrm{max}_{user}$ is the probability of the most likely user; $P_{realuser}$—current user probability.

If we apply it to the probabilities obtained, then DC will be approximately 25%. It is therefore clear that the approach described above is valid for DC estimation in PA. But the threshold for "responding" must be chosen above 25. Take 50, by analogy with the statistical approach. Thus, a very high percentage of recognition is not required for our problem. It is enough for the current user to simply be among the most likely.

Based on the results obtained, we conclude that the most appropriate NN architecture for user recognition on request was a multi-layer perceptron with one hidden layer. It is used in NN, which was developed for testing on the WSN system and trained for 60,000 records. For her, the number of successes in the control sample was 90%. Source code in high-level C# programming language for working with trained NN was generated by a code generator included in the STATISTICA Neural Networks package.

Based on the generated code, Visual Studio 2012 created a project based on the SQL Project template and implemented the KO4 scalar function for DC computation using the NA. The resulting assembly was published to an MS SQL server and integrated into the trigger described above, which is used to parse queries. Information about "dangerous" requests, as before, was displayed in the Windows Event Viewer window.

In Fig. 28 shows the result of the query for all four implemented DC estimation variants. It can be seen that at least one of the DCs obtained by the statistics corresponds to a more or less high NN estimation result. That is, the algorithm proposed to determine deviations is valid. Moreover, since it involved all the characteristics of the request, it showed high DCs for both types of offenses during the trial phase. Previously, it was stated that when manually editing data, a high deviation was shown, and when queries from a dedicated account and. And so he showed high deviations in all three cases. However, the approach has several disadvantages. Adaptation in it occurs at the expense of retraining NN on the schedule. This is not a big problem, but it should be noted that the implementation can be performed using more sophisticated NN architectures that adapt during their operation. A more serious problem that requires further research is the addition of new users to the WSN. In this case, new neurons should be added to the input layer and the NN architecture changed.

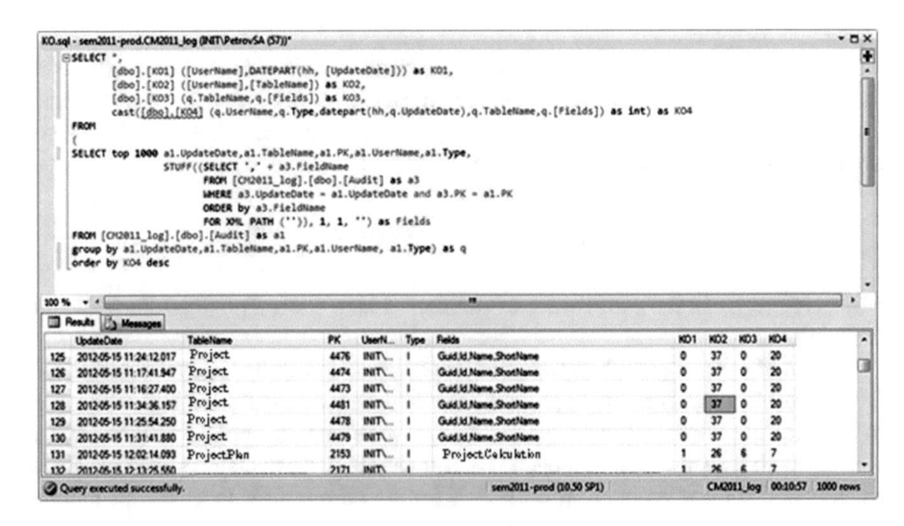

Fig. 28 A detailed analysis of the query classification process

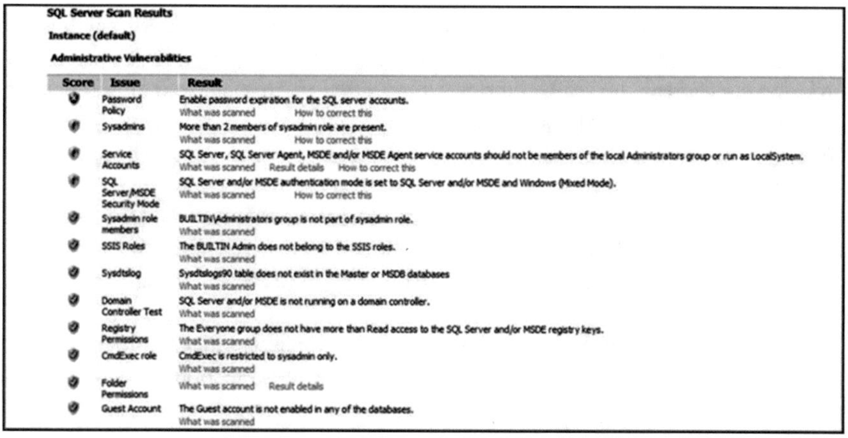

SQL Server Scan Results

Instance (default)

Administrative Vulnerabilities

Score	Issue	Result
	Password Policy	Enable password expiration for the SQL server accounts. What was scanned How to correct this
	Sysadmins	More than 2 members of sysadmin role are present. What was scanned How to correct this
	Service Accounts	SQL Server, SQL Server Agent, MSDE and/or MSDE Agent service accounts should not be members of the local Administrators group or run as LocalSystem. What was scanned Result details How to correct this
	SQL Server/MSDE Security Mode	SQL Server and/or MSDE authentication mode is set to SQL Server and/or MSDE and Windows (Mixed Mode). What was scanned How to correct this
	Sysadmin role members	BUILTIN\Administrators group is not part of sysadmin role. What was scanned
	SSIS Roles	The BUILTIN Admin does not belong to the SSIS roles. What was scanned
	Sysdtslog	Sysdtslogs90 table does not exist in the Master or MSDB databases What was scanned
	Domain Controller Test	SQL Server and/or MSDE is not running on a domain controller. What was scanned
	Registry Permissions	The Everyone group does not have more than Read access to the SQL Server and/or MSDE registry keys. What was scanned
	CmdExec role	CmdExec is restricted to sysadmin only. What was scanned
	Folder Permissions	What was scanned Result details
	Guest Account	The Guest account is not enabled in any of the databases. What was scanned

Fig. 29 Results of scanning SQL server using MBSA

And this will require retraining of the whole NA. Bypassing this problem will allow you to use as a result the classification of user roles or other more general clusters that can be assigned to users. But in this case, the results of the assessment may be worse. It all depends on the choice of characteristics, both for the input layer and for the output, and it is possible to implement it, knowing the specific conditions of the problem. For the problem discussed in this section, NN showed a good result, which is confirmed by the identified offenses.

In addition, SQL Server was scanned using Microsoft Baseline Security Analyzer to confirm the vulnerabilities. The scan results (Fig. 29) of SQL Server using MBSA showed that the policy of mandatory password updating is disabled on the target server, that is, the password is not limited. This is an indirect confirmation of the existence of identified vulnerabilities using dynamic security templates.

Overall, the results show that the proposed approach can be used to estimate DC. Given that the assumptions about connections in the original data were not made, it can be said that NN is more easily transferred to other PAs. In conclusion, it can be stated that the queries presented and implemented as DC SQL calculation algorithms confirm the possibility of detecting and evaluating deviation in WSN components.

5 Conclusion

For the proposed method for creating a static WSN security template, the ontology used to analyze real WSNs is formally described, the results of which confirm the effectiveness of the developed method.

The proposed multi-agent system was simulated for real WSN, confirming the effectiveness of the developed method for creating a dynamic WSN security template. Revealed "weak" server.

A multi-agent system has been developed to obtain a dynamic safety standard, the composition of agents, their relationship, and algorithms. Algorithms for calculating deviation coefficients for SQL-queries using statistical methods and neural network technologies have been developed. The developed model of a multi-agent system for a real corporate software system was tested, which confirmed the practical applicability of the developed method.

Implemented and implemented in a real enterprise software system intelligent security agent MS SQL, which uses the developed algorithms to estimate the deviation of SQL queries. Vulnerability in the functioning of the corporate software system has been identified.

An intelligent MS SQL protection agent is implemented and implemented in a real WSN, using developed algorithms to evaluate the rejection of SQL queries. Vulnerabilities in WSN functioning were revealed.

In future research, we will focus more on the very protocols of transmitting and encrypting data between sensors on wireless networks.

References

1. Wu J, Yuan Sh, Zhou G, Ji S, Wang Z, Wang Y (2009) Design and evaluation of a wireless sensor network based aircraft strength testing system. Sens J 9:195–210. https://doi.org/10.3390/s90604195
2. Somov A, Baranov A, Savkin A, Spirjakin D, Spirjakin A, Passerone R (2011) Development of wireless sensor network for combustible gas monitoring. Sens Actuators A Phys 171:398–405. https://doi.org/10.1016/j.sna.2011.07.016
3. Silvani X, Morandini F, Innocenti E (2015) Evaluation of a wireless sensor network with low cost and low energy consumption for fire detection and monitoring. Fire Technol 51:971–993. https://doi.org/10.1007/s10694-014-0439-9
4. Minor CP, Johnson KJ, Rose-Pehrsson SL, Owrutsky JC, Wales SC, Steinhurst DA, Gottuk DT (2010) A full-scale prototype multisensor system for damage control and situational awareness. Fire Technol 46:437–69
5. Garcia-Sanchez A-J, Garcia-Sanchez F, Garcia-Haro J (2011) Wireless sensor network deployment for integrating video-surveillance and data-monitoring in precision agriculture over distributed crops. Comput Electron Agric 75:288–303. https://doi.org/10.1016/j.compag.2010.12.005
6. Ehsan S, Bradford K, Brugger M, Hamdaoui B, Kovchegov Y, Johnson D, Louhaichi M (2012) Design and analysis of delay-tolerant sensor networks for monitoring and tracking free-roaming animals. IEEE Trans Wirel Commun 11:1220–27. https://doi.org/10.1109/TWC.2012.012412.111405
7. Umar MM, Khan S, Ahmad R, Singh D (2018) Game theoretic reward based adaptive data communication in wireless sensor networks. IEEE Access 6:28073–28084. https://doi.org/10.1109/ACCESS.2018.2833468
8. Bradford K, Brugger M, Ehsan S, Hamdaoui B, Kovchegov Y. (2011) Data loss modeling and analysis in partially-covered delay-tolerant networks. In: 2011 Proceedings of 20th international conference on computer communications and networks
9. Parra L, Karampelas E, Sendra S, Lloret J, Rodrigues PC (2015) Design and deployment of a smart system for data gathering in estuaries using wireless sensor networks. In: 2015 international conference on computer, information, and telecommunication systems. https://doi.org/10.1109/cits.2015.7297757

10. Kruger CP, Hancke GP (2014) Implementing the Internet of Things vision in industrial wireless sensor networks. In: 12th IEEE international conference on industrial informatics (INDIN), pp 627–632. https://doi.org/10.1109/indin.2014.6945586
11. Karim L, Anpalagan A, Nasser N, Almhana J (2013) Sensor-based M2M agriculture monitoring systems for developing countries: state and challenges. Netw Protoc Algorithms 5(3):68–86. https://doi.org/10.5296/npa.v5i3.3787
12. Chen B, Mao J, Guo N, Qiao G, Dai N (2013) An incentive detection mechanism for the cooperation of nodes selfish behavior in wireless sensor networks. In: 2013 25th Chinese control and decision conference, pp 4021–4024. https://doi.org/10.1109/ccdc.2013.6561653
13. Belej O (2020) The cryptography of elliptical curves application for formation of the electronic digital signature. In: Hu Z, Petoukhov S, Dychka I, He M (eds) Advances in computer science for engineering and education II. ICCSEEA 2019. Advances in intelligent systems and computing, vol 938. Springer, Cham, pp 43–57. https://doi.org/10.1007/978-3-030-16621-2_5
14. Belej O, Lohutova T, Sadeckii J (2020) The features of security of transfer and storage data for the Internet of Things in Cloud Database. In: Proceeding of the international scientific-practical conference on problems of infocommunications science and technology 2019, pp 7–15
15. Radivilova T et al (2018) Decrypting SSL/TLS traffic for hidden threats detection. In: Proceedings of the 2018 IEEE 9th international conference on dependable systems, services, and technologies (DESSERT). IEEE, pp 143–146. https://doi.org/10.1109/dessert.2018.8409116
16. Radivilova T, Kirichenko L, Ageyev D, Bulakh V (2019) Classification methods of machine learning to detect DDoS attacks. In: 2019 10th IEEE international conference on intelligent data acquisition and advanced computing systems: technology and applications (IDAACS). IEEE, Metz, France, pp 207–210. https://doi.org/10.1109/idaacs.2019.8924406
17. Kryvinska N (2008) An analytical approach for the modeling of real-time services over the IP network. Math Comput Simul 79(4):980–990. https://doi.org/10.1016/j.matcom.2008.02.016
18. Kryvinska N, Zinterhof P, van Thanh D (2007) An analytical approach to the efficient real-time events/services handling in converged network environment. In: Enokido T, Barolli L, Takizawa M (eds) Network-based information systems. NBiS 2007. Lecture notes in computer science, vol 4658. Springer, Berlin, Heidelberg
19. Mashkov OA, Sobchuk VV, Barabash OV, Dakhno NB, Shevchenko HV, Maisak TV (2019) Improvement of the variational-gradient method in dynamical systems of automated control for integrodifferential models. Math Model Comput 6(2):344–357. https://doi.org/10.23939/mmc2019.02.344
20. Laptiev O, Shuklin G, Savchenko V, Barabash O, Musienko A, Haidur H (2019) The method of hidden transmitters detection is based on the differential transformation model. Int J Adv Trends Comput Sci Eng 8(6):2840–2846. https://doi.org/10.30534/ijatcse/2019/26862019

Studying of Useful Signal Impact on Convergence Parameters of the Gradient Signal Processing Algorithm for Adaptive Antenna Arrays that Obviates Reference Signal Presence

Oleksandr Pliushch, Serhii Toliupa, Viktor Vyshnivskyi, and Anatolii Rybydajlo

Abstract Signal processing algorithms in adaptive antenna arrays are analyzed with respect to their application in mobile telecommunication systems. It is established that both in literature and in practice insufficient attention is paid to search for and design of such simple adaptation algorithms that do not require presence of the reference signal. A simple algorithm of antenna array adaptation was earlier deduced that operates in the absence of the reference signal. Its successful testing was carried out under assumptions that the useful signal in the adaptive antenna array does not have an impact on the adaptation process. This assumption is valid for radar applications with short duration useful signals but not necessarily applicable for telecommunication systems with continuous ones. To address these limitations, influences of the useful signal on the algorithm convergence parameters are studied. Computer simulation of adaptive antenna arrays with the designed algorithm with account of useful signal impact in different situations is performed. It is shown that with some adjustments of the constant that determines the adaptation step, the impact of continuous useful signals on the designed gradient algorithm convergence performance can be greatly mitigated.

Keywords Mobile telecommunication systems · Adaptive antenna arrays · Gradient adaptation algorithms · Computer simulation · Signal-to-interference ratio · Algorithm performance

---

O. Pliushch (✉) · V. Vyshnivskyi
State University of Telecommunications, Kyiv, Ukraine

S. Toliupa
Taras Shevchenko National University, Kyiv, Ukraine
e-mail: tolupa@i.ua

A. Rybydajlo
Center for Military and Strategic Studies of the National Defence University of Ukraine Named After Ivan Cherniakhovskyi, Kyiv, Ukraine

D. Ageyev et al. (eds.), *Data-Centric Business and Applications*, Lecture Notes on Data Engineering and Communications Technologies 69,
https://doi.org/10.1007/978-3-030-71892-3_11

1 Introduction

Adaptive antenna arrays (AAA) have been known to improve performance of different complex systems for decades [1–4]. Among their main applications, the most important are those in telecommunications and radars. Scientists and engineers in the field have designed a big range of pretty simple, nevertheless effective, algorithms for AAA convergence to the optimum weight vector. Careful analysis of these algorithms reveals that many of them necessitate creation of the reference signal, what imposes limitations on their practical applications [5–8]. This problem presents itself as an especially acute in the telecommunications field. In telecommunications, only those signals are transmitted that contain some sort of new information, hence engineers can only make-do with surrogate reference signals, which negates some of the advantages of AAA deployment.

To address this problem, the authors have developed a simple yet efficient gradient algorithm that obviates any presence of the reference signal [9]. In computer simulation, the algorithm has shown good performance in challenging interference environment and is considered as a viable option for application in telecommunications field, namely mobile telecommunications. Still, some unexplored areas of research remain, namely those relating to the impact of the continuous useful signal on the convergence of the algorithm. The problem is that one of the assumptions made before testing the algorithm is that the useful signal does not have any influence on its workings. This premise usually holds when it comes to radar applications, in which interfering noises are continuous while the useful signal is simply a pulse of short duration. In telecommunication environment, however, useful signals are continuous as well and can impact the performance of the algorithm.

Therefore, the aim of this paper is to study the effects of the continuous useful signal on the performance of the previously designed algorithm [9] and to find ways, if necessary, to lessen those effects to sustain good convergence characteristics obtained earlier.

2 Description of the Algorithm that Does not Require a Reference Signal and Its Computer Simulation

2.1 Presentation of the Algorithm

The algorithm was purposefully developed to eschew the requirement of creating the reference signal—the procedure entailing some difficulties in the telecommunication applications. This approach is proposed to be used in so-called signal aligned AAA. Functional diagram of this type of an array is presented in Fig. 1, which features a narrow-band version of the AAA [9].

Phase shifters deployed in each AAA channel play the role of attunement to the useful signal. The only information required about the useful signal is its arrival angle,

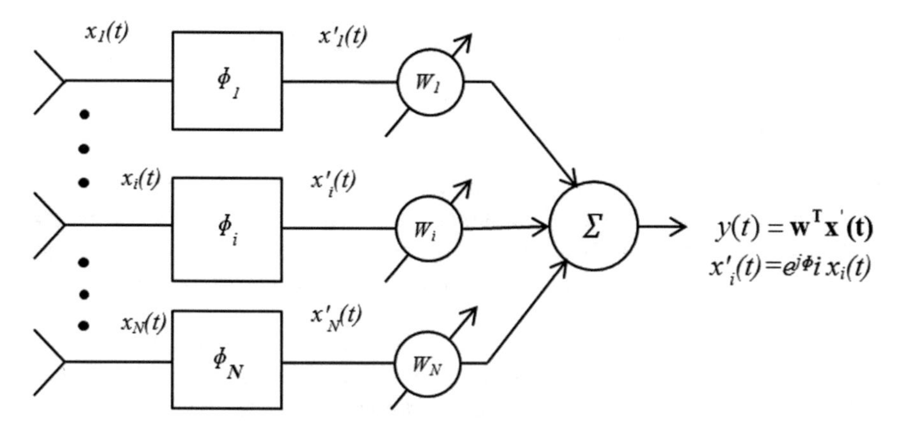

Fig. 1 Functional diagram of the narrow-band useful signal-aligned array

and it is presumed that the angle is known, what is a valid assumption. Such approach obviates the requirement of forming the reference signal in any explicit form. In Fig. 1, $x_i(t)$—denotes received signal in i element of the array, and w_i—weight coefficient in i element.

After passing the phase shifters, antenna elements transformed signal vector can be presented by the equation:

$$\mathbf{x}'(t)=\Phi\mathbf{x}(t), \tag{1}$$

where $\boldsymbol{\Phi}$ is a special diagonal matrix, diagonal elements of which are complex exponents with modules valued one and phase shifts related to the useful signal incidence angle:

$$\Phi = \begin{pmatrix} e^{i\Phi_1} & 0 & 0 \\ 0 & \ddots & 0 \\ 0 & 0 & e^{i\Phi_N} \end{pmatrix}.$$

Gradient algorithm for AAA that was designed by the authors for the array shown in Fig. 1 can be described as follows [9]:

$$\begin{aligned} \mathbf{w}(k+1) &= \mathbf{w}(k) - \Delta_s(\mathbf{x}(k)\mathbf{x}^{\mathbf{T}}(k)\mathbf{w}(k) - y(k)y(k) * \mathbf{1}) \\ w_i(k+1) &= \frac{w_i(k+1)}{\sum_{n=1}^{N} w_n(k+1)}, \end{aligned} \tag{2}$$

where N is the number of elements in the antenna array, vector $\mathbf{1} = [1, 1,\ldots,1^{\mathrm{T}}]$, T—denotes complex vector and matrix conjugation and transportation, * indicates scalar conjugation and Δ_s—the adjustment step size.

2.2 Computer Simulation of the Algorithm Without Impact of the Continuous Useful Signal

A well-tried and widely-used method of computer simulation in AAA is to describe internal noises of each antenna element as well as external interfering signals as those with the Gaussian distribution. This approach allows testing performance of an algorithm or a method for a wide range of applications. In addition, the performance of AAA in Gaussian environment can be used as the benchmark for other kinds of signals, not least with less random character.

Algorithm (2) was tested using linear narrow-band AAA with the following parameters: the number of elements in the AAA—9, distance between the elements—half the wavelength, relative power of internal noise in each antenna element—one unit.

Adaptation process was evaluated by observing signal-to-noise ratio (SNR) at the output of the array. Each complex noise signal (internal noise or interference) was presented by uncorrelated samples with normal distribution. It corresponds to the situation when samples are taken at a sampling rate equaling the channel bandwidth. Power of the noise signals was measured in relative unites with respect to the internal noise power. There were simulated three sources of interfering signals with angles of incidence −28.648, −57.296 and 57.296 degrees, with respective power levels 100, 200 and 100 units. Arrival angle of the useful signal was 0° and it had the power 100 units.

Potential signal-to-noise ratio for the situation under consideration, which is calculated using ideal covariance matrix by applying well-known formulas [5, 6], equals approximately 29.47 dBs.

With the step size fixed at 3.4636e−05, the SNR curve for this case is presented in Fig. 2. It is relevant to accentuate here that the useful signal does not have any impact on the convergence process, which can reflect the situation of the useful signal as a relatively short duration pulse.

Figures 3 and 4 demonstrate AAA gain coefficient or directivity patterns for, respectively, weight vector after adaptation and the optimal weight vector.

Analysis of data in Fig. 2 shows that algorithm (2) secures good convergence rate versus achievable SNR at the end of the adaptation process. Adaptation process is indicative of three distinct phases: first phase, which lasts approximately 70–80 iterations, whereby SNR quickly climbs to 22 dBs; second phase, in which SNR stays the same until the number of iterations comes to 5000; and third phase, ongoing up to about 10,000 iterations, after which stationary state is reached. The difference between potential SNR and that at the end of the adaptation process equals 2–3 dBs, which is an acceptable number. The SNR curve in Fig. 2 will be used as a bench mark to evaluate behavior of the algorithm (2) in the situations in which useful signal has an impact on the adaptation process.

Figures 3 and 4 demonstrate the fact that directivity pattern of the AAA at the end of the adaptation process closely matches the one obtained using optimal weight vector. It serves as an extra indication of the good performance of the algorithm (2).

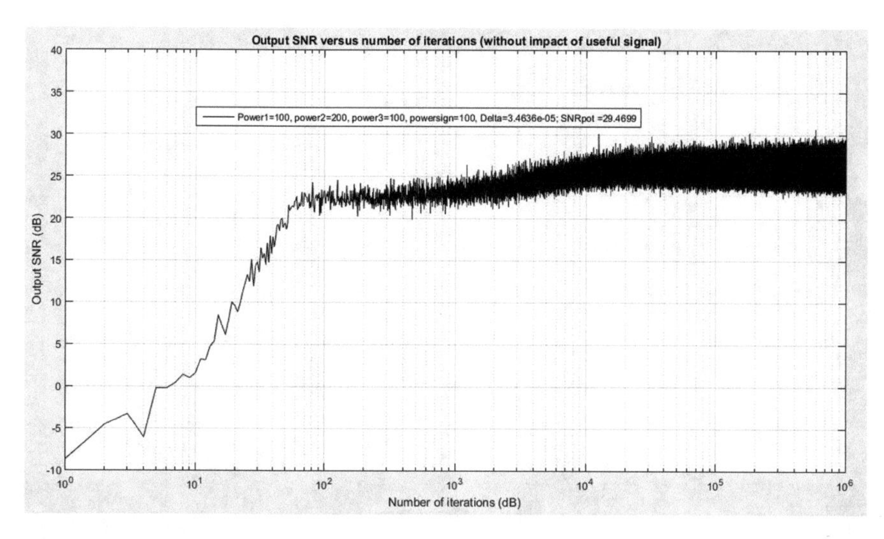

Fig. 2 Output SNR versus number of iterations for algorithm (2) with Gaussian probability distribution without impact of useful signal

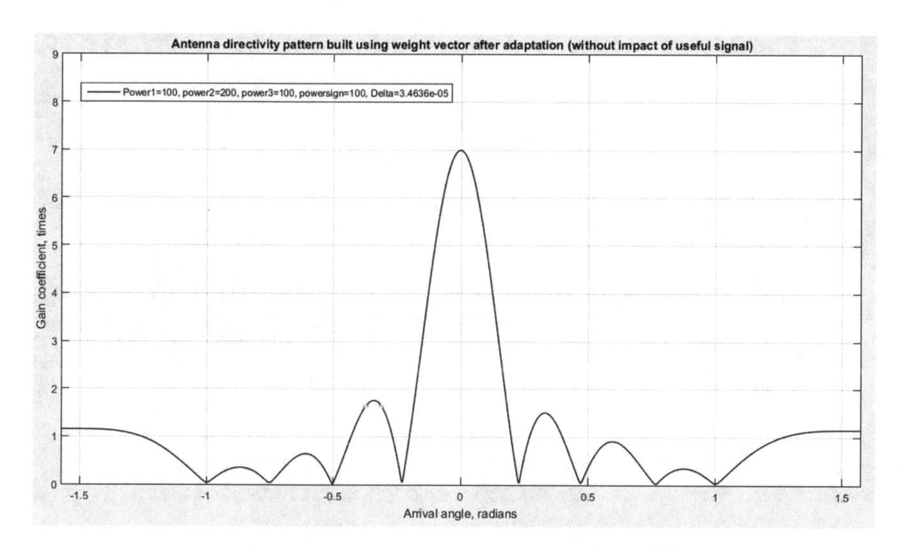

Fig. 3 Antenna directivity pattern built using weight vector after adaptation

2.3 *Computer Simulation of the Algorithm with Account of the Useful Signal Impact*

To assess the influence that the useful signal has on the adaptation process in the algorithm, useful signal was simulated as well as a noise with Gaussian distribution. This approach is viable because signals in modern mobile telecommunications due

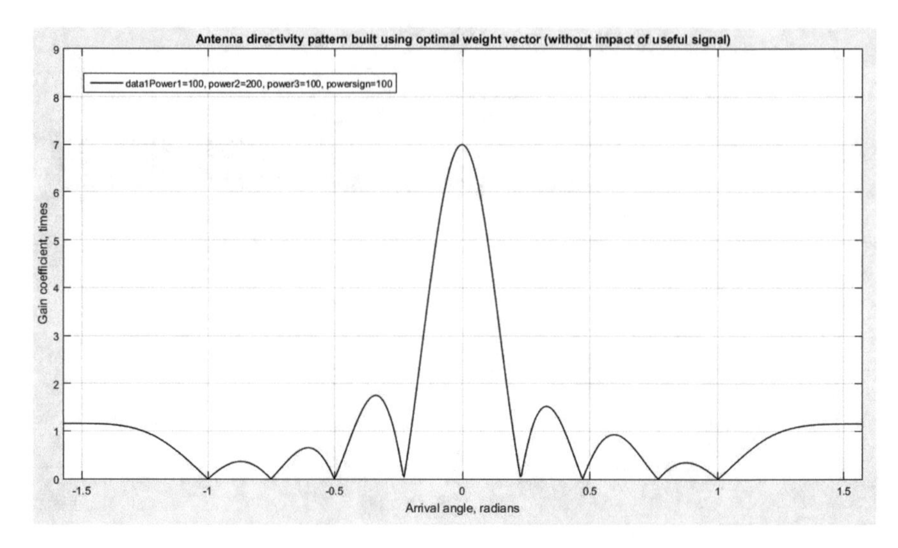

Fig. 4 Antenna directivity pattern built using optimal weight vector

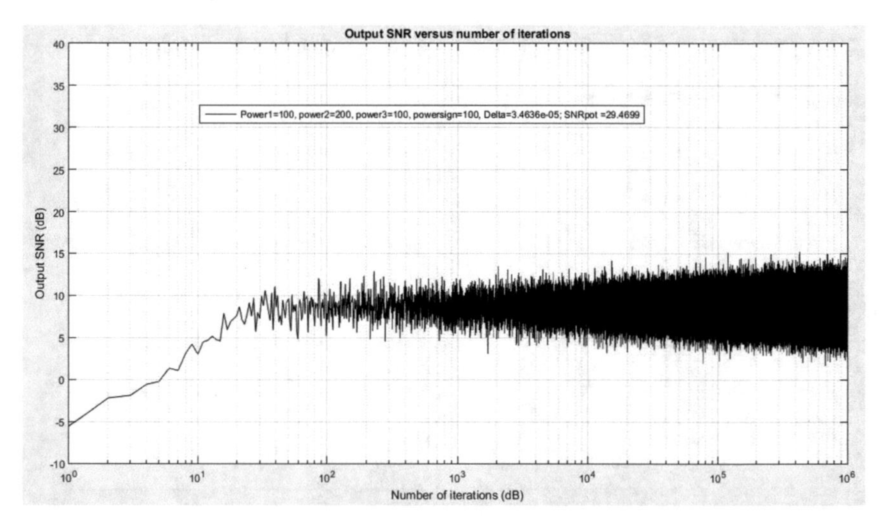

Fig. 5 Output SNR versus number of iterations for algorithm (2) with account of impact of useful signal with power 100 units

to unpredictable nature of the messages, use of complex higher-order modulations and different coding methods closely resemble a noise process.

For obtaining bigger picture, useful signal for the noise situation presented above was simulated at two power levels of 100 and 10 relative units.

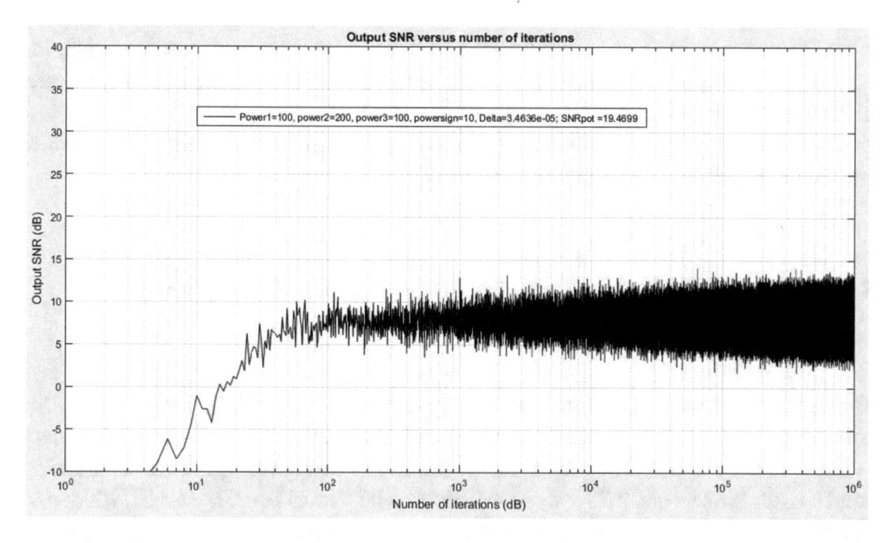

Fig. 6 Output SNR versus number of iterations for algorithm (2) with account of impact of useful signal with power 10 units

The AAA SNR curves for the case of useful signal influences on the adaptation process are presented in Fig. 5, for the relative useful signal power of 100 units, and in Fig. 6, for that of 10 unites.

Analysis of data in the above figures clearly shows that the useful continuous signal with Gaussian distribution considerably degrades performance of the algorithm (2). For example, for the case of relative useful signal power 100 units, the loss in achievable SNR reaches 19 dBs, while for the case of 10 units the loss comes to roughly 12 dBs. Clearly, this fact needs an explanation as well as a search for ways of redressing the situation.

3 Ways of Mitigating Continuous Useful Signal Impact on the Convergence Performance

3.1 Understanding the Influence of the Continuous Useful Signal and Searching for Possible Solutions

According to the algorithm (2), at each step of the adaptation the weight vector is normalized, that is, its module is set to 1. The phase shifters in Fig. 1 are not adjustable for a certain period of time and their phases are selected such that the useful signal always has the biggest gain in AAA and, as a result, it is never cancelled by the algorithm. Hence, useful signal adds to the internal noise at the output of the AAA and the resulting noise becomes much elevated. This higher noise translates into what is known as weight vector noise, and the bigger the weight vector noise, the greater the losses in achievable SNR. This is evidenced by the graphs in Figs. 5 and 6. That is the mechanism at the origin of the problem.

One of the ways of improving performance in the case of high weight vector noise is to reduce the step of adjustment size in (2).

3.2 *Computer Simulation of the Algorithm with Different Adjustment Step Values*

To assess the possibility of compensating losses due to weight vector noise, the computer simulation of the algorithm (2) with different values of the step of adaptation was carried out. The step size, or Δ_s, in (2) was given gradually decreasing values of 9.2362e−06, 2.7709e−06, 1.3854e−06, 5.5417e−07, and 2.7709e−07. The noise situation is the same as above and the useful signal power assumes the same relative values of 100 and 10 units.

Computer simulation results, which show dependence of SNR on number of iterations for different step sizes and useful signal powers, are presented in Figs. 7, 8, 9, 10, 11, 12, 13, 14, 15 and 16.

Analysis of the date clearly reveals some trends and behaviors that prove right the proposed method of mitigating losses due to weight vector noise.

Firstly, as the step size decreases, the loss in SNR goes down as well. For, example, in Fig. 7 loss in SNR as compared with the potential one is 5 dBs less than that in Fig. 5. With further decrease in the step size, this trend continues. In Fig. 9 losses in

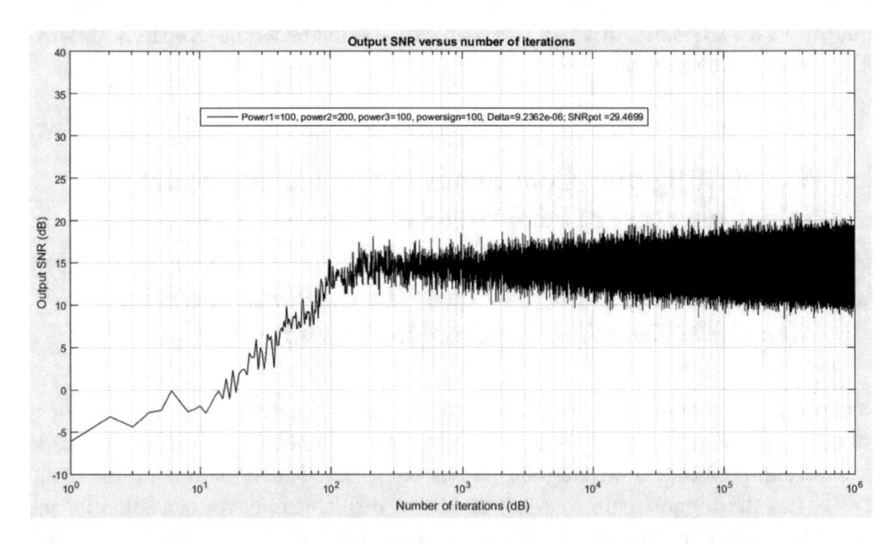

Fig. 7 Output SNR versus number of iterations for algorithm (2) with account of impact of useful signal with power 100 units and step size 9.2362e−06

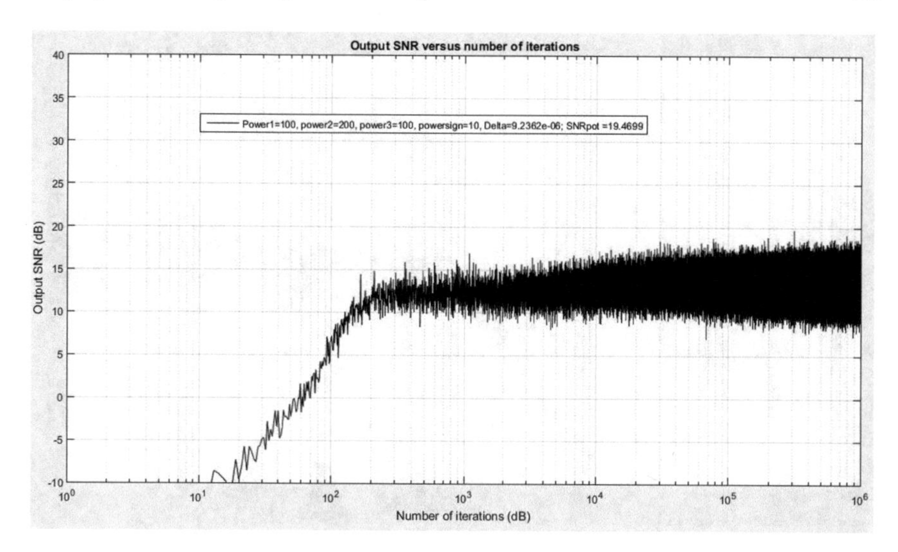

Fig. 8 Output SNR versus number of iterations for algorithm (2) with account of impact of useful signal with power 10 units and step size 9.2362e−06

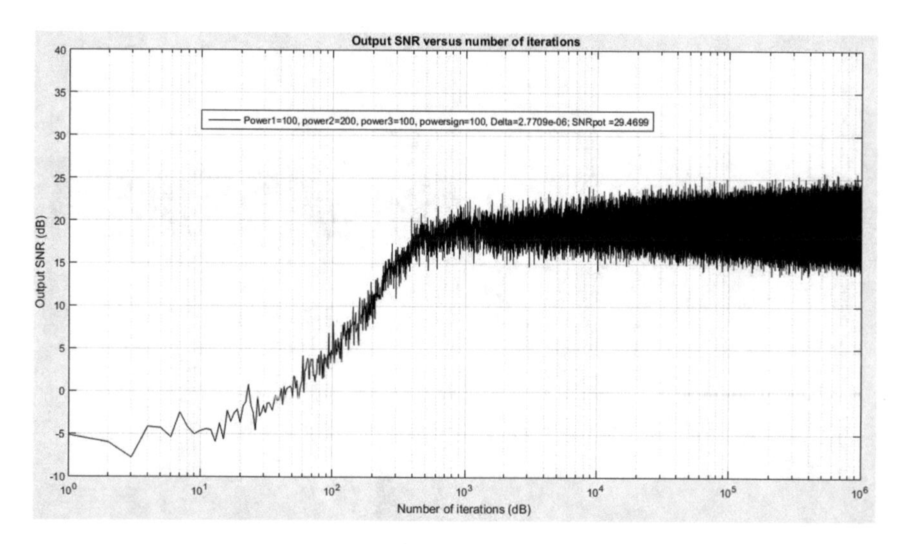

Fig. 9 Output SNR versus number of iterations for algorithm (2) with account of impact of useful signal with power 100 units and step size 2.7709e−06

SNR are yet again roughly 5 dBs less than those in Fig. 7. Overall, for relative power 10 in Fig. 10 the losses caused by the impact of the useful signal can be deemed already compensated by the way of reducing the step size. For the relative power of the useful signal equaling 100, compensation of the losses due to useful signal

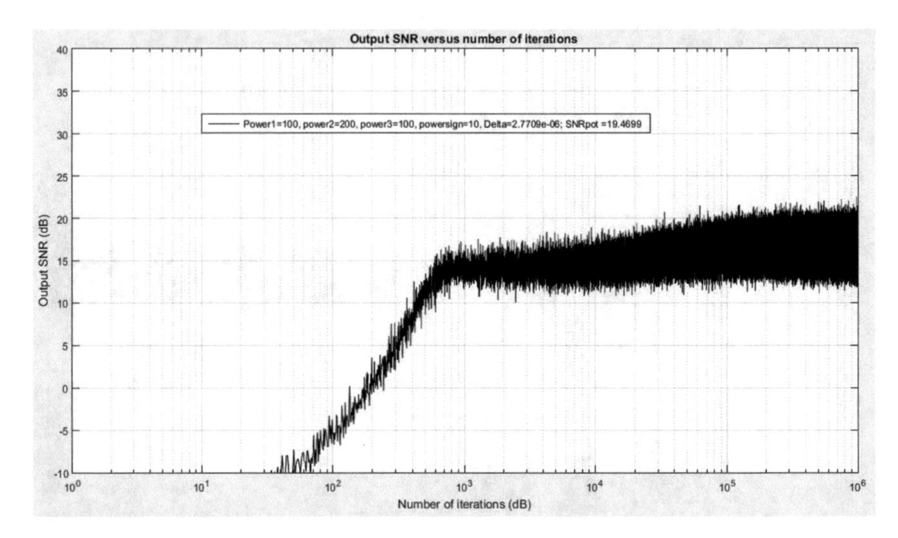

Fig. 10 Output SNR versus number of iterations for algorithm (2) with account of impact of useful signal with power 10 units and step size 2.7709e−06

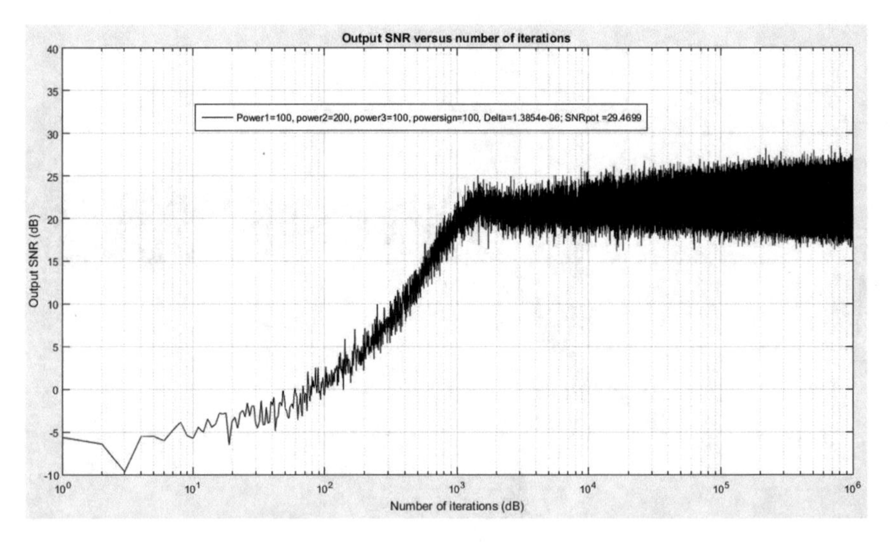

Fig. 11 Output SNR versus number of iterations for algorithm (2) with account of impact of useful signal with power 100 units and step size 1.3854e−06

power impact on the algorithm performance seems to be achieved much later as it is evidenced by the data in Fig. 15.

Secondly, the compensation of SNR losses shown above comes at a price. For example, the time of adaptation is increasing with every decrease in the step size. And

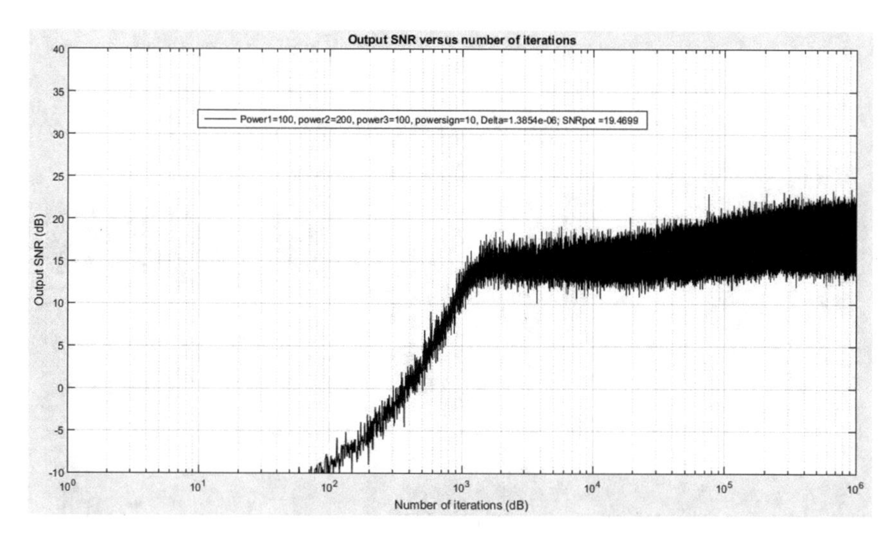

Fig. 12 Output SNR versus number of iterations for algorithm (2) with account of impact of useful signal with power 10 units and step size 1.3854e−06

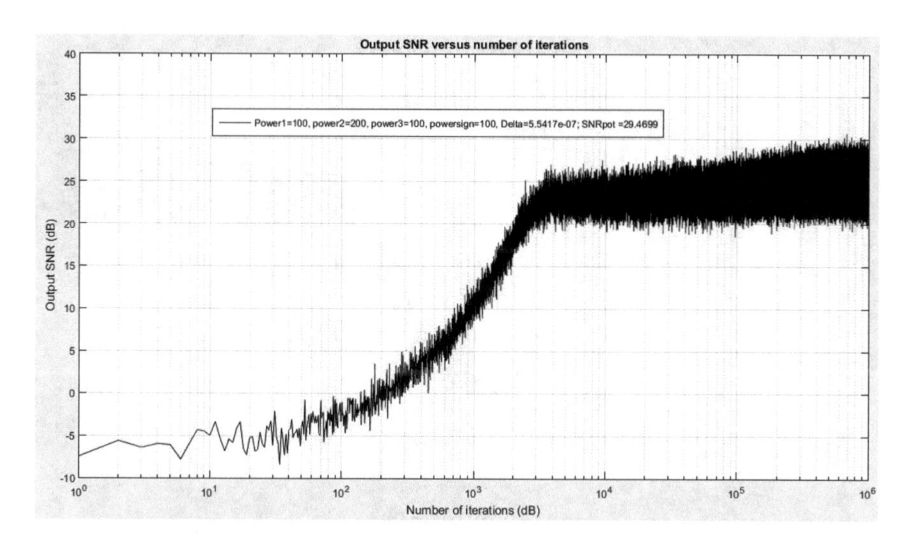

Fig. 13 Output SNR versus number of iterations for algorithm (2) with account of impact of useful signal with power 100 units and step size 5.5417e−07

this increase is rather considerable, from convergence time of roughly 90 iterations for the curve in Fig. 2, to 2000 iterations in Fig. 11.

Figures 17, 18, 19, 20, 21, 22, 23, 24, 25 and 26 demonstrate AAA directivity patterns at the end of the adaptation process for different step sizes and two values of useful signals powers, 100 and 10. In this case, the noise situations correspond to those used to build SNR curves above.

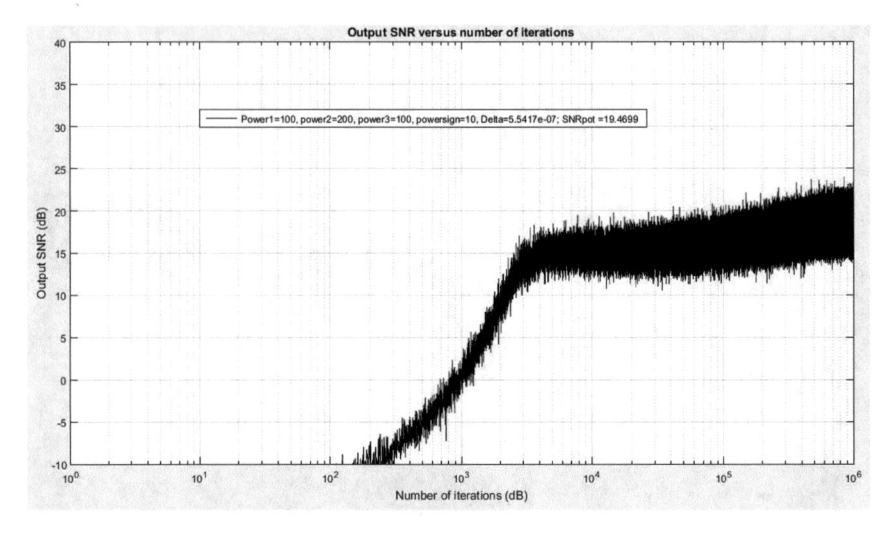

Fig. 14 Output SNR versus number of iterations for algorithm (2) with account of impact of useful signal with power 10 units and step size 5.5417e−07

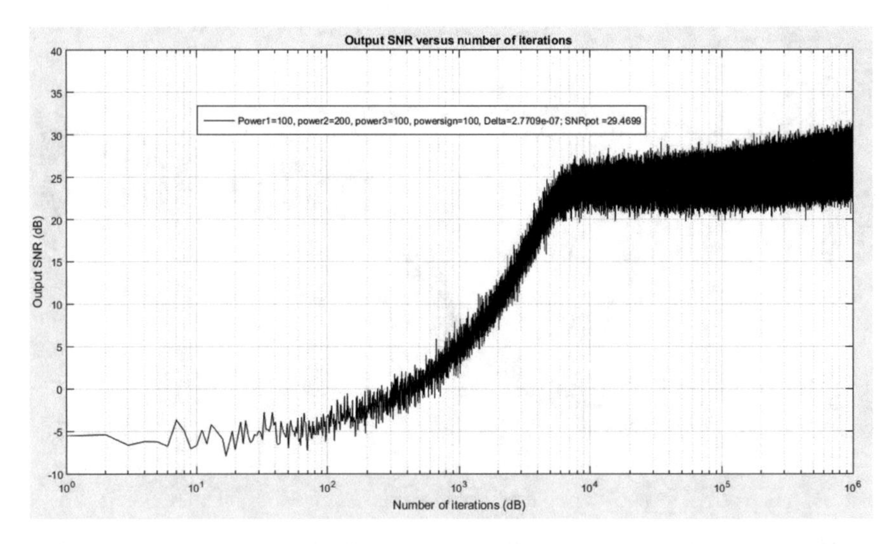

Fig. 15 Output SNR versus number of iterations for algorithm (2) with account of impact of useful signal with power 100 units and step size 2.7709e−07

Graphs in those figures confirm the finding that the lower SNR observed previously is reflective of elevated weight vector noises due to the fact that the algorithm (2) perceives useful signal as an added noise at the output of the AAA.

It can be deduced from the figures that at the higher step sizes the form of the directivity patterns loosely resembles that of the optimum one, as depicted in Fig. 4.

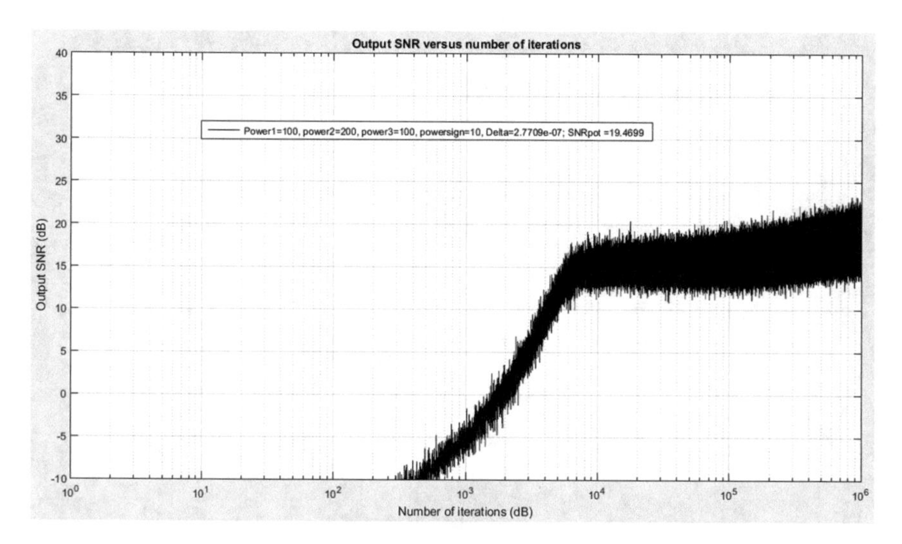

Fig. 16 Output SNR versus number of iterations for algorithm (2) with account of impact of useful signal with power 10 units and step size 2.7709e−07

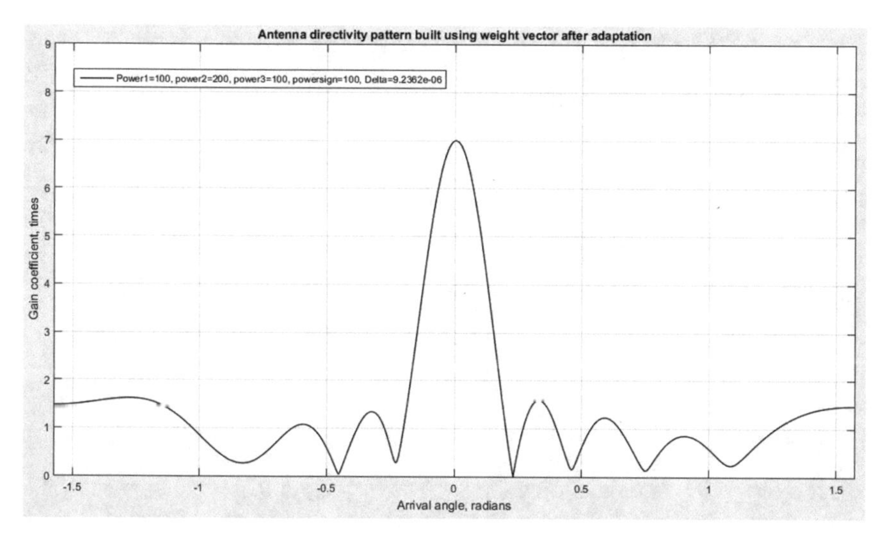

Fig. 17 Antenna directivity pattern built using weight vector after adaptation with account of impact of useful signal with power 100 units and step size 9.2362e−06

It is reflective of quicker convergence rate. At the same time, zeros in the directions of the interfering signals are not accurate, which leads to lower achievable SNR. Examples of this trend one can find in Figs. 17 and 19.

Conversely, if the step sizes are lower, the directivity patterns deviate more from the optimum one. But in this case zeroes in directions of interfering signals are much

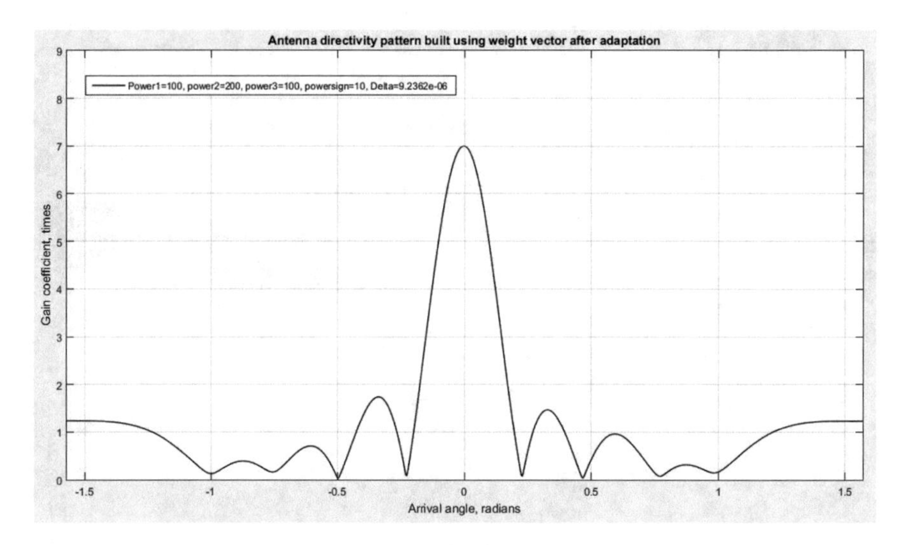

Fig. 18 Antenna directivity pattern built using weight vector after adaptation with account of impact of useful signal with power 10 units and step size 9.2362e−06

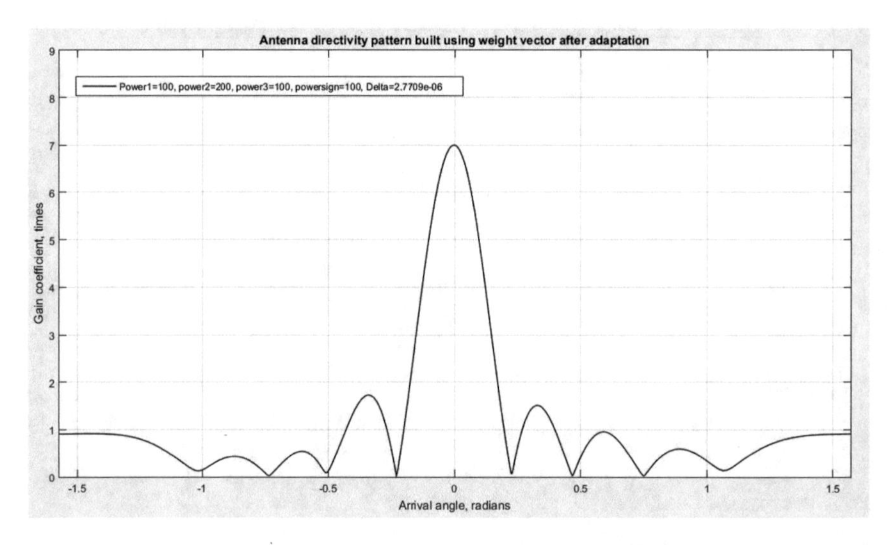

Fig. 19 Antenna directivity pattern built using weight vector after adaptation with account of impact of useful signal with power 100 units and step size 2.7709e−06

more accurate and this provides much higher SNR. This trend is evidenced by the graphs in Figs. 24, 25 and 26.

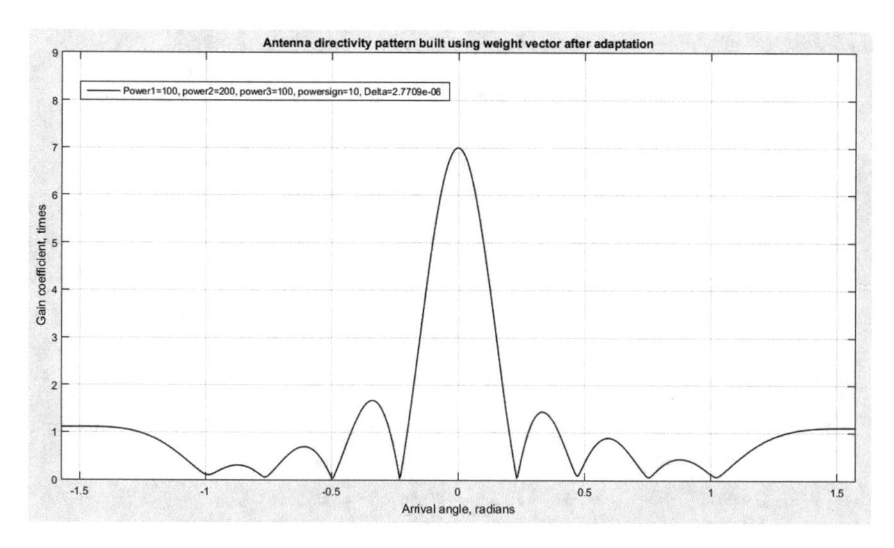

Fig. 20 Antenna directivity pattern built using weight vector after adaptation with account of impact of useful signal with power 10 units and step size 2.7709e−06

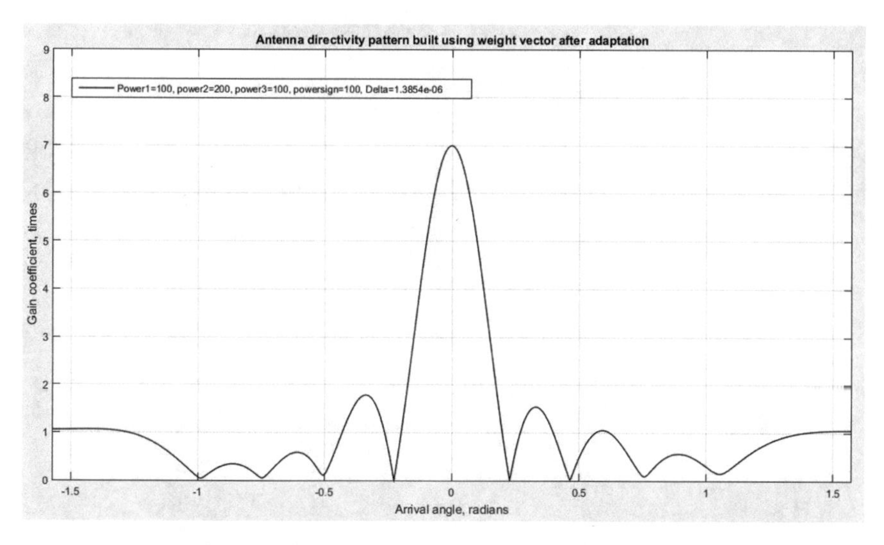

Fig. 21 Antenna directivity pattern built using weight vector after adaptation with account of impact of useful signal with power 100 units and step size 1.3854e−06

3.3 Discussion of the Research Results

The algorithm (2) has shown extremely good performance in the cases in which useful signal is considered as offering no influence on the adaptation mechanism.

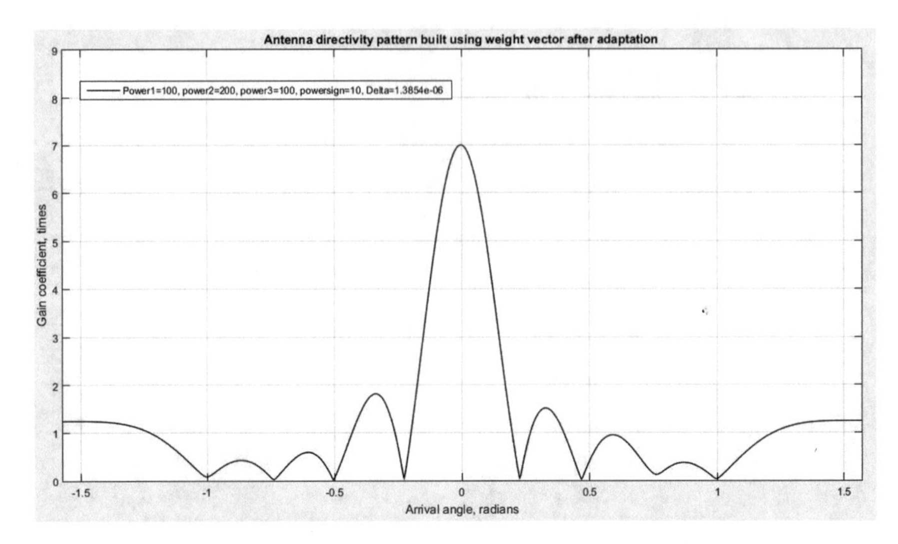

Fig. 22 Antenna directivity pattern built using weight vector after adaptation with account of impact of useful signal with power 10 units and step size 1.3854e

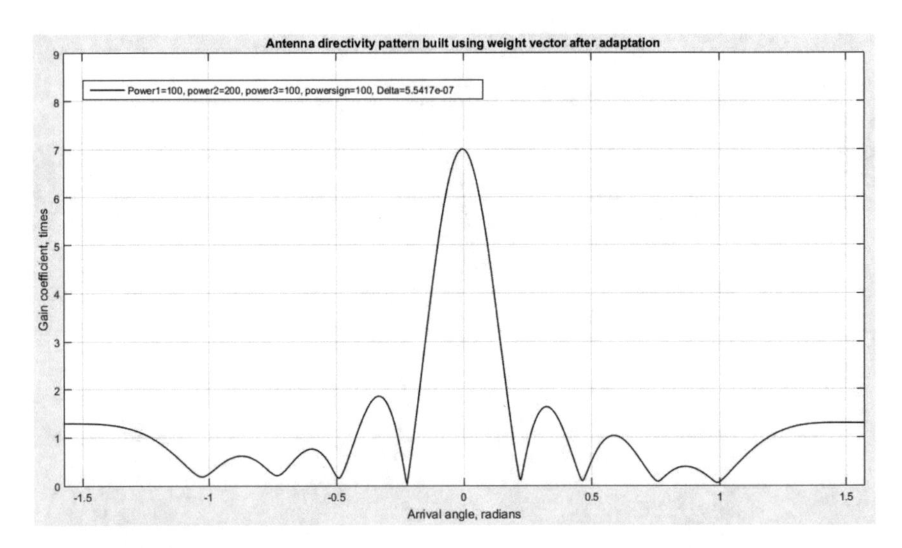

Fig. 23 Antenna directivity pattern built using weight vector after adaptation with account of impact of useful signal with power 100 units and step size 5.5417e−07

Such situations are common place in radar applications, in which useful signals are presented as relatively short pulses.

In telecommunications, this is not always the case, even to the contrary—useful signals are continuous in nature and are close in their structure to the randomly distributed ones.

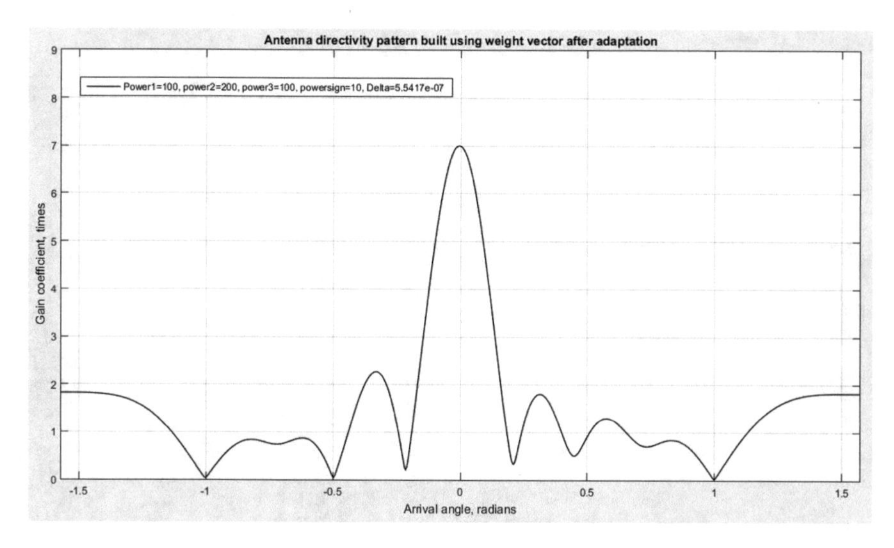

Fig. 24 Antenna directivity pattern built using weight vector after adaptation with account of impact of useful signal with power 10 units and step size 5.5417e−07

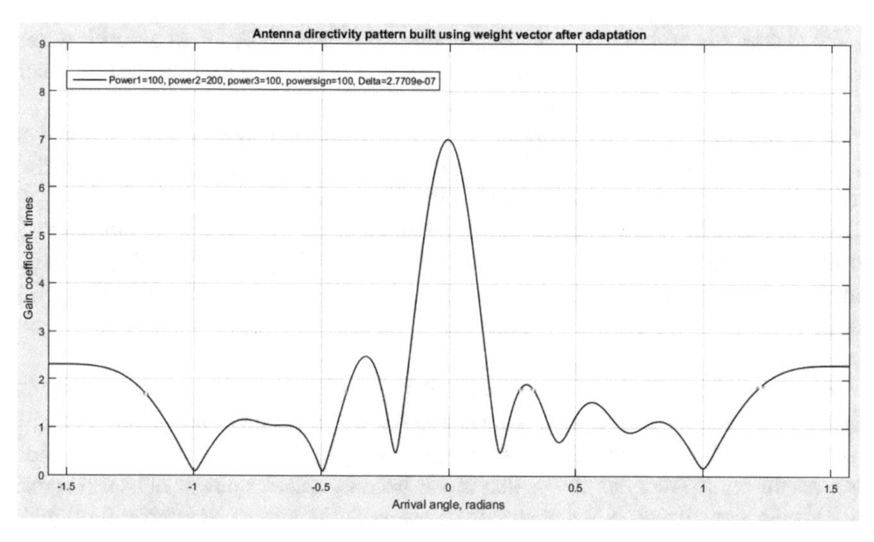

Fig. 25 Antenna directivity pattern built using weight vector after adaptation with account of impact of useful signal with power 100 units and step size 2.7709e−06

Computer simulation, performed in this paper, clearly shows that such useful signals can impact the workings of the algorithm (2) and greatly reduce its performance. The nature of this influence is such that it leads to the higher weight vector noises.

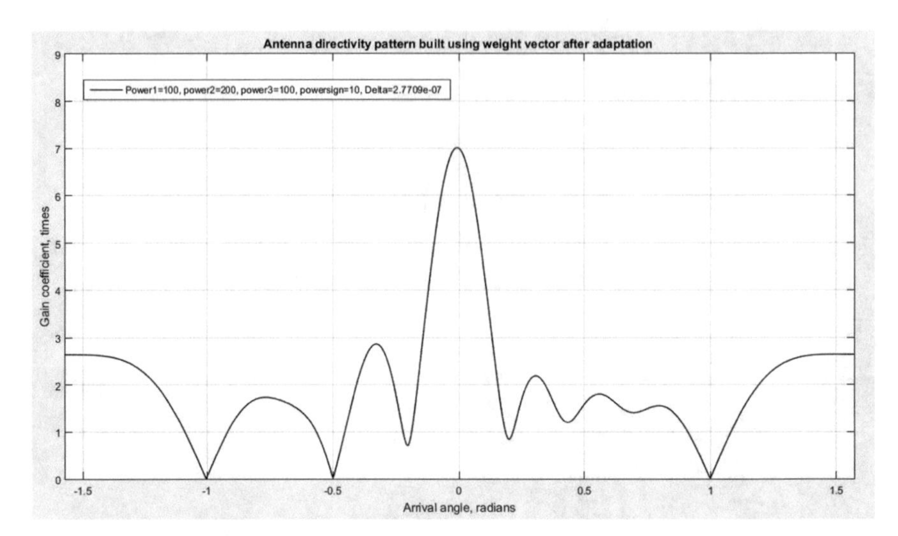

Fig. 26 Antenna directivity pattern built using weight vector after adaptation with account of impact of useful signal with power 10 units and step size 2.7709e−06

To tackle this problem, the paper proposes to adjust the gradient algorithm step size. This solution works well, but it comes at a price. The price is that of much slower convergence of the algorithm to the optimal solution.

Although this slower convergence seems to be still manageable, in the range of 2–3 ms for a 5 MHz channel, help comes from where the problem originated—continuous nature of useful signals in mobile telecommunications.

As it is well-known, in mobile telecommunications all incoming signals are kept to the minimum possible power levels, which are defined by acceptable bit error rate, error corrective coding and modulation type. This number is usually within the range of 5–15 dBs. Another important observation is that those interfering signals are many with low levels of power.

From the simulation, we can observe that the negative impact from the useful signal with the power10 relative units is much lower than the impact from that with 100 relative unites, and it can be compensated with much less increase in adaptation process duration. Analysis shows that this is because relative power of useful signal of 100 units constitutes 25% from relative power of 400 units of all interfering signals combined, while relative power of 10 units constitutes only 2.5%. This leads us to the important conclusion, that in real telecommunication environment with many small power interfering sources, the useful signal can constitute about fractions of a percent and this will greatly lessen the impact of the useful signal on the performance of the algorithm (2).

4 Conclusion

This paper addressed the shortcomings of the algorithm developed earlier. One of these shortcomings is the impact of the continuous useful signal on the performance of the algorithm in terms of convergence parameters.

Computer simulation helps to observe that continuous useful signal with random nature can greatly reduce achievable SNR due to increase in weight vector noise.

Corresponding reduction of the step size can redress the deficiency, but at the expense of much increased adaptation time.

The inference is made that for a typical telecommunication environment in which there are tens or even hundreds of low-power interfering sources, the problem of the described impact of the useful signal is much less significant and can be easily resolved by fine-tuning of the step size in the gradient algorithm. The accompanying increase in convergence time appears to be insignificant.

On balance, with account of the results of this research, algorithm (2) proves to be a good choice for deployment in telecommunications in 4th and 5th generations of mobile networks.

References

1. Mailloux RJ (2018) Phased array antenna handbook, 3rd edn. Artech House, Boston
2. Volakis J (2018) Antenna engineering handbook, 5th edn. McGraw-Hill Education, New York
3. Ageyev D, Al-Ansari A (2015) LTE RAN and services multi-period planning. In: 2015 Second international scientific-practical conference problems of infocommunications science and technology (PIC S&T). Kharkov, Ukraine: IEEE, pp 272–274. https://doi.org/10.1109/INFOCOMMST.2015.7357334
4. Loshakov V, Moskalets M, Ageyev D, Drif A, Sielivanov K (2021) Adaptive space-time and polarisation-time signal processing in mobile communication systems of next generations. In: Radivilova T, Ageyev D, Kryvinska N (eds) Data-centric business and applications. Lecture Notes on data engineering and communications technologies, vol 48. Springer, Cham. https://doi.org/https://doi.org/10.1007/978-3-030-43070-2_21
5. Grigoriev VA, Shchesnyak SS, Gulyushin VL, Raspaev YuA, Hvorov IA, Shchesnyak AS (2016) Adaptivnie Antennie Reshotki, Uchebnoye Posobie v Dvuch Chastyah. Chast 1. [Adaptive Antenna Arrays, Textbook in Two Parts, Part 1]. Universitet ITMO, Seint-Petersburg, Russia (In Russian)
6. Grigoriev VA, Shchesnyak SS, Gulyushin VL, Raspaev YuA, Hvorov IA, Shchesnyak AS (2016) Adaptivnie Antennie Reshotki, Uchebnoye Posobie v Dvuch Chastyah. Chast 2. [Adaptive antenna arrays, textbook in two parts, part 2]. Seint-Petersburg, Russia: Universitet ITMO (In Russian)
7. Kryvinska N, Zinterhof P, van Thanh (2007) New-emerging service-support model for converged multi-service network and its practical validation. In: First international conference on complex, intelligent and software intensive systems (CISIS'07). IEEE, pp 100–110. https://doi.org/10.1109/CISIS.2007.40

8. Kirichenko L, Vitalii B, Radivilova T (2020) Machine learning classification of multifractional Brownian motion realizations. In: CEUR workshop proceedings, vol 2608, pp 980–989
9. Pliushch OG (2019) Gradient signal processing algorithm for adaptive antenna arrays obviating reference signal presence. In: Presented at the IEEE international scientific-practical conference PIC S&T, Kyiv, Ukraine, 8–11 Oct 2019, Paper 190

Interference Immunity Assessment Identification Friend or Foe Systems

Iryna Svyd, Ivan Obod, and Oleksandr Maltsev

Abstract In the work, based on the consideration of the place and role of Identification friend or foe (IFF) systems in the airspace control system, it is shown that the principle of building modern IFF systems in the form of an asynchronous network for transmitting request and response signals and the implementation of the principle of servicing request signals based on an open single-channel queuing system with refuses, as well as the use of primitive coding of request signals and response signals do not allow an acceptable level of information support for the air control space and air traffic control in conditions of significant intensities of intra-systemic, as well as deliberate correlated and uncorrelated interference. A general description of the considering information systems is given and a brief description of the signals used in IFF systems is given. Based on the presentation of IFF systems in the form of two-channel systems for transmitting request and response signals, the noise immunity of aircraft re-sponders is evaluated under the action of request signals and intentional as well as unintended (intrasystem), correlated and uncorrelated interference in the request channel, which made it possible to evaluate the noise immunity of the entire IFF system in the form estimates of the probability of detection of airborne objects by the considering system.

Keywords Identification Friend or Foe · Aircraft responder · Request signal · Chaotic impulse noise · Air object

I. Svyd (✉) · I. Obod · O. Maltsev
Kharkiv National University of Radio Electronics, Nauky Ave. 14, Kharkiv 61166, Ukraine
e-mail: iryna.svyd@nure.ua

D. Ageyev et al. (eds.), *Data-Centric Business and Applications*, Lecture Notes on Data Engineering and Communications Technologies 69,
https://doi.org/10.1007/978-3-030-71892-3_12

1 Place and Role of Identification Friend or Foe Systems in the Information Support of an Airspace Control System

1.1 General Characteristics of Identification Friend or Foe Systems

The main source of air traffic data in the airspace control system [1–6] are IFF systems [7–10]. The IFF system provides independent monitoring of airspace, which involves the determination of the location of an air object (AO) using ground-based facilities. The use of an aircraft responder to generate response signals, which are necessary for assessing the location of AO in a ground requester, determines the belonging of the considered information tools to cooperative surveillance systems [11–15].

The informational task of identification systems is to determine the ownership of a detected AO based on the "friend or foe" attribute. IFF systems belong to the class of asynchronous two-way data transmission systems and consist of a number of transmitters and receivers that use different frequency ranges for receiving and transmitting. This allows us to conclude that IFF systems form a non-synchronous information network by the principle of construction. Each IFF systems is formed by two data transmission channels: a request channel and a response channel. The most vulnerable is aircraft responder (AR), which is an open single-channel queuing system with refuses, which causes significant shortcomings in its security [16–20]. This suggests that despite the use of modern IFF technology, aircraft recognition remains problematic even though a great deal of research effort has already been invested in this area. Indeed, the construction of AR according to the principle of an open single-channel queuing system with refuses presents wide opportunities for the interested party as unauthorized use of AR information, as well as suppressing them by setting up intentional interference of the required intensity [21].

IFF systems transmitters generate discrete signals $s_i(t)$, belonging to a finite set—an ensemble $\vec{S} = \{s_i(t)\}$; $i = 1, 2, \ldots, V$, and transmit them to the transmission line of request signals and response signals asynchronously, independently from each other, at the same time points determined by them. In this case, the condition $t_i \ll T_p$, is usually satisfied where t_i is the signal duration; $s_i(t)$is the repetition period of request signals. The use of a single channel for transmitting request signals, as well as the construction of an entire monitoring system based on the principle of an open queuing system with refuses, impedes the operation of such systems under the influence of intentional as well as intra-system interference [22].

At the same time, it should be noted that the work of IFF systems in the frequency band of data transmission of cooperative monitoring systems (1030 MHz—channel for transmitting request signals; 1090 MHz—channel for transmitting an information packet) leads to significant difficulties, which is noted, in particular, in [23–28]. This design of the system causes a significant density of intra-system interference in the request and response channels, which leads to a significant decrease in the noise immunity of the considering IFF systems. The papers presented show that the frequency band 1030/1090 MHz which is allocated for air traffic surveillance, and includes cooperative surveillance systems, the collision avoidance system and ADS-B systems are experiencing significant overloads and alternative methods for using variable query powers with respect to azimuth sectors and tracking data are presented aircraft to reduce spectrum overload. It is shown that a high probability of overlap leads to the fact that the 1090ES signal cannot be correctly decoded, and the use of the 1090 MHz aerial surveillance frequency band with an increasing number of aircraft, applications and equipment types can reach critical levels of intra-system interference. Data loss due to overlapping messages or distortion is to some extent acceptable in all protocols, but there is concern that this performance loss will soon become unacceptable when message density increases.

The networks of IFF systems are non-synchronous according to the construction principle, which means, firstly, the lack of synchronization in time of the radiation of individual requesters by aircraft responders and, secondly, the absence of any time synchronization between different Identification friend or foe systems.

In accordance with the principle of service of request signals, airborne responders of IFF systems refer to single-channel queuing systems (QS) with servicing of the first correctly received request signal. Therefore, according to the principle of construction, responders of IFF systems relate to QS with refuses. The essence of these systems is that when servicing a correctly received request signal, AR closes for a certain time, which is called the paralysis time t_p. The amount of time depends on the paralyzation mode IFF system (simulation resistant or non simulation resistant). The presence of AR paralysis time limits the throughput of both the aircraft responder and the Identification friend or foe system as a whole. Therefore, as a characteristic of the noise immunity of the IFF system, can use the probability of detection of an air object by the requester P_s, which is defined as the probability of receiving the required number of response signals to respond to requests from this requester. The capacity of the aircraft responder IFF system is characterized by the availability factor of the aircraft responder P_0, which is defined as the probability that the aircraft responder emits a response signal to a particular request signal of the requester.

Since the responders serve any correctly received request signal (and even generated by an interested party), they belong to open queuing systems. The presence of only one channel for servicing request signals and the possibility of paralyzing an aircraft responder when servicing request signals allows the aircraft responder to be classified as single-channel open queuing systems with refuses.

Thus, IFF systems are implemented on the principles of (Fig. 1), which largely determine the low noise immunity of the latter.

Given the above rationale, conclusions can be drawn that Identification friend or foe systems are a query-response data transmission system. Actually, they contain a channel for transmitting request signals and a channel for transmitting response signals. The principle of constructing the considering information systems, which is based on the principle of an open single-channel queuing system with failures, allows the interested party to use both unauthorized use of IFF systems airborne

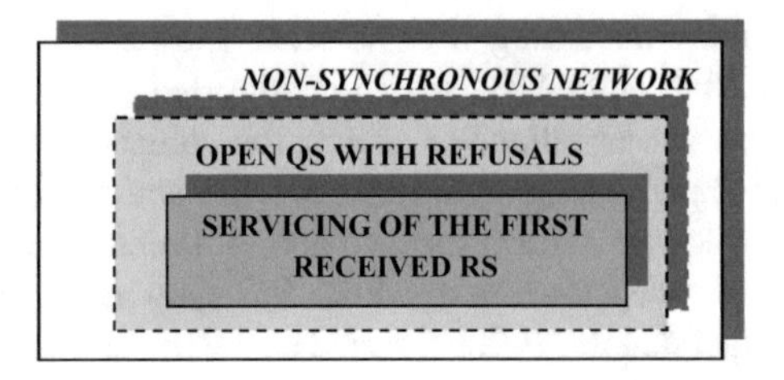

Fig. 1 Principles of IFF systems building

responders to obtain information and paralyze the latter by setting correlated (in the request channel) interference with necessary intensity [21]. Such a construction principle and a service signal request principle leads to low information security of the considered systems [28–33].

In addition, the choice of signals for encoding information in IFF systems was not entirely successful. Indeed, the use of primitive time-interval and positional codes [34–36], significantly reduced both the information capacity of data transmission channels and the noise immunity in general considered IFF systems [37–40].

1.2 Signals of IFF Systems and Their Mathematical Models

To solve the main information problem, the IFF system has the following operating modes—1, 2, 3, 4, and 5 [41], which are divided into non simulation resistant (1,2,3) and simulation resistant (4, 5) [41–44].

Mode 1 is designed to determine the state affiliation of air objects and is the main one when working in peacetime. The request signal (Fig. 2) is the simplest time-interval code [34] which includes two pulses of the mode code P1 and P3. The mode code distance T_k is 3 μs. Impulse P2 serves as a suppression of side lobes on request.

Mode 2 is designed for the individual identification of aircraft of military facilities or individual number of critical facilities. The code distance of the request signal (Fig. 2) $T_k = 5$ μs.

Mode 4 is designed for imitation-resistant identification of military facilities and is the main mode of operation in wartime.

In Fig. 3 shows the structures of the request signal, and Fig. 4 shows the response signals in the simulated-resistant Mode 4 mode.

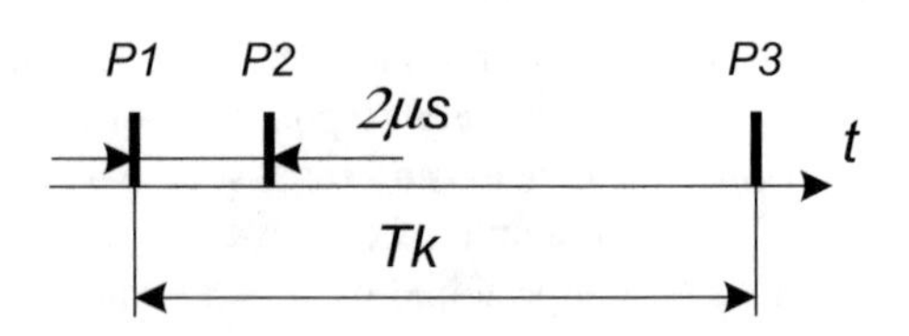

Fig. 2 Request signal of non-resistant operation modes

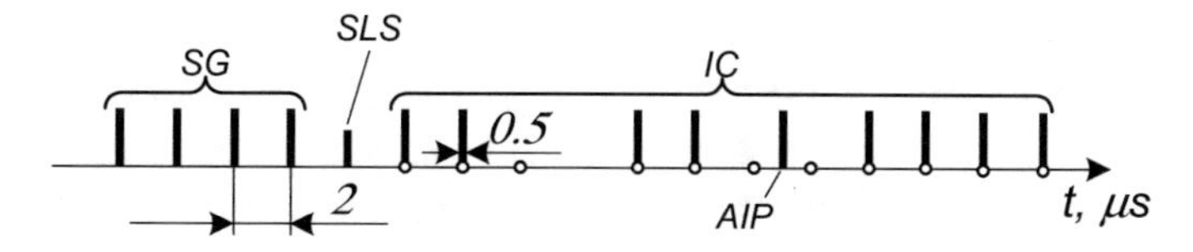

Fig. 3 Mode 4 request signal structure

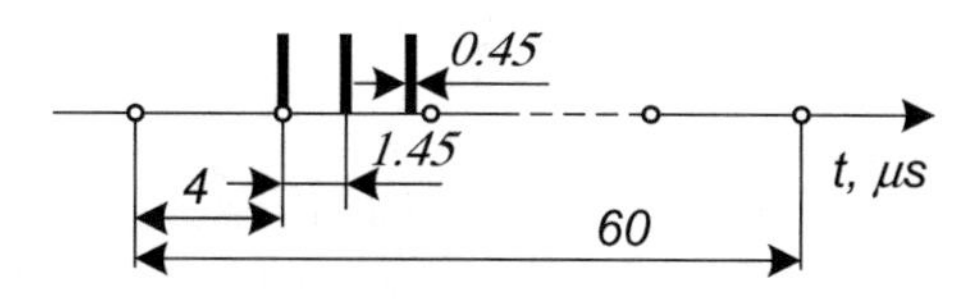

Fig. 4 Mode 4 response signal structure

Simulation resistant considered mode is provided by closed coding. The request signal includes a four-pulse synchronygroup (SG), a side lobe suppression pulse (SLS), and a 32-bit information code (IC). It is formed in the requester and processed in the responder by a crypto computer. In the absence of pulses at adjacent positions, an anti-interference pulse (AIP) is inserted. The response unchangeable three-pulse pulse-time code is supplemented by passive pause encoding. Response delays (16 values with a resolution of 4 μs) are determined by a cryptographic algorithm.

In all operating modes, when a P2 pulse is detected (this indicates that the signal is emitted from the side lobes of the requester's antenna), a response is not generated, and the aircraft responder is paralyzed by 35 μs, that is, it becomes unable to receive request signals [45].

When a request signal is received and the side lobe suppression signal is not detected, the aircraft responder is paralyzed for a considerable time, depending on the service of the request signals of simulation-resistant or nonresistant operating modes.

At present, NATO countries' aviation is being equipped with a new generation identification system Mk-12A, which additionally uses the simulation resistant mode of operation (Mode 5) [46–50]. The features of Mode 5 mode signals are the presence of a reference group of pulses (sync groups) with time-pulse modulation that determines the identification sub-mode, and frequency modulation with a continuous phase (CPFSK, MSK—Minimal Shift Keying) of preamble pulses and side-lobe suppression by a 16-bit Walsh sequence [51] used in decoding an information group.

The request signal of the M5 mode can be analytically presented as follows

$$S_{m5}(t) = A\sum_{n=0}^{3} s_1(t-t_n,k)p(t-t_n) + A_p s_1\big(t-t_p,k\big)p\big(t-t_p\big) + A\sum_{m=0}^{M-1} s_1(t-t_n,k)p(t-t_m), \quad (1)$$

where $p(t-t_n) = \{1|t_i \in [t_0 + iNT, t_0 + (i+1)NT]; 0|\, p(t-t_n) = \{1|t_i \in [t_0 + iNT, t_0 + (i+1)NT]; 0| \notin [t_0 + iNT, t_0 + (i+1)NT]\}; A$ and A_p pulse amplitudes of the preamble and SLS; t_n and t_p time moments of the appearance of the pulses of the preamble and SLS; t_m moments of occurrence of the.

mth pulse from M pulses of the information group $s_1(t,k) = \sum_{i=0}^{N-1} U_m \cos\left(2\pi f_s t + \frac{\pi b_i(k)}{2T} + \phi_i\right)$, $t_0 \le t \le t_0 + NT$ time-domain representation of the RQS impulse, frequency-manipulated 16-bit Walsh sequence; U_m signal amplitude; t_0 signal start time; $N = 16$ number of elements in the signal; T signal element (parcel) duration; f_s signal carrier frequency; $b_i(k) = \mathrm{sgn}(wal(k, \theta_i))$ sign of the ith element of the parcel in the time interval $[t_0 + iT,\ t_0 + (i+1)T]$; $wal(k, \theta)$ Walsh function under number k, which is defined on the interval $[0,\ NT]$; $\theta_i = t_i/NT = i/N, \theta_i \in [0, 1]$ dimensionless time at which moments values are formed $b_i(k) \in \{-1; 1\}$; $\phi_i = \frac{\pi}{2} - \sum_{j=1}^{i-1} b_j - \frac{(i-1)\pi}{2} b_i + \phi_s; \phi_s$ initial phase of the signal; modulation index $\beta = \Delta f T = 0.5$; $\Delta f-$ frequency spacing. The preamble pulses can be represented as follows

$$g_k(t) = \sum_{n=1}^{L-1} a_k(n) p(t - nT), \tag{2}$$

where $p(t)$ impulse of duration nT; $a_k(n),\ 0 \le n \le L$ code sequence that determines the placement of the preamble pulses in time.

Therefore, as the request and response signals in IFF systems, time-interval codes [42–46] and position codes are used, which are characterized on the one hand by simplicity and, on the other hand, by low noise immunity. Indeed, suppression by interference of at least one pulse of the request signals or the appearance of false pulses on the code intervals of the request (response) signals immediately leads to errors: omissions, false alarms of the first and second kind, or mess of the request signals.

The IFF systems request signal can be written as follows:

$$s_i(t) = U_i f_i(t) \cos(\omega_o t + \phi), \tag{3}$$

where $f_i(t)$ some binary coding sequence that takes values 1 or 0; "ones" correspond to impulses, and "zeros" correspond to pauses of the signal $s_i(t)$ and this sequence itself; ω_o carrier frequency, ϕ random initial phase of high-frequency filling, in the general case for each of the pulses of the coding sequence is different; U_i amplitude factor constant within one full realization of this signal. The entire ensemble of signals $\vec{S}$ contains K various signals (as a rule, according to the number of operating modes of IFF systems), which differ from each other only in the coding sequence $f_i(t)$, with a common carrier frequency ω_o. The condition for the adequacy of the request signals is expressed in that the pulse duration τ_i and the number of pulses n

must be the same in all request signals of the considering information systems. The number of pulses in the code of the request signal is called the code value [22, 34].

Let's evaluate the noise immunity of request signals, taking into account the different nature of intentional and unintentional interference that may be present in IFF systems.

The main harmful effects that reduce the noise immunity of the considered signals, and, consequently, the considered information tools in general, are interference correlated and uncorrelated with the existing set of signals. It should be noted that such interference occurs both in the IFF systems (intra-system) [52–57], and can be formed by the interested party (intentional) [51, 53, 58].

In case of asynchronous reception of request signals, the following main types of signal reception errors can occur [34]:

- request signal missing;
- false alarm of the first kind;
- false alarm of the second kind.

Thus, the noise immunity of IFF signal systems must be estimated by three error probabilities: P_p, F_1 and F_2.

Then it is necessary to assesse the noise immunity of request signal in Identification friend or foe systems, depending on the uncorrelated interference in the request channel.

We assume that the considered interference is stationary. It has no aftereffect, and is also statistically and structurally independent of the considered request signal. The request signal consists of n elementary pulses, has an uncertainty function R, and the detection threshold equals k.

In this situation, the probability of missing a signal, that is, the conditional probability that when receiving a request signal will be suppressed by an interference of at least $n - k + 1$ pulses, can be defined as:

$$P_p = \sum_{i=n-k+1}^{n} C_n^i P_{10}^i (1 - P_{10})^{n-i}, \tag{4}$$

where P_{10} the probability of suppression by the considered interference of at least one pulse of the request signal, which is determined by the type of interference and its intensity.

The probabilities of false alarms are a little more complicated. Indeed, for the fact of a fictitious alarm of the first kind, it is necessary that a false combination with k or more interference pulses arrive at the decoder input, and also that this erroneous combination must be decoded by the considered decoder.

Consequently, the probability of such event can be defined as

$$F_1 = \sum_{i=k}^{n} C_n^i P_{01}^i (1 - P_{01})^{n-i} P_d(i), \tag{5}$$

Where P_{01} is probability of a false pulse at the input of the decoder in the time interval $\delta\tau$, which is determined by the type and intensity of the interference; P_d the probability of decoding by the decoder the considered i—pulse combination.

In the calculation of the probability of false alarm of the 2nd kind we will determine the probability that the time for passing the signal through a decoder to happen at least one false alarm of the 2nd kind. In this case, the reciprocal will determine the conditional probability of the event, which consists in the fact that the arrival time of the received signal will be uniquely determined, with accuracy within the pulse duration.

Therefore, with such a formulation of the question, the probability of a false alarm of the second kind can be defined as

$$F_2 = 1 - \prod_{s=1}^{R} [1 - F_2(s)]^{M(s)}, \tag{6}$$

where $F_2(s)$ the probability of the formation and decoding of a false combination of the second kind, consisting of signal pulses that give one lobe of the uncertainty function, equal to s, and i interference pulses $(k - s < i < n - s)$

$$F_2(s) = \sum_{i=k-s}^{n-s} C_{n-s}^{i} P_{01}^{i} (1 - P_{01})^{n-s-i} P_d(i + 1), \tag{7}$$

where $M(s)$ is the number of side lobes of the uncertainty function equal to s.

Thus, the above expressions (4)–(6) for given values of P_{10}, P_{01} and the established upper limits of error probability P_p, F_1 and form a system of three equations with three unknowns—n, k, R. Having solved this expression, one can find rational values of these signal parameters, that is, such minimum n, maximum R and corresponding to them k $(R < k < n)$, at which the error probabilities do not exceed the given upper limits.

In real situations usually have to deal with the two varieties of uncorrelated interference: random impulse noise and fluctuation [54–57]. However, based on the feasibility and practicality, it can be shown that the main type of intentional interference to IFF systems can be chaotic impulse noise. Such interference, to a large extent, is present in the considered information systems in the form of intrasystem interference [23–26]. Let's determine the probabilities of events P_{10} and when chaotic pulse interference occurs in the request channel.

As a model of the flow of chaotic pulsed interference, the Poisson stream of single pulses is most often taken. (For impulse noise, the Poisson flux of single pulses is the same idealization as normal noise for fluctuation interference). In this case, the probability of the appearance of at least one pulse in the time interval t_0 is:

$$P = 1 - \exp(-\lambda_0 t_0), \tag{8}$$

where λ_0 is the intensity of chaotic impulse noise.

Therefore, the probability P_{01}, can be defined as

$$P_{01} = 1 - \exp(-\lambda_0 \tau_0). \tag{9}$$

The probability of suppressing one pulse of the interfering request signal can be determined from the following relation

$$P_{10} = 1 - \exp(-\lambda_0 \tau_p)[1 - \gamma(1 - \exp(1 - \lambda_0 \tau_0))], \tag{10}$$

where τ_p is the time of paralysis of the receiving device after passing through an impulse of interference; γ is a coefficient that determines the probability of interference suppression of the pulse of the received signal when it coincides in time with the interference pulse.

Consequently, the relation (10) takes into account both types of suppression occurring in systems with considered pulse signals: inertial and interferential. The first suppression is due to the presence in the receiver of elements with a non-zero sensitivity restoration time, and the second suppression is due to the interaction of the high-frequency fillings of the signal pulses and the considered interference when they coincide in time.

Coefficient γ depends on the ratio of amplitudes and phases of interfering oscillations. With a uniform distribution of the phase difference in the interval $[0; 2\pi]$ it is usually assumed that γ is 0.2 [34].

2 Immunity Assessment Identification Friend or Foe Systems

As noted above, systems noise immunity can be characterized by the probability of detection of air object IFF system P_s, which can be defined as

$$P_s = f(_0, \lambda_{CINa}, N, k), \tag{11}$$

where

$$P_0 = f(\lambda_{RS}, \lambda_{CIN}, \lambda_{RSs}, t_{p1}, t_{p2}) \tag{12}$$

is availability factor of the aircraft responder; $\lambda_{RS}(\lambda_{RSs})$ RS flow rate recovered from the main (side) lobe of the requester's antenna; $\lambda_{CIN}(\lambda_{CINa})$ chaotic impulse noise (CIN) intensity in the request (response) channel; $t_{p1}(t_{p2})$ AR at time paralyzation request signal (RS) radiation along the side (main) lobes of requester's antenna;

N total number of response signals in a packet of served request signals; k-digital threshold for decision making on detection (identification) of AO.

The choice of the considered quality indicator is consistent with the general tendency to solve the AO identification problem with its detection by both the primary radar and the IFF system, which involves the joint processing of data by the considered information tools [18–21, 40, 48–50].

The ground requester will receive a response signal from AR when two events occur simultaneously:

- AR will receive, correctly decode the request signal and emit a response signal (the probability of this event is nothing other than the AR availability factor P_0);
- The AR response signal will be received and detected by the terrestrial radio interrogator.

Next, we will consider the probabilities of these two events in the presence of interference and analyze the probability of the simultaneous execution of these two events.

2.1 *Interference Immunity Assessment for Aircraft Responders Identification Friend or Foe Systems*

Suppose that the general interference stream is formed by the RS stream of neighboring IFF systems, the stream of intentional correlated interference from the interested party, and the CIN stream (intentional and unintentional uncorrelated interference). We will carry out calculations for non-resistant and imitation-resistant operating modes of IFF systems based on the "friend or foe" principle and for existing detection algorithms of AO IFF systems.

The influence of the RS stream leads, to paralysis of AR for the time of paralysis, which is determined by the request type mode. We will note that when receiving RS on the mainly of the antenna pattern of the requester, the AR is completely paralyzed for the duration of the service. When RS is received from the side lobes of the antenna pattern of the requester, AR is paralyzed for the time between the RS pulse, the amplitude of which is remembered and the SLS pulse.

CINs (intentional or intra-systemic) affect the work of AR bilaterally:

- initially, it suppresses individual RS pulses, which makes it impossible to service the considered RS;
- then, it paralyzes AR through the formation of false RS (false alarm of the first (5) and second (6) kind).

Let's evaluate the AR availability factor under the influence of the indicated interference. Upon receipt of RS and CIN streams at the AR input, the responder will not generate a response signal if at least one of such adverse situations occurs:

- the RS of the considered requester is suppressed due to the formation of leading false RSs (false alarm of the first kind) from the CIN, which lead to the emission of a response signal or the operation of the side lobe suppression circuit (we denote the probability of this situation P_1);
- the request signal of the considered requester is suppressed due to the leading RS of the requester or requester of the interested party P_2);
- individual pulses of the request code of the considered requester are suppressed at a high frequency due to the coincidence in time of pulses of different RSs with unfavorable phase relations (probability P_3);
- the RSs of the considered requester are suppressed due to leading false RSs that are formed when the first RS pulse of the requester interacts with the leading (to the code base) CIN or RS pulses (false alarm of the second kind) and lead to the emission of a response signal or the operation of the side lobe suppression circuit (probability P_4);
- the RS of the considered requester is suppressed due to the appearance of a false suppression pulse at the signal position, which was formed from interference (probability P_5);
- the request signal is suppressed due to the operation of the time-selection circuit of the responders (probability P_6).

Let's determine the probabilities of these events under the assumption that the RS and CIN streams affect the request codes of this requester independently of each other and the number of sources forming the common RS stream is large enough to characterize the stream as Poisson [7–9].

We assume that the input AO receives:

- CIN stream with intensity λ_0;
- RS stream with intensity λ_1, which consists of the RS stream of neighboring requesters and the RS stream, intentional interference;
- RS flow, which lead to the trigger the circuit side lobe suppression of the antenna radiation pattern, intensity λ_2.

The combined action of the CIN and the RS stream leads to high frequency suppression of individual pulses of the RS stream at unfavorable phase relations, as a result of which the intensity of the RS stream decreases.

The probability that at least one CIN pulse coincides in time with the pulse of the RS stream and suppresses it is

$$P_p = \gamma\left(1 - e^{-\lambda_0 \tau_0}\right), \tag{13}$$

where γ is interference suppression coefficient, which determines the probability of interference suppression of the pulse of the received request signal when it coincides in time with the interference pulse.

Due to the high-frequency suppression, the RS flux intensity decreases, which cause the radiation response signal:

$$\lambda_1^1 = \lambda_1 (1 - P_p)^n, \tag{14}$$

and the intensity of the RS stream that trigger the circuit side lobe suppression:

$$\lambda_2^1 = \lambda_2 (1 - P_p)^n, \tag{15}$$

where n is the significance of the RS code.

The probability that at least one RS falls into the leading interval and suppresses the RS of this IFF systems due to the time of paralysis of the aircraft transponder in the mode when the response signal is emitted is determined respectively by CIN and RS stream as:

$$P_1^1 = 1 - e^{-\lambda_x t_1}, \tag{16}$$

$$P_1^2 = 1 - e^{-\lambda_1 t_1}, \tag{17}$$

where λ_x is the average number of false n-pulse codes that lead to the emission of a response signal, which can be determined from the following expression

$$\lambda_x = n \tau_0^{n-1} \lambda_0^{n-1} (1 - \tau_s / \tau_0), \tag{18}$$

where τ_s a given duration of the selection of pulses in time.

Consequently, the resulting probability of RS suppression of the considered requester of the system due to the paralysis of the responder upon emission, the response signal

$$P_1 = 1 - \prod_{i=1}^{2} \left(1 - P_1^i\right). \tag{19}$$

Hereinafter, the calculations are carried out under the condition that the intensity λ_2 of the RS stream emitted from the side lobes of the requester antenna radiation pattern is three times higher than the intensity λ_1 of the RS stream emitted from the main lobe of the requester antenna radiation pattern.

The probability P_2 that at least one RS falls into the leading interval and suppresses the RS of this IFF systems due to the paralysis time t_2 AR when the side lobe suppression circuit is triggered, from the CIN and from the RS stream is determined in accordance with:

$$P_2^1 = 1 - e^{-\lambda_0 t_2}, \tag{20}$$

$$P_2^2 = 1 - e^{-\lambda_2 t_3}. \tag{21}$$

The resulting probability of RS suppression of this IFF requester through paralysis of the responder when receiving RS along the side lobes of the requester antenna radiation pattern is

$$P_2 = 1 - \prod_{i=1}^{2} \left(1 - P_2^i\right). \tag{22}$$

The probability of suppressing one RS pulse of this requester due to coincidence with the pulses of the CIN and RS streams is:

$$P_{10} = \gamma\left(1 - e^{-\lambda_s \tau_0}\right), \tag{23}$$

where $\lambda_s = \lambda_0 + \lambda_1 + \lambda_2$ is the intensity of the total interference flow and RS.

Taking into account n RS pulses, the probability of suppressing the request signal is

$$P_3 = 1 - (1 - P_{10})^n. \tag{24}$$

The probability of RS suppression of this requester due to the appearance of leading false request codes generated as a result of the interaction of the first pulse of the request code and leading pulses of the RS stream, and lead to the emission of a response signal or triggering of the side lobe suppression circuit, is determined by the ratio

$$P_4 = (1 - P_{01})^n \left[1 - (1 - P_{10})^{n+1}\right]. \tag{25}$$

The second factor considers into account possible situations of the formation of false leading request codes: n request codes, which lead to the emission of a response code, and one code of the suppression signal, which leads to the triggering of the side-lobes suppression circuit.

Second order false alarm probability P_{01} is defined as

$$P_{01} = 1 - e^{-\lambda_s \tau_0}. \tag{26}$$

The probability P_5 of suppressing the request signal of this requester due to the appearance of a false suppression pulse, which was formed from interference, is determined by the formula

$$P_5 = (1 - P_{10})^n P_{01}^{n-1}. \tag{27}$$

The probability P_6 of RS suppression as a result of triggering of the AR time selection circuits is determined from the following relation

$$P_6 = 1 - e^{-2\lambda_s \tau_0}. \quad (28)$$

If the average number of RSs exceeds the permissible load AR λ_m, then the probability of a response during the operation of the load limitation circuit of an aircraft responder decreases and is $P_{AR} = \lambda_i/\lambda_3$, where $\lambda_3 = \lambda_1 + \lambda_2$.

Thus, the AR availability factor is:

When

$$\lambda_3 < \lambda_m \; P_0 = \prod_{i=1}^{6} (1 - P_i), \quad (29)$$

When

$$\lambda_3 > \lambda_m \; P_0 = P_{AR} \prod_{i=1}^{6} (1 - P_i). \quad (30)$$

Calculations based on the above expressions of non-resistant and simulated resistant operating modes are presented in Figs. 4 and 5, respectively. In this case, we considered that the intensity of the CIN stream $\lambda_0 = 0; 1 \cdot 10^4; 2 \cdot 10^4$, and the intensity λ_1 of the RS stream, which lead to the emission of the response signal, are five times less than the intensity λ_2 of the RS stream, which cause the side lobe suppression circuit to trigger.

From the above results, the following conclusions can be drawn:

- an increase in the RS flux intensity leads to a decrease in the availability factor of the aircraft responder during serving of both non-imitation resistant and imitation-resistant operating modes, which indicates a AR low noise immunity. Indeed, from Figs. 4 and 5, it can be seen that, with an RS flux intensity of 1600, it reduces the AR availability factor for the non-resistant mode to 0.72, and that of the resistant one to 0.65;

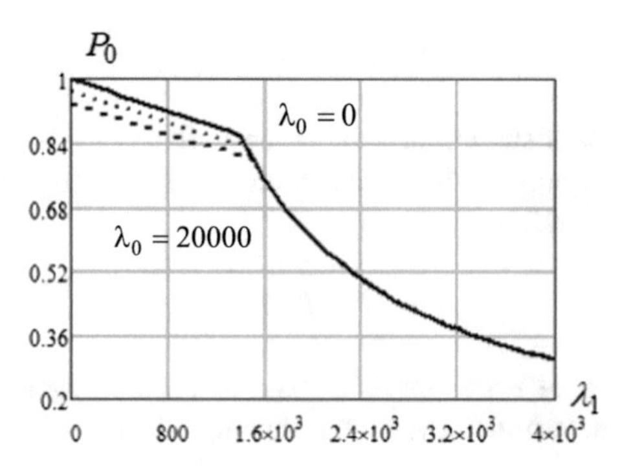

Fig. 5 Assessment of the availability factor AR (non-resistant operating modes)

- uncorrelated interference (CIN) has little effect on the AR availability factor;
- achievement of the maximum AR load for the non-resistant mode is achieved with $\lambda_1 = 1450$. For imitation-resistant mode, the maximum load AR is achieved for $\lambda_0 = 0$ at $\lambda_1 = 2100$; for $\lambda_0 = 10{,}000$ with $\lambda_1 = 2500$. With $\lambda_0 = 20{,}000$ the maximum load of AR is not achieved at all.

The above calculations provide an estimate of the AR noise immunity and assessment allows us to estimate the bandwidth of existing IFF systems and indicates the low efficiency of IFF systems when exposed to intentional correlated interference.

2.2 Immunity Assessment Identification Friend or Foe Systems

Suppose that the IFF systems request processing equipment implements a bilaterally algorithm for quasi-optimal detection of a packet of response signals, and the AR availability factor is constant throughout the entire packet of response signals. The IFF systems requestor receives independent response signal streams and, in the general case, a CIN stream of intensity λ_0. However, due to the fact that the interrogator's antenna is narrowly targeted, deliberate jamming on the response channel is not of interest.

Therefore, in the absence of interference on the response channel, the detection of AO IFF systems using the logic "k from M" can be determined from the following expression

$$P_s = \sum_{i=k}^{M} C_M^k P_0^i (1 - P_0)^{M-i}, \tag{31}$$

where $C_M^k = \frac{M!}{i!(M-i)!}$ binomial coefficients.

The dependence of the detection probability of air object Identification friend or foe systems on the flow rate RS, which lead to the radiation of the response signal, is posted on Fig. 6—for the non-resistant mode, and in Fig. 7—for the resistant mode.

During when carrying out the calculations, it was believed that the intensity of the RS stream, which leads to the emission of the response signal, is five times lower than the RS stream intensity, which lead to the triggering of the side-lobes suppression circuit. The logic of deciding whether an IFF aerial object was detected by the system (that is, deciding that it was a "friend") was $k/M = 8/15$.

The presented calculations show quite acceptable results of the IFF systems noise immunity assessment when operating in a non-imitation resistant mode of operation with a request signal stream intensity of up to 1600. However, with an increase in the flow of request signals to 2400, the probability of detecting an air object is only 0.5, which is unacceptable in essence.

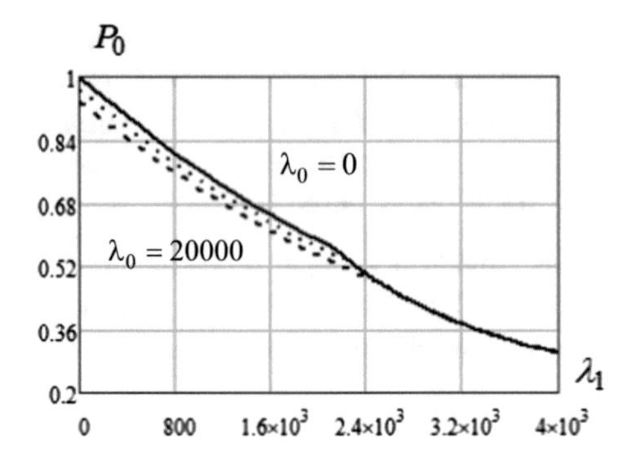

Fig. 6 Assessment of the availability factor AR (resistant operating modes)

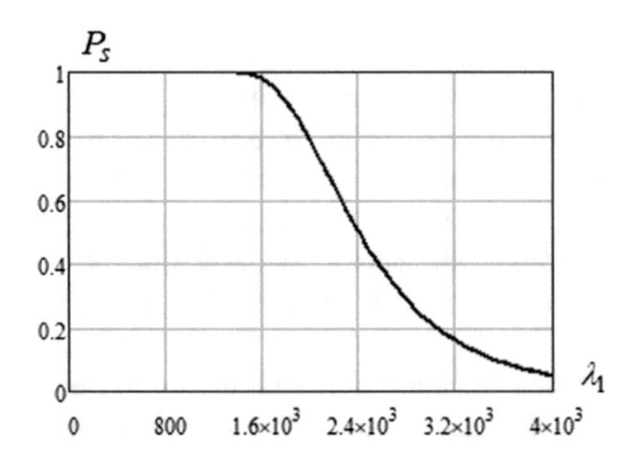

Fig. 7 IFF immunity assessment in non simulation resistant mode

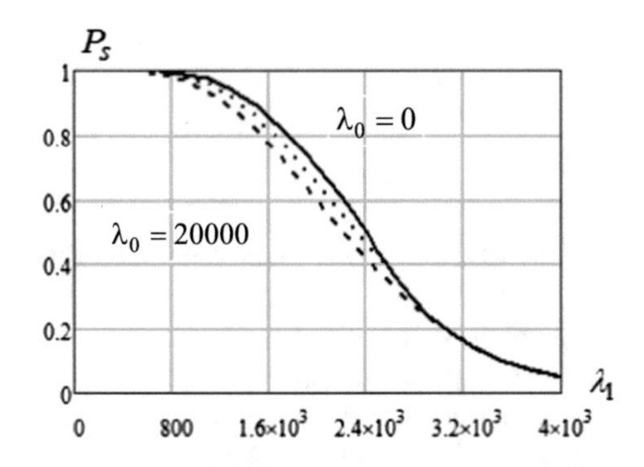

Fig. 8 IFF immunity assessment in simulation resistant mode

In the simulation resistant mode (Mod4), as shown in Fig. 6, the CIN significantly influences the noise immunity. This is caused by both a significant time base of the request signal in this mode and a significant number of pulses of the sync group and the information part in the form of a binary code. This is what stimulated the transition to the new Mod5 mode.

3 Conclusions

The paper discusses the principles of building existing Identification friend or foe systems. An assessment of their noise immunity is also given when the system is exposed to intentional and unintentional correlated and uncorrelated interference. The relationship between the noise immunity of the system and the quality of the information support of the system is shown. One of the versions for improving the quality of information support for Identification friend or foe systems was proposed under conditions of intense intrasystem and deliberate correlated and uncorrelated interference.

References

1. Stevens B, Lewis F, Johnson E (2016) Aircraft control and simulation: dynamics, controls design, and autonomous. Wiley
2. Air Force Doctrine Annex 3-52 Airspace Control (2012, October 26) CreateSpace Independent Publishing Platform
3. Air Force Doctrine Annex 3-52 Airspace Control (2019, September 30) Independently published
4. Bruckbauer B (2004) The airspace control system—a key to operational maneuver. Naval war coll newport ri joint military operations dept, Newport, Rhode Island
5. Buchanan T (1987) Hetactical air control system: its evolution and its need for battle managers. Research Report No. AU-ARI-87-1. Air University Press, Maxwell AFB, AL
6. Bloisi D, Iocchi L, Nardi D, Fiorini M, Graziano G (2012) Ground traffic surveillance system for air traffic control. In: 2012 12th international conference on ITS telecommunications. https://doi.org/10.1109/ITST.2012.6425151
7. Stevens M (1988) Secondary surveillance radar. Artech House, Norwood
8. NATO (2016) STANAG 4193 PT III. Technical characteristics of IFF MK XA and MK XII interrogators and transponders. Part III: installed system characteristics, 23 May 2016
9. NATO (2008) STANAG 4193 PT IV. Technical characteristics of IFF MK XA and MK XII interrogators and transponders. Part IV: technical characteristics of mode S in military interrogators and transponders, 3 Oct 2008
10. Schuck T, Shoemaker B, Willey J (2000) Identification friend-or-foe (IFF) sensor uncertainties, ambiguities, deception and their application to the multi-source fusion process. In: Proceedings of the IEEE 2000 national aerospace and electronics conference. NAECON 2000. Engineering tomorrow (Cat. No.00CH37093). https://doi.org/10.1109/NAECON.2000.894896
11. Ramasamy S, Sabatini R, Gardi A (2016) Cooperative and non-cooperative sense-and-avoid in the CNS+A context: a unified methodology. In: 2016 international conference on unmanned aircraft systems (ICUAS). https://doi.org/10.1109/ICUAS.2016.7502676
12. Svyd I, Obod I, Maltsev O, Vorgul O, Zavolodko G, Goriushkina A (2018) Noise immunity of data transfer channels in cooperative observation systems: comparative analysis. In: 2018 international scientific-practical conference problems of infocommunications. Science and Technology (PIC S&T). IEEE. https://doi.org/10.1109/INFOCOMMST.2018.8632019

13. Siergiejczyk M, Krzykowska K, Rosiński A (2014) Reliability assessment of cooperation and replacement of surveillance systems in air traffic. In: Proceedings of the ninth international conference on dependability and complex systems DepCoS-RELCOMEX. June 30–July 4, 2014, Brunów, Poland. https://doi.org/10.1007/978-3-319-07013-1_39
14. Obod I, Svyd I, Maltsev O, Vorgul O, Maistrenko G, Zavolodko G (2018) Optimization of data transfer in cooperative surveillance systems. In: 2018 international scientific-practical conference problems of infocommunications. Science and technology (PIC S&T). IEEE. https://doi.org/10.1109/INFOCOMMST.2018.8632134
15. Svyd I, Obod I, Maltsev O, Tkachova T, Zavolodko G (2019) Optimal request signals detection in cooperative surveillance systems. In: 2019 IEEE 2nd Ukraine conference on electrical and computer engineering (UKRCON), IEEE. https://doi.org/10.1109/UKRCON.2019.8879840
16. Pollack J, Ranganatha P (2018) Aviation navigation systems security: ADS-B, GPS, IFF. In: International conference on security & management, SAM'18, Las Vegas, Nevada, USA
17. Svyd I, Obod I, Maltsev O, Zavolodko G, Maistrenko G, Saikivska L (2019) Method of enhancing information security of requesting cooperative surveillance systems. In: 2019 IEEE international scientific-practical conference problems of infocommunications, science and technology (PIC S&T). https://doi.org/10.1109/PICST47496.2019.9061366
18. Strelnytskyi O, Svyd I, Obod I, Maltsev O, Voloshchuk O, Zavolodko G (2019) Assessment reliability of data in the identification Friend or Foe systems. In: 2019 IEEE 39th international conference on electronics and nanotechnology (ELNANO), IEEE. https://doi.org/10.1109/ELNANO.2019.8783397
19. Svyd I, Obod I, Maltsev O, Strelnytskyi O, Zubkov O, Zavolodko G (2019) Method of increasing the identification Friend or Foe systems information security. In: 2019 3rd international conference on advanced information and communications technologies (AICT), IEEE. https://doi.org/10.1109/AIACT.2019.8847853
20. Svyd I, Obod I, Maltsev O, Tkachova T, Zavolodko G (2019) Improving noise immunity in identification Friend or Foe systems. In: 2019 IEEE 2nd Ukraine conference on electrical and computer engineering (UKRCON), IEEE. https://doi.org/10.1109/UKRCON.2019.8879812
21. Obod I, Svyd I, Maltsev O, Bakumenko B (2020) Comparative analysis of noise immunity systems identification Friend or Foe. In: 2020 IEEE 40th international conference on electronics and nanotechnology (ELNANO). https://doi.org/10.1109/elnano50318.2020.9088856
22. Yarlykov, M., Chernyakov, M. (1979) Optimization Of Asynchronous Address Radio Communication Systems. Svyaz, Moscow.
23. Otsuyama T, Naganawa J, Honda J, Miyazaki H (2017) An analysis of signal environment on 1030/1090MHz aeronautical L-band systems. In: 2017 international symposium on antennas and propagation (ISAP). https://doi.org/10.1109/ISANP.2017.8228911
24. Otsuyama T, Honda J, Naganawa J, Miyazaki H (2018) Analysis of signal environment on 1030/1090MHz aeronautical surveillance systems. In: 2018 IEEE international symposium on electromagnetic compatibility and 2018 IEEE Asia-Pacific symposium on electromagnetic compatibility (EMC/APEMC). https://doi.org/10.1109/ISEMC.2018.8394048
25. Piracci E, Galati G, Petrochilos N, Fiori F (2009) 1090 MHz channel capacity improvement in the air traffic control context. Int J Microw Wirel Technol 1(3). https://doi.org/10.1017/s1759078709000191
26. EUROCONTROL (2006, July) CASCADE programme: 1090 MHz capacity study–final report, 2.7 edn
27. Chen Y, Lo S, Enge P, Jan S (2014) Evaluation & comparison of ranging using Universal Access Transceiver (UAT) and 1090 MHz Mode S extended squitter (Mode S ES). 2014 IEEE/ION Position, Location and Navigation Symposium—PLANS 2014. https://doi.org/10.1109/PLANS.2014.6851456
28. Li W, Kamal P (2011) Integrated aviation security for defense-in-depth of next generation air transportation system. In: 2011 IEEE international conference on technologies for homeland security (HST). https://doi.org/10.1109/ths.2011.6107860
29. Skaves P (2011) Information for cyber security issues related to aircraft systems. In: 2011 IEEE/AIAA 30th digital avionics systems conference. https://doi.org/10.1109/dasc.2011.6095968

30. De Cerchio R, Riley C (2011) Aircraft systems cyber security. In: 2011 IEEE/AIAA 30th digital avionics systems conference. https://doi.org/10.1109/dasc.2011.6095969
31. El-Badawy E, EL-Masry W, Mokhtar M, Hafez A (2010) A secured chaos encrypted mode-S aircraft identification friend or foe (IFF) system. In: 2010 4th international conference on signal processing and communication systems. https://doi.org/10.1109/icspcs.2010.5709756
32. Purton L, Abbass H, Alam S (2010) Identification of ADS-B system vulnerabilities and threats. In: Australasian Transport Research Forum 2010 Proceedings 29 September–1 October 2010, Canberra, Australia
33. De Cerchio R, Riley C (2012) Aircraft systems cyber security. In: 2012 integrated communications, navigation and surveillance conference. https://doi.org/10.1109/icnsurv.2012.6218454
34. Globus I (1972) Binary coding in asynchronous systems. Svyaz , Moscow
35. Belov A, Kozlov S, Korobkov A, Chabdarov S (2018) Multi-signal extraction of address flows in asynchronous impulse radio-systems. Vestnik KGTU im. A.N. Tupoleva, 74(4)
36. Tsikin I, Poklonskaya E (2017) Secondary surveillance radar signals processing at the remote analysis station. St. Petersburg State Polytechnical Univ J Comput Sci Telecommun Control Syst 10(2). https://doi.org/10.18721/JCSTCS.10205
37. Svyd I, Obod I, Maltsev O, Shtykh I, Zavolodko G, Maistrenko G (2019) Model and method for request signals processing of secondary surveillance radar. In: 2019 IEEE 15th international conference on the experience of designing and application of CAD systems (CADSM). https://doi.org/10.1109/CADSM.2019.8779347
38. Zhironkin S, Bliznyuk S, Kuchin A (2019) Jamming resistance of the inbound channel of an identification system with broadband signals and error control codes in the conditions of pulse noise and intra-system jamming. J Siberian Fed Univ Eng Technol. https://doi.org/10.17516/1999-494X-0166
39. Petrov A, Mikhalev V (2019) Bit-error rate in a digital data transmitting channel at chaotic impulse noise with random radio-pulse duration action. Syst Control Commun and Secur 3. https://doi.org/10.24411/2410-9916-2019-10303
40. Svyd I, Obod I, Maltsev O, Shtykh I, Zavolodko G (2019) Model and method for detecting request signals in identification friend or foe systems. In: 2019 IEEE 15th international conference on the experience of designing and application of CAD systems (CADSM). https://doi.org/10.1109/CADSM.2019.8779322
41. Malyarenko A (2007) Secondary radar systems for air traffic control and state recognition. HUVS, Kharkiv
42. Guo Y, Yang J, Guan C (2013) A mode 5 signal detection method based on phase and amplitude correlation. In: 2013 ninth international conference on natural computation (ICNC). https://doi.org/10.1109/ICNC.2013.6818164
43. Huan L, Feng Z, Dai L, Jian W (2015) One joint demodulation and despreading algorithm for MOD5. Open Autom Control Syst J 7(1). https://doi.org/10.2174/1874444301507010386
44. SBIR (2018) Mode 5 identification Friend or Foe (IFF) simulation and subsequent RF injection, 2nd ed. Department of Defense, Scalable Network Technologies, Inc., Culver City, CA, 12 Dec 2018
45. Lenshin A, Lebedev V (2015) Characteristics of detection of signals of imitostable regimes of identification systems. Dynamics of complex systems—XXI century, vol 9, issue no 1
46. Lenshin A, Lebedev V (2016) Algorithm for detecting signals of imitostable modes of identification. Telecommunications (7)
47. Leonardi M, Gerardi F (2020) Aircraft mode S transponder fingerprinting for intrusion detection. Aerospace 7(3). https://doi.org/10.3390/aerospace7030030
48. Obod I, Svyd I, Maltsev O, Zavolodko G, Pavlova D (2020) Optimization of data processing of primary radar systems. In: 2020 IEEE 40th international conference on electronics and nanotechnology (ELNANO). https://doi.org/10.1109/elnano50318.2020.9088842
49. Obod I, Svyd I, Maltsev O, Zavolodko G, Pavlova D (2020) Evaluation of measuring accuracy of the airborne object Azimuth when fusion the primary data radar observation systems. In: 2020 IEEE 15th international conference on advanced trends in radioelectronics, telecommunications and computer engineering (TCSET). https://doi.org/10.1109/TCSET49122.2020.235511

50. Pavlova D, Zavolodko G, Obod I, Svyd I, Maltsev O, Saikivska L (2019) Optimizing data processing in information networks of airspace surveillance systems. In: 2019 10th international conference on dependable systems, services and technologies (DESSERT). https://doi.org/10.1109/DESSERT.2019.8770022
51. Borisov V, Zinchuk V, Limarev A, Shestopalov V (2011) Interference immunity of radio communication systems with spectrum spreading by direct modulation by a pseudo-random sequence, 2nd edn. RadioSoft, Moscow
52. Obod, I. (1999) Integrated coordinate-and-time support for the address inquiry in the secondary radar systems. Telecommun Radio Eng 53(3). https://doi.org/10.1615/TelecomRadEng.v53.i3.100
53. Pankov V, Manezhkin A, Mytil V (2016) Evolution of aviation means of identification. RIO IPHVF RAS, Chernogolovka
54. Svyd I, Obod I, Maltsev O, Shtykh I, Maistrenko G, Zavolodko G (2019) Comparative quality analysis of the air objects detection by the secondary surveillance radar. In: 2019 IEEE 39th international conference on electronics and nanotechnology (ELNANO). https://doi.org/10.1109/ELNANO.2019.8783539
55. Obod I, Svyd I, Maltsev O, Vorgul O, Maistrenko G, Zavolodko G (2020) Optimization of the quality of information support for consumers of cooperative surveillance systems. In: Radivilova T, Ageyev D, Kryvinska N (eds) Data-centric business and applications. Lecture notes on data engineering and communications technologies, vol 48. Springer, Cham. https://doi.org/10.1007/978-3-030-43070-2_8
56. Obod I, Svyd I, Maltsev O, Zavolodko G, Pavlova D, Maistrenko G (2020) Fusion the coordinate data of airborne objects in the networks of surveillance radar observation systems. In: Radivilova T, Ageyev D, Kryvinska N (eds) Data-centric business and applications. Lecture notes on data engineering and communications technologies, vol 48. Springer, Cham. https://doi.org/10.1007/978-3-030-43070-2_31
57. Leonardi M, Di Gregorio L, Di Fausto D (2017) Air traffic security: aircraft classification using ADS-B message's phase-pattern. Aerospace 4(4). https://doi.org/10.3390/aerospace4040051.
58. Mantilla-Gaviria I, Galati G, Leonardi M, Balbastre-Tejedor J (2014) Time-difference-of-arrival regularised location estimator for multilateration systems. IET Radar Sonar Navig 8(5). https://doi.org/10.1049/iet-rsn.2013.0151

A Stand for Diagnosing the Durability of Infocommunication Equipment to the Effects of Powerful Ultra-Wideband Electromagnetic Pulses and Laser Radiation

Igor Shostko, Evgeniy Avchinnikov, and Yuliia Kulia

Abstract To check the stability of infocommunication equipment, which includes optico-electronic components to the effects of powerful electromagnetic radiation, special test stands are used. However, such stands are usually used when testing finished products to obtain a certificate. At the same time, there is often a need to assess the resistance to the effects of powerful electromagnetic radiation of individual modules, prototypes of radio electronic equipment at the stage of research and development. This requires a simple, inexpensive laboratory bench that allows preliminary testing. A laboratory bench has been developed for testing the resistance to the influence of powerful ultra-wideband electromagnetic pulses and laser radiation on infocommunication equipment at the stage of research and development. To reproduce the electric, magnetic and electromagnetic fields of ultra-wideband pulses, a pulse voltage generator and a field formation system were manufactured. The pulse voltage generator is built on the principle of the Arkadiev-Marx generator. The field formation system is an open strip line. The linear dimensions of the field formation system are taken based on the requirements for the dimensions of the required working volume with a uniform field. It is shown that the parameters of ultra-wideband electromagnetic pulses depend on the design features of the generating line of the generator. The parameters of the shaper design are optimized according to the criterion of achieving the maximum generator power. When observing the effects of powerful ultra-wideband electromagnetic pulses on the W922C camera as part of the Fortress W922C/R418 wireless video surveillance system and on the Nikon Coolpix digital camera, no significant deviations in their operation were found (the effects of short-term breakdown of the scan were observed during frame formation). The study of the resistance of optoelectronic devices to high-power radiation in the optical range

I. Shostko (✉) · Y. Kulia
Kharkiv National University of Radio Electronics, 14 Nauky Ave., Kharkiv, Ukraine
e-mail: ihor.shostko@nure.ua

Y. Kulia
e-mail: yuliia.kulia@nure.ua

E. Avchinnikov
Private Joint Stock Company "Scientific Research Institute of Laser Technology", 14 Nauky Ave., Kharkiv, Ukraine

D. Ageyev et al. (eds.), *Data-Centric Business and Applications*, Lecture Notes on Data Engineering and Communications Technologies 69,
https://doi.org/10.1007/978-3-030-71892-3_13

was carried out using an LGN-302 laser. The main attention was paid to the study of illumination by laser radiation of optoelectronic devices, taking into account the peculiarities of their design and lens adjustment modes, in which there is a significant loss of the information component obtained in real time. The limiting angles of side illumination are determined, when exceeding the observed illumination effects are not significant. In these cases, the illumination of a part of the image does not lead to the loss of its information component. The experiment involved the lenses Industar-22 and Industar-69, the W922C camera as part of the Fortress W922C/R418 wireless video surveillance system, and a Nikon Coolpix digital camera. For the tested optical and optoelectronic systems, the values of the limiting angles of side illumination did not exceed 10 … 15% of the angle of their field of view. During experiments, side illumination was significantly influenced by the presence of protective glass, which was located at a small distance from the lens and on which laser radiation is scattered. The consequence of this was a violation of the regular functioning of the optoelectronic device. Reduction of influence of laser emission side illumination is achieved during installation blend, polarizers.

Keywords Ultra-wideband electromagnetic pulse · Laser radiation · Information and communication systems · Video surveillance systems · Lateral illumination angle · Field of view of the lens

1 Introduction

The widespread use of microelectronics in modern infocommunication systems leads to an increase in the probability of failure of critical control and communication systems when exposed to an ultra-wideband electromagnetic impulses (UWB EMI) of electromagnetic factors of natural and technogenic origin (lightning discharges, static electricity discharges, electromagnetic fields of radio transmitting and radar stations, etc.). Under the influence of powerful UWB EMI, the performance of information and communication systems is interrupted (stopping tasks—the so-called "freeze", restarting the server, router) or the failure of individual system nodes. UWB low power EMP can interfere with radio receivers, for example, in a Wi-Fi network, connection time is increased, information transfer speed is reduced or subscribers become completely inaccessible. In video surveillance systems, the effect of powerful UWB EMIs is manifested visually from the failure of video synchronization to the complete cessation of the television system.

For equipment containing optico-electronic elements (video cameras), an additional threat is associated with exposure to powerful laser radiation.

Thus, under these conditions, the problems of protecting infocommunication systems, assessing their stability under the influence of UWB EMI [1–11] and laser radiation [12] become relevant.

To reproduce the impact of UWB EMI in our country and abroad, direct and indirect methods have been developed and used. Using direct methods, electromagnetic fields are reproduced whose parameters correspond to UWB EMI. Indirect methods are based on reproducing the calculated values of currents and voltages on the elements of the infocommunication system.

To test the stability of electronic equipment to the effects of high-power UWB EMI using the direct method, special stands is used. The main parts of the stand are a generator of powerful short pulses of nanosecond duration and transforming TEM camera system [13]. Such stand are designed and operated in the USA, Russia, Ukraine and other countries. However, such stands are usually used when testing finished products to obtain a certificate. At the same time, there is often a need to assess the durability to UWB EMI of individual modules, prototypes of electronic equipment at the stage of research and development work (R&DW). To do this, a simple, inexpensive laboratory stand is required, which allows conducting preliminary tests at the place of R&DW. For television equipment, it is necessary to further study the durability to laser radiation. Therefore, a laser will additionally be included in the stand.

Depending on the requirements for testing various radio systems, the stand can reproduce the magnetic and electric components of the field separately, or generate electromagnetic field strengths. To obtain electromagnetic fields, systems such as TEM-cameras or open strip lines are used, which ensure the formation of a plane electromagnetic wave with a ratio of amplitudes of electric and magnetic fields of 377 Ohms in their working volume.

In our stand, the field-forming TEM-camera system (limiting frequency 500 MHz) is made in the form of a segment of a symmetrical air stripe line with a constant characteristic durability, in which a transverse electromagnetic wave of the TEM type propagates. The electromagnetic field distribution in the regular part of the TEM-camera was calculated using the HFSS electrodynamic simulation program. This type of modeling installations is recommended by ISO 11,452–3 [14, 15] for reproducing electromagnetic fields and testing objects for electromagnetic durability to electromagnetic radiation (EMR).

2 Description of the Main Elements of the Stand

The stand for diagnosing the levels of durability of electronic equipment to the effects of UWB EMI is designed to study and test electronic equipment for durability to UWB EMI at the stage of R&DW.

It allows you to abandon field tests of electronic equipment (EE) associated with high economic costs, as well as to obtain multiple reproduction of the defeating factors EE. As a result of this, it is possible to more accurately determine the levels of electromagnetic durability of the test objects and their final "fine-tuning" according to the test results. The structural diagram of the stand is shown in Fig. 1. The appearance and arrangement of the elements of the stand is shown in Fig. 2.

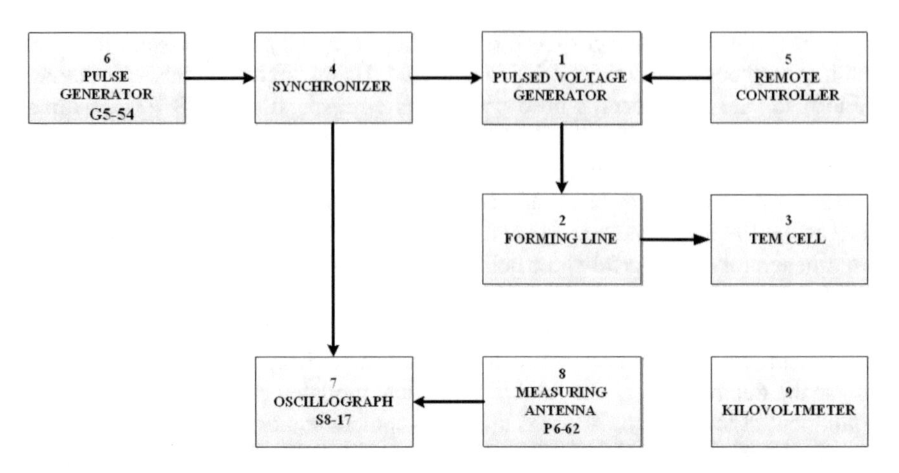

Fig. 1 The structural diagram of the stand

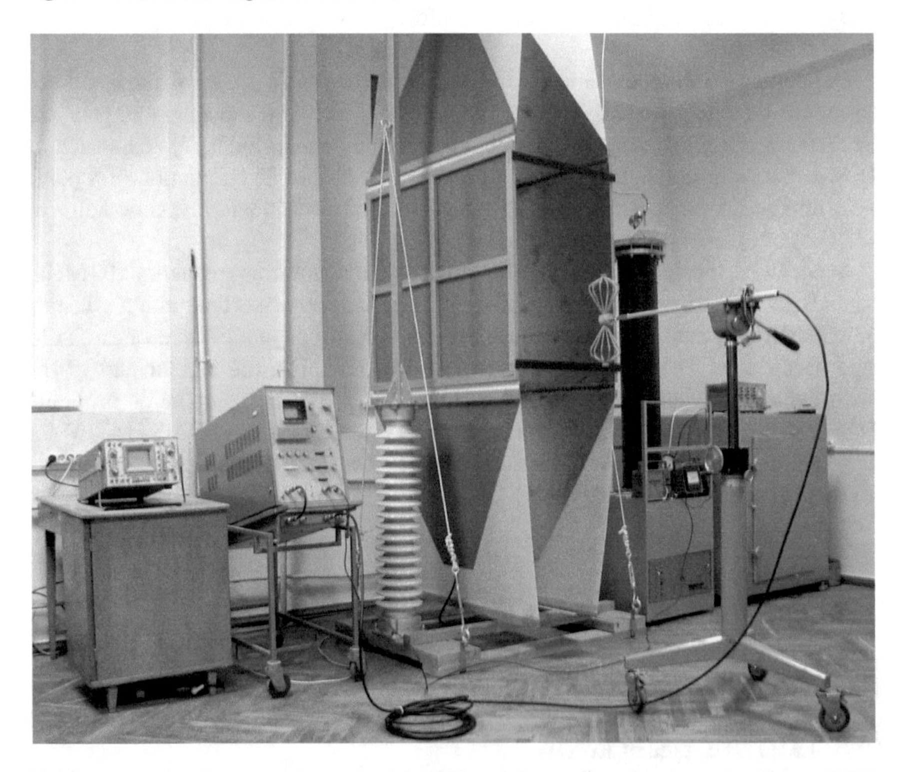

Fig. 2 A stand for diagnosing the levels of durability of electronic equipment to the effects of UWB EMI

The stand (Fig. 1) consists of the following parts:

1. a pulse voltage generator (PVG);
2. formative line;
3. TEM cell;
4. synchronizer;
5. remote control;
6. pulse generator G5-54;
7. S8-17 oscilloscope;
8. measuring antenna P6-62;
9. kilovoltmeter.

The pulse voltage generator consists of an Arkadyev-Marx generator and an adjustable high-voltage power supply. The Arkadyev-Marx generator (Fig. 3) is a system of 11 capacitors that are charged for a long time when connected in parallel and then, when the specified voltages reach the capacitors, are discharged when connected in series through spark gaps in a very short period of time. The capacitors are charged through resistors with a resistance of 33 kOhm from an adjustable high-voltage power supply voltage of 12–14 kV. The generator capacity in active mode is 62 pF, and the pulse energy is 0.54–0.74 J. The generated pulse has duration of

Fig. 3 Arkadyev-Marx generator and adjustable high-voltage power supply

0.5 μs. with an amplitude of 132–154 kV. This pulse is charged forming line 2, made of a segment of a coaxial cable, with a wave impedance of 50 Ohms.

For oscillography of the shape of the generated pulses, we used a S7-10B high-speed oscilloscope with a vertical deviation path bandwidth of 1200 MHz and a deviation coefficient of 63 mV/mm. To delay the onset of the appearance of the pulse on the screen relative to the Deployment start, a delay line of 120 ns made of coaxial cable was used. The input high-voltage resistive divider is assembled from high-voltage resistors KEV-5. The measurement error is not more than 20%, the coefficient of vertical deviation is 630 mV/cell, the sweep time is 200 ns/cell.

After the voltage in the forming line 2 reaches its maximum, a high-voltage arrester is triggered, as a result of which a high-voltage pulse of nanosecond duration, an amplitude of 66–77 kV, is formed at the output of the line. Next, the generated pulse enters the TEM cell 3, creating a uniform high-voltage field inside it. By putting various telecommunication equipment inside this cell, we study the effect of ultra-wideband nanosecond pulses on it.

TEM cell 3 (Fig. 1) is an air strip line of the type "Sandwich" with a rectangular cross-section loaded on a matched load with a resistance of 50 Ohms. Its size, excluding matching elements, is 1 m × 1.2 m × 0.6 m. The linear dimensions of the line are selected from the condition of providing a working volume with a uniform field sufficient to accommodate the test object in it. The working volume of the line is 1 m × 1.2 m × 0.3 m. The external overall dimensions of the line are 1.2 m × 1.4 m × 3.7 m.

The TEM cell (Fig. 4) consists of a regular segment of strip line 1 with pyramidal transitions 2 and 3 included at the input and output. Pyramidal transitions are used to coordinate the geometric dimensions of the regular part of the camera with coaxial connectors 4. In the middle part of the camera, where the mode a traveling wave emitting a wave in open space, field has a minimum non-uniformity and contains no longitudinal components, the test object is located 5.

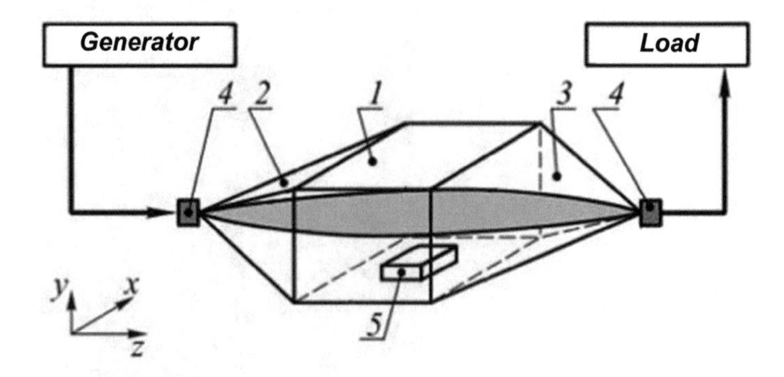

Fig. 4 Symmetric strip line (TEM cell)

3 The Principle of Operation of the Stand

The device under test is placed inside the TEM cell 3 (Fig. 1). The control signal is supplied from the remote control 5 to the pulse voltage generator 1. As a result, the high-voltage regulated power supply is turned on and the Arkadyev-Marx generator is charged to a voltage of 13 kV. The charge voltage is controlled by a kilovoltmeter 9. The pulse generator 6 outputs the trigger pulses to the input of synchronizer 4.

As a result of this, the synchronizer 4 generates an igniting impulse, which ensures the discharge of the Arkadyev-Marx generator to the forming line 2. After charging the forming line 2 to maximum voltage, the spark gap is triggered and it is discharged on TEM cell 3.

In addition, the synchronizer 4 generates a start trigger of the oscilloscope 7, which, together with the measuring antenna 8, serves to control the characteristics of the received pulse. Moreover, in the working volume of the TEM cell 3, a uniform electromagnetic field is formed with the specified amplitude-time parameters. After the discharge, the change in the parameters of the test device is investigated.

4 Shaper of Powerful Microwave Pulses of Nanosecond Duration

When the power of the generated pulse is large, there is a problem of how to transfer energy through a switch to a load with high efficiency in a short time [16–18]. Therefore, the actual development of new methods for commutation and shortening the pulse front to excite a TEM cell is relevant. The UWB EMI shaper is a segment of a coaxial line in which instead of a central conductor there is a multi-element adjustable spark gap Fig. 5.

An analog of this device is the design of multi-element spark gaps [19]. An equivalent shaper circuit is shown in Fig. 6, where: C_P and C_C are capacitances between the electrodes of the spark gaps and capacitances between the core of each section and the external conductor, K are the keys that simulate the successive breakdown of the spark gaps, R_n is the load resistance. For $C_C \gg C_P$, the main part of the voltage U_Σ of the high-voltage source applied to the multi-element spark gap will be in the first gap. This leads to the successive operation of m spark gaps (starting from the first) and the formation as a result of the redistribution of voltages on the remaining spark gaps, damped oscillation amplitude at load R_n.

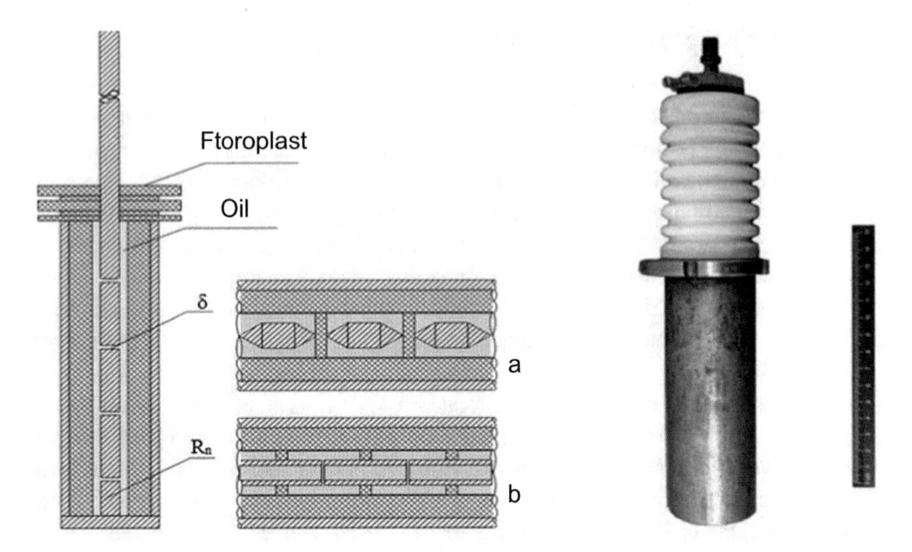

Fig. 5 Design and appearance of the pulse shaper. The principle of constructing the forming line: a—with solid-state spark gaps; b—with gas spark gaps; R_n—is the load resistance; δ—interelectrode gap

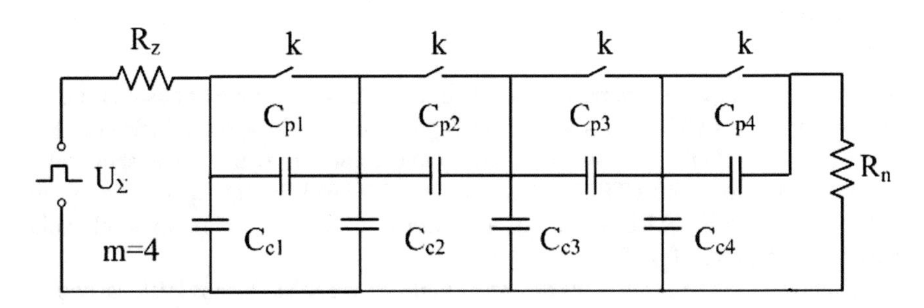

Fig. 6 The equivalent circuit shaper

5 Analysis of the Influence of Structural Features of the Shaper on the Parameters of Microwave Pulses

The value of the equivalent capacitance of the shaper C_F, after the breakdown of k spark gaps at $C_C \gg C_P$, will be determined by the value:

$$C_F = kC_C + \frac{C_C C_P}{C_C + C_P} \approx kC_C, \tag{1}$$

i.e., the shaper capacity, before the breakdown of the first arrester ($k = 1$), is almost equal to the capacity of the first section. As the breakdowns of the spark gaps, the capacity of the shaper will increase in proportion to the number of pierced spark gaps. The voltage on the capacitance:

$$U_C = \frac{1}{C_F}\int_0^t I(t)dt = U_\Sigma\left(1 - e^{-\frac{t}{\tau}} - \frac{t}{\tau}e^{-\frac{t}{\tau}}\right). \tag{2}$$

The voltage on the elements of the shaper [2]:

$$U_{k+1} = \frac{U_\Sigma^{Sh[(m-k)\gamma]}}{Sh[m\gamma]}, \tag{3}$$

where

$$\gamma = \ln\left(1 + \frac{C_C}{2C_P}\left(1 + \sqrt{\frac{4C_P}{C_C} + 1}\right)\right), k = 0, 1, 2, \ldots, m. \tag{4}$$

Voltage at the first gap and between the electrodes of the first spark gap:

$$\Delta U_1 = U_1 - U_2 = U_\Sigma\left(1 - \frac{e^{(m-1)\gamma} - e^{-(m-1)\gamma}}{e^{m\gamma} - e^{-m\gamma}}\right) \approx U_\Sigma(1 - e^{-\gamma}), \tag{5}$$

$$\Delta U_{cp} = U_\Sigma(1 - e^{-\gamma})\left(1 - e^{-\frac{t}{\tau}} - \frac{t}{\tau}e^{-\frac{t}{\tau}}\right) \tag{6}$$

The voltage distribution on the elements of the shaper can be controlled by changing the configuration of the multi-element spark gap: by choosing the material and thickness of the dielectric inserts in the spark gaps (Fig. 5, a), adjusting the pressure inside the shaper (Fig. 5, b), etc.

Equating the expression (6) to the value of the voltage static magnitude of the breakdown $\Delta U_{cp} = U_p$, we find the expression for determining the breakdown time interval t_p:

$$1 - e^{-\frac{t_p}{\tau}} - \frac{t_{PR}}{\tau}e^{-\frac{t_p}{\tau}} = \frac{U_p}{U_\Sigma(t_p)(1 - e^{-\gamma})} \tag{7}$$

Since U_p depends primarily on the size of the gaps and gas pressure in the discharge gaps, the value of t_p found from expression (7) will also be determined by these parameters. The amplitude of the generated microwave oscillations will be determined by the magnitude of the breakdown voltage of the spark gaps. It will increase with increasing gaps and gas pressure in the discharge gaps.

In order to maintain a high breakdown voltage and, therefore, a high power of the generated pulse at small gaps, it is necessary to switch to solid-state (membrane) dischargers. At small gaps, the breakdown field strength in solid dielectrics sharply increases with a decrease in the gap. By combining such parameters as the membrane thickness and the number of membranes in the multilayer construction of the discharger, it is possible to adjust the value of the breakdown voltage and the breakdown delay time for each of the dischargers.

Thus, an increase in the power of microwave pulses and a shortening of their duration are ensured by changing the profile of the multi-element core of the coaxial shaper design, by choosing the number of sections, the gap thickness or the number of dielectric membranes in the dischargers.

6 The Results of the Study of the Shaper of Microwave Pulses

The structure (length 24 cm, diameter 38 mm) of the pulse shaper is made and tested, which consists of four sections with an insulator made of fluoroplastic F_4. Copper tubes (length 1.5 cm, diameter 8 mm) are used for the inner core, which are fixed with ebonite washers. With the size of the gaps of 4 mm, the capacitance of the section and the gap: $C_P \approx 0, 05$ pF, $C_C \approx 0.9$ pF.

The voltage at the output of the power source is 100 kV, the inductance of the power source is 500 nH. Charging resistance $R_z = 1050$ Ohms. The momentum front is exponential.

The maximum value of the pulse power in the load depends on the delay time during when triggered each of the spark gaps. For the above parameters, the maximum pulse power was obtained at a breakdown delay time for each of the spark gaps $t_z \approx 5$ ns. The nature of the excited oscillations in each section of the shaper and the shape of the power pulse in the load depend on the delay time during the operation of each of the spark gaps. By changing the delay times of the spark gaps, you can change the shape of the pulse. The maximum power value is obtained when all the electrical parameters of the shaper and the power supply are matched. When optimizing the parameters of the shaper according to the criterion of achieving the maximum power of the generated pulse, it is necessary: to increase the inductance and container sections, reduce the capacity of the gaps.

The power and shape of the pulse at the output of the shaper also change depending on the values of the resistances R_z and R_n. A graph of the change in pulse power, for various resistance values R_n, is shown in Fig. 7. The maximum power value of R_n is obtained with a minimum resistance R_z. The optimal value of R_n is 80 Ohms.

The frequency response of the shaper changes during the triggering of the spark gaps. With increasing resistance R_z changes the amplitude frequency characteristic of the shaper. If R_z—are units of Ohm, then only frequencies coinciding with the resonant frequencies of the shaper are amplified from the pulse spectrum. With a resistance of the order of one kOhm, the amplitude frequency characteristic in the band up to 1 GHz becomes uniform, but the voltage gain decreases. Therefore, depending on the shape (spectrum) of the pulse at the output of the shaper, the maximum power in the pulse is achieved by adjustment R_z.

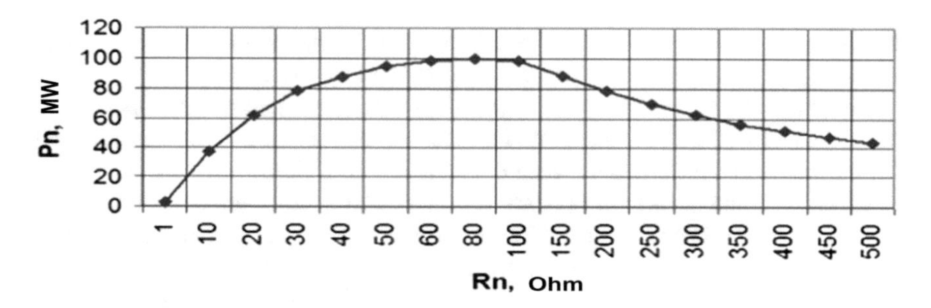

Fig. 7 The dependence of the maximum value of the pulse power on the resistance value Rn

7 Stand for Testing the Durability of Optical-Electronic Equipment to Laser Radiation

The stand for testing optical devices and optico-electronic devices is presented in Fig. 8.

The installation includes the LGN-302 laser which has the following characteristics:

- working gas mixture—helium–neon;
- wavelength of radiation—0.63 μm (beam color—red);
- radiation power—0.5 mW.

Other types of lasers can be used as a source of laser radiation.

The scattering lens is designed to simulate the discrepancy of laser radiation when passing through the atmosphere. The scattering lens was located near the LGN-302 laser. If there is a scattering lens at a distance of 2.2 m, the diameter of the laser beam increases from 0.5 to 25 mm. Experimental studies were performed both in the absence of a scattering lens and in its presence.

The supporting turning mechanism is intended for arrangement on it of the optico-electronic means which are investigated, change and measurement of an angle between an optical axis and a laser beam.

Experimental studies were performed according to the following method.

At the preparatory stage of experimental research, the direction of the laser beam and the device under study were coordinated. To do this, regardless of the presence

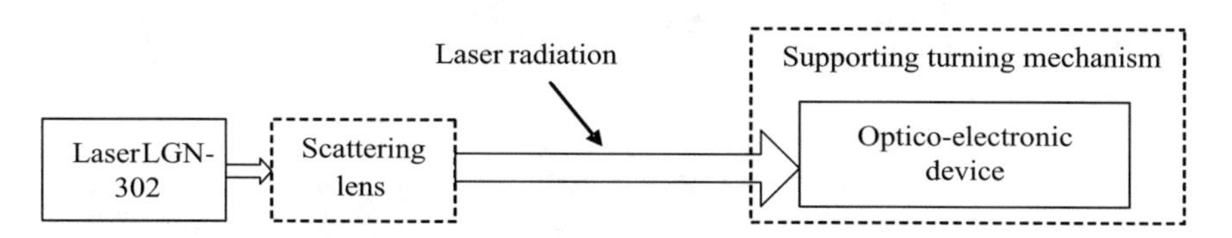

Fig. 8 Additional part of the stand for optical devices and optico-electronic devices

or absence of a scattering lens, the laser beam was directed to the central part of the objective, the focal plane of which was orthogonal to the laser beam.

Subsequently, by means of the supporting turning mechanism, the direction of the optical axis of the optoelectronic means relative to the laser beam was changed. For each position, the angle between the optical axis of the tool and the laser beam was measured and the effects of glare or lateral illumination were recorded.

Studies of the lateral angles of laser radiation of optical devices have shown that the angles of lateral illumination do not exceed 10…15% of the value of the field of view and depend on the design aperture of the lens, lens characteristics, the presence of the diaphragm. The most detailed information about the angles of lateral illumination of these lenses should be as a part of optico-electronic means.

8 Investigation of the Angles of Lateral Illumination by Laser Radiation of the Lens "Industar-22"

The Industar-22 lens is a standard lens with enlightened optics, made according to the Tessar scheme (four lenses in three groups—two lenses are glued), has an iris aperture, focal length—52 mm, field of view angle—46°.

In the Table 1 shows the results of the experiment.

The results show that for a given lens, the blinding of the receiving element (photographic film or semiconductor matrix, if any) can be carried out within the angles between the optical axis of the lens and the direction of the laser beam up to 20° (this angle is close to half the field of view lens, which is 23°). In this case, for small values of the angles between the optical axis of the lens and the direction of the laser beam (from 0° to 10°) can achieve maximum glare effects, which are

Table 1 The results of the impact of laser radiation on the lens "Industar-22" depending on the angle between the optical axis of the lens and the direction of the laser beam in the absence of a scattering lens and with an open aperture

The value of the angle, deg.	0	5	10	15
The results of exposure to laser radiation	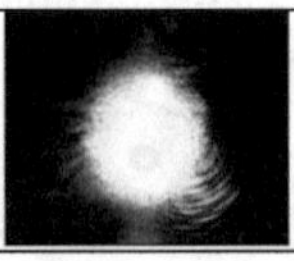	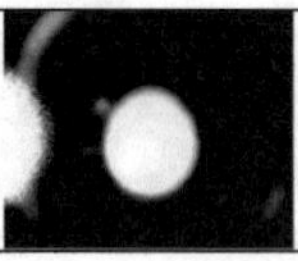	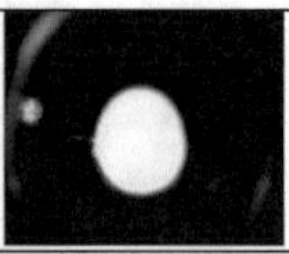	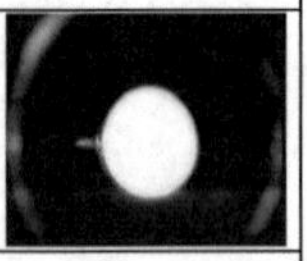
The value of the angle, deg.	20	25	30	35
The results of exposure to laser radiation	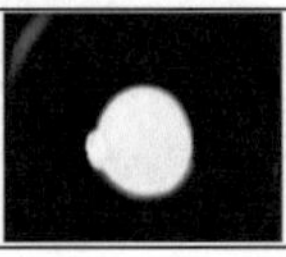	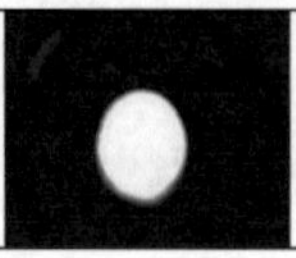		

associated with illumination of most of the receiving element, which eliminates the possibility of obtaining useful information photos or videos. Further increase in the angle between the optical axis of the lens and the direction of the laser beam leads to a decrease in the effects of glare, which are associated with a decrease in laser beam intensity due to increased reflection on lens surfaces and absorption by non-working (internal) surfaces. with a shift of the illumination area to the edge of the receiving element.

At a laser angle of 25°, there is a significant decrease in the intensity of the laser beam at the output of the lens, and at a further increase in the angle—the effects of laser radiation cease to be observed.

Thus, when a narrow (5 mm) laser beam hits the Industar-22 lens (15 mm diameter of the input lens) outside the field of view, the angles of lateral illumination can be up to 7°. In the Table 2 shows the results of the experiment.

With the diaphragm closed (the diameter of the open part is 3 mm), the blinding of the receiving element can be carried out within the angles between the optical axis of the lens and the direction of the laser beam up to 20°. At a laser beam angle of 25°, no laser beam is observed at the lens output.

Thus, for these conditions of experimental studies, the effects of lateral illumination are not observed. The laser beam, which is reflected (refracted) from the cylindrical non-working surfaces of the lens, the inner part of the lens body, is not transmitted by a closed aperture, which, in fact, acts as an internal aperture.

Similarly to the previous study, when the angle of influence of laser radiation is increased, the effects of glare within the field of view of the lens are reduced, and the area of illumination is shifted to the edge of the receiving element. In the Table 3 shows the results of the experiment.

Table 2 The results of the impact of laser radiation on the lens "Industar-22" depending on the angle between the optical axis of the lens and the direction of the laser beam in the absence of a scattering lens and a closed aperture

The value of the angle, deg.	0	5	10
The results of exposure to laser radiation	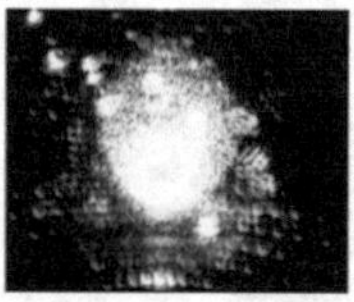	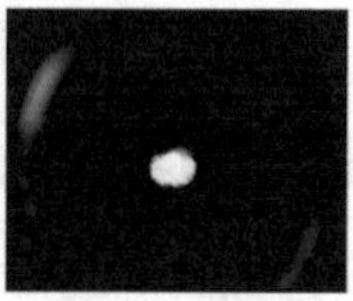	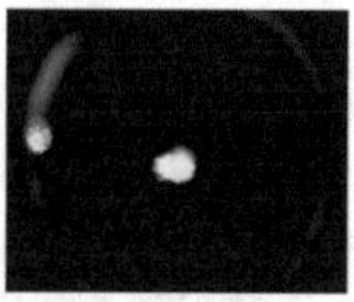
The value of the angle, deg.	15	20	25
The results of exposure to laser radiation	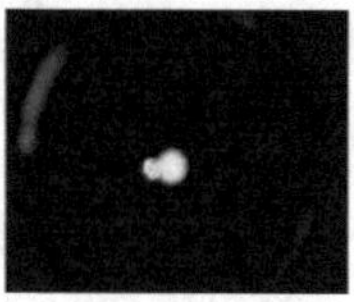	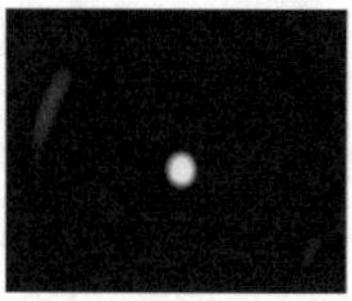	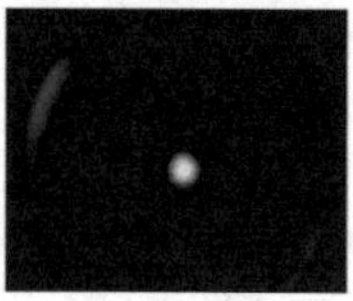

Table 3 The results of the effect of laser radiation on the lens "Industar-22" depending on the angle between the optical axis of the lens and the direction of the laser beam in the presence of a scattering lens and with an open aperture

The value of the angle, deg.	0	5	10
The results of exposure to laser radiation	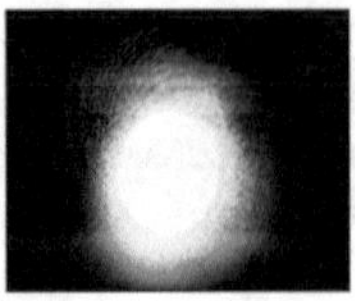	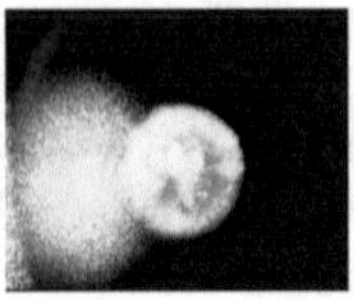	
The value of the angle, deg.	15	20	25
The results of exposure to laser radiation			
The value of the angle, deg.	30	35	40
The results of exposure to laser radiation			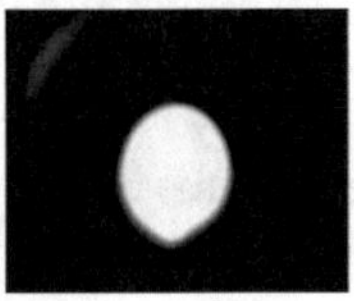

Unlike previous studies, the presence of a scattering lens provides an increase in the diameter of the laser beam from 5 to 25 mm at the lens inlet (at a distance of 2.2 m from the laser). The obtained diameter of the laser beam completely overlaps the diameter of the input lens, which is 15 mm.

From the analysis of the obtained results it is seen that for this lens at full overlap of the laser beam of the input lens area, the blinding of the receiving element can be carried out within the angles between the optical axis of the lens and the laser beam direction up to 35° (this angle exceeds half the field of view lens, which is 23°). Maximum glare effects, which are associated with illumination of most of the area of the receiving element, are achieved at small values of angles between the optical axis of the lens and the direction of the laser beam (from 0° to 15°), which eliminates the possibility of obtaining useful information from photographs or videos.

Further increase in the angle between the optical axis of the lens and the direction of the laser beam (up to 35°), leads to a significant reduction in the effects of glare, which are associated with reduced laser beam intensity due to increased reflection on lens surfaces and absorption by non-working (internal) the surfaces of the lens, as well as with the shift of the illumination area to the edge or beyond the receiving element. At angles between the optical axis of the lens and the direction of the laser beam of 40° or more, the effects of lateral illumination of the lens cease to be observed.

Thus, when the area of the input lens "Industar-22" is completely covered by the laser beam, the angles of lateral illumination can be up to 15°.

9 Investigation of the Angles of Lateral Illumination by Laser Radiation of the Lens "Industar-69"

The Industar-69 lens is a semi-format lens with enlightened optics, made according to the Tessar scheme (four lenses in three groups—two lenses are glued), has a focal length—28 mm, angle of view—56°. In the Table 4 shows the results of the experiment.

The analysis of the obtained results shows that for this lens the blinding of the receiving element can be carried out within the angles between the optical axis of the lens and the direction of the laser beam up to 25° (this value is close to half the field of view of the lens, which is 28°). Maximum glare effects, which are associated

Table 4 The results of the effect of laser radiation on the lens "Industar-69" depending on the angle between the optical axis of the lens and the direction of the laser beam in the absence of a scattering lens

The value of the angle, deg.	0	5	10
The results of exposure to laser radiation	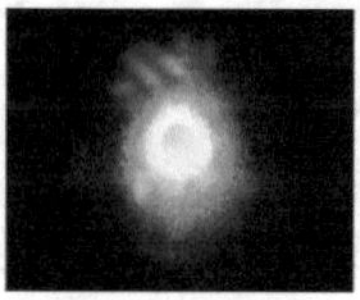	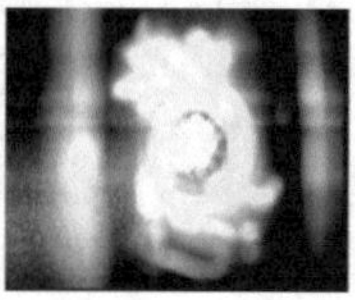	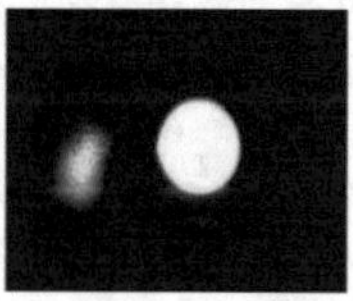
The value of the angle, deg.	15	20	25
The results of exposure to laser radiation	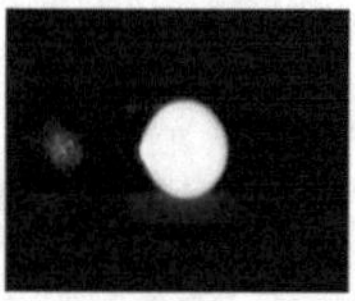		
The value of the angle, deg.	30	35	40
The results of exposure to laser radiation			Limited to the lens body

Table 5 The results of the impact of laser radiation on the lens "Industar-69" depending on the angle between the optical axis of the lens and the direction of the laser beam in the presence of a scattering lens

The value of the angle, deg.	0	5	10
The results of exposure to laser radiation	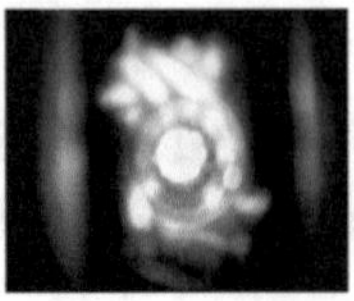	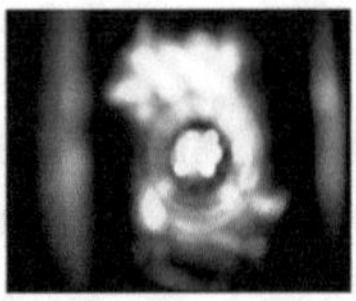	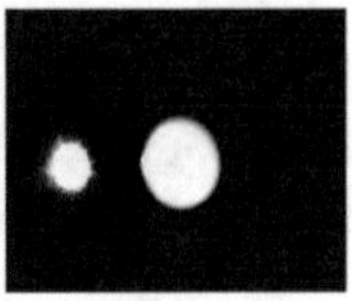
The value of the angle, deg.	15	20	25
The results of exposure to laser radiation			–

with illumination of most of the area of the receiving element, are achieved at small values of angles between the optical axis of the lens and the direction of the laser beam (from 0° to 15°), which eliminates the possibility of obtaining useful information from photographs or videos.

At a laser angle of 25°, there is a significant decrease in the intensity of the laser beam at the output of the lens, and at angles of 30° and 35°—the effects of laser radiation cease to be observed. In addition, at angles of 40° or more, the impact of the laser beam is limited to the lens case. Thus, when hitting a narrow (5 mm) laser beam into the lens "Industrial-69" (diameter of the input lens—8 mm) outside the field of view, the angles of lateral illumination can be up to 2°. In the Table 5 shows the results of the experiment.

The presence of a scattering lens provides a laser beam diameter of 25 mm at a distance of 2.2 m at the entrance of the lens and a complete overlap of the input lens with a diameter of 8 mm.

The analysis of the obtained results shows that when the laser beam completely covers the area of the input lens of this lens, the blinding of the receiving element can be carried out within the angles between the optical axis of the lens and the direction of the laser beam up to 15°. Maximum glare effects, which are associated with illumination of most of the area of the receiving element, are achieved at low values of angles between the optical axis of the lens and the direction of the laser beam (0°–10°), which eliminates the possibility of obtaining useful information from photographs or videos.

At laser exposure angles of 15° and 20°, there is a significant decrease in the intensity of the laser beam at the output of the lens, and at angles of 25° and above—the effects of laser radiation cease to be observed.

Thus, when hitting a laser beam whose diameter exceeds the diameter of the input lens, outside the field of view, the angles of lateral illumination of the lens "Industrial-69" are not observed. Unlike previous studies, in which the diameter of the laser beam was smaller than the diameter of the input lens of the lens, in this case should take into account the reduction (9.77 times) of the intensity of the laser beam passing through the lens in a proportion equal to the ratio of the plane of the laser beam after the scattering lens to the plane of the input lens.

10 Investigation of the Angles of Lateral Illumination by Laser Radiation of the W922C Camera as a Part of the Fortress W922C/R418 Wireless Video Surveillance Means

The W922C has a single-lens object-glass with a focal length of 4 mm and a field of view angle of 62°, equipped with a flat protective glass. In the Table 6 shows the results of the experiment.

Table 6 The results of the effect of laser radiation on the camera W922C depending on the angle between the optical axis of the camera lens and the direction of the laser beam in the absence of a scattering lens in a lighted room (infrared camera spotlight does not work)

The value of the angle, deg.	0	10	20
The results of exposure to laser radiation	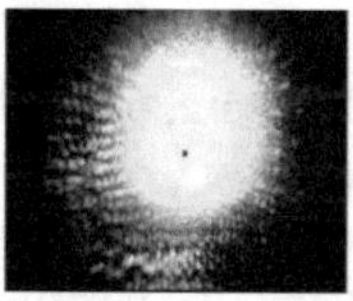	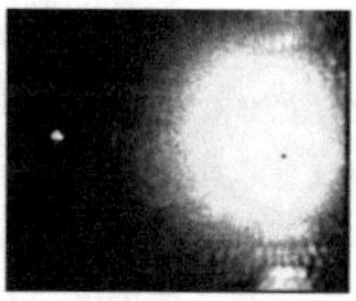	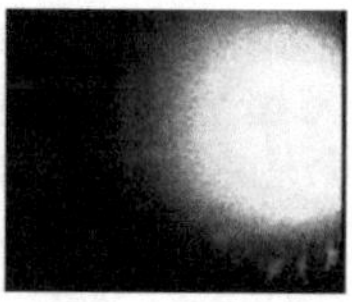
The value of the angle, deg.	30	40	50
The results of exposure to laser radiation		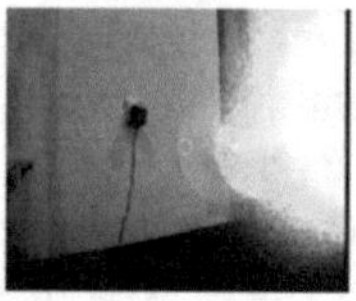	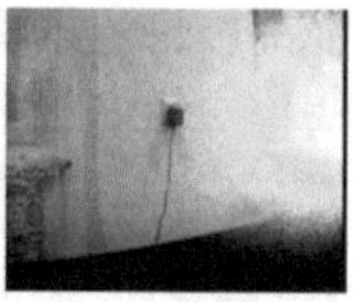
The value of the angle, deg..	60	70	80
The results of exposure to laser radiation			

The results show that the dazzle of the W922C is within the angles between the optical axis of the camera lens and the direction of the laser beam up to 80° (this value is more than twice the field of view of the camera lens, which is 31°). At the same time, for angles between the optical axis of the camera lens and the direction of the laser beam from 0° to 30°, maximum glare effects are achieved, which are associated with illumination of most of the receiving element, which eliminates the possibility of obtaining useful information from images or videos. Further increasing the angle between the optical axis of the camera lens and the direction of the laser beam to 80° gradually reduces the glare of the receiving element of the camera (most images or video recording are suitable for useful information). At an angle of 90° glare of the receiving element of the camera is not observed.

The blinding of the W922C's receiving element in such a wide range of angles is due to the presence of a flat protective glass, which is located directly in front of the camera lens at a short distance. It is this design feature of the camera that causes the reflection (refraction) of the laser beam on the surfaces of the protective glass and its impact on the receiving element.

Thus, when a narrow (5 mm) laser beam hits the W922C camera (input lens diameter—6 mm) outside the field of view, the angles of lateral illumination are up to 50°. In the Table 7 shows the results of the experiment.

The results obtained are similar to the previous ones. Under these conditions of experimental research (in a dark room), the sensitivity of the camera is automatically increased, which led to a greater manifestation of the effects of its blinding. In the Table 8 shows the results of the experiment.

The results show that in the presence of a scattering lens, the dazzle of the camera W922C is carried out within the angles between the optical axis of the camera lens and the direction of the laser beam up to 40° (this value is one third more than half the field of view of the camera lens 31°). In this case, for angles between the optical axis of the camera lens and the direction of the laser beam from 0° to 20°, maximum glare effects are achieved, which are associated with illumination of most or half the area of the receiving element, what eliminates the possibility of obtaining useful information from images or videos. Further increase of the angle between the optical axis of the camera lens and the direction of the laser beam to 50° gradually reduces the glare of the receiving element of the camera (most of the images or videos are suitable for useful information). At an angle of 60° blinding of the receiving element of the camera ceases to be observed.

In the presence of a scattering lens, the range of angles at which blinding of the receiving element of the camera W922C is observed is smaller than in the absence of a scattering lens. This is due to a reduction (17.4 times) in the intensity of the laser beam passing through the camera lens in a ratio equal to the ratio of the plane of the laser beam after the scattering lens at the camera input to the plane of the input lens. In the Table 9 shows the results of the experiment.

The results show that in the presence of a scattering lens and in a darkened room (camera sensitivity is increased) dazzle of the W922C camera is carried out within the angles between the optical axis of the camera lens and the direction of the laser beam to 40° (this value is one third more than half field of view of the camera lens, which

Table 7 The results of the effect of laser radiation on the camera W922C depending on the angle between the optical axis of the camera lens and the direction of the laser beam in the absence of a scattering lens in a darkened room (with the camera's infrared spotlight on)

The value of the angle, deg.	0	10	20
The results of exposure to laser radiation	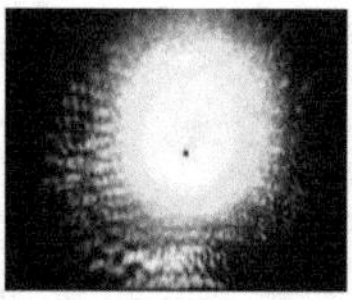	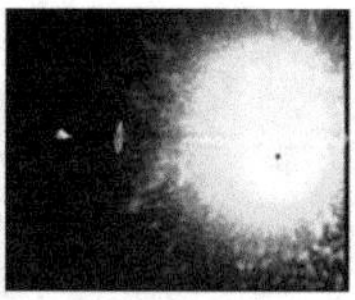	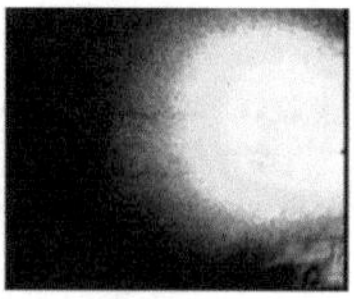
The value of the angle, deg.	30	40	50
The results of exposure to laser radiation			
The value of the angle, deg.	60	70	80
The results of exposure to laser radiation			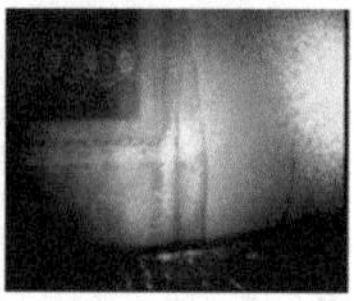

Table 8 The results of the effect of laser radiation on the camera W922C depending on the angle between the optical axis of the camera lens and the direction of the laser beam in the presence of a scattering lens in a lighted room (infrared searchlight does not work)

The value of the angle, deg.	0	10	20
The results of exposure to laser radiation	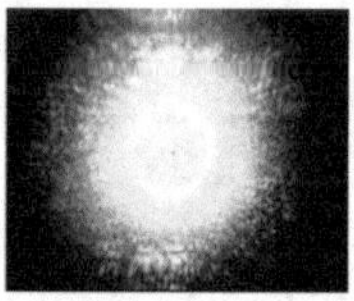	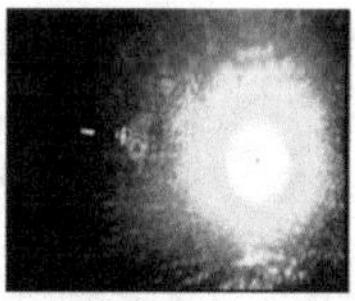	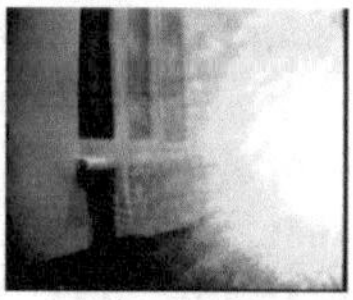
The value of the angle, deg.	30	40	50
The results of exposure to laser radiation		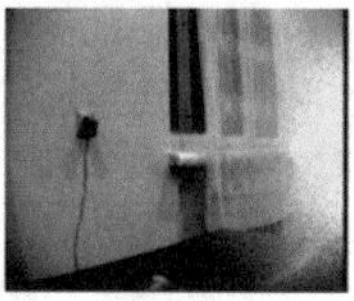	

Table 9 The results of the effect of laser radiation on the camera W922C depending on the angle between the optical axis of the camera lens and the direction of the laser beam in the presence of a scattering lens in a darkened room (with the camera's infrared spotlight on)

The value of the angle, deg.	0	10	20
The results of exposure to laser radiation	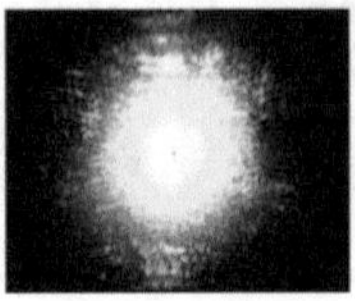	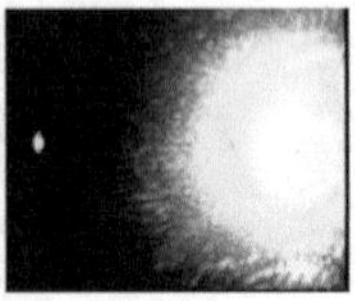	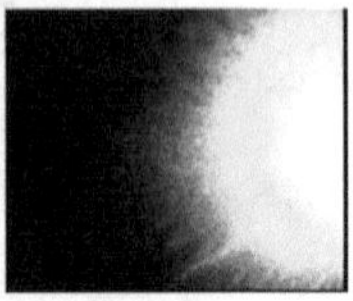
The value of the angle, deg.	30	40	50
The results of exposure to laser radiation	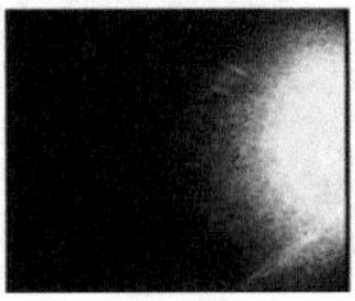	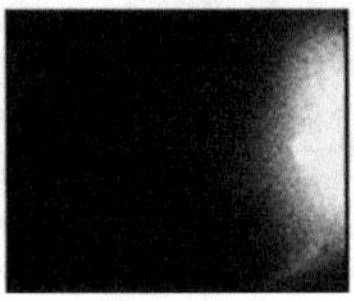	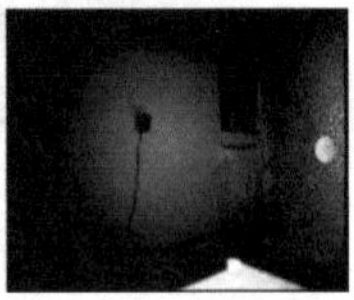

is 31°). Within these angles, the maximum blinding effects are maintained. Further increase of the angle between the optical axis of the camera lens and the direction of the laser beam to 50° gradually reduces the glare of the receiving element of the camera (most of the images or videos are suitable for useful information). At an angle of 60° blinding of the receiving element of the camera ceases to be observed.

11 Research of Angles of Lateral Illumination by Laser Radiation of the Nikon Coolpix Digital Camera

The Nikon Coolpix digital camera has a NIKKOR multi-lens lens with a variable focal length of 4.5 to 99 mm, which corresponds to the field of view angles for a 1/2.3″ matrix from 81° to 4.5°, respectively. In the Table 10 shows the results of the experiment.

The results show that the dazzling of the Nikon Coolpix digital camera is carried out within the angles between the optical axis of the camera lens and the direction of the laser beam up to 40° inclusive (this value coincides with half the field of view of the camera lens, which is 40.5°). At the same time, for angles between the optical axis of the camera lens and the direction of the laser beam from 0° to 10°, maximum blinding effects are achieved, which are associated with illumination of most of the receiving element, which eliminates the possibility of obtaining useful information from images or videos. Further increasing the angle between the optical axis of the camera lens and the direction of the laser beam to 40° gradually reduces the glare of the camera receiving element (most pictures or videos are suitable for

Table 10 Results of the effect of laser radiation on a Nikon Coolpix digital camera depending on the angle between the optical axis of the camera lens and the direction of the laser beam in the absence of a scattering lens and without zooming (corresponding to a field of view angle of 81°)

The value of the angle, deg.	0	5	10
The results of exposure to laser radiation	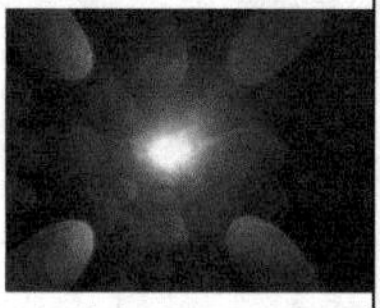		
The value of the angle, deg.	20	30	40
The results of exposure to laser radiation	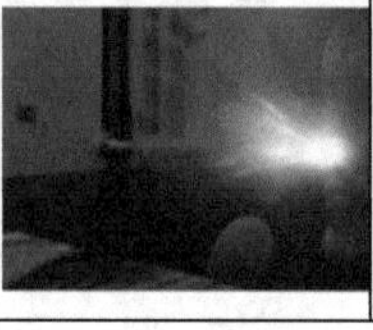		

useful information). At angles greater than 40°, glare of the receiving element of the camera is not observed.

Thus, when a narrow (5 mm) laser beam hits a Nikon Coolpix digital camera (input lens diameter 20 mm, which is limited by a rectangular window 20 × 17 mm), which operates in the absence of image magnification, outside the field of view angles lateral illumination are absent. In the Table 11 shows the results of the experiment.

The results show that the dazzling of the Nikon Coolpix digital camera is carried out within the angles between the optical axis of the camera lens and the direction of the laser beam up to 5° inclusive (this value is twice more than half the field of view of the camera lens, which is 2.25°). When the angle between the optical axis of the camera lens and the direction of the laser beam increases by more than 5°, glare from the camera receiving element ceases to be observed.

Table 11 Results of the effect of laser radiation on a Nikon Coolpix digital camera depending on the angle between the optical axis of the camera lens and the direction of the laser beam in the absence of a scattering lens and the maximum approximation of the image (corresponding to a field angle of 4.5°)

The value of the angle, deg.	0	5	10
The results of exposure to laser radiation			

The result obtained at an angle of 5° has a speckle structure. This is probably due to the reflection of a laser beam on the non-working surfaces of the lens, which increases the length when the image magnification function is introduced, or for other reasons.

Thus, when a narrow (5 mm) laser beam hits a Nikon Coolpix digital camera (input lens diameter 20 mm, which is limited by a rectangular window 20 × 17 mm), which operates in the mode of maximum image magnification, outside the field of view of the side lighting angles are approximately up to 3°. In the Table 12 shows the results of the experiment.

The results show that in the presence of a scattering lens, the Nikon Coolpix digital camera can be blinded within the angles between the optical axis of the camera lens and the direction of the laser beam up to 40° (this value coincides with half the field of view of the camera lens, which is 40,5°). At the same time, for angles between the optical axis of the camera lens and the direction of the laser beam from 0° to 10°, maximum glare effects can be achieved, which are associated with illumination of most of the receiving element, which eliminates the possibility of obtaining useful information from images or videos. Further increasing the angle between the optical axis of the camera lens and the direction of the laser beam to 40° gradually reduces the glare of the camera receiving element (most pictures or videos are suitable for useful information). At angles greater than 40°, glare of the receiving element of the

Table 12 Results of the effect of laser radiation on a Nikon Coolpix digital camera depending on the angle between the optical axis of the camera lens and the direction of the laser beam in the presence of a scattering lens and without zooming (corresponding to an angle of view of 81°)

The value of the angle, deg.	0	5	10
The results of exposure to laser radiation			
The value of the angle, deg.	15	20	25
The results of exposure to laser radiation			
The value of the angle, deg.	30	35	40
The results of exposure to laser radiation			

Table 13 Results of the effect of laser radiation on a Nikon Coolpix digital camera depending on the angle between the optical axis of the camera lens and the direction of the laser beam in the presence of a scattering lens and the maximum image approximation (corresponds to a field of view angle of 4.5°)

The value of the angle, deg.	0	5	10
The results of exposure to laser radiation			

camera is not observed. Comparison of these results with the results obtained without a scattering lens, indicates a significant dependence of the effects of blinding on the intensity of the laser beam. In this case, the intensity of the laser beam at the input of the camera lens is less than 1.44 times compared with the results obtained in the absence of a scattering lens.

Thus, when hit by a wide (25 mm) laser beam in a Nikon Coolpix digital camera (diameter of the input lens—20 mm, which is limited by a rectangular window 20 × 17 mm), which works in the absence of image magnification, out of sight lateral illumination angles missing. In the Table 13 shows the results of the experiment.

The results obtained are similar to the previous ones carried out in the absence of a scattering lens.

12 Conclusions

A laboratory bench was developed for measuring the durability to UWB EMI of electronic equipment. The boundary frequency of 500 MHz pulsed electromagnetic fields is limited by the size of the manufactured field-forming system. The electromagnetic field distribution in the regular part of the TEM camera was calculated using the HFSS electrodynamic simulation program.

The problem of optimizing the design parameters of the shaper by the criterion of achieving maximum generator power has been solved. To increase the generator power, it is proposed to use a new method of excitation of microwave pulses of nanosecond duration, which is based on the property of charge redistribution in a coaxial line with a multi-element core and a change in the arrester switching delays.

The influence of the multi-core core profile of the coaxial design shaper and the correction of the switching time of the spark gap on the pulse parameters was revealed.

Additionally resistance was evaluated television equipment under the action of laser radiation. Analysis of the results showed that the value of the angle of lateral illumination depends on the design of the lens. The greater the angle of lateral illumination, the greater the field of view of the lens, the greater the angle between the optical axis of the lens and the tangent to its body (the structural surface of the outer lens). For lenses and camcorders, the values of the lateral illumination angles did not exceed 10...15% of the value of the field of view.

As a result of the experiment, it was found that the lateral illumination is significantly affected by the presence of protective glass, which is located at a short distance from the lens and on which the scattering of laser radiation. Due to this, there is a violation of the normal functioning of the optoelectronic means in fact within the anterior hemisphere.

The effect of lateral illumination by laser radiation of optical means can be significantly reduced by installing blends, polarizers.

Further research in this area is to study the stability of television facilities in the conditions of laser illumination of different ranges and their combination with UWB EMI.

References

1. Prather WD, Agee FJ, Baum CE et al (1999) Ultra-wideband sources and antennas. In: Heyman E, Mandelbaum B, Shiloh Y (eds) Ultra-Wideband, Short-Pulse Electromagnetics 4, Plenum Publishers, pp 119–130
2. Lehr JM, Baum CE, Prather WD et al (1998) Ultra-wideband transmitter research. IEEE Trans Plazma Sci 26(3)
3. Prather WD, Baum CE, Agee FJ et al (1997) Ultrawide band sources and antennas: present technology, future challenges. Baum et al (eds) Ultra-wideband, short-pulse electromagnetics, 3 edn. Plenum Press, N.Y., pp 381–389
4. Grekhov IV, Kardo-Sysoev AF (1979) Subnanosecond current drops in delayed breakdown of silicon p-n junctions. Sov Tech Phys Lett 5(8)
5. Bertoni HL, Carin L, Felsen LB (eds) (1993) Ultra-wideband, short—pulse electromagnetics. Plenum Press
6. Podosenov SA, Sokolov AA (1995) Linear two-wire transmission line coupling to an external electromagnetic field, Part I: theory. IEEE Trans Electromagn Compat 37(4):559–566
7. Podosenov SA, Sakharov KYu, Svekis YG, Sokolov AA (1995) Linear two-wire transmission line coupling to an external electromagnetic field, Part II: specific cases, experiment. IEEE Trans Electromagn Compat 37(4):566–574
8. MEK 61000-1-3 (2002) Elektromagnitnaya sovmestimost (EMS). Ustoychivost k elektromagnitnomu impulsu vyisotnogo yadernogo vzryiva (EMI VYV). Vozdeystvie SSHP EMI na oborudovanie i sistemyi grajdanskogo naznacheniya (Resistance to the electromagnetic impulse of a high-altitude nuclear explosion (EMI HANE). Impact of UWB EMI on equipment and systems for civilian use)
9. Yakushin SP, Vedmidskiy AA (2003) Analiz metodov rascheta vzaimodeystviya SSHP EMI s elementami TKS (Analysis of methods for calculating the interaction of UWB EMI with elements of TCS) In: Kechieva JIN (ed) pp 17–32
10. Myirova JIO, Voskobovich VV (2004) Vozdeystvie sverhshirokopolosnogo impulsnogo elektromagnitnogo izlucheniya na tehnicheskie sredstva (The impact of ultra-wideband pulsed electromagnetic radiation on technical means) Tehnologii EMS, № 3(10), pp 25–30
11. Voskobovich VV (2004) Aktualnost i sovremennoe sostoyanie problemyi zaschityi tehnicheskih sredstv ot sverhshirokopolosnyih impulsov bolshoy moschnosti (The relevance and current state of the problem of protecting technical equipment from ultra-wideband high-power pulses) Tehnologii EMS, № 3, pp 17–24

12. Yu K, Shostko I, Avchinnikov E, Tevyashev A, Neofitnyi M (2019) Experimental researches on determination of angles of side illumination by laser radiation of optical devices and opto-electronic devices. Paper presented at the "Problems of Infocommunications. Science and Technology" (PICS&T-2019) International Scientific-Practical Conference, Kyiv, pp 319–323
13. Kravchenko VI (2010) Molniya. Elektromagnitnyie faktoryi i ih porajayuschee vozdeystvie na tehnicheskie sredstva (Lightning. Electromagnetic factors and their damaging effects on technical means) Harkov, Izd-vo NTU «HPI», 292 p
14. ISO 11452–3 Component test methods for electrical disturbances from narrowband radiated electromagnetic energy. Transverse electromagnetic (TEM) cell
15. Crawford ML (1974) Generation of standard EM fields using TEM transmission cells. Paper presented at the IEEE Transactions on Electromagnetic Compatibility, Vol. EMC-16, №4, 1974. pp 189–195
16. Burtsev VA, Kalinin NV, Luchinskiy AV (1963) Electric explosion of conductors and its application in electro installations, Energoatomizdat, p 288
17. Mesyats GA, Vorobyov GA (1963) Technique for formation of high-voltage nanosecond pulses, Gosatomizdat, p 167
18. Mick J, Kregs J (1960) Electrical breakdown in gases. Publishing Foreign Literature, p 605
19. Mesyats GA (1974) Generation of high-power nanosecond pulses. Radio, Sov, p 256

Method for Planning SAN Based on FTTH Technology

Iskandar Saif Ahmed Al-Vandavi, Mykola Moskalets, Kateryna Popovska, Dmytro Ageyev, and Yana Krasnozheniuk

Abstract Analytical models for determining the length of an optical cable based on symmetric graphical models with one-way and two-way cable laying topology in an urban area with a base of potential customers of a subscriber access network uniformly distributed over the square area are considered, which allow optimizing the economic costs of deploying an access network. An access network model has been developed, which allows to consider the selection of access technology, operating and capital costs. An optimization procedure has been developed that is aimed at minimizing the objective function according to the criterion of the cost of an FTTH network deploying, taking into account the costs of purchasing network elements and deploying street optical cable infrastructure. An optimization problem has been formulated and solved that allows minimizing the cost of a passive optical (PON) access network, considering the number of optical splitters, floor splitters, and ONU subscriber units. The solution to this problem is presented in general form, which allows to adapt it for any set of these network elements. The developed methodology for calculating the cost of the designed access network includes the total capital and operating expenditures for the purchasing, installation and maintenance of network elements presented in approximate prices. The calculation of expenditures on a typical cable structure is shown on a specific example. The presented methodology also allows to take into account the costs of deploying cable infrastructure and linear structures.

Keywords Fiber to the home · Subscriber access network · Passive optical networks · Optic fiber

I. S. A. Al-Vandavi (✉) · M. Moskalets · K. Popovska · D. Ageyev · Y. Krasnozheniuk
Kharkiv National University of Radio Electronics, 14 Nauky Ave, Kharkiv, Ukraine
e-mail: aleksandrua@icloud.com

M. Moskalets
e-mail: mykola.moskalets@nure.ua

K. Popovska
e-mail: kateryna.popovska@nure.ua

Y. Krasnozheniuk
e-mail: yana.krasnozheniuk@nure.ua

D. Ageyev et al. (eds.), *Data-Centric Business and Applications*, Lecture Notes on Data Engineering and Communications Technologies 69,
https://doi.org/10.1007/978-3-030-71892-3_14

1 Introduction

The role of access in the overall architecture of modern telecommunication networks is steadily increasing. This is primarily due to the introduction of new services, such as video telephony, HDTV, games and various options for virtual reality. All this significantly increases the requirements for link bandwidth in the "last mile" section. In Ukraine, as in other countries, the number of subscribers with high-speed access is steadily increasing [1–6]. At the same time, there was a trend towards a gradual decrease in the share of xDSL and a gradual increase in the number of connections according to the FTTB and FTTH scenarios.

Currently, almost all access network solutions involve the use of fiber-optic transmission systems at various sites.

The main options for organizing existing and future access networks according to the FTTH scenario are:

- PTP where each CPN is connected by a separate fiber from the OLT;
- P2MP using PON technology where CPN is connected by a separate fiber, but in the distribution or trunk sections, one fiber is shared by several CPNs;
- AON where each CPN is connected by a separate fiber from the active removal located in linear structures.

Currently, the PON direction is actively developing with the separation of the PON-WDM wavelength [7–10]. With the introduction of PON-WDM, the optical link utilization coefficient and the specific effective transmission bandwidth significantly increase (Fig. 1).

The effectiveness of using information technology is largely determined by the level of accessibility for the population and industry of various services and content [6, 8, 11–13]. To do this, telecom operators should plan the development of telecommunication networks, taking into account the display of the list and types of services on the plane of the network characteristics of transport networks they need.

According to [13–15], all services can be divided into seven main groups:

1. Communication services.
2. Computer and storage.
3. Entertainment.
4. Health care provision.
5. Education and distance learning.
6. Providing a "smart" home.
7. Commercial and etc.

Parameters of subscriber access networks (SAD) in the subscriber's premises form a group of services that can be provided.

Thus, the results of planning and modernizing the SAD directly determine the economic efficiency and profit of telecom operators.

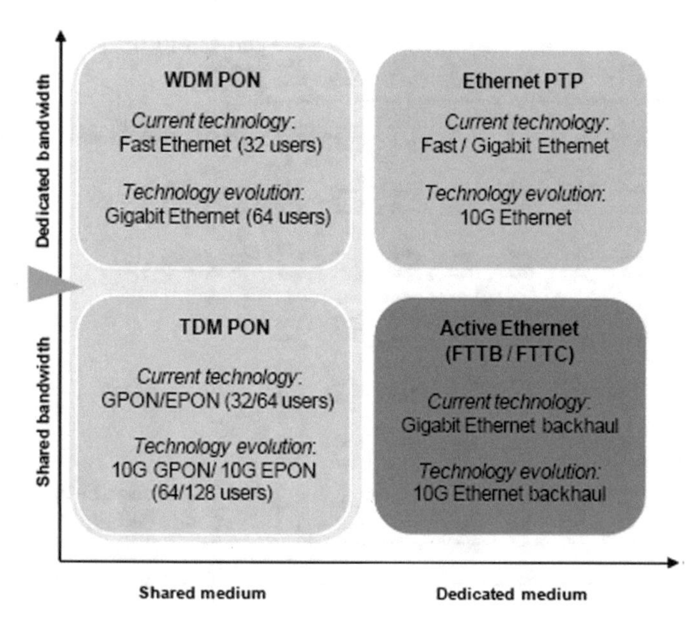

Fig. 1 Comparison of the main ways to implement the FTTH scenarios

In this regard, the aim of this work is to develop an access network model with a different architecture, which allows taking into account the choice of access technology, operating and capital costs and is focused on maximizing economic efficiency in general.

2 Analytical Models for Deployment and Determining the Length of Optical Fiber

In this subsection, various analytical models based on symmetric graphical models with a potential customer base evenly distributed over the square area are studied in detail (Fig. 2).

One side of the square contains n houses, and the area contains n^2 houses. The distance between the two houses is l. When considering only connection points in houses, the longest horizontal or vertical distance between the two most remote houses is $(n-1)l$. The longest horizontal or vertical distance is nl. Surface area does not exceed n^2l. The central office (CO) is always located in the center of the square.

In the following, we will use the terms: length of cable installation, L, and length of optical fiber F.

A fully deployed FTTH scenario involves deploying an optical cable in the areas between the houses and the central station, as a rule, in cable duct on one or both

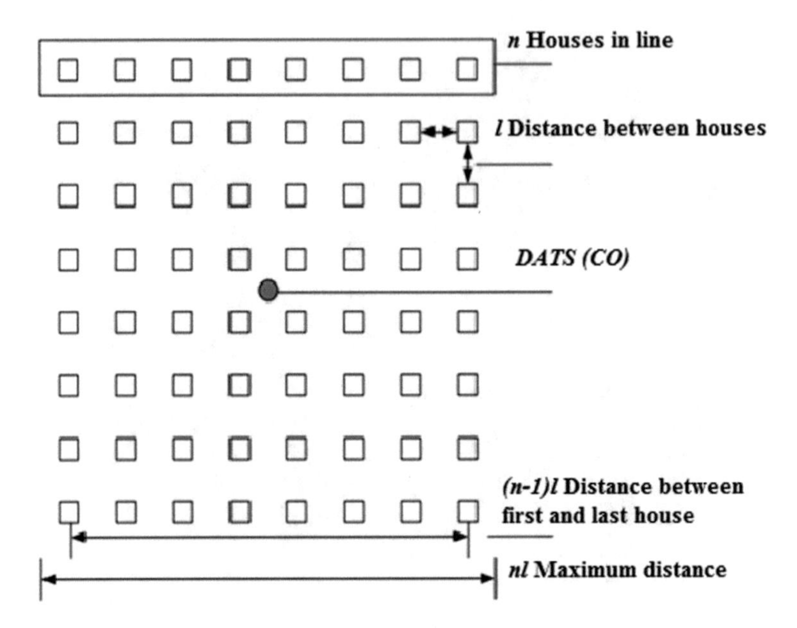

Fig. 2 Schematic representation of the logical structure, location parameters, and analytical determination of optical fiber length

sides of the street. Thus, we can distinguish a one-way and two-way street analytical model of installing and determining the length of the optical cable.

2.1 *One-Way Street Model for Determining the Optical Cable Length*

In a one-way street model, the cable runs on one side of the street and connects the houses on both sides, while in the calculation model the cable runs in the middle of the street. With this structure, we can combine all the houses by 2, as shown in Fig. 3.

To connect all pairs of houses from 2 adjacent sides of the street, we use the installation length nl. Moreover, we obtain $n/2$ of such rows. To connect the coupled pairs of houses in one fully connected street, we use the length of $(n - 1)l$ in each $n/2$ adjacent row. Finally, the connection to the central office takes place through a street that has a length of $(n - 2)l$. The combination in the presented way gives the installation length:

$$\mathrm{L} = \frac{n^2 l}{2} + \frac{n(n-1)l}{2} + (n-2)\mathrm{l} = \left(n^2 + \frac{n}{2} - 2\right)l. \tag{1}$$

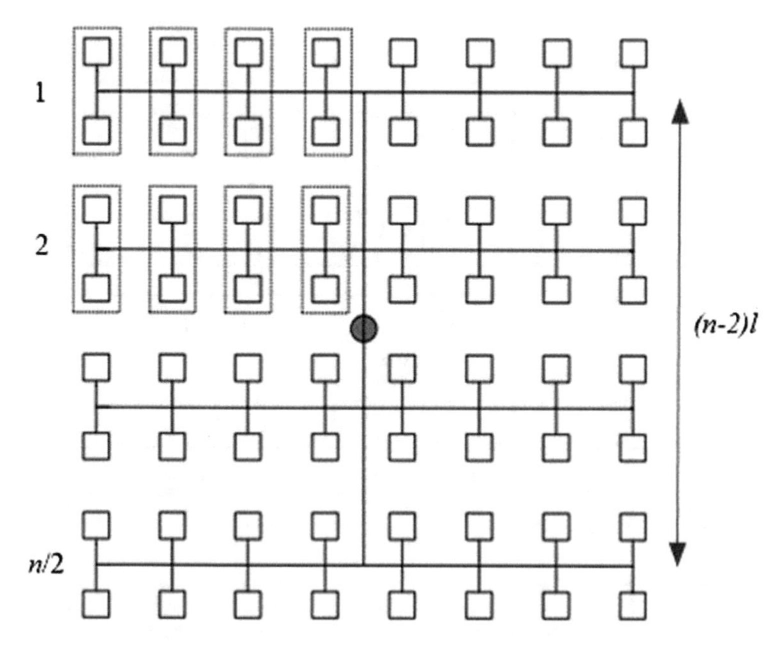

Fig. 3 Schematic representation of a one-way street model

2.2 *Two-Way Street Model for Determining the Optical Cable Length*

When using a two-way street model, the cable is deployed on each side of the street, which reduces the distance to the installation point of the equipment and the number of cable intersections in the street (Fig. 4).

For the presented structure, we use groups of two houses. However, in this case the neighboring houses are not directly connected to each other, since there is no intersection of streets with a width of w. Under this installation, the connection distance is $(l - w)$, the number of adjacent houses is $n^2/2$. Given the connection in the rows, we need an additional installation length on both sides of the street. In all cases, except for the upper part of the street in the upper row and the lower side of the lower row street, the installation length is $(n - 1)l$ minus the width of the street w, which does not intersect. In the case of the upper and lower rows of the scheme, the width of the street is not taken into account. In total, we get $n/2$ streets and n sides of the streets. Horizontal streets are spaced at a distance of $2l$, the length of the side that connects the two streets is $2l - w$. Given the number of streets and connections on both sides, the length of the connecting cable is $2(n/2 - 1)(2l - w)$. Thus, the total cable installation length can be represented as:

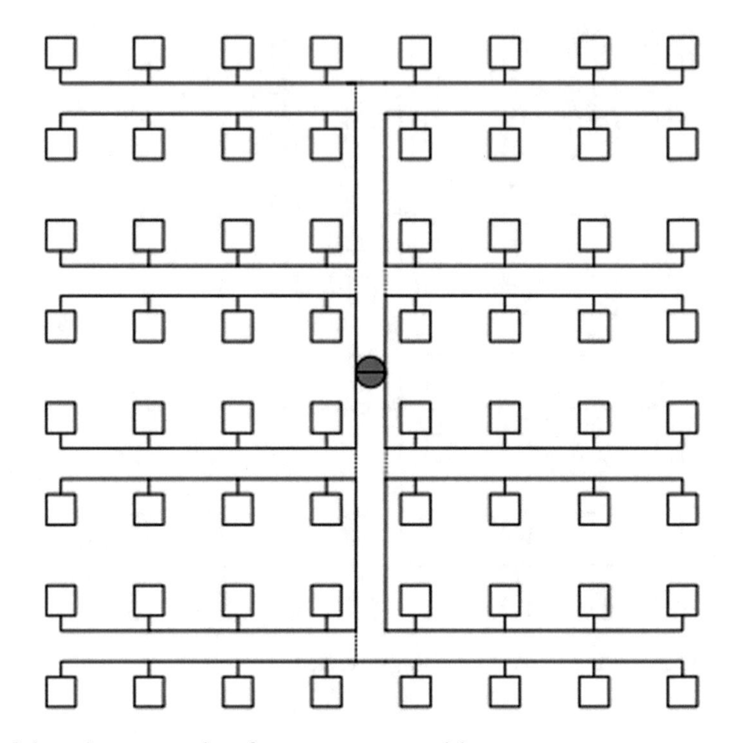

Fig. 4 Schematic representation of a two-way street model

$$
\begin{aligned}
L &= \frac{n^2}{2}(l - w) + [(n - 2)((n - 1)l - w) + 2(n - 1)l] \\
&\quad + 2(n/2 - 1)(2l - w) \\
&= \left(\frac{3n^2}{2} + \mathrm{n} - 4\right)l + \left(\frac{n^2}{2} + 2n - 4\right)w
\end{aligned} \tag{2}
$$

3 Selection Criteria for FTTH Network Deployment Zones

When designing the FTTH network deployment zones, the existing lowland of the residential area for cabinet areas (CAs) and direct power supply zones (DPZs) of the existing telephone network of the community is taken as a basis [13–16]. Areas of service providing on the FTTH technology are determined in a residential area on the territories with predominantly multi-story buildings (5 floors and more). Such zones are created only provided the availability of ADSL subscribers in existing buildings:

- in direct power supply zones—more than 96;
- in cabinet areas of existing and new buildings, where more than 48 subscribers of FTTH access network can be expected.

Table 1 The number of buildings that can connect up to one trunk fiber-optical cable (FOC) depending on the maximum number of entrances in the house

The number of buildings in the service area of one trunk cable	Number of entrances in a group of buildings			
	House 1	House 2	House 3	House 4
1	≤16	–	–	–
2	≤12	≤4	–	–
2	≤8	≤8	–	–
3	≤8	≤4	≤4	–
4	≤4	≤4	≤4	≤4

In the given zones, groups of buildings with a high concentration of ADSL service subscribers of Ukrtelecom OJSC are determined according to the principles:

- the total number of apartments in the group of buildings is up to 500;
- in cabinet areas—more than 48 ADSL service subscribers for 500 apartments;
- in the group there can be up to 4 closely located buildings. The number of buildings and the maximum number of entrances are determined according to Table 1.

The total number of entrances cannot exceed 16 entrances in a group.

The initial selection of the distribution zone for deploying FTTH technology is made taking into account the density of subscribers in each trunk direction.

Density of subscribers in the district automatic telephone station (DATS) service area:

$$D_{\text{ens}_{\text{Sub}_{\text{CO}}}} = \frac{\text{CO}_{\text{Sub}}}{S_{\text{CO}}}, \tag{3}$$

where CO_{Sub} is the number of subscribers serviced by DATS; S_{CO} is the subscriber distribution area of DATS.

To account for the existing ADSL subscriber base, we introduce the following indices of constants and variables:

$i \in I$	Distribution zone index
$j \in J$	house index in the distribution zone
a_{ij}	the number of ADSL subscribers in the ith house of the jth distribution zone
c_{tarif}	transition plan when connecting to an optical access network
t	time interval for evaluating access network planning
Ψ_{ij}	indicator of the connection between the ith distribution zone and the jth house $0 \leq \Psi_{\text{ij}} \leq 1$

When deploying the FTTH network, it is necessary to maximize the objective function:

Table 2 FOC capacity depending on the number of apartments in the distribution area

Zone type	The number of apartments in the service area of one trunk cable			
	<150	<250	<350	<500
CA	–	–	–	144 OFs
DPZ	48 OFs	72 OFs	96 OFs	144 OFs

$$\max \sum_{i \in I} \sum_{j \in J} t \cdot c_{tarif} a_{ij} \Psi_{ij}, \tag{4}$$

which takes into account the reduction in the total cost due to reconnecting the existing ADSL subscriber base to the new pricing plan for the fiber optic network.

4 Stating the Problem of Trunk Segment Optimization in FTTH Network

In the process of deploying a subscriber access network according to the FTTH scenario, station cable sleeves should provide mutual connection of one station and one line cable. The cable sleeve is located in a cable shaft or station jointing chamber. The trunk cable must provide the connection of the required number of apartments in the service area. Given the linear supply of the optic fiber (OFs), the capacity of the trunk cable is determined according to the Table 2.

The main distribution sleeve is located in the inspection device of the telephone duct (since it is impossible to locate the cable sleeve in the existing jointing chamber, it is located in an additional separate jointing chamber) or in the basement of a residential building and should ensure the interconnection of one main cable to 1–4 optical distribution cables).

The cable sleeve should not be installed if the trunk cable is connected to one house with more than 350 apartments. In this case, the main cable is mounted in the house distribution sleeve.

To optimize the access network according to the cost criterion, when deploying the FTTH scenario on the trunk segment, we introduce the following additional indices, parameters, and variables:

$k \in K$	Cable type index
V_k	the number of k-type cable OFs
α	cost of the OLT port
β	cost of laying or drawing optical fiber, UAH/km
θ_k	cost of the k-type optical cable
l_i	distance from the DATS to the ith distribution zone

(continued)

(continued)

x_0, y_0	coordinates of the DATS location
x_i, y_i	location coordinates of the *i*th distribution zone, $i \in I$
Φ_i	indicator of laying the FOC to the *i*th distribution zone. It receives the value 1 if the cable is laid, and 0 otherwise.

The planned number of active OLT ports at the connection stage is defined as:

$$N_{\mathrm{OLT}} = \mathrm{F_{CO}} \cdot \mathrm{Pl_{CO}}, \tag{5}$$

where F_{CO} is the number of households in the DATS distribution area; $\mathrm{Pl_{CO}}$ is the planned value of the total number of households, in percent.

It is necessary to minimize the objective function:

$$\min\left(\alpha \sum_{i \in I} \Phi_i V_{\mathrm{ik}} + \beta \sum_{i \in I} \Phi_i l_i + \sum_{i \in I} \sum_{k \in K} \Phi_i l_i \theta_k\right). \tag{6}$$

The following constraints have been introduced.

The number of optical fibers in the trunk cables is greater than or equal to the number of OLT ports:

$$N_{\mathrm{CO}} \leq \sum \Phi_i V_{\mathrm{ik}}, \tag{7}$$

Only one cable of the *k*th type is laid in the *i*th distribution zone;

$$\sum_{k \in K} \Phi_i \leq 1, \tag{8}$$

The smallest distance to the *i*th distribution point;

$$l_i = \sqrt{(x_i - x_0)^2 + (y_i - y_0)^2}, \forall i \in I, \tag{9}$$

Distance to the *i*th distribution point, taking into account the rotation of the cable;

$$l_i = |x_i - x_0| + |y_i - y_0|, \forall i \in I. \tag{10}$$

The last constraint gives a more accurate estimate of the cable length when using the coordinate system on the network planning map.

5 Stating the Problem of Distribution Segment Optimization in FTTH Network

The distribution segment of the FTTH network is part of the access network, which consists of an optical distribution cable and house distribution boxes. Distribution cable must provide the connection for the required number of apartments in the service area of the specified cable. Thus, given the above:

l_{ij}	Distance from the *i*th distribution area to the *j*th house
s_{ij}	the number of households in the *j*th house of the *i*th distribution zone
ν	the planned number of connected users of the total number of apartments in the house

It is necessary to minimize the objective function:

$$\sum_{i \in I} \sum_{j \in J} \sum_{k \in K} \Psi_{ij} l_{ij} \theta_k, . \tag{11}$$

Constraints:

The number of optical fibers in the *k*-type cable is greater than or equal to the planned number of connections in the house;

$$\upsilon \cdot s_{ij} \leq \Psi_{ij} V_k, \tag{12}$$

The total number of optical fibers in the distribution area is greater than or equal to the number of OLT ports:

$$N_{OLT} \leq \sum_{j \in J} \Psi_{ij} V_k, \tag{13}$$

If a cable enters the distribution zone, then at least one cable must go out:

$$\Phi_i \leq \Phi_i \sum_j \Psi_{ij} \tag{14}$$

6 Stating the Problem of Optimizing a Passive Optical Access Network

The subscriber segment of the PON is part of the access network, which is deployed inside the house from the splitter in the house or in a separate entrance to the user's apartment and ends with a subscriber outlet module and a fiber-optic connecting cord (patch cord) in the user's apartment.

The subscriber segment is of more interested in the PON optimization process regarding the location of optical splitters.

The inner house part consists of a fiber-optic cable for interfloor laying, a distribution box on the top floor, floor branching devices and a subscriber outlet module. Fibers on each floor can branch from the fiber-optic interfloor cable and be deployed to the place where the subscriber outlet is installed.

To formulate a PON optimization problem, we introduce the following sets of indices:

S	Index of a set of optical splitters
T_S	set of splitter types
U	index of the ONU set

Parameters:

x_0, y_0	DATS location coordinates
x_i^0, y_i^0	location coordinates of the ith ONU, $i \in U$
Δ	large value necessary for integer linear programming
T_k	the total number of outgoing ports of the k-type splitter
α	cost factor or OLT weight
β_i	cost factor or weight of each outgoing port of the k-type splitter
γ	factor of the cost of laying or broaching optical fiber, UAH/km
θ	cost factor of the optical cable, UAH/km
$l_{\max}^{\text{total}}$	total maximum PON transmission distance between OLT and ONU, generally less than 100 km
$l_{\max}^{\text{diff}}$	maximum differential distance between different ONUs within the same PON network, generally less than 20 km

Solution variables:

x_i^S, y_i^S	Coordinates of the ith splitter location, $i \in S$
Φ_i	indicator of the use of the ith splitter. It receives the value 1 if the splitter is used, otherwise it is equal to 0
Ψ_i^j	indicator of the connection between the jth ONU and the ith splitter. It receives the value 1 if the ONU connects to the splitter, otherwise it is equal to 0
π_i^k	indication function of the type of the ith splitter. It receives the value 1 if the splitter is of the kth type, otherwise it is equal to 0
l_i^s	distance from the ith splitter to the central OLT
l_i^j	distance from the jth ONU to the ith splitter

(continued)

(continued)

$l_i^{\max}$	maximum distance from the ONU to OLT of the *i*th PON, $i \in S$
$l_i^{\min}$	minimum distance from the ONU to OLT of the *i*th PON, $i \in S$
τ_i, ζ_i^j	additional binary variables for the "if" and "then" states in the optimization model

It is necessary to minimize the total cost of deploying an access network when connecting all ONUs to a central OLT, taking into account restrictions including the maximum transmission distance, maximum differential distance, and optical separation coefficient:

$$\min\left(\alpha\sum_{i\in S}\Phi_i + \beta\sum_{i\in S}\sum_{k\in T_S}\Phi_i T_k \pi_i^k + (\gamma+\theta)\left(\sum_{i\in S}\Phi_i l_i^S + \sum_{i\in S}\sum_{j\in U}\Psi_i^j l_i^j\right)\right). \quad (15)$$

Constraints:

$$\sum_{i\in S}\Psi_i^j = 1, \forall j \in U \quad (16)$$

$$\Delta\cdot\Phi_i \geq \sum_{j\in U}\Psi_i^j, \forall i \in S \quad (17)$$

$$\sum_{j\in U}\Psi_i^j \leq \sum_{k\in T_S} T_k\pi_i^k, \forall i \in S \quad (18)$$

$$\sum_{k\in T_S}\pi_i^k - 1 \leq \Delta\cdot\tau_i, \forall i \in S, \quad (19)$$

$$1 - \sum_{k\in T_S}\pi_i^k \leq \Delta\cdot\tau_i, \forall i \in S, \quad (20)$$

$$\Phi_i \leq \Delta\cdot(1-\tau_i), \forall i \in S, \quad (21)$$

$$l_i^S = \sqrt{\left(x_i^S - x_0\right)^2 + \left(y_i^S - y_0\right)^2}, \forall i \in S, \quad (22)$$

$$l_i^j = \sqrt{\left(x_i^S - x_j^0\right)^2 + \left(y_i^S - y_j^0\right)^2}, \forall i \in \mathrm{S}, \forall j \in U, \quad (23)$$

$$\left(l_i^S + 1_i^j\right) - l_i^{\max} \leq \Delta\cdot\zeta_i^j, \forall i \in \mathrm{S}, \forall j \in U, \quad (24)$$

$$l_i^{\min} - \left(l_i^S + 1_i^j\right) \leq \Delta\cdot\zeta_i^j, \forall i \in \mathrm{S}, \forall j \in U \quad (25)$$

$$\Psi_i^j \leq \Delta \cdot \left(1 - \zeta_i^j\right), \forall i \in S, \forall j \in U, \quad (26)$$

$$l_i^{\max} \leq l_{\max}^{\text{total}}, \forall i \in S, \quad (27)$$

$$l_i^{\max} - l_i^{\min} \leq l_{\max}^{\text{diff}}, \forall i \in S \quad (28)$$

In the presented model, the total cost consists of three parts. The first part is the cost of OLT, since each splitter corresponds to a PON that needs OLT. The second part is the cost of splitters and the last part includes the cost of laying optics and the cost of optical cables.

The constraint (16) corresponds to the fact that each ONU must be connected to the splitter. The constraint (17) determines when the ith splitter should be installed and depends on whether there are ONU connected to the splitter. If such ONUs are available, then the splitter is installed and it is not installed otherwise. The constraint (18) selects the type of splitter, taking into account the sufficient number of outgoing ports for connecting the ONU.

Constraints (19)–(21) provide the state "if"–"then". If the ith splitter is selected, then there should be only one type of splitter associated with it. More specifically, if $\Phi_i = 1$ (that is, the ith splitter is selected), then the value is $\tau_i = 0$, which subsequently leads to $\sum \pi_i^k = 1$, taking into account the constraints (20), (21) (that is, only one type of splitter for the ith point). The constraint (22) calculates the distance from the splitter to the central OLT. The constraint (23) calculates the distance from the splitter to the ONU. These two constraints are nonlinear.

Constraints (24)–(26) find the maximum and minimum distances between the ONU and OLT, respectively, within each PON. The constraints are also based on "if"–"then" conditions. If $\Psi_i^j = 1$ (i.e., the jth ONU refers to the ith PON or splitter), then the value $\zeta_i^j = 0$ is set, thus the constraints (24), (25) take the form $l_i^S + 1_i^j \leq l_i^{\max}$ and $l_i^{\min} \leq l_i^S + 1_i^j$, respectively.

The constraint (27) provides the condition of not exceeding the maximum PON transmission distance. The constraint (28) provides the condition for satisfying the maximum differential distance between different ONUs within the same PON network.

7 Calculation of Cost for Designing an Access Network

The basic principle in calculating capital costs when creating a Subscriber Access Network (SAN) is considered in [17–25], which can be defined as:

$$\sum_c [d_c/g_c] \cdot p_c \quad (29)$$

where:

d_c—the required number of elements of the installed equipment with type c;
g_c—c-type equipment item;
p_c—price per item of c-type equipment.

Approximate European prices for FTTH scenario equipment are shown in Table 3.

General economic costs for the SAN will include the cost of equipment and linear constructions:

$$C = C_{Equipment} + C_{cableplant} \tag{30}$$

Based on the specifics of the SAN under the FTTH scenario, equipment costs include:

Table 3 Approximate prices for FTTH technology components

	Equipment	Price (€)	Type
Inside plant	ODF	1500	Passive optics
	Rack	1500	Passive mechanic
	Shelf	800	Passive mechanic
	OLT Card	5000	Active optics
	Control card	3000	Active electronics
	Patch cable	10–20	Passive optics
Outside plant	Cable	0.125/m 1/m 2.5/m	Passive optics
	Ducts	5/m	Passive mechanic
	Connector	5–10	Passive optics
	Splitter	12.5 per split	Passive optics
	Flexibility point	5000	Passive mechanic
	Aerial/façade deployment	10/m	Manual labor
	Trenching crossing	20–50/m ×6	Manual labor
Customer premises	Bender in wall	10	Passive mechanic
	ONTP + ONT	150–300	Active optics

$$C_{Equipments} = \sum_{PONs} \left(C_{OLT} + C_{Splitter} \right) + \sum_{Customers} C_{ONU} \tag{31}$$

A feature of calculating the cost of optical linear structures is the division into one-time expenditures and costs required for connecting subscriber ports.

$$C = C_0 + \sum C_v$$

where:

C_0 is one-time costs in the construction of linear structures;
C_v is the additional costs required to connect subscribers.

We can consider one-time costs as those costs to create a trunk section (feeder), additional to connecting subscribers at the distribution network level (distribution).

$$C_{Cableplant} = C_{feed} + C_{dist}$$

$$C_{feed} = \sum_{e \in Feeder} \left(C_0^{feed} + \sum C_v^{feed} \right)$$

$$C_{dist} = \sum_{e \in Dist} \left(C_0^{dist} + \sum C_v^{dist} \right)$$

Therefore, the general costs for linear constructions are:

$$C_{Cableplant} = \sum_{e \in Feeder} \left(C_0^{feed} + \sum C_v^{feed} \right) + \sum_{e \in Dist} \sum C_v^{dist} \tag{32}$$

The list of capital expenditures for hardware and software components:

$$\mathrm{CapEx} = \mathrm{C_{HW}} + \mathrm{C_{SW}} + \mathrm{C_{CW}} + \mathrm{C_{OTHER}} \tag{33}$$

where:

C_{HW} is the hardware;
C_{SW} is the software;
C_{CW} denotes installation and construction;
C_{OTHER} denotes other expenditures.

The list of equipment costs consists of network equipment and ancillary equipment (C_{GenHW}). Network equipment includes OLT, switches, OC, splitters, cabinets, as well as equipment for connecting a subscriber [26–28].

Ancillary equipment consists of air conditioning systems, power systems, measuring and other equipment.

$$C_{HW} = C_{NetHW} + G_{GenHW} \tag{34}$$

where:

C_{HW}—cost of equipment;
C_{NetHW}—cost of network equipment;
G_{GenHW}—cost of ancillary equipment.

When calculating software costs, both one-time costs and regular payments depending on the number of connected subscribers or a software type are considered.

Installation costs are highly dependent on status and specifics of laying OC in linear structures. In particular:

- use of existing channels;
- construction of new cable ducts using earthworks;
- laying of lines in the form of air communication lines.

$$\begin{aligned} Ccw = {} & N_{CO} \cdot C_{CO-Inst} + N_{StreetCab} \cdot C_{StreetCab-inst} + N_{SplitCab} \cdot C_{SplitCab-Inst} \\ & + N_{HE} \cdot C_{HE-Inst} + L_{Duct} \cdot C_{Ducts} + \sum_{FiberType\ t} (L_{Fiber}(t) \cdot C_{Fiber-Inst}(t)) \end{aligned} \tag{35}$$

where:

N_{CO}	Number of installation sites for station equipment
$C_{CO-Inst}$	cost of preparing an installation site for station equipment
$N_{StreetCab}$	number of distribution cabinets for the installation of equipment
$C_{StreetCab-Inst}$	cost of installing a distribution cabinet
$N_{SplitCab}$	number of boxes for installing splitters
$C_{SplitCab-Inst}$	cost of installing the box for splitters
N_{HE}	number of house entries
$C_{HE-Inst}$	cost of installing house entries
L_{Duct}	length of the cable duct or air communication lines
C_{Ducts}	cost of building cable ducts or air communication lines
L_{Fiber}	length of the fiber cable
$C_{Fiber-Inst}$	cost of laying the fiber cable.

Additional costs, such as transportation and project management costs (as a percentage of the cost of equipment) are:

$$C_{OTHER} = C_{Car} \cdot N_{CO} + P_{PM} + C_{NetHW} \tag{36}$$

Working expenditures are divided into two parts: fixed and variable

$$OpEx = OpEx_{Fixed} + OpEx_{Variable} \tag{37}$$

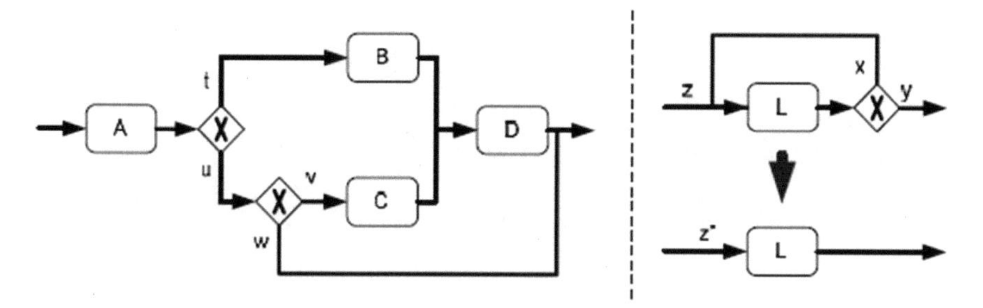

Fig. 5 The general principle of using the BPMN language

Fixed operating expenditures are divided into:

- rental of the site;
- salary to staff;
- other costs (network security, insurance, payment of licenses, training);

$$\mathrm{OpEx}_{\mathrm{Fixed}} = \mathrm{O}_{\mathrm{SiteRental}} + \mathrm{O}_{\mathrm{Salaries}} + \mathrm{O}_{\mathrm{OtherFixed}} \tag{37}$$

Electricity expenditures for power supply of active equipment:

$$\mathrm{OpEx}_{\mathrm{Variable}} = \mathrm{O}_{\mathrm{Energy}}$$

$$O_{\mathrm{Energ}}\mathrm{y} = \mathrm{O}_{\mathrm{kWh}} \cdot 365 \cdot 24 \cdot \sum_{\mathrm{AreasA}} \sum_{\mathrm{HWTypest}} (N_t(A) \cdot P_t)$$

Operating expenditures, OpEX, are conveniently formalized using the Business Process Modeling Notation (BPMN) language.

The general principle of using the BPMN language is illustrated in Fig. 5.

Total expenditures of OpEX will be:

$$\mathrm{Cost}_{\mathrm{tot}} = \mathrm{Cost}_A + \mathrm{t} \cdot \mathrm{Cost}_B + \mathrm{u} \cdot v \cdot \mathrm{Cost}_C + (\mathrm{t} + \mathrm{u} \cdot v) \cdot \mathrm{Cost}_D \tag{38}$$

In addition, in the case of a loop emerging, the probability of an event is calculated as

$$z' = \mathrm{z} \cdot \left(1 + \mathrm{x} + \mathrm{x}^2 + \ldots\right) = \mathrm{z} \cdot \sum_{\mathrm{i}=0}^{\infty} x' = \frac{z}{1 - x} \tag{39}$$

The general installation procedure includes the basic operations explained in Fig. 6.

The installation process consists of four basic blocks:

- creating a common FTTH infrastructure (FTTH Deployment);

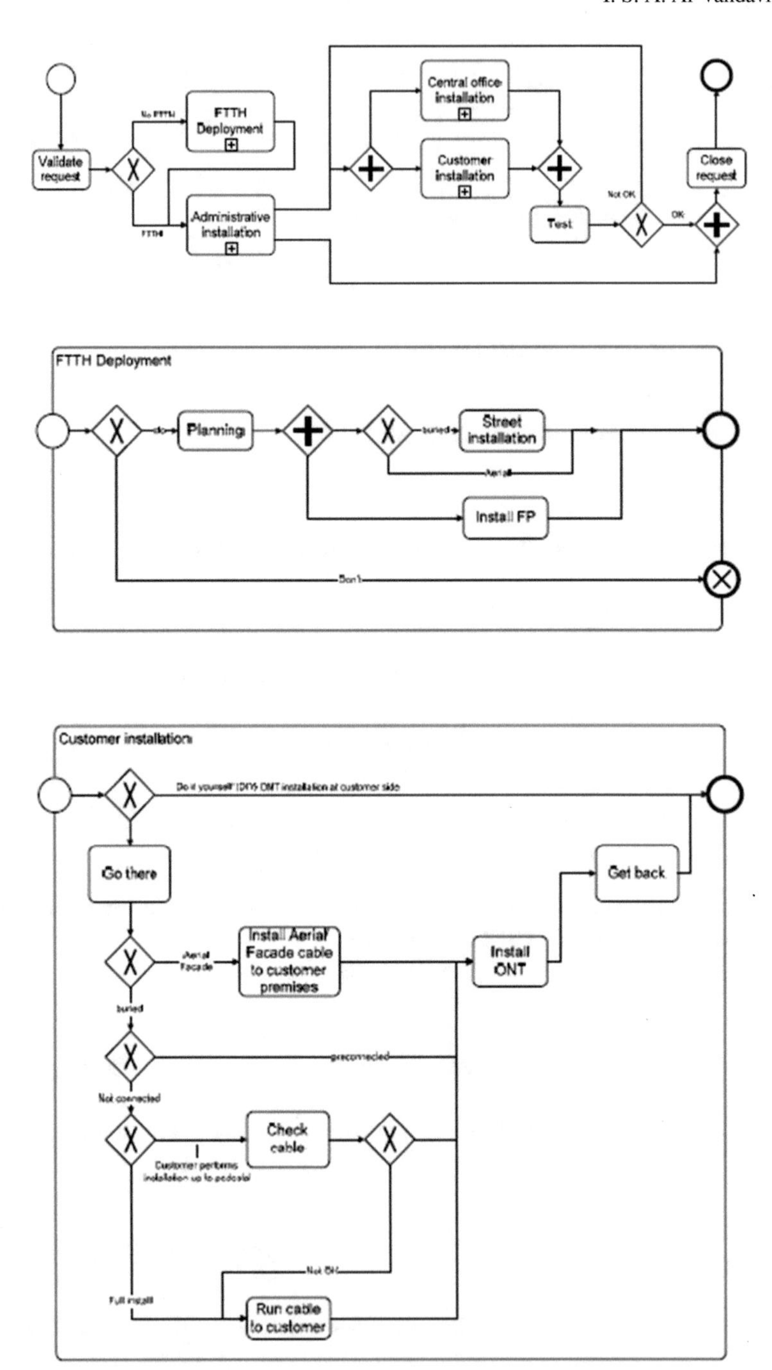

Fig. 6 OpEX Process Model Blocks for FTTH Technology

- preliminary preparation for installation of subscriber equipment (Administrative Installation);
- installation of subscriber equipment (Customer Installation);
- installation of the connection on the station side (Central Office Installation).

8 Conclusions

In order to apply to a typical urban development, analytical models for determining the lengths of optical fibers have been developed, which allow optimizing the economic costs of deploying an access network.

An optimization procedure has been developed that focuses on minimizing the objective function according to the criterion of the cost of deploying an FTTH network.

The problem of minimizing the cost of the FTTH network distribution segment is formulated in the form of a separate optimization subtask, which takes into account the number of OLT ports, the type and minimum cable length, the distribution density of subscribers of the existing ADSL base necessary to solve the problem.

With regard to PON technology, the optimization problem is formulated and solved, taking into account the number of optical splitters, floor splitters and ONU subscriber devices. The problem is solved in the general form, which allows it to be solved for any set of specified network elements.

A methodology has been developed for calculating the cost of the designed access network, taking into account the total capital CAPEX and operational OPEX costs for the buying, installing and maintaining network elements presented in approximate prices.

A specific calculation of costs on a typical structure is given (Table 3). The presented technique also allows to take into account the costs of deploying cable infrastructure and linear structures.

References

1. Ageyev, D., et al.: Method of self-similar load balancing in network intrusion detection system. In: 2018 28th International Conference Radioelektronika (RADIOELEKTRONIKA), pp. 1–4. IEEE (2018). https://doi.org/10.1109/radioelek.2018.8376406
2. Ageyev, D., et al.: Classification of existing virtualization methods used in telecommunication networks. In: Proceedings of the 2018 IEEE 9th International Conference on Dependable Systems, Services and Technologies (DESSERT), pp. 83–86 (2018)
3. Ageyev, D.V., Ignatenko, A.A., Wehbe, F.: Design of information and telecommunication systems with the usage of the multi-layer graph model. In: Proceedings of the XIIth International Conference The Experience of Designing and Application of CAD Systems in Microelectronics (CADSM), pp. 1–4. Lviv Polytechnic National University, Lviv-Polyana, Ukraine (2013)

4. Kryvinska, N.: Converged network service architecture: a platform for integrated services delivery and interworking. In: Electronic Business Series, vol. 2. International Academic Publishers, Peter Lang Publishing Group (2010)
5. Kryvinska N (2008) An analytical approach for the modeling of real-time services over IP network. Math. Comput. Simul. 79(4):980–990. https://doi.org/10.1016/j.matcom.2008.02.016
6. Barabash, O.V., Open'ko, P.V., Kopiika, O.V., Shevchenko, H.V., Dakhno, N.B.: Target programming with multicriterial restrictions application to the defense budget optimization. Adv. Military Technol. **14**(2), 213–229 (2019). https://doi.org/10.3849/aimt.01291
7. Radivilova, T., Kirichenko, L., Ageiev, D., Bulakh, V.: The methods to improve quality of service by accounting secure parameters. In: Hu, Z., Petoukhov, S., Dychka, I., He, M. (eds) Advances in Computer Science for Engineering and Education II. ICCSEEA 2019. Advances in Intelligent Systems and Computing, vol 938. Springer, Cham (2020)
8. Ageyev, D., et al.: Provision security in SDN/NFV. In 2018 14th International Conference on Advanced Trends in Radioelecrtronics, Telecommunications and Computer Engineering (TCSET), pp. 506–509. IEEE (2018). https://doi.org/10.1109/tcset.2018.8336252
9. Kirichenko, L., Radivilova, T., Bulakh, V.: Classification of fractal time series using recurrence plots. In: 2018 International Scientific-Practical Conference Problems of Infocommunications. Science and Technology (PIC S&T), pp. 719–724. IEEE (2018). https://doi.org/10.1109/infocommst.2018.8632010
10. Ivanisenko, I., Kirichenko, L., Radivilova, T.: Investigation of self-similar properties of additive data traffic. In: Proceedings of the International Conference on Computer Sciences and Information Technologies, CSIT 2015, pp. 169–171. IEEE, Lviv, Ukraine (2015). https://doi.org/10.1109/STC-CSIT.2015.7325459
11. Radivilova, T., Hassan, H.A.: Test for penetration in Wi-Fi network: attacks on WPA2-PSK and WPA2-enterprise. In: 2017 International Conference on Information and Telecommunication Technologies and Radio Electronics (UkrMiCo), pp. 1–4. IEEE (2017)
12. Geilhardt, F.: Migration guidelines for DSL from operator's view. MUSE, Deliverable DB2.3 version 2.0., p. 64 (2005)
13. EURESCOM Project P-709: Planning of Optical Network. EURESCOM, Deliverable 3, p. 38 (2000)
14. EURESCOM Project P-614: Evaluation of broadband home networks for residential and small business users. EURESCOM, Deliverable 11, p. 94 (1998)
15. Prat, J.: Next-Generation FTTH Passive Optical Networks, p. 224 (2008)
16. Ims, L.A., Myhre, D., Olsen, B.T.: Investment costs of broadband capacity upgrade strategies in residential areas. In: Proceedings of GLOBECOM 98, vol. 6, pp. 3153–3158 (1998)
17. Ims, L.A., et al.: Key factors influencing investment strategies of broadband access network upgrades. In: Proceedings of ISSLS'98, Venice, p. 246 (1998)
18. Stordahl, K. et al.: Evaluating broadband strategies in a competitive market using risk analysis. In: Proceedings of Networks, Sorrento, p. 567 (1998)
19. Minoux M (1999) Network synthesis and optimum network design problems: models, solution methods and applications. Networks 19:313–360
20. Monath, T.: Towards multi-service business models. MUSE, Deliverable DA1.1, version 1.0., p. 76 (2004)
21. Monath, T.: Techno-economics for fixed access network evolution scenarios. MUSE, Deliverable DA3.2p version 1.0., p. 87 (2005)
22. Kind M (2008) Economical guidelines and decisions in BB access (results of use cases). MUSE, Deliverable DA3.3 version 1.0., p 123
23. Olsen, B.T.: OPTIMUM—a techno-economic tool. Telektronikk (2/3), 239–250 (1999)
24. Claunir, P.: Dimensioning of multilayer optical networks (2009)
25. Barabash, O.V., Musienko, A.P., Sobchuk, V.V., Lukova-Chuiko, N.V., Svynchuk, O.V.: Distribution of values of cantor type fractal functions with specified restrictions. In: Sadovnichiy, V.A., Zgurovsky, M.Z. (eds.) Contemporary Approaches and Methods in Fundamental Mathematics and Mechanics. Understanding Complex Systems. Springer, Cham (2021). https://doi.org/10.1007/978-3-030-50302-4_21

26. Matuszewski, J.: Application of clustering methods in radar signals recognition. In Proceeding of 2018 International Scientific-Practical Conference Problems of Infocommunications, Science and Technology (PIC S&T), Kharkiv, Ukraine, pp. 745–751. https://doi.org/10.1109/infocommst.2018.8632057
27. Osadchuk, V.S., et al.: Experimental research and simulation of microwave oscillator based on structure of static inductance transistor with negative resistance. In: 2010 20th International Crimean Conference "Microwave and Telecommunication Technology", Sevastopol, Ukraine, pp. 13–17 (2010). https://doi.org/10.1109/crmico.2010.5632543
28. Toliupa, S., Nakonechnyi, V., Tereikovskyi, I., Tereikovska, L., Korystin, O.: One-periodic template marks model of normal behavior of the safety parameters of information systems networking resources. In: 2019 IEEE International Scientific-Practical Conference Problems of Infocommunications, Science and Technology (PIC S&T), pp. 764–768. IEEE (2019)

Estimation of Signal Parameters Using SSA and Linear Transformation of Covariance Matrix or Data Matrix

Volodymyr Vasylyshyn

Abstract The joint application of singular spectrum analysis (SSA) approach (basic variant or adaptive variant) and unitary transformation of the forward-backward data matrix obtained after SSA technique or forward-backward version of the covariance matrix is proposed for improvement of estimation performance of the signal parameters. The unitary transformation reduces the computational complexity and improves the threshold performance of spectral analysis performed by subspace-based techniques. Performance improvement can be explained by forward–backward averaging effect that has a place when performing the unitary transformation. This averaging effectively doubles the number of samples. The proposed approach can be characterized by reduced computational load such as the computations with real-valued numbers are performed after unitary transformation. Unitary Root-MUSIC is mainly used for the simulation. The unitary formulation of ESPRIT is obtained for the problem of estimation of signal component frequencies. The possible applications of considered approach in the communication systems (including channel estimation, speech processing, automatic modulation classification and so on) are considered. Simulation results are presented to demonstrate the improved performance of subspace-based techniques when using proposed approach.

Keywords Adaptive singular spectrum analysis · Singular value decomposition · Unitary transformation · Superresolution methods

1 Introduction

The performance of the most methods of signal processing, parameter estimation, detection, modern methods of spectral analysis (SA), and recognition of signals depends on signal-to-noise ratio, number of samples, angular (frequency) distance between sources and many other factors [1–10]. The application of subspace-based methods of SA requires the low-rank model of the considered system. Such structure has a place in the number of applications including array signal processing. However,

---

V. Vasylyshyn (✉)
Kharkiv National Air Force University, Kharkiv, Ukraine

D. Ageyev et al. (eds.), *Data-Centric Business and Applications*, Lecture Notes on Data Engineering and Communications Technologies 69,
https://doi.org/10.1007/978-3-030-71892-3_15

in the case of frequency estimation, system identification, speech processing the application of the window to the received data permits to receive low-rank data structure. Data model in vector form transformed from an initial process in scalar form is sometimes named as windowed data models [1, 2, 9].

The windowing or embedding step is also a basic step of one of noise reduction methods- singular spectrum analysis (SSA) method [10–17]. The mathematical basis of this method is widely used SVD (singular value decomposition) of data matrix (Fig. 1).

The spectral or eigenvalue decomposition (EVD) of covariance matrix (CM) of data can be applied in many cases instead of SVD. These two decompositions are related with independent component analysis (ICA), principal component analysis (PCA). They are used in the blind source subspace decompositions, spectral analysis,

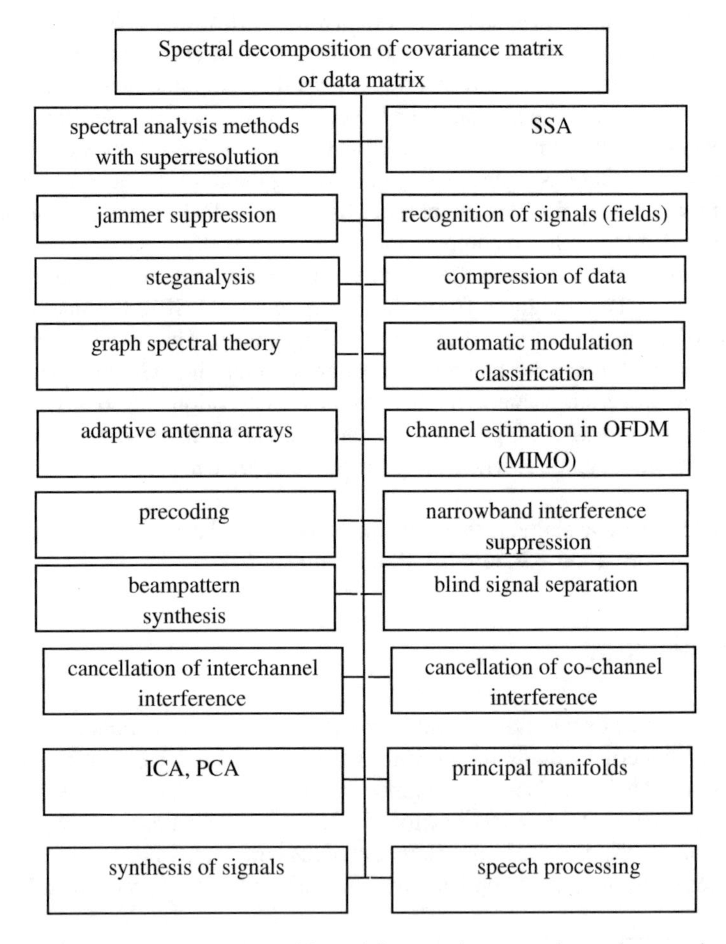

Fig. 1 Applications of eigendecomposition of the covariance matrix or data matrix spectral decomposition

precoding (beamforming), channel estimation in Orthogonal Frequency Division Multiplexing (OFDM), multiple input-multiple output system (MIMO) systems and so on. Furthermore, these decompositions also can be used for spectrum sensing in cognitive radio networks, automatic classification of signals with digital modulation, detection in wireless sensor networks, extraction of signals, signal separation, neural networks [1–18].

ICA can be seen as extension of PCA which searches the linear transformation that minimizes the statistical dependence between components of initial random vector [2, 14]. The several new terms are related with PCA including principal manifolds, principal curves and principal graphs [5].

The term field in Fig. 1 can be understood as the image. SVD and eigenimage term are widely used in digital image processing for a long time [18].

The theory of superresolution despite of many famous and fundamental results is still developing. It is explained by new terms such as spectral theory of operators (usual and self-adjoint), graphs, random matrix theory, compressing sensing [4, 7, 19–22]. Furthermore, the recent developed mathematical theory of superresolution is of interest [23].

The term "spectral analysis" is often related with Fourier transform. However, the spectral decomposition of a time series can be realized in other orthonormal bases. The basis of eigenvectors of CM, Walsh function basis, Haar function basis, Vilenkin–Chrestenson function basis are the most famous bases. In the case of SSA such basis functions as temporal empirical orthogonal functions (EOFs) (or corresponding temporal principal components (PCs)) are considered.

It is known that the attainability of resolution higher than the Rayleigh limit requires high signal-to-noise ratios [2, 6]. Application of different de-noising approaches (filtering, SSA (singular spectra analysis), wavelet) or approaches which allow to increase the number of samples (surrogate data technology, bootstrap, permutation and so on) [2, 10–17, 19–21, 24–27] is useful. Furthermore, the exploitation of structure of CM can be useful in many situations [2, 19, 21, 25, 28–30].

However, the advantages obtained using spectral decomposition of CM or data matrix is attained by quite high computational load because of the necessity of calculation of the EVD or SVD. The calculation of eigenvalues can be performed using QR decomposition (Givens rotations), Lanczos algorithm and many others algorithms [2, 16, 17]. Furthermore, LU decomposition, Cholesky decomposition, Gram-Schmidt orthogonalization, Householder transformation (reflection) are the examples of orthogonal transformations widely used in signal processing.

The CORDIC (Coordinate Rotation Digital Computer) method, pipelining of computation using array of digital signal processors, fast subspace decomposition techniques can be used in this case [2, 17, 31].

It is known the relationship between the eigenanalysis and Fourier analysis [25–27]. In the case of uniform linear array the spatial correlation matrix is Toeplitz and Hermitian and ideal signal vectors are equal to discrete Fourier transformation (DFT) vectors. The eigenvectors of such matrix approach the DFT vectors as the size of matrix grows. Circulant matrices belong to the class of Toeplitz-Hermitian matrices. The eigenvectors of circulant matrices are DFT vectors.

Furthermore, additional reduction of computational load will be received in the case of preprocessing related with linear transformation of data matrix (or CM) [2, 19, 28–30]. The most known examples are beamspace transformation and unitary transformation. They allow transforming the complex-valued matrix with some properties in real one. Furthermore, the number of data vectors is increased nearly twice.

It is known, that using the prior information about specificity of the CM structure could significantly improve the analysis precision and processing efficiency and reduce the size of the parameter vectors [28].

The widely known and used in practice specificity of CM is persymmetry of symmetric real CM and Hermitian CM. The persymmetry means the CM symmetry relative to the secondary diagonal. The quantity of parameters (CM elements) which determine the persymmetric CM (PCM) is approximately twice less as compared with CM of general form [2, 19, 21, 29].

The persymmetric matrices occur often in the antenna array with central symmetry, communication systems and other systems. There are a number of problems when the solution includes the eigenvectors of persymmetric matrices [25]. The using persymmetry property after SSA was considered in [19, 21].

Filtering properties and frequency interpretations of SSA are considered in [12–16, 32, 33]. Filtering interpretation of SSA in [32] was performed using the eigenvector properties of persymmetric matrix.

Besides of persymmetry the Toeplicity is one of structural property. Toeplitz SSA is based on the assumptions about stationarity of time-series. It should be noted that in the case of non-stationary time-series the difference between the Basic and Toeplitz approaches of SSA is insignificant [12].

The joint using of SA methods with superresolution and SSA was discussed in [13–16, 19, 21, 34]. It was confirmed by several authors that performance of SA methods is improved after application SSA method.

Additional ways of improvement of the subspace-based method performance include high-order statistics, expected likelihood approach, joint estimation strategy, pseudo-noise resampling (based on perturbation theory), surrogate data technology. Application of these techniques improves the performance of SA at the threshold signal-to-noise ratios (i.e. reduces the threshold effect) [35]. This effect has a place for important in the practice situations with small signal-to-noise ratio (SNR). The same effect appears when number of snapshots becomes small.

Specific attention should be pointed to the application of preprocessing of data matrix (or CM of observations) by linear transformation [2, 19, 28–30, 36]. Reduction of the complexity of subspace-based approaches and improvement of their performance after unitary transformation causes the combination of the advantages of unitary transformation approach and SSA with aim of improvement of the performance of SA performed by subspace-based methods.

The unitary transformation was applied to the Root-MUSIC in [29]. In this paper the results of performance analysis of the Unitary Root-MUSIC were presented. Unitary ESPRIT for direction of arrival estimation (DOA) with antenna array was proposed in [36].

The implementation of the preliminary linear transformation of forward-backward data matrix formed after SSA step or forward-backward CM was proposed in [19]. The short explanation of proposed idea was performed based on Unitary Root-MUSIC. The modification of the SSA method which is based on the results of the random matrix theory (G-analysis) proposed in [4, 16, 21, 34] is used in the paper.

Generalization of Unitary ESPRIT method for frequency estimation is also proposed in the paper. The simulation results are presented for Unitary Root-MUSIC and Unitary ESPRIT. The possible generalizations of the proposed approach to the case of speech processing, problem of signals recognition and other problems are discussed in the paper in shortly.

2 Signal Model and SSA Method

The proposed technique can be adapted for many variants of signal models. The simple model is used in the paper. However, it is used in many practical cases such as amplitude and phase estimation approach for synthetic aperture radar, weather radar, MIMO communication system and so on, where the problem of frequency estimation arises [1, 2, 9, 21, 37].

The observed signal $y(n)$ is scalar-valued. It is the sum of V harmonic components and white Gaussian noise with zero-mean [2, 9, 19, 37]

$$y(n) = \sum_{\upsilon=1}^{V} \alpha_\upsilon e^{j(\omega_\upsilon n + \varphi_\upsilon)} + e(n) = s(n) + e(n), \tag{1}$$

where $n = 1, \ldots, N$, N is the sample number, α_υ is the amplitude, $j = \sqrt{-1}$. The frequencies (which are assumed to be distinct) can be defined as $\omega_v = 2\pi f_{v_n} = 2\pi f_v / f_s$, $\omega_v \in [0, \pi)$, where f_s is the frequency of sampling. Moreover, φ_υ is the initial phase of the corresponding component. Our goal is frequency estimation (i.e. estimation of ω_υ, $\upsilon = 1, \cdots, V$) using the obtained data sequence.

The frequencies ω_υ, $\upsilon = 1, \cdots, V$ and amplitudes are modeled as deterministic quantities. The initial phases φ_v are independent random variables which are uniformly distributed on $[-\pi, \pi]$. The complex white noise term e(n) has variance σ^2.

As in the case with one snapshot (one time sequence) the first step of SSA is the forming the trajectory matrix [10–17, 21]

$$\mathbf{Y} = \begin{bmatrix} y(1) & \cdots & y(K-1) & y(K) \\ y(2) & \cdots & y(K) & y(K+1) \\ \vdots & \vdots & \vdots & \vdots \\ y(m) & \cdots & y(N-1) & y(N) \end{bmatrix}. \tag{2}$$

Here the number of rows is defined by parameter of embedding m (named also as embedding window) and the number of columns is $K = N - m + 1$. Matrix $\mathbf{Y}$ is a Hankel matrix. In the case of antenna array signal processing such presentation is possible in the case of single snapshot. The obtained data matrix is named as smoothed data matrix.

The columns of the trajectory matrix can be written as

$$\mathbf{y}(n) = [y(n) \ldots y(n_m)]^T, \tag{3}$$

where $n = 1, \ldots, K$, $n_m = n + m - 1$, $()^T$ is the transpose. These columns can be referred as snapshots (by analogy to antenna array signal processing).

The data matrix $\mathbf{Y}$ can be considered as multivariate data with K observations $\mathbf{y}(n)$.

The Takens' theorem is used in the embedding approach [11]. It explains how to embed and reconstruct the time series.

Low-rank matrix representation of the model is used by subspace methods. The guarantee that rank of observation CM will be equal to component number is provided by the segmentation of the input sequence.

Size of window influences on performance of SSA approach [14–16, 21] and mentioned approaches related with use of one snapshot in antenna array. Window size is usually selected as $m > V$ for the signal-subspace methods.

Data model in the matrix form can be presented as [2, 9, 21]

$$\mathbf{y}(n) = \mathbf{B}\mathbf{x}(n) + \mathbf{e}(n), \tag{4}$$

where the components of $V \times 1$ vector $\mathbf{x}(n) = [x(n) \ldots x(n_m)]^T$ are $x(n) = \alpha_v \exp(j(\omega_v n + \varphi_v))$, the $m \times V$ matrix $\mathbf{B}$ consists of columns $a(\omega_v) = [1 \exp(j\omega_v) \ldots \exp(j(m-1)\omega_v)]^T$. The vector form of additive noise is $\mathbf{e}(n) = [e(n) \ldots e(n_m)]^T$.

The windowed sequence CM is given by [2, 9]

$$\mathbf{R} = \sum_{v=1}^{V} \alpha_v^2 \mathbf{a}(\omega_v)\mathbf{a}^H(\omega_v) + \sigma^2\mathbf{I} = \mathbf{B}\mathbf{S}\mathbf{B}^H + \sigma^2\mathbf{I}. \tag{5}$$

Here the CM of signal $\mathbf{S} = diag(\boldsymbol{\alpha})$, where $\boldsymbol{\alpha} = [\alpha_1^2, \ldots, \alpha_V^2]^T$ and $()^H$ is the Hermitian transpose operator.

The spectral representation of CM $\mathbf{R} = E\{\mathbf{y}(n)\mathbf{y}^H(n)\}$ can be written in the following form [2, 38]

$$\mathbf{R} = \mathbf{U}\boldsymbol{\Sigma}\mathbf{U}^H, \tag{6}$$

where $\mathbf{U} = [\mathbf{u}_1 \cdots \mathbf{u}_m]$ is the matrix of orthonormal eigenvectors (EVs) associated with eigenvalues $\lambda_1 > \lambda_2 > \ldots > \lambda_V \geq \lambda_{V+1} = \ldots = \lambda_m$, which are the elements of $m \times m$ diagonal matrix $\boldsymbol{\Sigma}$. It is related with SVD of $\mathbf{Y}$

In order to show spectrum of CM we perform simulation where signal was considered as consisting of two harmonic components with frequencies $f_1 = 0.3$ Hz and $f_2 = 0.32$ Hz and equal power. The amplitudes of signal components were equal to 1.12, $N = 64$. The window size was selected as $m = 18$, $m = 26$, $m = 34$.

Spectrums of eigenvalues of signal covariance matrix (i.e. without noise) versus the window size are presented in Fig. 2.

We can see from Fig. 2 that the eigenvalues corresponding to signal significantly differ from the rest eigenvalues.

The results of simulations for the situation with noise are illustrated in Fig. 3 (only the first 15 eigenvalues are shown). Here we can see that difference between the signal and noise eigenvalues is not too big. The threshold also has a place. However, it should be noted that in the case as SNR decreasing the second eigenvalue can becomes approximately equal to noise subspace eigenvalues.

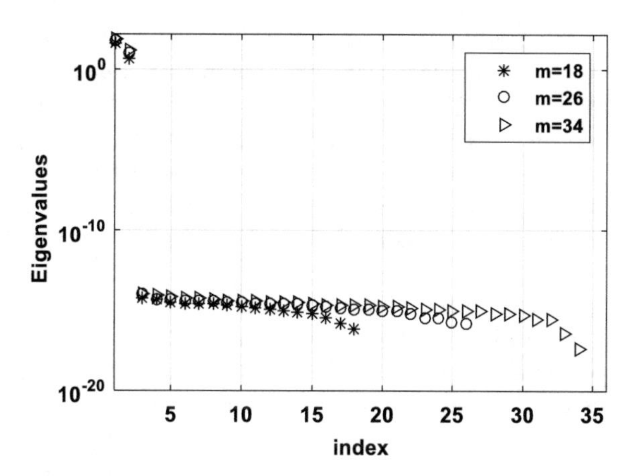

Fig. 2 Spectrums of covariance matrix in the absence of noise

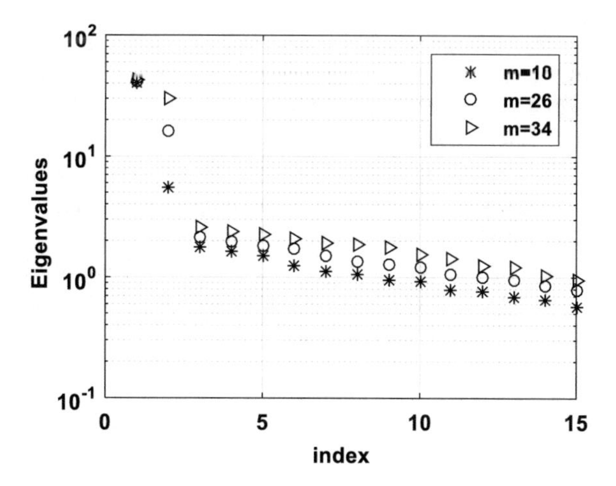

Fig. 3 Spectrum of covariance matrix in the occurrence of noise, SNR = 5 dB

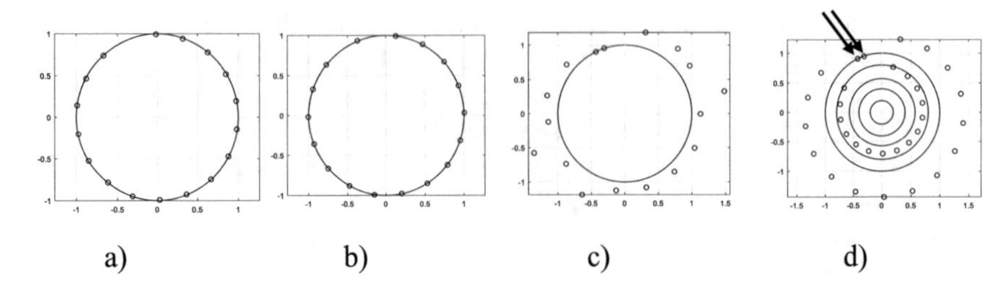

a) b) c) d)

Fig. 4 Roots of eigenfilters and root-music polynomial: **a** for first EV; **b** for second EV; **c** for third EV; **d** roots of Root-MUSIC

The eigenvalues can be traditionally considered as observation for detection purposes when minimum descriptive length (MDL), Akaike information criterion (AIC), Bayesian information criterion are used [2, 16].

It is possible to consider the signal-subspace eigenvector components as coefficients of filter impulse response with finite-impulse response. This filter is named as eigenfilter [2, 16, 32, 33]. The eigenfilters constructed from the eigenvectors of persymmetric matrix were investigated.

Here we consider the eigenfilters constructed from the eigenvectors of CM with general form. The SNR was 5 dB, the frequencies of components were the same as one for Fig. 1. Results of simulation are plotted in Fig. 4.

Analysis of the Fig. 4 shows that the use of the third eigenvector (corresponding to noise subspace) allows to estimate the frequencies of signal components (Fig. 4c). This possibility can be viewed from the comparison of Fig. 4c, d, where the roots of polynomial of Root-MUSIC are presented. The positions of roots corresponding to frequency estimates are indicated by arrows.

The application of the last noise sunspace eigenvector corresponds to Pisarenko method. It should be noted, that Root-MUSIC is an extension of Pisarenko approach [2].

Furthermore, the locations of roots of eigenfilters corresponding to eigenvectors of signal subspace do not indicate directly the positions of sources. However, the signal subspace eigenvectors (or singular vectors) are used in precoding performed in MIMO, OFDM and other communication systems.

It is necessary to mention separately also such terms as spectral portrait, spectral radius, Gerschgorin circle, dichotomy of the spectrum of a matrix, ε-spectrum that are related with considered problem [4, 6].

We define the vector $\mathbf{v}_i$ as $\mathbf{v}_i = \mathbf{Y}^H\mathbf{u}_i/\lambda_i$, $i = 1, \ldots, d$. Here $d = \max\{i \in \{1, \ldots, m\} | \lambda_i > 0\}$ is the rank of $\mathbf{Y}$. Spectral decomposition (SVD) of the matrix $\mathbf{Y}$ can be presented as

$$\mathbf{Y} = \sum_{i=1}^{d} \sqrt{\lambda_i}\mathbf{u}_i\mathbf{v}_i^H = \mathbf{Y}_1 + \ldots + \mathbf{Y}_d, \tag{7}$$

where $\mathbf{Y}_i$ is the biorthogonal matrix of unit rank . It should be noted that $\mathbf{v}_i$ is the i-th eigenvector of $\mathbf{Y}^H\mathbf{Y}$. The set $(\sqrt{\lambda_i}, \mathbf{u}_i, \mathbf{v}_i)$ is named as i-th eigentriple (in the case of EVD we have only two parameters- eigenvectors and eigenvalues, i.e. eigenpair) [7]. The eigenvalue λ_i characterizes the contribution of $\mathbf{Y}_i$ in the decomposition (7).

The linear and nonlinear weighting can be used together with SVD [18]. The case of linear weighting is given as

$$\mathbf{Y} = \sum_{i=1}^{d} \rho_i \sqrt{\lambda_i} \mathbf{u}_i \mathbf{v}_i^H \tag{8}$$

where ρ_i is the weighting coefficient. The nonlinear weighting can be defined as

$$\mathbf{Y} = \sum_{i=1}^{d} \lambda_i^{\beta/2} \mathbf{u}_i \mathbf{v}_i^H \tag{9}$$

Equation (9) can be used for "low-pass" and "high-pass" nonlinear filtering (the situation $\beta > 1$ has a place when the larger eigenvalues are dominate and the situation $\beta < 1$ is correct when smaller eigenvalues are dominate). In the case $\beta = 0$ the weighting is similar for all singular values.

Eigentriple grouping is very important step of basic SSA. The index set $I \subset \{i = 1, \ldots, d\}$ is divided into several subsets [12, 14, 15]. These subsets are disjoint. As a result, the subset of $\mathbf{Y}_i$ (described in eq. 7) is selected. The separability of SSA is depends on this step.

The problem of separation of eigenvector subspaces is important in addition to signal-subspace methods [2]. The general recommendations on grouping step are presented in [15].

In order to reconstruct the denoised data matrix the truncated SVD is used [2, 14–16]. However, this matrix is not Hankel matrix. Such problem is solved by applying Hankelization operator. Hankelization is performed by diagonal averaging.

The averaging of CM elements has a place in the antenna array signal processing when spatial smoothing or variants of the procedure is performed [2, 21].

Truncated SVD (TSVD) is also a well-known regularization technique. Besides of truncation the clipping procedure is known. In the clipping SVD transformation the clipping operation performed on the singular values is sometimes referred as "inverted clip".

Simplification of the averaging step of SSA can be attained using discrete Fourier transformation and convolution theorem [17]. Furthermore, the Toeplitz structure can be applied instead of Hankel structure.

Let $y_f(n)$ be the filtered sequence (i.e. sequence obtained at last step of SSA). Therefore, the initial time sequence is given by

$$y(n) = y_f(n) + \varepsilon_r(n), \tag{10}$$

where $\varepsilon_r(n)$ is the residual series. Based on $y_f(n)$ we form the data matrix $\mathbf{Y}_{filt.}$.

In the practice, the estimate of CM (i.e. sample CM) should be computed. The sample CM can be given as

$$\widehat{\mathbf{R}} = \frac{1}{K}\mathbf{Y}\mathbf{Y}^H = \frac{1}{K}\sum_{n=1}^{K}\mathbf{y}(n)\mathbf{y}^H(n). \tag{11}$$

The features of unitary transformation and joint application of the unitary variant of superresolution method and SSA-like methods are considered in the next section. In most cases the Unitary Root-MUSIC is used. However, Unitary MODE, Unitary FINE, Unitary Min-Norm and other methods can be used. Furthermore, Unitary ESPRIT is described later. It can be realized using least squares (LS), total LS, structured total LS method.

3 Joint Application of the SSA and Unitary Root-MUSIC

The typical preprocessing approaches are the unitary transformation, beamspace transformation, precoding (widely used in communication systems with MIMO and OFDM), spatial smoothing, orthogonalisation of data (Gram-Schmidt preprocessor), preconditioning [2, 19].

The bijective mapping allows to map the persymmetric (centro-Hermitian) matrices to the set of real matrices of the same size $\psi : \mathbf{M} \mapsto \mathbf{Q}_m\mathbf{M}\mathbf{Q}_K$ by corresponding matrices [2, 36]. Here $m \times m$ matrix $\mathbf{Q}_m$ and $K \times K$ matrix $\mathbf{Q}_K$ are the unitary matrices, $m \times K$ matrix $\mathbf{M}$ is the complex persymmetric (centro-Hermitian) matrix. The unitary matrix [2, 19, 29, 30, 36]

$$\mathbf{Q}_{2\tilde{m}+1} = (1/\sqrt{2})\begin{bmatrix} \mathbf{I}_{\tilde{m}} & \mathbf{0} & j\mathbf{I}_{\tilde{m}} \\ \mathbf{0}^T & \sqrt{2} & \mathbf{0}^T \\ \widetilde{\mathbf{I}}_{\tilde{m}} & \mathbf{0} & -j\widetilde{\mathbf{I}}_{\tilde{m}} \end{bmatrix} \tag{12}$$

which is column conjugate symmetric matrix, can be selected when m is odd ($m = 2\tilde{m}+1$, where $\tilde{m} = fix(m/2)$. For the case when m is even we can form the unitary matrix based on (12) by dropping the central column and row

$$\mathbf{Q}_{2\tilde{m}} = (1/\sqrt{2})\begin{bmatrix} \mathbf{I}_{\tilde{m}} & j\mathbf{I}_{\tilde{m}} \\ \widetilde{\mathbf{I}}_{\tilde{m}} & -j\widetilde{\mathbf{I}}_{\tilde{m}} \end{bmatrix}. \tag{13}$$

Here $\mathbf{0}$ is the vector formed from zeros, $\widetilde{\mathbf{I}}_m$ is the $m \times m$ exchange matrix

$$\widetilde{\mathbf{I}}_m = \begin{bmatrix} 0 & 0 & 1 \\ 0 & \cdot^{\cdot^{\cdot}} & 0 \\ 1 & 0 & 0 \end{bmatrix}. \tag{14}$$

The real-valued CM obtained in the result of mapping is given by [19, 28, 36]

$$\mathbf{R}_u = \tfrac{1}{2}\mathbf{Q}^H(\mathbf{R} + \widetilde{\mathbf{I}}\mathbf{R}^*\widetilde{\mathbf{I}})\mathbf{Q}, \tag{15}$$

where $(\cdot)^*$ is the complex conjugate operator. The distribution problem of the real-valued CM is discussed in [28]. The matrix in parentheses is the forward-backward averaged CM. Forward-backward averaging on the data matrix level (the CM level is described by (15)) can be presented as [2, 21, 36]

$$\mathbf{Y}^{fb} = [\mathbf{Y}\ \ \widetilde{\mathbf{I}}_m\mathbf{Y}^*\ \ \widetilde{\mathbf{I}}_K]. \tag{16}$$

where $K \times K$ matrix $\widetilde{\mathbf{I}}_K$ is the exchange matrix. Forward-backward averaging uses a symmetry in the data to create an additional set of "virtual" snapshots.

The data after SSA transformed by analogy to (16) have a form

$$\mathbf{Y}^{fb}_{filt.} = [\mathbf{Y}_{filt.}\ \widetilde{\mathbf{I}}_m\mathbf{Y}^*_{filt.}\widetilde{\mathbf{I}}_K]. \tag{17}$$

Eigendecomposition of $\mathbf{R}_u$ is given by

$$\mathbf{R}_u = \mathbf{E}_{sun}\mathbf{\Lambda}_{sun}\mathbf{E}^{\mathrm{H}}_{sun} + \mathbf{E}_{nun}\mathbf{\Lambda}_{nun}\mathbf{E}^{\mathrm{H}}_{nun}. \tag{18}$$

Columns of $m \times \mathrm{V}$ matrix $\mathbf{E}_{sun}$ are the orthonormal eigenvectors corresponding to the V signal eigenvalues containing in $\mathbf{\Lambda}_{\mathrm{sun}}$. In addition, the columns of $m \times (m - \mathrm{V})$ matrix $\mathbf{E}_{nun}$ are the noise-subspace eigenvectors.

However, in the practice the forward-only (traditional) estimate of CM is not persymmetric. The estimate of the persymmetric CM $(1/2)(\widehat{\mathbf{R}} + \widetilde{\mathbf{I}}\widehat{\mathbf{R}}^*\widetilde{\mathbf{I}})$ is the forward-backward variant of CM The real-valued sample CM $\widehat{\mathbf{R}}_u = \frac{1}{2}\mathbf{Q}^H(\widehat{\mathbf{R}} + \widetilde{\mathbf{I}}\widehat{\mathbf{R}}^*\widetilde{\mathbf{I}})\mathbf{Q}$ or $\widehat{\mathbf{R}}_u = \mathrm{Re}(\mathbf{Q}^H\widehat{\mathbf{R}}\mathbf{Q})$ can be used for frequency estimation [2].

Unitary transformation can be also applied to forward-backward data matrix $\mathbf{Y}^{fb}_{filt.} = [\mathbf{Y}_{filt.}\ \widetilde{\mathbf{I}}_m\mathbf{Y}^*_{filt.}\widetilde{\mathbf{I}}_K]$. The real-valued data matrix can be obtained as

$$\mathbf{Y}_{ext} = \mathbf{Q}^H_m[\mathbf{Y}_{filt.}\ \widetilde{\mathbf{I}}_m\mathbf{Y}^*_{filt.}\widetilde{\mathbf{I}}_K]\mathbf{Q}_{2K} = \mathbf{Q}^H_m\mathbf{Y}^{fb}_{filt.}\mathbf{Q}_{2K} \tag{19}$$

The size of corresponding matrices that perform mapping corresponds to the size of matrices $\mathbf{Y}^{fb}_{filt.}$ and $\widetilde{\mathbf{I}}_K$. After the unitary transformation the methods of SA can be used in usual way.

Unitary Root-MUSIC is the result of application of unitary transformation to Root-MUSIC. It gives the estimates of the frequencies via rooting the polynomial of degree $2(m - 1)$ [29]

$$P_{urm}(z) = \mathbf{a}^T(z^{-1})\mathbf{Q}\widehat{\mathbf{E}}_n\widehat{\mathbf{E}}_n^T\mathbf{Q}^H\mathbf{a}(z), \tag{20}$$

where $z = \exp(j\omega)$, $\mathbf{a}(z) = [1, z, \ldots, z^{M-1}]^T$.' Taking into account that $\mathbf{E}_n\mathbf{E}_n^H = (\mathbf{I} - \mathbf{E}_s\mathbf{E}_s^H)$ the following equation for Unitary Root-MUSIC can be obtained

$$P_{urm}(z) = \mathbf{a}^T(z^{-1})\mathbf{Q}(\mathbf{I} - \widehat{\mathbf{E}}_s\widehat{\mathbf{E}}_s^T)\mathbf{Q}^H\mathbf{a}(z). \tag{21}$$

Such representation is appropriate for the case with small number of harmonic components of the signal (i.e. only signal eigenvalues and eigenvectors are to be calculated).

In respect to unit circle the roots of $P_{urm}(z)$ occur in mirrored pairs (similar to Fig. 4d). The phases of V roots (z_v, $v = 1, \ldots V$) closest to the unit circle (selected from $m - 1$ roots situated inside unit circle) are used for the frequency estimation

$$\hat{f}_v = f_s \arg(z_v)/2\pi. \tag{22}$$

The computational complexity of Unitary Root-MUSIC is reduced approximately by a factor of four in comparison with complex Root-MUSIC.

Combined application of adaptive SSA [21] and Unitary Root-MUSIC can be described by the sequence of the steps:

(1) perform the preprocessing by the adaptive SSA. Obtain the matrix $\mathbf{Y}_{filt.} = \sum_{i=1}^{\widehat{V}} (\widehat{\mu}_i - \widehat{\sigma}_n)\hat{\mathbf{u}}_i\hat{\mathbf{v}}_i^H$, where $\widehat{\sigma}_n$ is the estimate of noise standard deviation;
(2) form the forward-backward matrix $\mathbf{Y}_{filt.}^{fb} = [\mathbf{Y}_{filt.} \;\; \widetilde{\mathbf{I}}_m\mathbf{Y}_{filt.}^*\widetilde{\mathbf{I}}_K]$, perform the unitary transformation (mapping) of $\mathbf{Y}_{filt.}^{fb}$ or calculate $\widehat{\mathbf{R}}_u$;
(3) calculate the SVD of mapped $\mathbf{Y}_{filt.}^{fb}$ or EVD of the $\widehat{\mathbf{R}}_u$;
(4) find the polynomial roots of Unitary Root-MUSIC. Calculate the frequency estimates according to (22).

The application of denoising by SSA-like methods and unitary transformation to spectral analysis can be considered as the one of possible ways of performance improvement of different areas in signal processing (Fig. 5). Additional directions of application of the proposed approach are also presented [39–43].

Let us consider the obtaining the Unitary ESPRIT method for frequency estimation. In some cases the post correlation can be performed before unitary transformation.

4 Unitary ESPRIT for Frequency Estimation

The Unitary ESPRIT is the famous method of direction of arrival estimation that can be used for frequency estimation and having number of advantages including reduced computation complexity, improved performance as compared to basic ESPRIT and

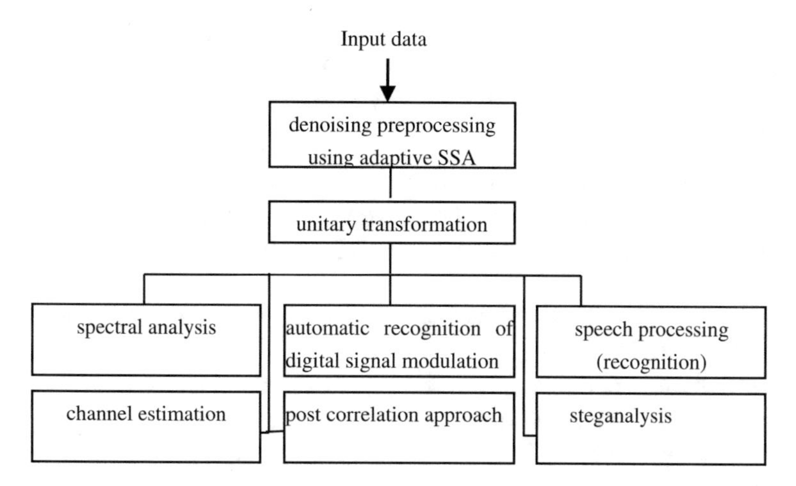

Fig. 5 Examples of SSA applications

so on. Unitary ESPRIT for antenna array signal processing with spatial frequency estimation is obtained in [36]. However, the generalization to the case of frequency estimation should be performed.

The fundamental equation in ESPRIT-like methods is the shift invariance relationship which for considered case can be presented as

$$\exp(j\omega)\mathbf{J}_{1\tilde{m}}\bar{\mathbf{a}}(\omega) = \mathbf{J}_{2\tilde{m}}\bar{\mathbf{a}}(\omega). \tag{23}$$

where the selection matrices $\mathbf{J}_{1\tilde{m}}$ and $\mathbf{J}_{2\tilde{m}}$ are used. Here $\mathbf{J}_{1\tilde{m}} = [\mathbf{I}_{\tilde{m}\times\tilde{m}} \ \ \mathbf{0}_{\tilde{m}\times 1}]$ and $\mathbf{J}_{2\tilde{m}} = [\mathbf{0}_{\tilde{m}\times 1} \ \ \mathbf{I}_{\tilde{m}\times\tilde{m}}]$, $\bar{\mathbf{a}}(\omega) = \exp(-j\omega/2)\mathbf{a}(\omega)$.

It should be noted that shift operator is very important operator in digital signal processing related, for example, with discrete Fourier transformation (DFT).

The matrices $\mathbf{K}_1 = \mathrm{Re}(\mathbf{Q}_{m-1}^H\mathbf{J}_2\mathbf{Q})$ and $\mathbf{K}_2 = \mathrm{Im}(\mathbf{Q}_{m-1}^H\mathbf{J}_2\mathbf{Q})$ are the selection matrices when using unitary transformation. Equation (23) can be transformed to the following form

$$\exp(j\omega/2)\mathbf{P}\mathbf{d}(\omega) = \exp(-j\omega/2)(\mathbf{P})^*\mathbf{d}(\omega), \tag{24}$$

where $\mathbf{P} = \mathbf{K}_1 - j\mathbf{K}_2$. Simple manipulations by analogy with [36] give the following invariance relationship

$$tg(\omega/2)\mathbf{K}_1\mathbf{d}(\omega) = \mathbf{K}_2\mathbf{d}(\omega). \tag{25}$$

The frequencies of signal components are calculated based on the eigenvalues of operator Ξ determined from equation $\mathbf{K}_1\mathbf{E}_{sun}\Xi = \mathbf{K}_2\mathbf{E}_{sun}$. The columns of $\mathbf{E}_{sun}$ are eigenvectors of $\mathbf{R}_u$ corresponding to V signal subspace eigenvalues. These comments and the Eq. (25) give the basis for Unitary ESFRIT. The equation $\mathbf{K}_1\mathbf{E}_{sun}\,\Xi = \mathbf{K}_2\mathbf{E}_{sun}$ can be solved by LS, total LS or structured total LS method.

Unitary ESFRIT method for frequency estimation can be presented as the sequence of steps:

(1) calculate the spectral decomposition of matrix $\mathrm{Re}(\mathbf{Q}^H\widehat{\mathbf{R}}\mathbf{Q})$;
(2) estimate the number of signal components $\widehat{V}$ and form $M \times \widehat{V}$ matrix $\widehat{\mathbf{E}}_{sun}$ from the eigenvectors of $\mathrm{Re}(\mathbf{Q}^H\widehat{\mathbf{R}}\mathbf{Q})$ corresponding to $\widehat{V}$ signal subspace eigenvalues;
(3) calculate $(M - 1) \times M$ selection matrices $\mathbf{K}_1 = \mathrm{Re}(\mathbf{Q}_{m-1}^H\mathbf{J}_2\mathbf{Q})$ and $\mathbf{K}_2 = \mathrm{Im}(\mathbf{Q}_{m-1}^H\mathbf{J}_2\mathbf{Q})$;
(4) solve the equation $\mathbf{K}_1\widehat{\mathbf{E}}_{sun}\,\Xi = \mathbf{K}_2\widehat{\mathbf{E}}_{sun}$ and find the eigenvalues of Ξ;
(5) calculate the estimate of frequencies $\widehat{\omega}_i = 2arctg(\widehat{\zeta}_i), i = 1, \ldots, \widehat{V}$, where $\widehat{\zeta}_i$ are the eigenvalues of Ξ.

5 Simulation Results

The complex data were simulated according to (1) and consisted of components with frequencies $f_1 = 0.3$ Hz and $f_2 = 0.311$ Hz. In the simulation the input sequence length $N = 64$. The power of components was the same and the frequency separation was less than Rayleigh resolution limit corresponding to $N = 64$ (1/N = 0.015). The each simulated point was obtained based on $L = 1000$ independent simulation runs. Root-mean square error (RMSE) of frequency estimation was calculated as in [19, 21].

The performance of usual Root-MUSIC and unitary variant, Root-MUSIC with preprocessing by SSA (Root-MUSIC with SSA) was compared. Moreover, usual Root-MUSIC with preprocessing by adaptive SSA method (Root-MUSIC with adaptive SSA) and unitary variant with the same preprocessing (Un. Root-MUSIC with adaptive SSA) were considered. Figure 6 shows the RMSE's of frequency estimation versus the window size.

SNR was fixed for Fig. 6. From Fig. 6 we can see that unitary transformation of CM improves the performance of Root-MUSIC. Furthermore, the performance of Unitary Root-MUSIC with adaptive SSA is better for all values of segment size.

Similar simulations for high SNR (SNR = 15 dB) are also performed. Corresponding dependences are presented in Fig. 7. The difference in performance becomes less evident. Furthermore, for such value of SNR the performance of usual Root-MUSIC with adaptive SSA at big values of segment size is better than one of Unitary Root-MUSIC with adaptive SSA.

It should be noted that in the situations when the length of the input sequence is greater as compared to value in the paper the window size considered in the paper is also greater. The typical examples can be speech processing, channel state estimation and so on [40–42]. Furthermore, the feature in this case is bigger number of signal component. The Pearson correlation coefficient, distance between the vectors,

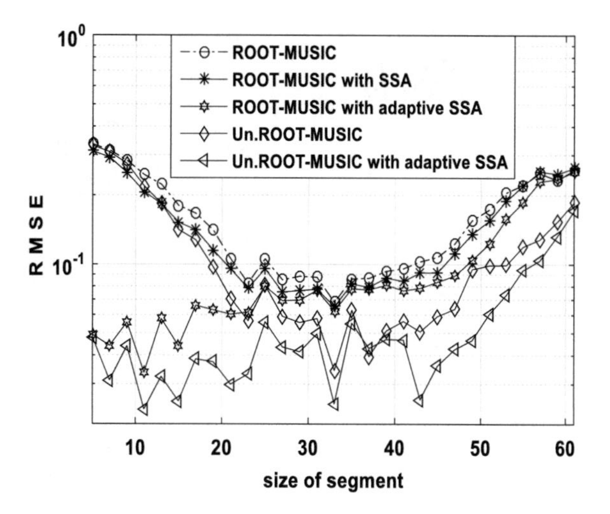

Fig. 6 RMSEs versus the size of window, SNR = 4 dB

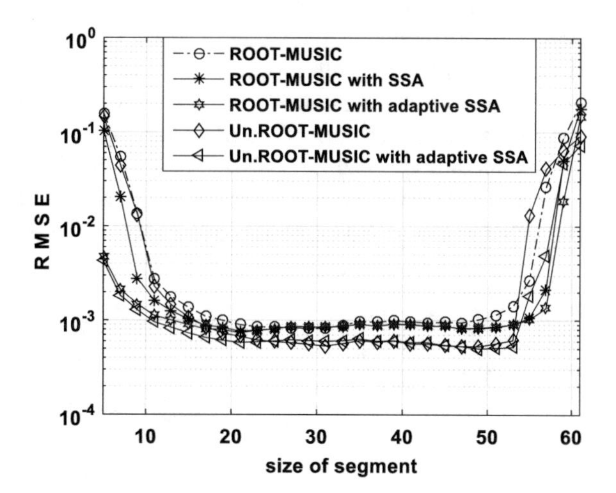

Fig. 7 RMSEs versus the size of window, SNR = 15 dB

Kaiser's rule, phase portraits can be applied for selection of the eigenvectors (EOFs or PCs) on the reconstruction stage.

The following simulation was performed based on the analysis of Fig. 6. The segment size was selected as $m = 15$.The results are illustrated in Fig. 8.

Figure 8 shows that for used conditions of simulations the performance of proposed approach (Unitary Root-MUSIC with adaptive SSA) is better as compared to performance of the remaining methods especially at low SNRs. Moreover, the proposed approach has a superior asymptotic performance as compared with Root-MUSIC with adaptive SSA.

Furthermore, in order to check the efficiency of proposed idea with ESPRIT method we compare the performance of traditional ESPRIT, unitary ESPRIT, ESPRIT with preprocessing by SSA method (ESPRIT with SSA), unitary ESPRIT

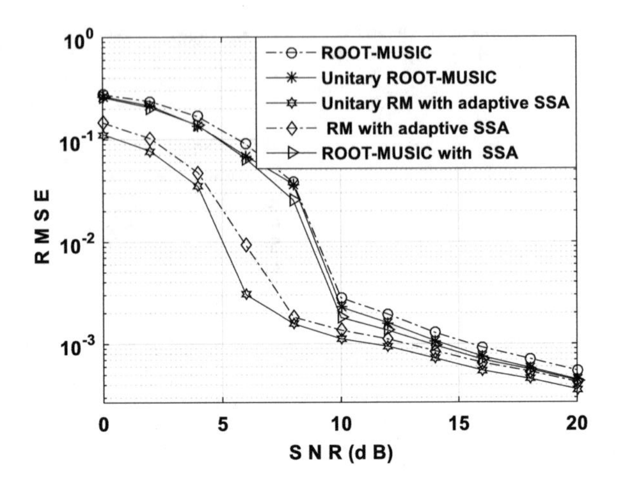

Fig. 8 Experimental RMSEs versus the SNR, $m = 15$, $\Delta f = 0.011\,\text{Hz}$

after SSA (Un. ESPRIT with SSA). Additionally, unitary ESPRIT with adaptive SSA method with different estimates of noise variance including usual one and proposed in [21] (Un. ESPRIT with adaptive SSA and Un.ESPRIT with adaptive SSA and usual estimate (UE)) are considered. The usual estimate is obtained by averaging the noise subspace eigenvaues. The corresponding results are presented in Fig. 9. The LS version of traditional ESPRIT and unitary variant was used, $m = 15$.

Figure 9 shows that performance of proposed technique with Unitary ESPRIT is also better as compared to other ESPRIT-like methods.

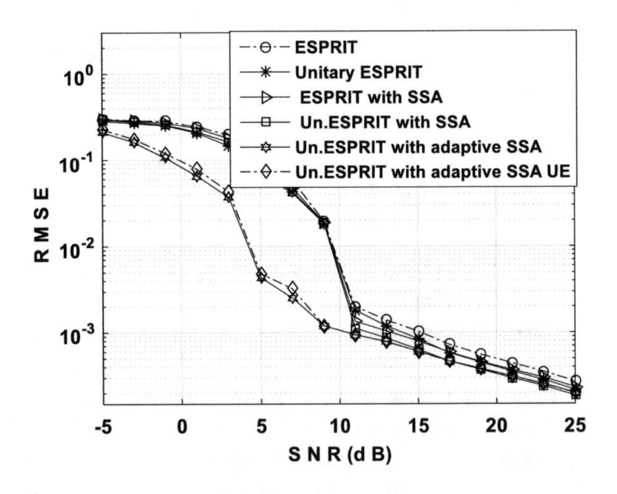

Fig. 9 RMSEs versus the SNR for ESPRIT-like methods, $\Delta f = 0.011\,\text{Hz}$

6 Conclusion

Unitary transformation method which is the example of linear transformation of covariance matrix with structural properties is applied to the output data of complex SSA method. The application of the transformation to forward-backward data matrix or covariance matrix allows obtaining corresponding real-valued matrices and improvement of performance of subspace-based methods. The unitary versions of Root-MUSIC and ESPRIT were used. However, Unitary Root-Min-Norm, Unitary matrix pencil method, Unitary MODE can also be used. Furthermore, the Unitary ESPRIT is obtained for the frequency estimation problem. Adaptive SSA method is used as preprocessing step before unitary transformation.

The possible applications of considered approach are the channel estimation in OFDM-like systems, channel sounding problem, speech enhancement, steganography, recognition of signals and fields (speech, digital modulation and so on) [1, 2, 8]. The additional performance improvement of proposed technique in some applications can be obtained using results obtained in [3, 20, 42–44].

It is of interest to use the components of graph signal processing, high-order SVD to extend the possible applications and possibilities of proposed approach. Furthermore, it is of interest to use singular value adaptation, consider the influence of SVD clipping. Furthermore, it is necessary find the ways to stop the use of SSA in certain application at high SNR. The investigation of the efficiency of application of the unitary transformation before SSA must be performed.

References

1. Proakis G, Salehi M (2008) Digital communications. Fifth edn, McGraw-Hill
2. Trees HLV (2002) Optimum array processing. Part IV of detection, estimation and modulation theory. Wiley–interscience
3. Michałowski Tadeusz (ed) (2011) Applications of matlab in science and engineering. Tech, RijekaIn
4. Bai Z, Fang Z, Liang Y-C (2014) Spectral theory of large dimensional random matrices and its applications to wireless communications and finance statistics. Random matrix theory and its applications. World scientific publishing
5. Gorban A, Kegl B, Wunsch D, Zinovyev A (eds) (2007) Principal manifolds for data visualisation and dimension reduction. Springer, Berlin
6. Percival DB, Walden AT (1993) Spectral analysis for physical applications: multitaper and conventional univariate techniques. Cambridge Univ Press, Cambridge
7. Trefethen LN, Embree M (2005) Spectra and pseudospectra. The behavior of non normal matrices and operators. Princeton University Press, Princeton and Oxford
8. Van Huffel S (1993) Enhanced resolution based on minimum variance estimation and exponential data modeling. Sig Process 33(3):333–355
9. Kristensson M, Jansson M, Ottersten B (2001) Further results and insights on subspace based sinusoidal frequency estimation. IEEE Trans SP 49(12):2962–2974
10. Cadzow JA (1988) Signal enhancement—a composite property mapping algorithm. IEEE Trans ASSP 36:49–62
11. Broomhead D, King G (1986) Extracting qualitative dynamics from experimental data. Phys D 20:217–236. https://doi.org/10.1016/0167-2789(86)90031-X

12. Vautard R, Yiou P, Ghil M (1992) Singular-spectrum analysis: a toolkit for short, noisy chaotic signals. Phys D 58:95–126. https://doi.org/10.1016/0167-2789(92)90103-T
13. Penland C, Ghil M, Weickmann KM (1991) Adaptive filtering and maximum entropy spectra with application to changes in atmospheric angular momentum. J Geophys Res 96(D12):22659–22671. https://doi.org/10.1029/91JD02107
14. Sanei S, Hassani H (2016) Singular spectrum analysis of biomedical signals. CRC Press, London
15. Golyandina N, Zhigljavsky A (2013) Singular spectrum analysis for time series. Springer
16. Vasylyshyn VI (2014) The signal preprocessing with using the SSA method in the spectral analysis problems. Appl Radio Electron 14(1):43–50 (in Russian)
17. Korobeynikov A (2010) Computation- and space-efficient implementation of SSA. Stat Interface 3:357–368
18. Andrews HC, Patterson CL (1976) Singular value decompositions and digital image processing. IEEE Trans Acoust Speech Sig Process 24:26–53. https://doi.org/10.1109/TASSP.1976.1162766
19. Vasylyshyn V (2019) Estimation of signal parameters using SSA and unitary root-music. Paper presented at 2019 international scientific-practical conference problems of infocommunications, science and technology, Kyiv, Ukraine, 2019. https://doi.org/10.1109/PICST47496.2019.9061460
20. Belda J, Vergara L, Gl Safont et al (2019) A New surrogating algorithm by the complex graph fourier transform (CGFT). Entropy 21(759):1–18
21. Vasylyshyn V (2021) Adaptive complex singular spectrum analysis with application to modern superresolution methods. In: Radivilova T et al (eds) Lecture notes on data engineering and communications technologies. Data-Centric business and applications, vol 48. Springer, Switzerland, pp 1–20. https://doi.org/10.1007/978-3-030-43070-2_3
22. Malioutov DM, Cetin M, Willsky AS (2005) A sparse signal reconstruction perspective for source localization with sensor arrays. IEEE Trans Signal Process 53(8):3010–3022
23. Candes EJ, Fernandez-Granda C (2014) Towards a mathematical theory of super-resolution. Commun Pure Appl Math 67(6):1–48
24. Kostenko PYu, Vasylyshyn VI (2015) Enhancing the spectral analysis efficiency at low signal-to-noise ratios using the technology of surrogate data without the segmentation of observation. Radioelectron Commun Syst 58:75–84. https://doi.org/10.3103/S0735272715020041
25. Cantoni A, Butler P (1976) Properties of the eigenvectors of persymmetric matrices with applications to communication theory. IEEE Trans Commun 24:804–809
26. Makhoul J (1981) On the eigenvectors of symmetric Toeplitz matrices. Proc Acoust Speech Signal Process 29:868–872
27. Reddi SS (1984) Eigenvector properties of Toeplitz matrices and their application to spectral analysis of time series. Signal Process 7:45–56. https://doi.org/10.1016/0165-1684(84)90023-9
28. Lekhovytskiy DI (2016) To the theory of adaptive signal processing in systems with centrally symmetric receive channels. EURASIP J Adv Signal Process 33(1):1–11. https://doi.org/10.1186/s13634-016-0329-z
29. Pesavento M, Gershman AB, Haardt M (2000) Unitary root-music with a real-valued eigendecomposition: a theoretical and experimental performance study. IEEE Trans SP 48(5):1306–1314. https://doi.org/10.1109/78.839978
30. Huarng K-C, Yeh C-C (1991) A unitary transformation method for angle-of-arrival estimation. IEEE Trans signal Process 39(4):975–977
31. Liu KR, Yao K (1992) Multiphase systolic algorithms for spectral decomposition. IEEE Trans SP 40(1):190–201
32. Harris TJ, Yuan H (2010) Filtering and frequency interpretations of singular spectrum analysis. Phys D 239:1958–1967. https://doi.org/10.1016/j.physd.2010.07.005
33. Hansen PC, Jensen SH (1998) FIR filter representations of reduced-rank noise reduction. IEEE Trans Signal Process 46:1737–1741. https://doi.org/10.1109/78.678511
34. Vasylyshyn VI (2018) Frequency estimation of signals by Esprit method using SSA- based preprocessing. Paper presented at the UkrMiCo, Odessa, Ukraine, 10–14 Sept 2018. https://doi.org/10.1109/UkrMiCo43733.2018.9047555

35. Thomas JK, Scharf LL, Tufts DW (1995) The probability of a subspace swap in the SVD. IEEE Trans Signal Process 43(3):730–736
36. Haardt M, Nossek JA (2005) Unitary ESPRIT: how to obtain increased estimation accuracy with a reduced computational burden. IEEE Trans Acoustic Speech Signal Process 43(5):1232–1242. https://doi.org/10.1109/78.382406
37. Li J, Stoica P (1996) An adaptive filtering approach to spectral estimation and SAR imaging. IEEE Trans Signal Process 44(6):1469–1484. https://doi.org/10.1109/78.506612
38. Gerbrandson JJ (1981) The relationships between SVD, KLT and PCA. Pattern Recoflnifion 14(6):375–381
39. Kostenko PYu, Vasylyshyn V, Barsukov A et al (2017) Nonparametric estimate of multiplicity of the signal phase-shift keying. Paper presented at 4th international scientific-practical conference "Problems of infocommunications. science and technology" (PICS&T-2017), Kharkiv, Ukraine, 10–13 Oct 2017
40. Vasylyshyn V, Barsukov O, Bekirov A et al (2020) The use of the "Caterpillar" method for the tasks of noise filtering in the voice range. Paper presented at the 2020 IEEE 40th international conference on electronics and nanotechnology (ELNANO), Kyiv, Ukraine, 2020. https://doi.org/10.1109/ELNANO50318.2020.9088853
41. Vasylyshyn V (2020) Channel estimation method for OFDM communication system using adaptive singular spectrum analysis. Paper presented at the 2020 IEEE 40th International conference on electronics and nanotechnology (ELNANO), Kyiv, Ukraine, 2020. https://doi.org/10.1109/ELNANO50318.2020.9088787
42. Narsimha B, Reddy KA (2018) Multi-scale singular spectrum analysis for channel estimation of OFDM transceiver system. Paper presented at the IEEE recent advances in intelligent computational systems RAICS 2018, Thiruvananthapuram, India, 6–8 Dec 2018. https://doi.org/10.1109/RAICS.2018.8634904
43. Elango GA, Sudha GF, Francis B (2017) Weak signal acquisition enhancement in software GPS receivers. Pre-filtering combined postcorrelation detection approach. Appl Comput Inf 13 (1):66–78. https://doi.org/10.1016/j.aci.2014.10.002
44. Volosyuk VK, Kravchenko VF, Kutuza BG, Pavlikov VV (2015) Review of modern algorithms for high resolution imaging with passive radar. Paper presented at 2015 international conference on antenna theory and techniques (ICATT), Kharkiv, 2015. https://doi.org/10.1109/ICATT.2015.7136779

Method of Creating a Passive Optical Network Monitoring System

Liubov Tokar and Yana Krasnozheniuk

Abstract The necessity of monitoring a passive optical network (PON) is discussed. The architecture and topologies of PONs are considered, their advantages and disadvantages are analyzed. The equipment management tools for creating a monitoring system are analyzed. The protocol and control information database were selected using the MIB-I, MIB-II, RMON MIB standards. It is proved that the selection of SNMP for management is conditioned by its simplicity and efficiency, as well as the ability to unifiedly control equipment of various manufacturers. SNMP features are analyzed. The general procedure for creating a monitoring system is formulated. It is shown that for high-quality network monitoring it is necessary to implement periodic polling of all available OLTs and tracking important indicators: detecting ONUs on OLTs that are not included in the ONU database, or registered simultaneously on two OLTs. Tools and technologies for creating a PON monitoring system (client and server technologies) are analyzed. Algorithms have been developed that allow processing and outputting OLT data structured according to PON specifics: an algorithm for adding a new OLT and obtaining information about ONUs; an algorithm for displaying information about ONUs in real time; and an algorithm for displaying information about client devices in real time. It is shown that in the process of monitoring a PON network, an important role is played by data obtained in real time—on demand. A PON monitoring system database has been developed with the allocation of the necessary set of domain objects. Its entities and relationships have been determined; and a database scheme as well as code examples have been presented.

Keywords Monitoring · Passive optical network · Database

L. Tokar (✉) · Y. Krasnozheniuk
Kharkiv National University of Radio Electronics, 14 Nauky Avenue, Kharkiv, Ukraine
e-mail: liubov.tokar@nure.ua

Y. Krasnozheniuk
e-mail: yana.krasnozheniuk@nure.ua

D. Ageyev et al. (eds.), *Data-Centric Business and Applications*, Lecture Notes on Data Engineering and Communications Technologies 69,
https://doi.org/10.1007/978-3-030-71892-3_16

1 Introduction

Network control is an important function; therefore, it is often separated from other functions of control systems and implemented by special means. Constant monitoring of a network is necessary to maintain it in operating condition.

The first stage of control is monitoring. At this stage, the procedure of collecting primary data about the network is performed. Monitoring tasks are solved by software and hardware meters, testers, network analyzers, built-in monitoring tools for communication devices, as well as control system agents. The stage of monitoring the network status will reduce the time spent on troubleshooting network problems and ensure continuous and high-quality network operation.

The technology of passive optical networks (PONs) is based on a tree-like fiber-cable architecture with passive optical splitters on the nodes [1]. In this regard, the PON architecture has the necessary efficiency of expanding network nodes and bandwidth, depending on the present and future needs of subscribers.

As part of the PON architecture, there is a main Optical Line Terminal (OLT). It has both Ethernet ports for connecting uplink channels, and output optical ports for transmitting and receiving information to a set of Optical Network Terminals (ONTs) of subscriber devices called Optical Network Units (ONU). The number of ONTs connected to one OLT depends on the power and maximum speed of the transceiver equipment [2]. The main elements of the PON architecture are shown in Fig. 1.

The direct flow at the optical signal level is broadcast. Each ONT reads address fields and extracts from this common flow a part of information designated only for it. All ONTs transmit in the reverse flow at the same wavelength and use the Time Division Multiple Access (TDMA) concept. To exclude the intersection of signals from different ONTs, each of them has its own individual data transfer schedule taking into account a correction for delay associated with the removal of this ONT from the OLT.

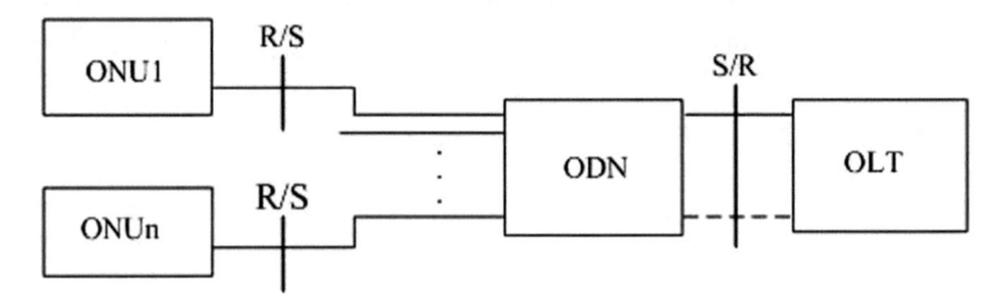

Fig. 1 Basic elements of the PON architecture

1.1 Analysis of Passive Optical Network Topologies

Let us consider main topologies for building optical access networks [3]:

The point-to-point (P2P) topology shown in Fig. 2 does not impose restrictions on the network topology; it can be implemented for any network standard as well as for non-standard solutions.

From the point of view of security and protection of the transmitted information in the P2P connection, maximum security of subscribers' nodes is provided. Since optical cable needs to be laid individually to the subscriber, this approach is the most costly.

1. Central OLT is a device installed in the central office; it receives data from the backbone networks via Service Node Interfaces (SNIs) and forms a direct (downsward) flow to the subscribers' nodes by a PON tree;
2. Optical network terminal (ONT) has, on the one hand, subscriber interfaces, and on the other hand, an interface for connecting a central OLT of the PON; data from the OLT is received by the ONT and transmitted to subscribers through user network interfaces (UNIs), forming a reverse (upward) flow.

The topology of a "tree with passive optical branching", or "point-to-multipoint" (P2MP) is shown in Fig. 3.

Any fiber-optic segment of the tree architecture, covering dozens of subscribers, is connected to one port of the central node; optical splitters are installed in the intermediate nodes of the tree. In general, a splitter has M input and N output ports.

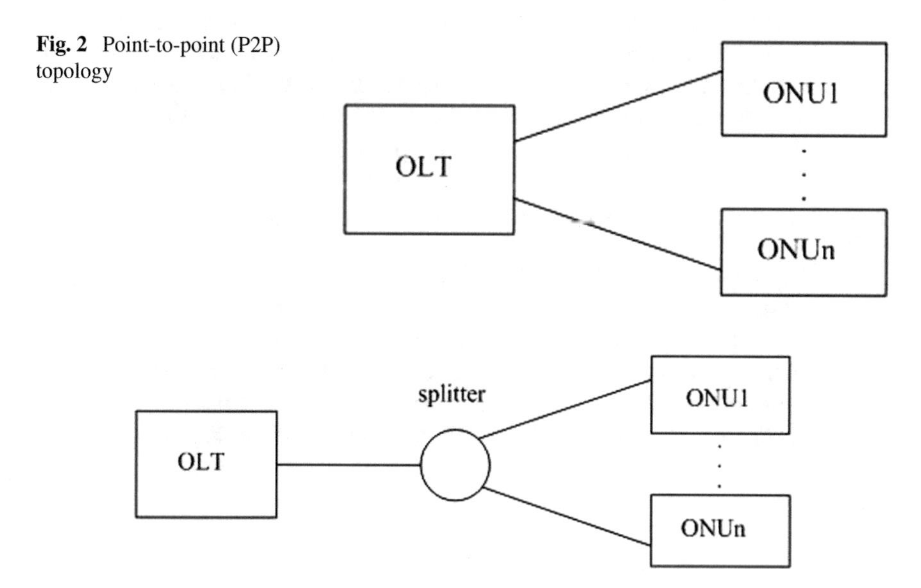

Fig. 2 Point-to-point (P2P) topology

Fig. 3 Point-to-multipoint (P2MP) topology

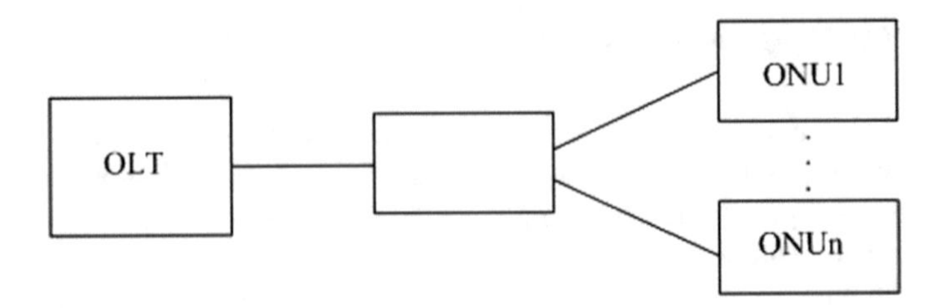

Fig. 4 «Tree with active nodes» topology

Most often, 1xN splitters with one input port are used. 2xN splitters are used in reservation systems with a single fiber.

The network of this topology is more cost-effective than the P2P topology, due to the reduction in the total length of the optical fibers, since only one fiber is used from the central node to the splitter.

The advantages of this topology are:

- no active intermediate nodes;
- saving of optical transceivers in the central node;
- fiber saving;
- easy connection of new subscribers.

The P2MP tree topology allows to optimize the installation of optical splitters based on the actual location of subscribers, the costs of laying optical cable and cable network exploitation.

The tree topology with active nodes (Fig. 4) is a cost-effective solution from the point of view of fiber usage; it complies with the Ethernet standard with a hierarchy of speeds from the central node to subscribers with 1000/100/10 Mbit/s (1000 Base-FL).

However, in each node of the tree, an active device is necessarily installed (in relation to IP networks, it is a switch or router). Optical Ethernet access networks that primarily use this topology are relatively inexpensive. Their disadvantage is the presence of active devices, which require individual power supply, on the intermediate nodes.

Ring topology—the "Ring" (Fig. 5) is more often used in telecommunication networks using the technology of Synchronous Digital Hierarchy (SDH).

It provides for the laying of two optical cables, the information is transmitted by two fibers simultaneously in different directions, which ensures high reliability. However, to connect new subscribers it is necessary to break the ring and insert additional segments. A network with such a topology is difficult to grow.

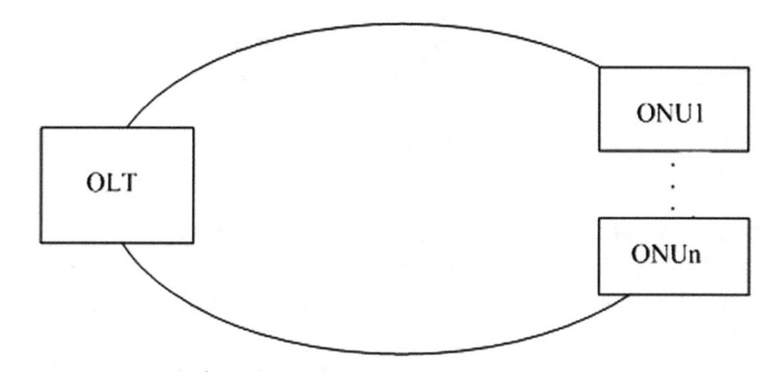

Fig. 5 Ring topology

1.2 Management Tools for Monitoring Equipment in Passive Optical Networks

1.2.1 Purpose and Concepts of SNMP

Simple Network Management Protocol (SNMP) is designed to remotely control network elements: switches, routers, gateways, servers, etc. It allows network specialists or automated systems to obtain information about the operation of network elements and change their configuration. The protocol allows unified management of equipment from various manufacturers operating under different OS and at different layers of the OSI model [4]. The SNMP architectural model is a collection of network management stations and controlled network elements. SNMP is used to exchange information between network management stations and network elements. Programs, the so-called managers, are implemented at network management stations; they monitor and control network elements. A software agent is implemented in the network elements. An agent in SNMP is a processing element that provides managers located at network management stations with access to the values of Management Information Base (MIB) variables and thereby enables them to implement device control and supervision functions.

The main control operations are performed by the manager, and the SNMP agent most often performs a passive role, passing to the manager the value of the accumulated statistical variables on demand. At the same time, the device works with minimal support for the management protocol and uses almost all of its computing power to perform its basic functions as a router, bridge or hub, and the agent collects statistics and values of the device status variables and transfers them to the manager of the control system [5].

The simplicity of SNMP is largely determined by the simplicity of the Management Information Bases, especially their first versions: MIB-I and MIB II. In addition, SNMP itself is also quite simple. The tree structure of the MIB contains the required (standard) subtrees, and it may also contain private subtrees that allow the manufacturer of intelligent devices to control any specific device functions based on specific MIB objects.

In management systems built on the basis of SNMP, the following elements are standardized [6]:

- protocol of interaction between the agent and the manager;
- language for describing MIB models and SNMP messages is the abstract syntactic notation language—ASN.1 (standard ISO 8824: 1987, recommendations ITU-TX.208);
- several specific MIB models (MIB-I, MIB-II, RMON, RMON 2), the object names of which are registered in the ISO standard tree.

SNMP and the closely related SNMP MIB concept have been developed to control Internet routers as a temporary solution. But the simplicity and effectiveness of the solution ensured the success of this protocol, and today it is used in the management of almost any type of equipment and software for computer networks. Although there is a steady trend of using ITU-T standards in the area of telecommunication network management, which includes the CMIP protocol, there are quite a few examples of successful use of SNMP management. SNMP agents are integrated into analog modems, ADSL modems, ATM switches, etc.

1.2.2 SNMP Basics

SNMP is a request-response protocol, that is, for each request received from a manager, an agent must transmit a response. A feature of the protocol is its extreme simplicity—it includes only a few commands:

1. *Get-request* command is used by the manager to get the value of any object from the agent by its name;
2. *GetNext-request* command is used by the manager to retrieve the value of the next object (without specifying its name) when sequentially viewing the table of objects;
3. using *Get-response* command, the SNMP agent passes the response to Get-request or GetNext-request commands to the manager.
   Set command is used by the manager to change the value of any object. Using Set command, the device itself is controlled. The agent must understand the meaning of the values of the object that is used to control the device. Based on these values, the agent performs a real control effect—disconnecting a port, assigning a port to a specific VLAN, etc. Set command is necessary to set up the condition under which the SNMP agent must send an appropriate message to the manager. In addition, the response to events such as: agent initialization, agent restart, disconnection, connection recovery, incorrect authentication and loss of the nearest router, is determined. If any of these events occurs, the agent initiates an interrupt;

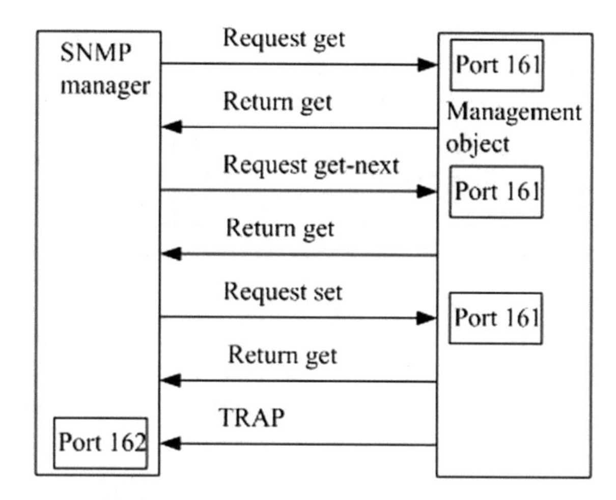

Fig. 6 Request/response SNMP scheme

4. *Trap* command is used by the agent to inform the manager about an emergency;
5. SNMP v.2 adds *GetBulk* command to this set, which allows the manager to get several variable values in one request [7]. Figure 6 shows the SNMP requests/response scheme.

1.2.3 SNMP Message Format

SNMP supports the data transfer between agents and a station that controls the network. SNMP uses the UDP datagram transport protocol, which does not provide reliable message delivery.

SNMP is often considered only as a solution for managing TCP/IP networks. Although SNMP most often works on UDP (it can also work on TCP), it can work on transport network protocols of the OSI stack as well as on MAC layer protocols.

SNMP messages, unlike messages of many other communication protocols, do not have fixed field headers. An SNMP message consists of an arbitrary number of fields, and each field is previously described by its type and size. Any SNMP message consists of three main parts: the protocol version (*version*); the community identifier (*community*), which is used to group devices controlled by a specific manager; and the *data area*, which actually contains the protocol commands described above, object names and their meaning.

The data area is divided into Protocol Data Unit (PDU). Figure 7 shows the format of SNMP messages embedded in UDP datagrams.

Below are the meanings of each field in the message:

1. the "version" field contains a value equal to the SNMP version number minus one;
2. the "password" field ("community"—defines an access group) contains a sequence of symbols, which is a pass during the interaction of the manager and a control object. Typically, this field contains a 6-byte line called "public";

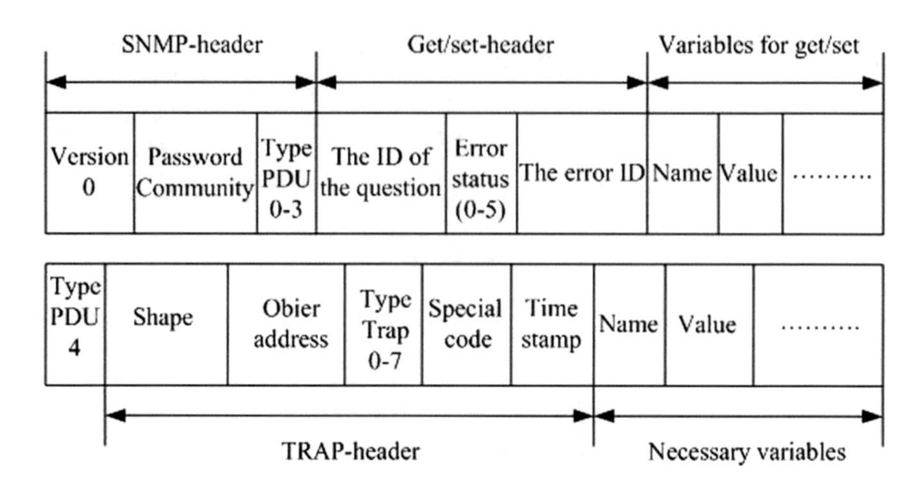

Fig. 7 Format of SNMP messages embedded in UDP datagrams

Table 1 Error status and their description

Error status	Error name	Description
0	Noerror	Everything is alright
1	Toobig	Object cannot wrap a response in one message
2	Nosuchname	Unknown variable specified in operation
3	Badvalue	Set command used an invalid value or incorrect syntax
4	Readonly	Manager tried to change the constant
5	Generr	Other errors

3. for GET, GET-next and SET requests, the value of the request identifier is determined by the manager and returned by the control object in the GET response, which allows pairing requests and responses;
4. the "error status" field is characterized by the number forwarded by the control object, as shown in Table 1.
5. the "error index" field characterizes which of the variables this refers to, and is a variable pointer and is set by the non-zero control object for Badvalue, Readonly and Nosuchname errors. For the TRAP operator (type PDU = 4), the message format is changed. The TRAP types table is presented below (Table 2).
6. The "time label" field contains the number of hundredths of a second (the number of ticks) from the moment the control object is initialized.

1.2.4 SNMP Disadvantages

SNMP is the basis for many management systems, although it has several fundamental flaws, which are listed below:

Table 2 TRAP types

TRAP type	TRAP name	Description
0	Coldstart	Setting the initial status of an object
1	Warmstart	Recovering the initial status of an object
2	Linkdown	Interface has shut down. The first variable identifies the interface
3	Linkup	Interface is turned on. The first variable identifies the interface
4	Authenticationfailture	Message was received from the manager with an invalid password (community)
5	EGPneighborloss	Partner disconnected. The first variable identifies the IP address of the partner
6	Enterprisespecific	Information on TRAP is contained in the "special code" field

1. lack of means of mutual authentication for agents and managers. The only means that could be attributed to authentication is the use of the so-called "community string" in messages. This string is transmitted over the network in an open form in the SNMP message and serves as the basis for dividing agents and managers into "communities", so the agent interacts only with those managers that specify the same character string in the community string field and the string stored in agent memory. This is certainly not a way of authentication, but a way of structuring agents and managers. SNMP v. 2 was supposed to eliminate this drawback, but as a result of disagreements between the developers of the standard, new authentication means are optional, although they appeared in this version;
2. operation through the unreliable UDP protocol. This is how the vast majority of SNMP agent implementations work, which leads to loss of alarm messages (trap messages) from agents to managers, which can lead to poor control. Correcting the situation by switching to a reliable transport protocol with the establishment of connections can lead to loss of communication with the huge number of built-in SNMP agents that are available in the installed equipment in the networks. (CMIP initially works on the reliable OSI stack transport and is not affected by this flaw). Management platform developers are trying to overcome these shortcomings. For example, the HP 0 V Telecom DM TMN platform, which is a platform for developing multi-level management systems in accordance with TMN and ISO standards, has a new SNMP implementation that organizes reliable messaging between agents and managers by independently organizing retransmission of SNMP messages in case of loss.

1.2.5 Management Information Base

Today, there are several standards of management information bases for SNMP. The main ones are the MIB-I and MIB-II standards, as well as the RMON (MIB remote control base version). In addition, there are standards for specific MIB devices of a specific type (for example, MIB for hubs or MIB for modems), as well as private MIBs of specific equipment manufacturers. The original MIB-I specification defined only read operations on variable values. Changing or setting values for an object are part of the MIB-II specification.

The MIB-I version (RFC 1156) defines 114 objects that are divided into 7 groups.

1. System—general data about a device (for example, manufacturer identifier, time of the last system initialization);
2. Interfaces—parameters of the device network interfaces (for example, their number, types, exchange rates, maximum packet size);
3. Address Translation Table—description of the correspondence between network and physical addresses;
4. Internet Protocol—data related to the IP protocol (IP gateways, hosts, statistics on IP packets);
5. ICMP—data related to the ICMP control message exchange protocol;
6. TCP—data related to the TCP, UDP—data related to the UDP (the number of transmitted, received and false UPD datagrams);
7. EGP—data related to the Exterior Gateway Protocol used by the Internet for the routing information exchange protocol (number of messages received with and without errors).

From this list of variable groups, it is clear that the MIB-I standard was developed with a strict focus on managing routers that support the TCP/IP stack protocols.

In MIB-II version (RFC 1213), the set of standard objects was significantly expanded (up to 185), and the number of groups increased to 10. The following are included in the number of objects describing each specific device interface:

ifType	Type of protocol that supports the interface. This object takes the values of all standard data link protocols, for example rfc877-x25, ethemet-csmacd, iso88023-csmacd, iso88024-tokenBus, iso88025-tokenRlng, etc.
IfMtu	Maximum packet size of the network layer that can be sent
ifSpeed	Interface bandwidth in bits per second (100 Fast Ethernet)
ifPhysAddress	Physical address of the port
ifAdminStatus	The desired port status
up	Ready to transmit packets
down	Not ready to transmit packets
testing	In test mode
ifOperStatus	Actual current status of the port same values as ifAdminStatus
ifInOctets	Total number of bytes received by this port from the moment of the SNMP agent last initialization, including service ones

(continued)

(continued)

iflnUcastPkts	Number of packets with an individual interface address delivered to the upper layer protocol
iflnNUcastPkts	Number of packets with the broadcast or multicast interface address delivered to the upper layer protocol
ifInDiscards	Number of packets that were received by the interface turned out to be correct, but were not delivered to the upper layer protocol, most likely due to overflow of the packet buffer or for another reason
ifin Errors	Number of received packets that were not transmitted by the upper layer protocol because of errors detected in them

As can be seen from the description of the MIB-II objects, this database does not provide detailed statistics on typical Ethernet frame errors. Moreover, it does not reflect the change in characteristics over time, which is often of interest to the network administrator. These restrictions were the new MIB standard—RMON MIB, which is specifically focused on collecting detailed statistics over the Ethernet protocol, in addition to supporting such an important function as constructing time dependencies on statistical characteristics by the agent.

2 Stating the Problem of Creating a System for Monitoring and Assessing PON Status

For the monitoring system to work, it periodically polls all available OLTs and monitors important indicators: detecting ONUs on OLTs that are not included in the ONUs database, or registered simultaneously on two OLTs. SNMP management was used to develop an algorithm for adding a new OLT and obtaining information about ONUs.

In management systems based on SNMP, the following elements are standardized: an agent-manager interaction protocol, a language for describing MIB models and SNMP messages (ASN.1 is an abstract syntactic notation language [8]). In addition, several specific MIB models (MIB-I, MIB-II, RMON, RMON 2) are standardized, objects' names of which are registered in the ISO standards tree, as mentioned above [9].

Thus, information about the number and status of ONUs during the process of adding a new OLT should contain: a list of SNMP interfaces connected by ONUs in the PON interface, PON interface numbers, ONU uplink port status and its traffic, status of other ONU ports.

The general procedure for creating a monitoring system consists of certain separate tasks:

- study of the subject area and existing monitoring tools for passive optical networks;
- analysis and comparison of existing monitoring tools for passive optical networks;

- study of requirements for a software product that would satisfy the needs of the system administrator and reduce the time spent on troubleshooting network problems, and thus ensure continuous and high-quality network operation;
- analysis of existing tools for the implementation of software applications, selection of the most rational and optimal set of tools and libraries;
- implementation of a software solution for collecting network equipment data, data analysis and display.

2.1 Tools and Technologies for Creating a Passive Optical Network Monitoring System

2.1.1 Client Technologies

HTML is the fundamental, basic technology of the Internet. Everything that a user sees on the website from the font, the background of the drop-down menu, the slider, was created using the three basic tools—the HTML language, cascading style sheets (CSS) and JavaScript (JS) scenarios.

HTML forms the structure of the page (skeleton), CSS defines its appearance, and JS gives the dynamics. Hypertext Markup Language (HTML) is the main building tool for web pages used to create and visualize web pages. It defines the structure and describes the content of the web page in the structured form.

An HTML page is an ordinary text document that uses special operators—tags or another name for descriptors to indicate in which form a text or another element will be displayed in a browser window. Examples of such operators are <img>, <title>, <p>, <div>, <table>, etc.

HTML allows to create text blocks on a website page, add images to them, organize tables, add sound to the website design, organize hyperlinks with transition to other sections of the website or Internet resources, and put all these elements together. Using HTML, we can create both a static and dynamic site. Pages created with HTML only have .html extensions.

A hyperlink is a basic functional element of an HTML document that implements a link between a specific object of a web page and another object. For a hyperlink, either a fragment of text or a graphic object can be used, and the connection itself can be established both between objects of one site and between objects that are located on different Internet sites [10].

HTML is a language that is only interpreted; therefore, in order to execute the code, it does not need to be compiled. The language interpreter is embedded in the browser and it "compiles" the code directly when a document is being opened. If an error is detected in the page code, the interpreter usually does not give an appropriate warning, but simply ignores the error, which can ruin the appearance of the loaded page. To prevent this, developers should be careful when writing HTML code and carefully test the results of their work.

CSS (Cascading Style Sheets) is a technology for describing the appearance of a document created using HTML.

CSS is used to assign certain features to elements of an HTML page: color, font, layout on the page, etc. Before CSS appeared, the design of elements was indicated directly in the HTML code of a page. However, with the emerging of CSS, a fundamental separation of the structure and description of a document became possible. Due to this distribution, it became possible to easily apply a single design style for several pages of the site, as well as quickly change this design [10].

CSS benefits:

1. use of several page design options for different viewing devices, for example, for a computer, tablet or phone;
2. reducing the loading time of site pages by transferring data description rules to a separate CSS file. In this case, the browser only downloads the document structure and the data contained on the page. A CSS file with the rules for describing this data is downloaded by the browser only once and stored in the browser cache;
3. simplicity of subsequent design changes. There is no need to fix every page, just change a few rules in the CSS file;
4. additional design options. For example, using CSS rules, we can apply text wrapping around a certain block or make the menu fixed in a certain place when scrolling.

JavaScript ("JS") is a complete dynamic programming language that, when applied to an HTML document, can provide dynamic interactivity on websites. JavaScript has extremely many implementations. JavaScript is a fairly compact and flexible language. The developers have provided a wide variety of tools that complement the core of the JavaScript language and open up a huge amount of additional functionality with minimal effort. Among them:

- program interfaces (APIs), for browsers—APIs built into browsers that provide functionality such as dynamic HTML creation and the use of CSS styles, the collection and processing of video streams from a user's webcam, generation of 3D graphics and audio samples;
- third-party APIs that allow developers to integrate the functionality of other providers, such as Twitter or Facebook, into their sites;
- third-party frameworks and libraries that can be applied to HTML to speed up the creation of sites and applications.

An important feature of JavaScript is object orientation. The programmer has access to numerous objects, such as documents, hyperlinks, forms, frames, and more. Objects are characterized by descriptive information (properties) and possible actions (methods) [11].

2.1.2 Server Technologies

Server programming is used to process user actions on dynamic complex projects such as search engines, email, forums, online stores, etc.

In these cases, the browser receives data from the visitor and sends it to the web server, which:

- accepts a request for downloading files (web pages, style sheets, graphic images, films, sounds, archives, executable files, etc.);
- searches for files on the disks of the server computer;
- forwards tasks to the appropriate programs that perform additional actions on files;
- generates results of processing server programs in HTML code;
- sends a generated web page to the browser in HTML code.

To perform the tasks of the web server, special programs are used that work together with the web server on the same server computer. They are called server programs, do not have a user interface and only "communicate" with the web server, receive user input from it and return the result to it. In this they are fundamentally different from client programs that work directly with the user. There are also multi-functional programs that perform functions not only from a web server, but also from FTP or a mail server.

Pages that are formed by server programs are called dynamic, in contrast to static pages in the .html format.

Server programs are divided into:

1. web server extension (ISAPI format applications, Apache modules, etc.). These are server programs built into the web server;
2. active server pages (ASP, PHP, etc.). These are ordinary static web pages, saved as files, which, in addition to the usual HTML code, contain commands processed either by the web server itself or its extensions;
3. server-side scripts written in an interpreted language (PHP, Python, Java, Ruby). Scenarios of data processing for various user actions on the site.

A web server extension is a type of server program that represents a library which implements the logic of a server program. Such libraries are embedded in the web server program and work as its integral part.

The Microsoft Internet Information Web Server extension is created in the form of DLLs (the extension is in the format of ISAPI (Internet Server Application Programming Interface). The format of the Apache extension modules is called Apache modules.

The goal of web server extensions is to conserve system resources. To process all user data sets, only one extension instance is launched, which consumes significantly less resources than many running programs.

Active server pages are regular web pages containing specific server scripts (program code in HTML code) and are executed by the web server itself or its extensions. The advantages of active server pages are ease and speed of writing, and ease of debugging.

Server-side scripts are similar to active server pages, however, they are "clean" program code without HTML fragments. The code interpreter is introduced as a web server extension. Scripts are usually written in the programming languages PHP, Java, Perl, Python, and Ruby. In fact, we can write scripts in any programming language for which there is an interpreter.

Server-side scripts consume significantly more system resources than active pages, because a separate copy of the interpreter is launched to process each user data set, and the interpreter, in turn, spends a lot of resources for processing the script. And yet, despite this, scripting is the most popular way to create server programs.

Ruby is an interpreted, fully object-oriented programming language with clear dynamic typing. The language is highly effective in program development and incorporates the best features of Perl, Java, Python, Smalltalk, Eiffel, Ada and Lisp. Ruby combines Perl-like syntax with the object-oriented approach of the Smalltalk programming language. Also, some features are borrowed from the Python, Lisp, Dylan, and CLU programming languages. The multi-platform implementation of the Ruby language interpreter is distributed as free software. The source code for the project is distributed under the BSD ("2-clause BSDL") and "Ruby" licenses, which refers to the latest version of the GPL and is fully compatible with GPLv3.

Ruby is an object-oriented programming language [12]. Each data type is an object, including types and classes, which in many other languages are implemented as primitives (such as "integer" or "null"). Each function is a method.

Ruby does not support multiple inheritance, but instead has a powerful Mix In mechanism. All classes (directly or through other classes) are derived from the Object class, therefore, any object can use the methods defined in it (for example, class, to_s, nil?). Procedural style is also supported, but all global procedures are implicitly private methods of the Object class.

Ruby features:

1. concise and simple syntax; there is often influence of Hell, Eiffel and Python;
2. can handle exceptions in the Java and Python style;
3. allows to reassign operators;
4. it is a fully object-oriented programming language. All data in Ruby are objects in the Smalltalk sense. The only exception is control constructions, which in Ruby, unlike Smalltalk, are not objects. For example, the number "1" is an instance of the Template: RDoc class. We can also add methods to a class and even to a specific instance during program execution;
5. multiple inheritance is not supported, however, the concept can be used instead.
6. contains an automatic garbage collector. It works for all Ruby objects, including external libraries;
7. creating extensions for Ruby in C is very simple due to garbage collection as well as a simple and convenient API;
8. supports loops with full binding to variables;

9. supports a block of code (the code is taken in {…} or do … end). Blocks can be used in methods or turned into loops;
10. integer variables in Ruby are automatically converted between the Fixnum (32-bit) and Bignum (more than 32 bits) types depending on their value, which allows to perform integer mathematical calculations with infinite accuracy;
11. does not require preliminary declaration of variables, although for the interpreter it is desirable the variable to be assigned an empty value nil (then the interpreter knows that the identifier points to a variable, and not to the name of the method). The language uses simple conventions to define the visibility scope. Example: just var is a local variable; @var is an instance variable (member or field of a class object); @@var is a class variable; $var is a global variable;
12. many programming patterns are directly implemented in the Ruby language, for example, a "singleton" can be implemented by adding the necessary methods to one specific object;
13. can dynamically load extensions, if the operating system allows it;
14. transferred to many platforms. The language was developed on GNU/Linux, but it works on many versions of Unix, DOS, Microsoft Windows (partly, Win32), Mac OS, BeOS, OS/2, etc.

Ruby on Rails is an object-oriented software framework for creating web applications written in the Ruby programming language. Ruby on Rails provides a Model-View-Controller framework for web applications and also integrates them with a web server and database server.

Ruby on Rails defines application development principles:

- applications should not define their own architecture, since they use a ready-made Model-View-Controller framework;
- Ruby language allows the use of notation, which is easy to read, to determine the semantics of web applications (such as relationships between tables in a database);
- Ruby on Rails provides reuse mechanisms to minimize code duplication in a web application (don'T Repeat Yourself principle);
- by default, Convention over configuration are used, which are typical for most web applications. An explicit configuration specification is needed only in non-standard cases.

The main components of Ruby on Rails applications are a *model*, a *view* and a *controller*. Ruby on Rails uses the REST style of building web applications.

The *model* provides the rest of the application components with an object-oriented display of data (such as a product catalog or list of orders). Model objects can load and save data in a database, as well as implement business logic. The ActiveRecord library is used by default in Rails 3 and above to store model objects in a relational DBMS. The competing analogue is DataMapper. There are plugins for working with non-relational databases, for example Mongoid for working with MongoDB.

The *view* creates a user interface using data received from the controller. The view also passes the user's request for data manipulation to the controller (as a rule, the view does not directly modify the model). In Ruby on Rails, the view is described using ERB templates. They are HTML files with optional Ruby code snippets (Embedded Ruby or ERb). The result generated by the embedded Ruby code is included in the template text, after which an HTML page is returned to the user. In addition to ERB, it is possible to use about 20 more template engines, including Slim, Haml.

The *controller* in Rails is a set of logic that runs after a server receives an HTTP request. The controller is responsible for calling the methods of the model and starts the formation of the view. The correspondence of the Internet address with the action of the controller (route) is set in the config/routes.rb file.

The controller in Ruby on Rails is a class inherited from ActionController:: Base. The open methods of the controller are the so-called actions. An action often corresponds to a separate view. For example, at the request of the user admin/list, the list class AdminController method will be called and then the list.html.erb view will be used.

2.1.3 Data Base

PostgreSQL is an object-relational database management system (ORDBMS) based on POSTGRES version 4.2. PostgreSQL is an open source product that is a descendant of the original code written in Berkeley. PostgreSQL supports most of the SQL standard and offers many modern features [13]:

- complex requests;
- foreign keys;
- representations (views);
- transactional integrity;
- multi-version concurrency control.

PostgreSQL supports most of the features of the SQL standard: 2011, ACID-compatible and transactional (including most DDL statements) avoids the blocking problem using the parallel access Multi Version Concurrency Control (MVCC) mechanism, provides immunity to dirty reading; manages complex SQL queries using a variety of indexed methods that are not available in other databases. It has updating materialized representations, triggers, foreign keys. It supports functions and procedures, and other extension options, and has many extensions written by third parties [14]. In addition to the ability to work with the main proprietary and open source databases, PostgreSQL supports migration from them, through its great support for the SQL standard and affordable migration tools.

Branded extensions in databases such as Oracle can be emulated with built-in and third-party open source compatibility extensions.

PostgreSQL is cross-platform and it runs on many operating systems, including Linux, FreeBSD, macOS, Solaris, and Microsoft Windows. Starting with Mac OS X 10.7 Lion Server, PostgreSQL is the standard default database, and PostgreSQL client tools come with the desktop version. The vast majority of Linux distributions have

Table 3 PostgreSQL 9.4.5 limitations

Maximum database size	No restrictions
Maximum table size	32 TB
Maximum record size	1.6 TB
Maximum field size	1 GB
Maximum records in a table	No restrictions
Maximum fields per record	250–1600 depending on field types
Maximum indexes in a table	No restrictions

PostgreSQL available in supported packages. This is free and open source software. Currently (version 9.4.5), there are limitations in PostgreSQL (Table 3).

The advantages of PostgreSQL are:

- high-performance and reliable mechanisms of transaction and replication;
- an extensible system of built-in programming languages, and there is also support for loading C-compatible modules;
- support of many programming languages: C\C++, Java, Perl, Python, Ruby, ECPG, Tcl, PHP and others;
- inheritance;
- easy extensibility.

2.1.4 List of Libraries and Solutions

1. *Trailblazer* provides new high-level abstractions for Ruby frameworks. It gently enforces encapsulation and an intuitive structure code, and also gives an object-oriented architecture. Its principles are:
   - all business logic is enclosed in operations (service objects);
   - additional validation objects (Reform or Dry-validation) in the operation deserialize and validate data entry;
   - policies block unauthorized users from starting an operation;
   - controllers instantly delegate everything into operations;
   - models exclusively define associations and areas. No business code, validation or callbacks should be found here.
2. *Puma* is a simple and fast server for Ruby/Rack applications. Puma is intended for use in the development process, and in the work environment;
3. *Sorcery* is a library for implementing authentication, authorization and user profile processing;
4. *Ruby-snmp* is a library, which implements the Simple Network Management Protocol (SNMP). It is implemented in pure Ruby, so there are no dependencies on external libraries such as Net-SNMP.

2.2 *Algorithmic Support*

There are algorithms, which have been developed for PON documenting and convenient monitoring of its effective operation. These algorithms allow processing and output of OLT data, which are structured according to the specifics of PON and the OLT model, in real time [15].

2.2.1 Algorithm for Adding a New OLT and Getting Information About ONUs

The block diagram of the algorithm for adding a new OLT and obtaining information about ONU is shown in Fig. 8 [16].

At the first stage, important data about a new OLT is received from the user: OLT IP addresses, locations, community name, coordinates.

In the second step, an SNMP request is sent to verify that the OLT is online and responds to the request. After this request, all information about this OLT is written to the database, but if the OLT is offline, then the "offline" flag is added. If the OLT is online, an SNMP request is sent to get all the information on the ONUs that belong to this OLT.

The third stage allows to create an ONU array, in the cycle of which a check is performed for each element in the array (for each ONU). After that, an SNMP request is sent to get all the information about a specific ONU and the ONU is written into the database.

2.2.2 Real-Time ONU Information Display Algorithm

An important role in the network monitoring process is played by data obtained in real time—on demand. Figure 9 shows a block diagram of an algorithm for obtaining information about a specific ONU [16].

The presented algorithm displays information about the signal levels of OLT-ONU and ONU-OLT, as well as distance data of OLT-ONU.

The first step is to obtain a unique ONU identifier from the HTML request. After that, this ID contains information about this ONU in the database and the OLT to which this ONU belongs.

At the second stage, having the IP address of the OLT, an SNMP request is made to find an OLT interface to which this ONU connects.

At the third stage, an SNMP request is made to a specific OLT interface and a check is made on whether the ONU online. If the ONU is offline, then the data about this ONU is viewed in the database and displayed to the user with a note that the ONU is offline. If the ONU is online, then an SNMP request is made for the signal level from OLT to ONU and from ONU to OLT, and then the next SNMP request is

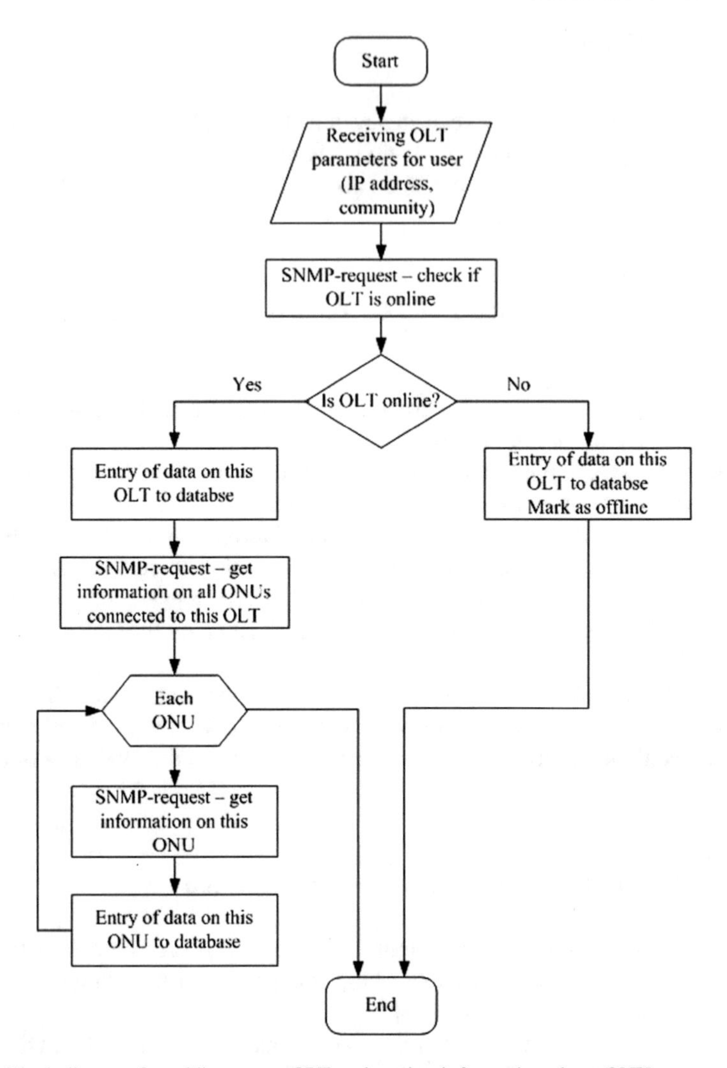

Fig. 8 Block diagram for adding a new OLT and getting information about ONU

made to find the distance from OLT to ONU. If there has been a change in data, the new data is written to the database and displayed to the user.

2.2.3 The Algorithm for Displaying Information About Client Devices in Real Time

Figure 10 shows the block diagram of the algorithm for obtaining information about connected devices on the client side [16].

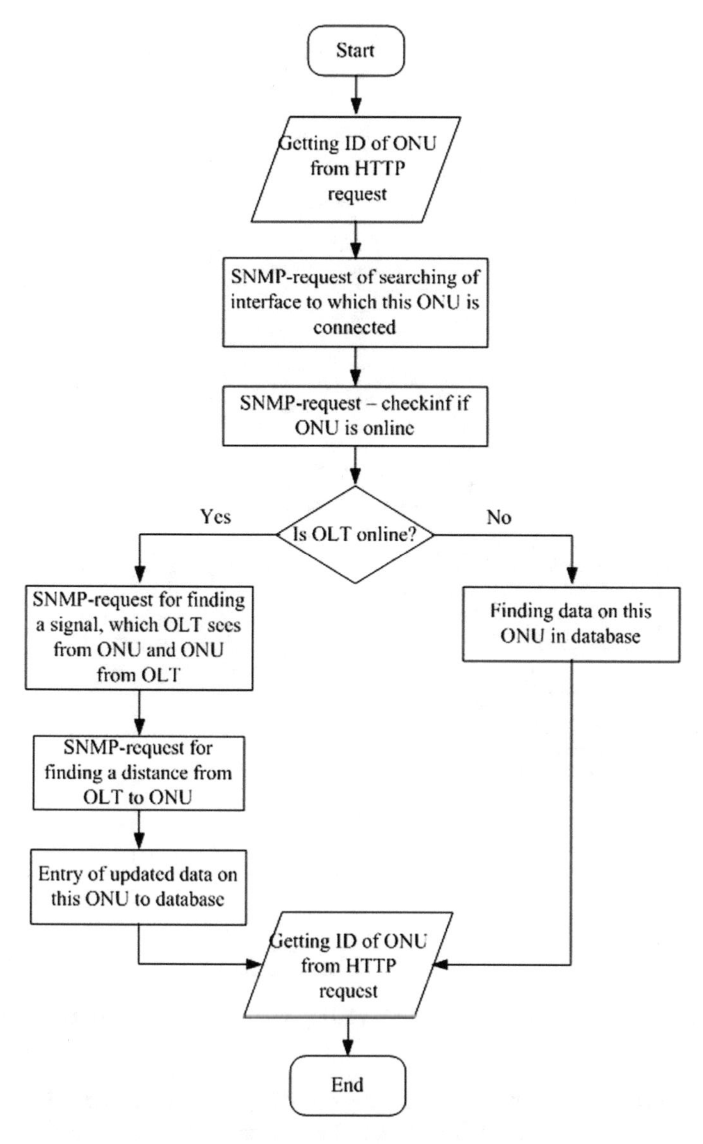

Fig. 9 Block diagram of the algorithm for displaying information about ONU in real time

At the first stage of the algorithm, a unique ONU identifier should be obtained from the HTML request. After that, the database contains information about this ONU, and then the OLT, to which this ONU belongs, is located.

The second stage is aimed at implementing the following SNMP requests:

- a request to find an OLT interface to which this ONU connects;
- a request for a specific OLT interface and checking whether the ONU is online.

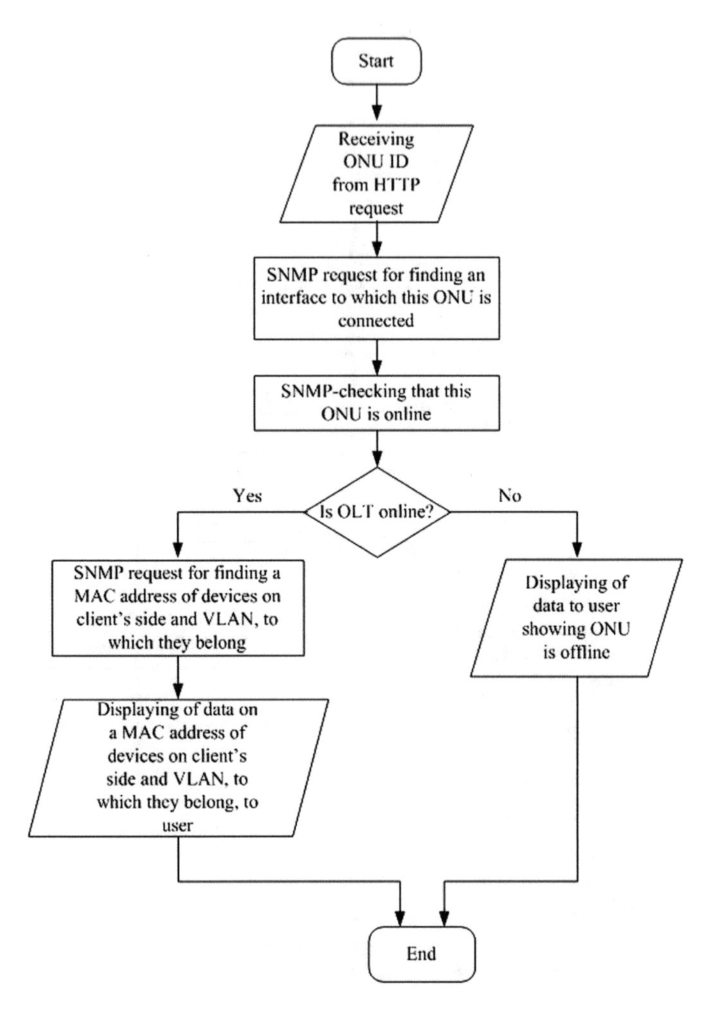

Fig. 10 Block diagram of the algorithm for displaying information about client devices located

If the ONU is offline, then this information is displayed to the user. If the ONU is online, then an SNMP request is first made to find the MAC addresses of the devices on the client side and the VLAN to which they belong, and then this data is displayed to the user.

3 Creation of a Passive Optical Network Monitoring System

3.1 Basic Software Requirements

Creating software is a process of determining the architecture, components, interfaces of other system characteristics and the final result.

Let us dwell in more detail on the software development processes, which together should provide a path from understanding the needs of a customer to transferring the ready product. On this path, a number of specific tasks are distinguished [17]:

- definition of requirements;
- operation and maintenance of a ready software system
- engineering requirements;
- analysis of requirements.

The stages of development, implementation and testing include collection and analysis of customer requirements by the responsible party and their presentation in a notation, which is understandable for both the customer and the responsible party. Typical scheme of the above tasks sequence is given in Fig. 11.

Figure 11 shows the so-called waterfall or cascade model, in which the assumption is true that each of the operations will be performed so carefully that after its completion and transition to the next operation, a return to the previous one is not necessary [18]. Such a model has the main drawback, consisting in the inability to repeatedly refine and change during development or after its completion and testing.

When changing the conditions of the system use or repeatedly returning to the stage of formulating requirements, a spiral development model is valid from any stage of operation, in which each coil of the spiral corresponds to one of the development versions (Fig. 12). At each stage of development, the need for changes is analyzed,

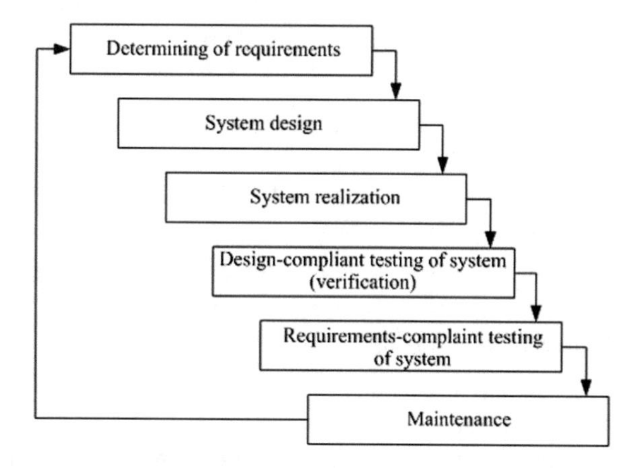

Fig. 11 Waterfall model of the software system life cycle

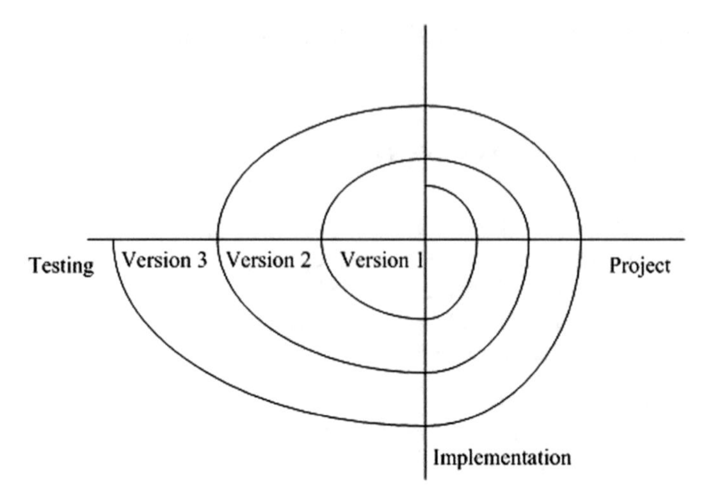

Fig. 12 Spiral model of development life cycle

and the introduction of changes at any stage necessarily begins with the introduction of changes to previously fixed requirements [18].

The process of formulating system requirements consists of several subprocesses:

- determining the goals and objectives of the system, which are formulated by the customer;
- development of general knowledge on the customer's problem area;
- study of departmental customer standards regarding organizational requirements, functioning environment of a future system, its performance and resource capabilities.

The product of the requirements development process is an informal description of these requirements. This description is the input to the next requirements engineering process—analysis of requirements. The first step in the analysis is the classification of requirements. Many of the requirements collected can be distributed between the two main categories:

- those that reflect the capabilities that the system must provide are called *functional requirements*;
- reflecting restrictions associated with the functioning of the system, called *non-functional requirements.*

There are several classes of non-functional requirements that are essential for most software systems, expressing limitations relevant to many problem areas. Among them are the following:

- confidentiality requirements;
- fault tolerance;
- number of clients simultaneously having access to the system;

- security requirements;
- waiting time for the system response;
- performing qualities of the system (limitations on memory resources, system response speed, etc.).

Functional requirements are associated with the semantic features of the problem area within which development is carried out. The product of the analysis process is a constructed model of the problem, focused on its understanding, which the responsible party must achieve before starting system design.

3.2 *Database Design for PON Monitoring System*

Creating a database should begin with its design (development). As a result of the design, the structure of the base should be determined, that is, the composition of the tables, their structure and logical connections. The structure of a relational table is determined by the composition of the columns, their sequence, the data type of each column and their size, as well as the table key.

3.2.1 System Analysis of the Subject Area

From the point of view of designing a database in the framework of system analysis, it is necessary to carry out a detailed description of the objects in the subject area and the real relationships between the described objects [19].

The functional approach of system analysis implements the principle of moving "from tasks" and is applied when the functions of a certain group of individuals and task complexes are known in advance, and a database is created to serve the information needs of them. In this case, one can clearly distinguish the minimum required set of domain objects that should be described.

For monitoring a passive optical network, the following are distinguished:

1. User—a user of the system that will create an OLT;
2. OLT—the main object of the system, which belongs to the User who created it and can have many ONUs that are connected to it;
3. ONU—a secondary system object, which belongs to the OLT to which it is connected and has many records about its characteristics in Signals History;
4. Signals History—a table for recording the history of signals that were received from the ONU.

3.2.2 Database Entities and Relationships Between Them

A *User entity* is a system user who logs into the system and creates an OLT. The main attributes of the User entity and their data types are shown in Table 4.

Table 4 The main attributes of the User entity and their data types

Attribute	Data type	Description
id	UUID (universally unique identifier)	Unique identifier of the object
email	String	Email address of the user, unique in the system
first_name	String	Username
last_name	String	Last name of the user
crypted_password	String	Encrypted password
salt	String	"Salt" to decrypt the password
created_at	Timestamp	Profile creation date
updated_at	Timestamp	Profile update date

User has the following relationships:

- one-to-many relationship with the OLT entity (i.e., one User has many OLTs);
- OLT has a one-to-one relationship with User (i.e., one OLT belongs to one User).

An *OLT entity* is the primary object of the system. The main attributes of the OLT entity and their data types are shown in Table 5.

OLT has the following relationships:

- one-to-many relationship with ONU entities (i.e., one OLT has many ONUs);
- ONU has a one-to-one relationship with the OLT (i.e., one ONU belongs to one OLT);

Table 5 The main attributes of the OLT entity and their data types

Attribute	Data type	Description
id	UUID (universally unique identifier)	Unique identifier of the object
user_id	String	Unique identifier of the user who owns the OLT
ip	String	OLT IP address
place	Text	Location OLT
read_only_community	String	Name of the community that has the right to read data
write_community	String	Name of the community that has the right to write data
last_activity	Timestamp	Date when OLT was last online
latitude	String	Latitude
longitude	String	Longitude
comment	Text	Comment
sfp_count	Integer	Sfp number

Table 6 The main attributes of the ONU entity and their data types

Attribute	Data type	Description
id	UUID (universally unique identifier)	Unique identifier of the object
olt_id	String	Unique identifier of the OLT to which the ONU belongs
interface_name	String	Name of the OLT interface to which the ONU is connected
mac_address	String	MAC address ONU
last_activity	String	Date when ONU was last online
latitude	String	Latitude
longitude	String	Longitude
comment	Text	Commentary
pwr	String	Signal strength OLT
deleted	Boolean	Flag whether ONU has been deleted

- one-to-one communication with User (i.e., one OLT belongs to one User).

An ONU entity is a secondary entity of the system. The main attributes of the ONU entity and their data types are shown in Table 6.

The ONU has the following relationships:

- one-to-one relationship with the OLT entity (i.e., one ONU belongs to one OLT);
- one-to-many relationship with the Signals History entity (i.e., one ONU has many Signals History);
- Signals History has a one-to-one relationship with the ONU. (i.e., one Signals History belongs to one ONU).

The Signals History entity is a table for recording the history of signals received from the ONU. The main attributes of the Signals History entity and their data types are shown in Table 7.

Signals History has the following relationships:

- one-to-one relationship with the ONU entity (i.e., one Signals History belongs to one ONU);

Table 7 The main attributes of the Signals History entity and their data types

Attribute	Data type	Description
id	UUID (universally unique identifier)	Unique identifier of the object
onu_id	String	Unique identifier of the ONU to which this signal belongs
pwr	String	Signal strength OLT
created_at	Timestamp	Record date
updated_at	Timestamp	Record update date

Fig. 13 Database diagram

- ONU has a one-to-many relationship with the Signals History entity (i.e., one ONU has many Signals History).

  Figure 13 shows a diagram of the entire database and relationships between tables

3.2.3 Code Examples

Creating an OLT. The controller receives parameters from a user, invokes the operation, and passes all the parameters to it. After the operation is completed, the controller is redirecting to other pages.

Figure 14 shows the part of the controller code that is responsible for invoking operations and redirecting.

The operation code, which is responsible for checking the input data, creating an OLT, writing to the database and finding information about the ONU, is shown in Fig. 15

The function that makes an SMNP request that OLT is on-line is shown in Fig. 16.

```
1 def create
2   run Olt::Create, params: params.merge(user: current_user) do |op|
3     flash[:success] = "Your OLT was successfully created!"
4     redirect_to show_olt_path(op.model) and return
5   end
6   flash.now[:error] = @form.errors.full_messages
7   render action: :new
8 end
9
```

Fig. 14 Controller code

```
1  class Olt
2    class Create < Trailblazer::Operation
3      include Model
4      model Olt, :create
5
6      contract do
7        property :user_id
8        property :ip
9        property :place
10       property :read_only_community
11       property :write_community
12       property :last_activity
13       property :latitude
14       property :longitude
15       property :comment
16       property :sfp_count
17     end
18
19     def process(params)
20       user = params[:user]
21       contract.user_id = user.id
22       validate(params['olt']) do |f|
23         f.save
24         alive = SnmpRequest.check_olt_alive(model)
25         model.last_activity = alive ? Time.now.utc : Time.new(2000, 1).utc
26         model.save
27         if alive
28           Onu::CollectAllOnusInfo.(olt: model)
29         end
30       end
31     end
32   end
33 end
```

Fig. 15 Operation of creating an OLT

```
1  def self.check_olt_alive(olt)
2    manager = SNMP::Manager.new(:host => olt.ip, :port => DEFAULT_SNMT_PORT, :community => olt
       .read_only_community)
3    alive = false
4    response = manager.walk("1.3.6.1.2.1.1.3") { |vb| alive = vb.value.present?}
5    return alive
6  end
7
```

Fig. 16 The function that makes an SMNP request to check whether an OLT is on-line

A function that makes an SMNP request to find the distance from the OLT to the ONU is given in Fig. 17.

```
1  def self.get_distance_to_onu_by_interface(olt, onu_int)
2    manager = SNMP::Manager.new(:host => olt.ip, :port => DEFAULT_SNMT_PORT, :community => olt.read_only_community)
3    response = manager.get(["1.3.6.1.4.1.3320.101.10.1.1.27." + onu_int.to_s])
4    distance = []
5    response.each_varbind {|vb|  distance.push(vb.value)}
6    distance.first
7  end
```

Fig. 17 A function that makes an SMNP request to find the distance from the OLT to the ONU

4 Conclusions

It is shown that the performance of PON elements is conditioned by timely monitoring using effective protocols and procedures. The architecture of a passive optical network is considered. To solve the problem of PON monitoring, the main variants of topologies, their advantages and disadvantages are analyzed.

For remote control of network elements, obtaining information about the operation of network elements and changing their configuration, equipment management tools are considered. The stages of the monitoring system using SNMP, a database of control information based on the MIB-I, MIB-II, RMON MIB standards are justified. It is shown that the use of SNMP to develop the ONU status algorithm with the basic MIB-I and MIB-II standards for control information databases is focused on obtaining information on the operation of network elements and collecting detailed statistics.

The general procedure for creating a monitoring system, consisting of individual specifics tasks, is formulated. Tools and technologies for creating a monitoring system for a passive optical network (client and server technologies) have been analyzed. The necessity of choosing a PostgreSQL database is justified; its advantages are shown. The libraries and solutions used for designing a monitoring system within the selected database are presented. There have been developed the algorithms that allow processing and outputting OLT data structured according to the PON specifics: an algorithm for adding a new OLT and obtaining information about ONUs; an algorithm for displaying information about ONUs in real time; an algorithm for obtaining information about connected devices on the client side.

Based on the proposed algorithms, a database was developed for the PON monitoring system with the allocation of the necessary set of domain objects, its entities and relationships were determined, a database diagram and code examples were given.

References

1. Ubaidullaev RR (2000) Fiber-optic networks. ECO-TRENDZ, Moscow, 267 p
2. Gaskevich E, Ubaidullaev R (2002) PON—broadband multiservice access network. TeleMultiMedia 2(12):29–32
3. Olifer VG (2006) Computer networks. Principles, technologies, and protocols. Peter, Saint Petersburg, 992 p
4. Khakimov AKh, Baranov SA (2016) Application of the SNMP protocol in the radio receiver control system proceedings of the north caucasus branch. Moscow Tech Univ Commun Inform 1:261–264
5. Golubintsev AV, Myasnikova AI, Legkov KE (2015) Architectural principles of automated systems organization management of special-purpose infocommunication networks. High-Tech Technol Space Earth Expl 7(4):16–23
6. Faraonov AV, Mikhailov RV (2020) Analysis of the SNMP protocol. World Science: Problems and Innovations, collection of articles of the XLI International scientific and practical conference, Publishing House: Science and Education, Penza, pp 21–24

7. How SNMP works (2014) https://technet.microsoft.com/en-us/library/cc783142(v=ws.10).aspx
8. Specification of abstract syntax notation one (ASN.1) (1988) ITU-T recommendation X-208, 74 p
9. Sokolov SA, Stockipny AL (2005) Multi-agent system architecture for the solution tasks for monitoring and analyzing the telecommunications network. Sistemi Obrobki Informatsii, Kharkiv 8(48):150–155
10. Hogan B (2014) HTML5 and CSS3. Web development based on new generation standards, 2nd edn. SPb, Petnr, 320 p
11. Flanagan D (2018) JavaScript. Detailed guide. SPb, Symbol, 1080 p
12. Olsen R (2011) Eloquent Ruby. Addison–Wesley Professional, New York, 448 p
13. Morgunov EP (2018) SQL. Basic course. Postgres Professional, Moscow, 257 p
14. Murav'ev KA, Terekhov VV (2017) Software and hardware complex for monitoring distributed telecommunications systems. In: Proceedings of the international symposium "reliability and quality", Moscow, vol 1, pp 324–329
15. Wilson E (2002) Monitoring and analysis of networks. Methods for detecting faults. Lori, Moscow, 386 p
16. Derevianko LS, Tokar LO, Krasnozheniuk YO (2019) Approach to algorithmic control of optical network nodes in monitoring tasks. Problemi telekomunìkacìj 1(24):113–122
17. Vigers K (2016) Development of software requirements. BHV-Petersburg, Saint Petersburg, 736 p
18. Popular Software Development Life Cycles (2018) https://training.qatestlab.com/blog/technical-articles/popular-software-development-life-cycles
19. Bogachev KG, Subbotin DV, Ivanin AN (2017) Statement of the problem of forming a control system for telecommunications equipment of various nomenclature used on special-purpose communication networks. Almanac Mod Sci Educ Tambov 4–5(118):24–29

Statistical Analysis and Optimization of Telecommunications Company Operating Business Processes

Olga Malyeyeva, Viktor Kosenko, Yurii Parzhyn, and Viktoriia Nevliudova

Abstract The object of research in this paper is the operational business processes of a telecommunications company. To analyze the implementation of business processes, the correspondences between the processes and the elements of the organizational structure that will allow controlling the correctness of the processes are described their correspondence to real goals, objectives and performers. The use of cluster analysis as a classification method is proposed. As a result, the entire set of processes is divided into several clusters based on the indicator of the excess of the work execution time, which makes it possible to make individual decisions on the optimization of work from each cluster. The overall runtime of business processes by network planning methods is adjusted. As a result, the manager has the opportunity to assess how productive and optimal the operational activities of this company are and to make the most profitable solution option in the field of telecommunications services management. The proposed approach allows improving the efficiency of the company's operations based on the optimization of business processes.

Keywords Telecommunication company · Business processes · Cluster analysis · Optimization · Planning

---

O. Malyeyeva (✉)
National Aerospace University "Kharkiv Aviation Institute", Kharkiv, Ukraine
e-mail: o.maleyeva@khai.edu

V. Kosenko · V. Nevliudova
State Enterprise "Kharkiv Scientific-Research Institute of Mechanical Engineering Technology", Kharkiv, Ukraine
e-mail: viktor.kosenko@nure.ua

V. Nevliudova
e-mail: d_tapr@nure.ua

Y. Parzhyn
National Technical University "Kharkiv Polytechnic Institute", Kharkiv, Ukraine

D. Ageyev et al. (eds.), *Data-Centric Business and Applications*, Lecture Notes on Data Engineering and Communications Technologies 69,
https://doi.org/10.1007/978-3-030-71892-3_17

1 Introduction

Competition in the field of telecommunications services necessitates the search for ways to optimize the company's activities. It is possible to change the situation for the better by making the management of the company flexible and mobile, responsive to the changes in the external environment and the internal state of the organization, that is, process-oriented. Telecommunications companies (TCC) can achieve this goal by optimizing and reengineering core business processes (BP). Reengineering here means the use by the management of a company of a set of methods for planning and optimizing a business in accordance with its goals. The main object of study in this work is operating BP TCC. Since the company's profitability and efficiency largely depend on its internal organization, first it is necessary to optimize the internal structure and processes.

2 Analysis of Publications and the Formulation of the Research Problem

Managing business processes of a company that is adequate to objective conditions is a creative task that does not boil down to using typical management structures (or systems) that have proven themselves in certain fairly limited business conditions [1, 2].

For effective management of business processes it is necessary to conduct measures for their optimization. The construction of a process-oriented management system of TCC involves the management cycle processes that consist of the three main elements [3]:

- process description,
- automation (regulation) of the process,
- performance monitoring and process evaluation.

The introduction of a full cycle of business process management and maintaining its efficiency allows for a significant increase in the efficiency of the company [4]. Therefore, the description and automation of the company's business processes will not bring the expected benefits, until effective mechanisms and controls are established for the flow and assessment of BP [5].

When analyzing management structures as a subject of research, it is necessary to use systemic and process approaches [6, 7]. Under the organizational design refers to a set of procedures deployed in time, allowing to form a specifically targeted system of the company. Such an approach is the basis of generally accepted standards of business modeling (for example, IDEF methodology) and structural analysis and design methodology, the basis of functional cost analysis and is implemented in a number of software systems (ARIS, IDEF/Design, Rational Rose, SAP R/3, etc.) [8, 9].

In General, a telecommunications company is an organization that, from the point of view of a systematic approach, has the properties of openness, reality, complexity, integrity, polystructure, variability and hierarchy. In addition, the organization is also a "living" system. It has its own boundaries and functions in its environment. Thus, the organization is a spatio-temporal structure of a set of internal and external factors that interact with each other in order to obtain the maximum qualitative and quantitative results in the shortest time and with minimal resources.

Any description of the company's activities begins with the subjective identification of individual business processes [10]. Definition of business processes of the company, their description, analysis and optimization—is the work on the organization of effective activity of the company, to improve its competitiveness [11].

Taxonomy of business processes in the field of telecommunications is described in the standard eTom [12], which defines the structure of BP, using their hierarchical decomposition. At the top level of the eTom model, there are three main BP groups: "Strategy, infrastructure and product" (BP managers), "Operational processes" (as a part of the main BP) and "Enterprise management" (providing BP).

When analyzing the company's activities, imitational modeling of a power supply is effective. This type of modeling [13, 14] includes almost all other types of modeling and provides an opportunity to conduct a comprehensive and complete assessment of the system under study. As a preparatory stage for software modeling, an analysis of the requirements for the BP system [15], algorithmization, selection of the development environment and mathematical apparatus [6, 16] is carried out. The simulation method is suitable for the study of stochastic systems, random processes.

In this paper, the use of cluster analysis is proposed as a method of analysis and classification of business processes. As a result, it becomes possible to divide the set of BP into several enlarged clusters according to the average time of work execution and to take separate optimization solutions for the work from each cluster.

To analyze, classify and optimize the business processes of the TCC, it is necessary to solve a number of tasks:

1. To investigate the structure of the organizational support of TCC BP.
2. To process statistical data obtained empirically or through or based on process modeling.
3. To analyze the obtained statistical data and determine the necessary actions aimed at optimizing the implementation of operational BP.

3 Structure of the Organizational Support of Business Processes TCC

The initial data for the analysis of the activities of the TCC is an infological model of business processes containing the following data:

(1) information about BP of the company;

(2) company BP algorithms;
(3) the organizational structure of the company;
(4) a list of company employees;
(5) data required for modeling:

- the probability of distribution of incoming work,
- laws of distribution, labor input of executed works,
- the performance ratios for each employee.

Before planning the scope of changes, you should assess whether the company is ready to introduce changes. One of the ways to analyze BP is to present the company's work in various conditions, to take into account the BP's performance and employee productivity, to imitate the company's work for a limited period. By modeling, the determination of the average workload of employees and the average execution time of the BP [17]. The analysis will make it possible to foresee possible problems in the company that may arise when the organizational structure or activity volumes change, to optimize the time spent on operational BPs, to minimize the time for performing individual tasks.

Based on the main tasks assigned to the TCC, the following can be attributed to the main BP of the TCC.

1. Development of telecommunication (TC) services.
2. Operating BP.

   2.1 Development and maintenance of communications.
      2.1.1. Quality management services.
      2.1.2. Registration and Troubleshooting.
   2.2 Sale of TC services.
      2.2.1. Registration of new subscribers.
      2.2.2 Registration of applications for connection.
      2.2.3 Connection of various services.
   2.3. Customer Service.
      2.3.1. Change of services.
      2.3.2. Disable services.

In the course of structural analysis of BP, its main constituent units (sub processes, main types of work, individual work) are distinguished. In relation to the company, we can distinguish internal and external works. Their main difference is that internal work requires time and resources of the company, while external work requires only time to wait for their work. Depending on the method of implementation, they can be divided into groups:

- automatic—triggers that are executed instantly and do not require resource expenditures;
- managed—work performed for a certain period, usually requiring labor costs.

The company carries out a structural division of work. There is a distribution and accounting of work, their synchronization and the choice of algorithm for implementation. To analyze the implementation of operational business processes, it is necessary to describe the correspondence between the company's BP and the elements of the organizational structure, which will allow controlling the correctness of processes, their correspondence to real goals, objectives and performers. Table 1 presents the BP responsibility matrix for the TCC departments [18].

For BP analysis, first of all, it is necessary to determine the key indicators that assess their effectiveness. Thus, business process management includes the following steps:

- comprehensive assessment of the company,
- building a power supply system,
- determination of BP performance indicators,
- development of a system of measures for the implementation of adjustments.

Optimization of business processes will reduce time costs and operating expenses, as well as improve the manageability of the company and the quality of customer service, in other words, to achieve target performance.

4 Processing of Statistical Data on Business Processes

An important step in the analysis of business processes is the processing of statistical data obtained in the course of the company's activities. The initial statistics contain the following information:

- process number,
- process performer,
- start date of the process,
- end date of the process,
- standard deadline,
- data of the employee assigned to this process.

Note that statistics can be obtained in the following ways:

- according to the results of BP modeling,
- according to the results of reporting on the proceeding processes.

From the point of view of information processing, the source of its receipt does not matter, since it is reduced to the same type of form (tabular). The tables contain data on the duration of execution in hours, the number of incoming processes, the percentage of processes exceeding the standard deadlines, the percentage of the excess of the standard periods for the processes in question, the minimum and maximum values for the specified simulation period, etc.

As a result of processing the above data at the output, we obtain statistical information for each process:

- the number of processes submitted for execution for each calendar year,

Table 1 Matrix of BP company responsibility

	Marketing department	Sales Department	Department of tariff policy	Brand management Department	Customer care Department	Service department	Product and service development Department	Capital construction department	Bookkeeping	Financial department	Information Technology Department	The HR Department	Corporate governance Department	Labor protection service	Secretariat	Legal department
The main processes																
Development of TC services				+	+		+									
Development and maintenance of communication facilities.						+										
Sale of TC services.		+	+													
Customer service.					+											
Supporting processes																
IT-support and communication											+					
Legal support																+
Security support													+			
Construction of communication facilities								+								
Accounting									+							
Management processes																
Strategic management							+									
Financial management										+						
Marketing management	+			+												
Personnel management												+		+		
Corporate management													+		+	
Network development						+										

- the standard deadline for processes,
- average actual deadline,
- the average rate of time exceeding the process for the selected billing year,
- the average ratio of the number of processes exceeding the standard deadline to the normative period,
- a process executed with a maximum delay in relation to other processes,
- information about the employee assigned to the process at the slowest performance,
- process executed in the shortest possible time in relation to other processes,
- information about the employee assigned to the process at the fastest execution.

The following is the output for a set of all processes involved in the processing:

- the number of one type of processes received to perform for each calendar year,
- average actual deadline,
- the average rate of exceeding the execution time of the processes for the selected settlement year in relation to the standard deadline for implementation,
- the dynamics of the execution of processes of the same type during the calculation period.

Figure 1 shows the integration scheme of the system for the preparation of statistical data [18].

Consider the method of modeling the company's activities in the implementation of the main BP. Upon receipt of the application in the TCC, its type is determined and the appropriate algorithm of actions is selected. Simulated bids are generated randomly within an established time frame [19]. There is a call to the database that stores the algorithms for the implementation of the BP, and the algorithm that corresponds to the type of application is selected. Further, in accordance with the

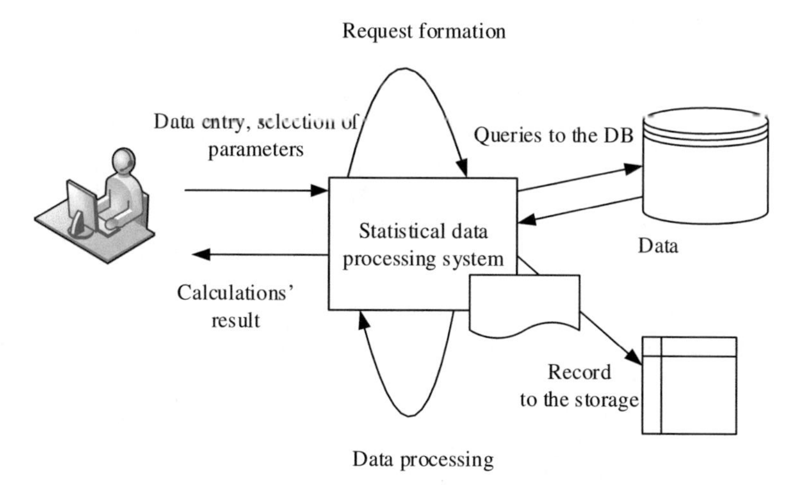

Fig. 1 Integrated scheme of statistical data processing system

selected algorithm, the implementation of the BP takes place: the application goes through all stages, they are going through the departments of the company in reality. In this case, time keeping is kept. Depending on the initial criteria and the selected algorithm, the processing of applications may take different times [20]. When the execution algorithm ends, the application is considered to be completed.

During the operation of the system there may be queues for the execution of applications at different stages (the queue of outstanding applications in specific departments of the company). After the application is completed at this stage, its place is released and immediately deals with the next application—from the queue. After the time has expired, the system provides the result of working with information on the number of completed applications and applications that are at different stages of implementation. The data obtained provide an opportunity for subsequent statistical analysis and classification of BP.

The protocol, which reflects the course of the simulation, contains the following data:

- identification number of the application for the process,
- process identification number,
- date of receipt of the application for the process,
- date of the process,
- a list of operations that were performed in this process
- list of employees who performed this process at different stages.

For each of the operations the following information is filled:

- identification number of the transaction.
- name of operation,
- standard deadline for the operation,
- the complexity of the operation,
- type of transaction.

Statistical data on the implementation of processes include the minimum, average and maximum deadlines for the execution of the selected process, the number of processed processes, the standard execution time, the percentage of excess depending on the number of processes, as well as the percentage of excess depending on the execution time (Table 2). Only the years for which information is available for the process in question will be included in the resulting table.

Table 2 Sample of statistics

Year	Amount	Average term, h	Expected term, h	Min term, h	Maxterm, h	Overhead (amount), %	Overhead (duration), %
2016	17	1326,235	72	11	6578	47	1793
2017	40	41,650	72	3	240	20	32

In the case of a comprehensive analysis of all processes, the following data will be recorded in the file: the percentage of each process performed by a year depending on the normative period of the process execution, the minimum value of the percentage exceeding the calculated one, the average value exceeding one percent, their ratio is different to a friend, as well as kurtosis for each process (Table 3).

The transition diagram of the statistical data processing process described above is shown in Fig. 2 [18].

5 Classification of Business Processes by Cluster Analysis Method

Cluster analysis is necessary for classifying information; with its help, it is possible to structure variables in a certain way. In this paper, we classify processes using the k-means method by building a hierarchical tree with the aim of developing recommendations for managing the timing of the processes for each group.

Cluster analysis, in contrast to most mathematical and statistical methods, does not impose any restrictions on the type of objects under consideration and allows us to consider many basic data of a virtually arbitrary nature. The solution of the cluster analysis problem is a partition that satisfies some optimality criterion. This criterion may be some kind of functional expressing the desirability levels of various partitions and groupings [21].

Let us consider the k-means method, which allows us to build k-clusters located at the greatest possible distances from each other.

Clustering was performed in two stages.

1. The initial distribution of the objects in clusters.

For the analysis, the initial number of clusters was chosen equal to three. The whole set of processes should be divided into three groups depending on the time of their execution. The first group includes processes that do not require optimization or changes; the second group includes processes with intermediate indicators. This group requires optimization, but the level of violations is not critical, in some processes the indicators are violated by a subjective factor (for example, the poor performance of duties by an individual employee). In addition, the third group, respectively, includes BP, significantly violating the time limits set by the rules. These processes are unstable or with large exceedances of standards and require rapid reengineering and optimization.

2. Iterative process.

Cluster centers are computed and then considered subordinate middle clusters. After each iteration, the objects are redistributed. The process of calculating the centers and the redistribution of objects continues until one of the following conditions is fulfilled:

Table 3 A fragment of statistical data for processes

Business process	BP number	2013	2014	2015	2016	2017	Min overhead	Average overhead	Average/min overhead	Kurtosis
…	…	…	…	…	…	…	…	…	…	…
Phone- IP Connection	38	3632	1590	100	0	0	100	1669.7	16.69	1.31
Acceptance into operation.	59	0	0	0	900	652	652	776	1.19	-0.93
TV optics	24	999	0	0	642	157	157	1195	7.61	3.54
DNS-domen	3	172	514	1376	295	120	110	431.16	3.91	3.98
DNS-hosting	4	33	22	2816	583	407	22	1990.1	90.46	3.67
DNS-mail_srv	5	146	0	409	3240	270	146	859.5	5.88	5.75
…	…	…	…	…	…	…	…	…	…	…

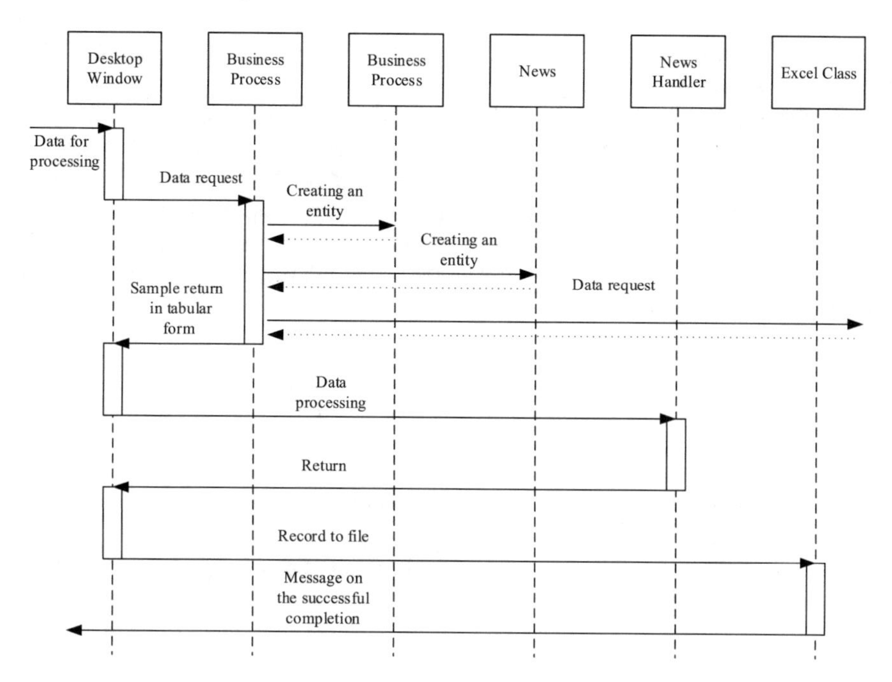

Fig. 2 Transition diagram of the statistical data processing system

- cluster centers have stabilized,
- the number of iterations is equal to the maximum number of iterations.

To assess the correctness of the initial choice of clusters, a hierarchical method was used, the essence of which is to build a hierarchy of clusters. To organize the data, an agglomerating strategy was used, which is a bottom-up approach: each observation starts in its own cluster, and the pairs of clusters are combined, providing a higher order hierarchy. The advantages of the method are that the agglomerative clustering method does not pre-determine the number of classes, is well applicable for classifying sets of not very large volume, and, in practice, usually leads to better results than clustering methods based on downward classification. Based on the observations, a dendrogram is constructed that describes the closeness of individual points and clusters to each other, graphically presents the sequence of combining (dividing) clusters.

When each object is a separate cluster, the distances between these objects are determined by the selected measure. The distance between the clusters is taken as the increment of the sum of squares of the distances of the objects to the centers of the clusters, obtained because of their combination. Unlike other cluster analysis methods for estimating the distances between clusters, here the methods of analysis of variance are used. At each step of the algorithm, these two clusters are combined, which lead to a minimal increase in the objective function, i.e. intragroup sum of squares. This method is aimed at combining closely located clusters and “seeks” to create clusters of small size.

In the first step, all objects are considered as separate independent clusters consisting of only one element. The distance between all possible pairs of objects using one or another metric is calculated. The Ward method was used to calculate the distance. In this case, the signs are quantitative, the components of the vector of observations are homogeneous in their physical meaning, and they are all equally important from the point of view of resolving issues relating to the assignment of an object to a particular class [22]:

$$\mathrm{d}(\mathrm{i},\mathrm{i}') = \sqrt{\sum_{j=1}^{k}\left(\mathrm{i}_s^{(\mathrm{j})} - \mathrm{i}_s'^{(\mathrm{j})}\right)^2}, \quad (1)$$

where $i_s^{(j)}$—normalized value of the *j*th characteristic of the *i*th object ($j \in J$);

k—number of characteristics,

The normalization of the values of the characteristics of objects is carried out according to the formula:

$$\mathrm{i}_s^{(\mathrm{j})} = \frac{i^{(j)} - \mathrm{i}_{\mathrm{min}}^{(\mathrm{j})}}{\mathrm{i}_{\mathrm{min}}^{(\mathrm{j})} - \mathrm{i}_{\mathrm{max}}^{(\mathrm{j})}}, \quad (2)$$

where $i^{(j)}$—value of the *j*th characteristic of the *i*th object ($j \in J$);

- $\mathrm{i}_{\mathrm{min}}^{(\mathrm{j})}$—minimum value of the *j*th characteristic;
- $\mathrm{i}_{\mathrm{max}}^{(\mathrm{j})}$—maximum value of the *j*th characteristic;
- In the method of agglomerative hierarchical clustering using the average connection, the following notation is used:
- $X(I)$—hierarchical classification (this is the set of non-empty subsets of the set I, partially ordered by the inclusion relation of sets);
- $T(X(I))$—the set of terminal classes of hierarchy;
- $M(X_0), M(X_1), \ldots, M\left(X_{|I|-1}\right)$—the sequence of nested partitions;
- $\nu(a)$—class a level index (stratification index or class diameter);
- $\delta_0\left(\{i\}, \{i'\}\right), \delta_1\left(\{i\}, \{i'\}\right), \ldots, \delta_{h-1}\left(\{i\}, \{i'\}\right)$—distance between classes;
- h—step;
- N—a non-terminal class;
- $Si(a)$—the set of classes located directly under the class *a;*
- $|I|$—the number of elements of the set *I;*
- $A(N) = i$—successor node *N*;
- $B(N) = i'$- the second successor of the node *N*;
- $P(N) = 2$—the number of elements of the node *N*;
- s_h and s_h'—two classes of $M(X_{h-1})$ at which the minimum value of distances δ_{h-1} is realized.

It is necessary to build a sequence of partial hierarchies. $X_0, X_1, \ldots, X_{|I|-1}$.

Consider a table of distances $d(i, i')$ calculated from the source table, using the selected metric. It is assumed that:

$$X_0 = X_0(I) = T(X(I)) = \{\{i\}; i \in I\},$$

$$M(X_0) = X_0(I) = T(X(I)) = \{\{i\}; i \in I\}, \quad (3)$$

$$\nu(\{i\}) = 0, \quad (i \in I).$$

Distances between single-element classes should be equal to the distances between elements:

$$\delta_0(\{i\}, \{i'\}) = d(i, i'), \quad (i, i' \in I). \quad (4)$$

The first step ($h = 1$) defines the minimum value δ for X_0. Let us say that this minimum value is reached on a pair of single-element classes $\{i\}$, $\{i'\}$. The first node with a serial number $|I| + 1$ is formed, so that $N = |I| + 1$:

$$a_1 = \{i, i'\}, \; Si(a_1) = \{i, i'\}, \; |a_1| = 2,$$

$$M(X_1(I)) = M(X_0(I)) \cup \{a_1\} \backslash \{i\} \backslash \{i'\}, X_1 = X_1(I) = X_0 \cup a_1 \quad (5)$$

$$\nu(a_1) = \min\{\delta_0(i, i') : i \neq i', \; i, i' \in M(X_0)\} = \nu(N).$$

Finally, the distances between all the classes of the new partition indicated by $M(X_1)$ are calculated. As the partition is obtained from $M(X_0)$ by combining two classes, you must use the distance value between two subsets of elements to recalculate the distances. Then you can calculate the distance between the new merged class and the other classes $\delta_1(a_1, t)$, $(t \in M(X_1))$.

In this case, the characteristic distance between two subsets of elements is the average distance between clusters. The distance between the new class and other classes $\delta(a, b)$ is calculated using the average coupling method:

$$\delta(a, b) = \sum_{j=1}^{k} \{d(i, i'); \; i \in a, \; i' \in b\} / |a||b|. \quad (6)$$

The distance between the two classes is defined as the average of the original distances between the elements belonging to these two classes. The recurrent formula has the following form:

$$\delta_h(t, s_h \cup s'_h) = (|s_h|\delta_{h-1}(t, s_h) + |s'_h|\delta_{h-1}(t, s'_h)) / (|s_h| + |s'_h|), \quad (7),)$$

at $t \neq s_h \neq s'_h;\ t,\ s_h,\ s'_h \in M(X_{h-1})$,

$$\delta_h(t,\ t') = \delta_{h-1}(t,\ t'), \quad (8)$$

at $t \neq t' \neq s_h \neq s'_h;\ t,\ t',\ s_h,\ s'_h \in M(X_{h-1})$.

Now we define the recurrence formulas for $h = Z$. The sequence of nested hierarchies is known X_{h-1}, as well as the top $M(X_{h-1})$. Recurrent formulas should be based only on information related to $M(X_{h-1})$. We get:

$$N = |I| + h, a_h = s_h \cup s'_h,\ Si(a_h) = \{s_h,\ s'_h\},$$

$$X_h(I) = X_{h-1}(I) \cup a_h,\ M(X_h(I)) = M(X_{h-1}(I)) \cup \{a_h\} \backslash \{s_h\} \backslash \{s'_h\} \quad (9)$$

$$\nu(a_h) = \min\{\delta_{h-1}(s,\ s');\ s \neq s',\ s,\ s' \in M(X_{h-1})\},\ |a_h| = |s_h| + |s'_h|.$$

When recalculating distances for $\delta_h(t, a_h),\ t \in M(X_h)$ the following values are used: $\delta_{h-1}(t,\ s_h),\ \delta_{h-1}(t,\ s'_h),\ \delta_{h-1}(s_h,\ s'_h),\ \nu(s_h),\ \nu(s'_h),\ |s_h|,\ |s'_h|,\ \nu(t),\ |t|$.

At the last step, when $h = |I| - 1$, it remains to combine only two classes in order to obtain the whole set I. In this case:

$$N = 2|I| - 1, a_h = I = s_h \cup s'_h,\ |a_h| = |s_h| + |s'_h| = |I|,$$

$$X_h = X_{|I|-1} = X(I), M(X_h) = \{I\}, \quad (10)$$

$$\nu(a_h) = \nu(I) = \delta_{h-1}(s_h,\ s'_h).$$

Thus, all objects become members of the same cluster.

In order to classify business processes, a two-stage cluster analysis is performed: the initial data are first evaluated on one basis and divided into clusters. Then there is a clustering of the same source data, but on a different basis. The resulting clusters based on the two stages are combined into a final cluster. Such an approach makes it possible to take into account the influence of various factors on the performance of individual BPs during the period of time under consideration, including the subjective and objective components. It becomes possible to concentrate management on a group of processes that are problematic in terms of specified requirements. At the same time, there is a significant similarity of processes in the considered parameter.

We present the results of the first stage of cluster analysis by the criterion of the ratio of the maximum execution time of processes to the average time of its execution. This coefficient (kurtosis) allows us to determine how great the difference is between the average process execution time and the maximum time that, in the case of a large difference, could occur for reasons that can be considered force majeure (because they occurred once and not repeated more, but in the calculation of statistics may distort the calculation). Under force majeure factors, you can mean a failure in the production process, economic factors, mistakes of the responsible employee, etc.

The first stage of cluster analysis is carried out using the k-means method to divide all processes into three clusters. The procedure is carried out iteratively to achieve a set number of clusters.

Table 4 presents the parameters of kurtosis for each of the clusters.

Successful partition was obtained after the third iteration (Table 5). The conductivity is achieved by the criterion of small size or no change in the position of the cluster centers.

Automatic processes (triggers) did not take part in the calculations since no duration was specified for them. Such works were considered not valid. The resulting table contains information about the belonging of the works to the clusters and their distance from the center of the cluster.

Works belonging to cluster 3 are considered to have an average level of criticality, the work of cluster 2 is identified as the most critical (it is for them that a control decision should be made), the work of cluster 1 is noncritical.

The following are the distances between the final centers of the clusters (Table 6), which are maximally balanced for a certain number of clusters. The values of F-statistics (Table 7) serve as an indicator of the correctness of the splitting; the level

Table 4 The parameters of the kurtosis of clusters

	Cluster		
	1	2	3
Kurtosis	−1.4961	5.3842	1.2952

Table 5 Iteration history

Iteration	Cluster center changes		
	1	2	3
1	0.459	0.591	0.376
2	0.029	0.024	0.029
3	0.002	0.001	0.002
4	0.000	3.783E−005	0.000
…	…	…	…
13	1.776E−015	0.000	1.599E−014
14	0.000	0.000	1.110E−015
15	0.000	0.000	0.000

Table 6 Distances between the final cluster centers

Cluster	1	2	3
1		2935	3365
2	2935		6300
3	3365	6300	

Table 7 The results ANOVA

SS effect	Pf effect	SS error	Pf error	F	p
228,046	2	0.565	48	403,898	0.000

of significance confirms the hypothesis of the significance of differences between clusters.

In total, 73 processes were processed during the cluster analysis (Table 8). 22 processes were automatic and did not participate in the division into clusters.

The second stage of the analysis is the clustering of processes according to the objective factor. To this end, at the stage of preparing statistical data, the ratio of the average time of work execution to the minimum time of their performance was determined.

This coefficient makes it possible to determine how great the difference is between the average time of the process and the minimum time, which, in the case of a small difference, indicates that the process requires adjustments.

Since there is a probability of distortion of the average time of the process as a result of the influence of external factors (force majeure), for the second stage the method of hierarchical trees is more effective.

In this example, the established boundaries of the number of clusters were 3 (the initial number of clusters) and 10 (since a more detailed partition exceeds the square of the minimum limit). In the analysis, the Ward method was applied, the essence of which is that, first, in clusters for all existing observations, the average values of individual variables are calculated. Then, the Euclidean distances from the individual observations of each cluster to the average values are calculated. In one new cluster, those clusters are combined, the combination of which results in the smallest increase in the total sum of distances.

In an iterative procedure, starting with five clusters, only the fragmentation of individual clusters occurred with the rest stabilized. Thus, the optimal number of clusters in this example is five.

The processes that are part of cluster 1, require optimization in the first place, the processes that are part of cluster 2, have variable indicators and are noncritical from the point of view of the need to make adjustments. The processes of the remaining clusters 3, 4, 5 include the highest values of the reduction factors for the duration of their execution, and, therefore, are the least critical.

Table 8 The number of cases in each cluster

Cluster	1	15
	2	24
	3	12
Valid		51
Missing values		22

Imagine the overall results of the results of the two stages of the analysis (Table 6). All processes are divided into clusters according to various criteria. To begin with, we will single out the processes that require optimization by the objective factor: this subgroup will include all processes except those belonging to clusters 3, 4, 5. These are 31 processes. For the remaining 42 processes, we will take into account the results of the analysis on the subjective factor, i.e. it is possible to exclude the clusters 3, 4, 5 following the results of the second stage of the analysis from the considered total set of processes that require optimization, reducing the value of the resulting cluster for them. That is if the process is in a cluster with the highest criticality, but it belongs to one of the clusters 3, 4, 5 from the second stage of the analysis, it automatically moves to the cluster with lower criticality.

The result is a separation of processes depending on the level of severity, which is represented by the values of the last column of Table 9.

As a result, the processes were classified as follows: cluster 1 includes processes that have a maximum "strength" margin, compared with other processes, i.e. in the queue for reengineering they can be placed at the end; cluster 2 includes processes that, on the contrary, are the most critical among the existing ones and should be considered when solving the reengineering problem in the first place; cluster 3 includes the remaining processes that are included in the intermediate group as having ambiguous indicators when evaluated by two criteria. Processes from cluster 3 are the most mobile group, due to which the size of two extreme clusters can vary, which should be determined directly by the person making the control decision.

6 Optimization of the Terms of Business Processes Execution by the Network Planning Method

After the labor intensity of the work has been determined, a group of "critical" processes has been identified from the point of view of possible exceeding the deadlines; the total runtime of the BP TCC has been adjusted. The Network planning methods are used for this [23, 24].

We will use the critical path method with the construction of the vertex-event model. Business processes are represented as arcs between nodes of the network graph. To build a network model of work we use the data from Table 10. We also define the set of immediately preceding and immediately following works.

Here are the main stages of network planning.

1. The calculation of the earliest possible date for the execution of the works.

The event corresponds to some node and represents the start of one or more activities. Let $T_j(E)$ is the earliest possible date of occurrence of the jth event, $j = \overline{1, n}$, where n is the number of events (nodes) in the network.

Note that $T_j(E)$ is the earliest possible completion date for all work that is suitable for the jth node. Let d_{ij} is the duration of the processes connecting two events (nodes) i and j.

Table 9 Fragment of the resulting clustering processes

Business process	Business process number	Clusters		
		Subjective factor	Objective factor	Result
...	...	...	...	...
Phone- IP Connection	38	3	3	1
Disable temporarily	43	3	1	3
HW	12	3	2	3
IP-net	2	3	2	3
Tariff change	39	3	2	3
WiPLL-2 connection	40	3	3	1
Reissue	37	3	2	3
Acceptance into operation	59	1	1	3
Tech-Wi	33	1	1	3
Phone- IP Connection	48	1	1	3
port	10	1	2	1
Agreement	26	1	1	3
Connection	7	1	2	1
Change of parameters	28	1	1	3
Additionally	21	1	1	3
Testing	31	1	2	1
DNS-mail_srv	5	2	2	2
Changing hosting settings	49	2	4	3
...	...	...	...	...

The earliest possible time of occurrence of each event is calculated as the length of the longest path from the initial to the given event. Suppose that r paths lead from the initial event (1st) to the jth event. Let us denote these paths $P_1, P_2, \ldots, P_r$. Each path corresponds to a certain measure, equal to the sum of the durations of all work on the path. Therefore,

$$T_j(P_k) = \sum_m \sum_p d_{mp}, \quad m, \ p \in P_k, \ k = \overline{1, r}, \quad j = \overline{1, n} \tag{11}$$

Thus, the longest path from the initial node to the jth node is defined as

Table 10 Processes in the «vertex—event» format

Processes		Directly preceding processes	The duration of the process, in days
A	Calculation	–	1
B	Appointment IP	–	3
C	Invoice payment	A	1
D	Billing	B, C	0
E	Port-w	D	3
F	Integration	E	1
G	Service activation	E, F	21
H	Confirmation	E, G	1
I	Start date	E	3
J	Signing of the act	I	1
K	Cancel enable	I	0
L	Registration of non-compliance	J, K	3

$$T_j(E) = \max_k [T_j(P_k)],\ \ j = \overline{1, n}, \tag{12}$$

where the maximum is taken over all paths connecting nodes i and j.

Let the longest path to the first node is zero $T_1(E) = 0$.

Let us now consider all the processes leading to the subsequent event. We calculate for each such process the time equal to the earliest possible date of occurrence of the previous event plus the duration of the process. The earliest possible date of occurrence of the event in question is the maximum of these two different periods of time. The earliest possible date of occurrence of the ith event is determined by the formula:

$$T_j(E) = \begin{cases} 0, & j = 1; \\ \max\limits_{i<j}[T_j(E) \mid d_{ij}], & 2 \leq j \leq n \end{cases}, \tag{13}$$

where the maximum is taken for all work completed in the jth node and exiting any previous ith node. For a network with n events, the calculations continue until the earliest possible time for the nth event to occur is determined.

2. Calculating the latest acceptable deadline to complete the work.

Let $T_i(L)$ is the latest date of occurrence of the ith event that does not affect the completion time of the whole process, that is, the latest date of completion of all works going to the ith node. Starting from the nth event, we move in the opposite direction through each previous event. To ensure that the critical path is not exceeded, you must start the procedure by equating the latest acceptable end of the event with the earliest possible end date of the process. Therefore,

$$T_n(L) = T_n(E). \tag{14}$$

The latest allowable date of occurrence of any ith event immediately preceding the nth event is determined by the formula

$$T_i(L) = T_n(L) - \min[d_{in}], \quad i \leq (n-1) \tag{15}$$

To calculate the latest time of occurrence of any ith event (i < n), consider all the processes coming from this event. Calculate for each such process the latest date of occurrence of all subsequent events and subtract the duration of the process. The lowest value is the latest date of the ith event:

$$T_i(L) = \begin{cases} T_n(E), & i = n \\ \min\limits_{j>i}[T_j(L) - d_{ij}, & 1 \leq i \leq n-1 \end{cases} \cdot \tag{16}$$

The minimum is taken for all jth events connected to the ith event of the work (i, j). Calculations are performed until the latest date of the initial event is determined.

3. The definition of reserve time and critical path.

Indicate S_i as a time reserve for the ith event. Then

$$S_i = T_i(L) - T_i(E). \tag{17}$$

Zero-time events are on a critical path.

Let ES_{ij} be the earliest possible time to start the process between events (i, j). Since the work cannot begin before the onset of the preceding event, we have

$$ES_{ij} = T_i(E), \tag{18}$$

the earliest possible deadline for finishing the processes (i, j):

$$EF_{ij} = T_i(E) + d_{ij} = ES_{ij} + d_{ij}. \tag{19}$$

Let us denote LF_{ij} as the latest acceptable end date of the process (i, j). Since the work can be completed no later than the maximum permissible period of occurrence of the subsequent event j, we have

$$LF_{ij} = T_j(L). \tag{20}$$

The latest allowable start date of the process (i, j) can be calculated by the formula:

$$LS_{ij} = LF_{ij} - d_{ij}. \tag{21}$$

4. Time reserve calculation.

We define four indicators of the time reserve: total, free, non-dependent and guaranteed.

The total time reserve TF_{ij} for the process (i, j) represents the maximum duration of the process delay (i, j), which does not cause delays in the implementation of the entire operating cycle:

$$TF_{ij} = LS_{ij} - ES_{ij} = LF_{ij} - EF_{ij}. \tag{22}$$

The free time reserve FF_{ij} differs from the total in that it characterizes the time that does not affect the delay of subsequent work:

$$FF_{ij} = T_j(E) - EF_{ij}. \tag{23}$$

The independent reserve of time is determined by the formula:

$$IF_{ij} = \max\begin{cases} 0 \\ T_j(E) - [T_i(L) + d_{ij}] \end{cases}. \tag{24}$$

Guaranteed reserve of time is determined by the formulas:

$$SF_{ij} = LF_{ij} - \left[T_i(L) + d_{ij}\right] = T_j(L) - \left[T_i(L) + d_{ij}\right]. \tag{25}$$

We get Table 11, which shows the time reserves for each process.

The given characteristics of network models allow us to evaluate the duration of the business processes of the TCC.

The next step is to build a calendar schedule of operational activities of the TCC. The availability of resources should be taken into account when planning the calendar, as it may not be possible to carry out some work at the same time due to limitations

Table 11 Time reserves for the processes

Process i	Total time reserve, TF_i	Free time reserve, FF_i	Independent time reserve, IF_i	Guaranteed time reserved, SF_i
A	1	2	0	0
B	2	3	3	0
C	5	6	3	6
D	2	4	1	0
E	3	5	3	3
F	6	8	0	9
G	8	9	3	9
H	9	10	1	12
I	9	11	1	12
J	10	12	21	13
K	4	7	3	1
L	7	12	1	4

Table 12 Critical BP paths

Number	Path of the events	Path of the works
1	1 -> 2 -> 3 -> 5 -> 6 -> 8 -> 9 -> 10 -> 12	A -> B -> E -> C -> F -> G -> H -> J
2	1 -> 2 -> 3 -> 5 -> 6 -> 8 -> 9 -> 11 -> 10 -> 12	A -> B -> E -> C -> F -> G -> I -> N -> J

related to staffing, equipment and other types of resources. In this respect, the full time reserves of non-critical work are valuable. By shifting non-critical work within its full time reserve, it is possible to achieve a reduction in the maximum resource requirement and to equalize the need for resources throughout the execution of all processes. This means that a more or less permanent staff of performers, compared with the case when the needs of performers (and other resources) change dramatically during the transition from one time interval to another, can perform the work.

Based on the specified start and end dates, the schedule of processes can be converted into a real time scale in the form of a calendar plan.

According to the results of the analysis of the BP "Connecting Phone—IP", its critical paths are presented in Table 12.

We will analyze the work with the cost of increasing (decreasing) the duration of their implementation. In the case of this BP without acceleration, taking into account the regulatory deadlines for the implementation of the process, its cost is 190 UAH; the duration of the critical path is 34 days. The standard period for the fulfillment of the "Phone-IP Connection" BP is a period of 7 days, it is necessary to accelerate the execution by 27 days. Reduction of the period can only be achieved by reducing the time needed to complete work belonging to the critical path, and reducing the time it takes to do any work on the BP requires additional costs. Therefore, by reducing the execution time of those critical works that require the least additional costs, it is possible to achieve a minimum increase in the cost of the whole complex of works on the provision of services. After the acceleration of the BP, its cost will be 445 UAH, the duration of the critical path is 7 days.

Network graph for this case is shown in Fig. 3.

On the basis of the calculations it is possible to build graphs of the relationships of the requirements and time frames (Figs. 4 and 5), according to which the manager will be able to determine the optimally acceptable to the company ratio "time–value" as a way to optimize the process. Thus, it is possible to determine the most optimal reduction in terms of operational business processes, which will not lead to a significant increase in costs.

7 Conclusion

The proposed approach to optimizing the business processes of a telecommunications company is intended to facilitate the process of evaluating the effectiveness of operating activities by visualizing and analyzing it over a certain period of time. Methods

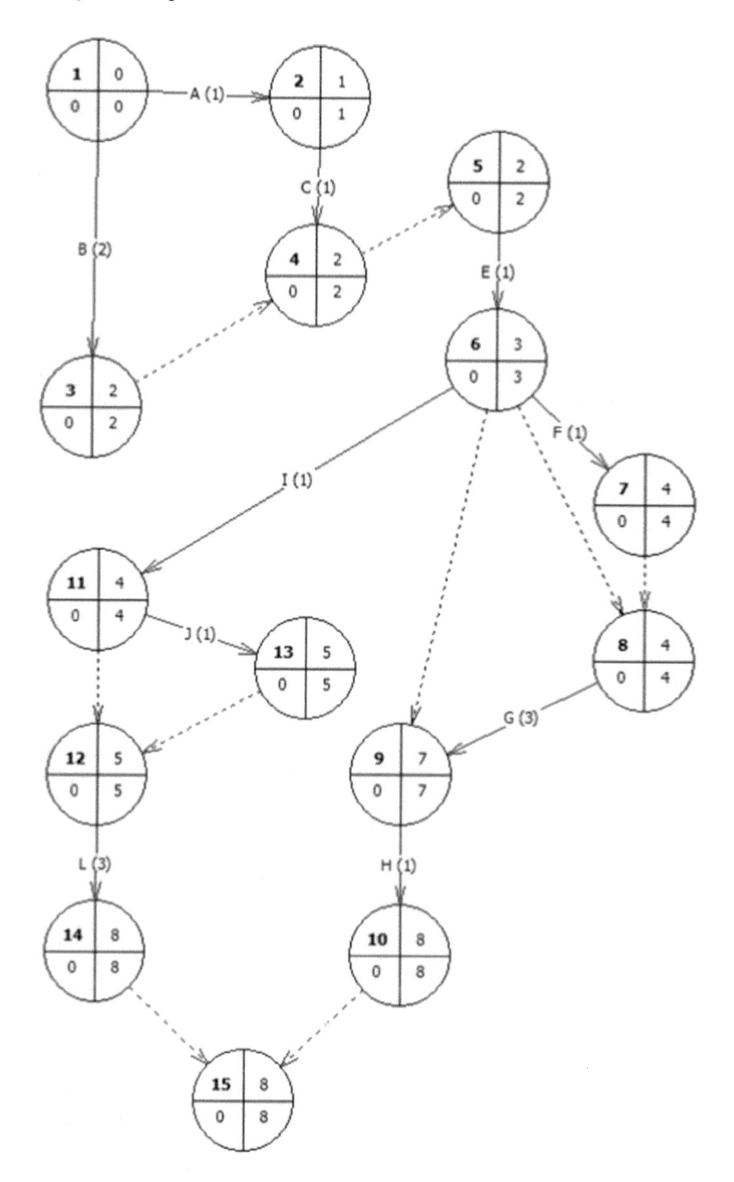

Fig. 3 Modified network graph

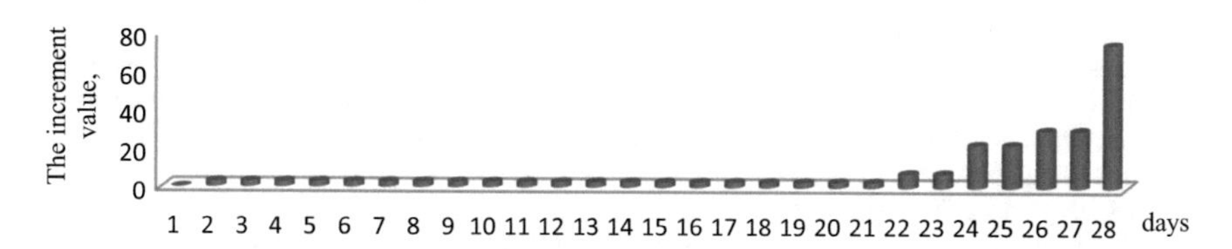

Fig. 4 The increment value of BP

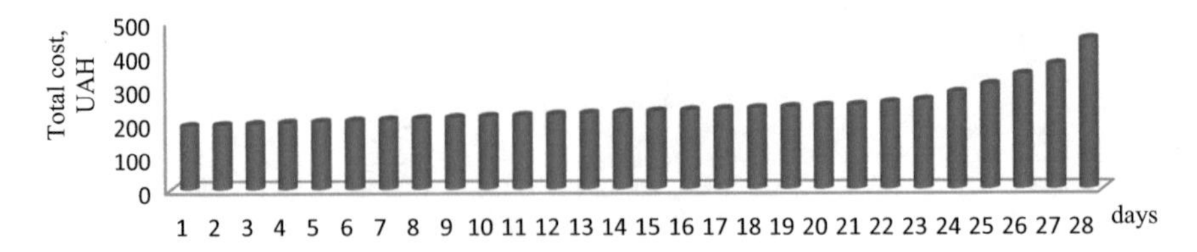

Fig. 5 Total cost of BP

of statistical analysis and network planning allow you to optimize the processes occurring in the company. The application of the developed concept contributes to improving the efficiency of TCC activity by simulating the interaction of technical and organizational structures and improving the individual characteristics of each of them. As a result, the manager has the opportunity to assess how productive and optimal the organizational structure of this company is. The approach allows saving many resources on finding and making the most profitable decision in the field of telecommunications services management.

References

1. Kryvinska N (2008) An analytical approach for the modeling of real-time services over IP network. Math Comput Simul 79(4):980–990. https://doi.org/10.1016/j.matcom.2008.02.016
2. Kirichenko L, Alghawli ASA, Radivilova T (2020) Generalized approach to analysis of multifractal properties from short time series. Int J Adv Comput Sci Appl 11(5):183–198. https://doi.org/10.14569/IJACSA.2020.0110527
3. Andersen B (2005) Biznes-protsessy. Instrumenty sovershenstvovaniya [Business processes. Improvement tools]. Standards and Quality, Moscow
4. Tahara K et al (2005) Comparison of "CO_2 efficiency" between company and industry. J Clean Prod 13(13–14):1301–1308
5. Ageyev D, Ali A-A, Nameer Q (2015) Multi-period LTE RAN and services planning for operator profit maximization. In: 2015 13th international conference on the experience of designing and application of CAD systems in microelectronics (CADSM). IEEE, pp 25–27. https://doi.org/10.1109/cadsm.2015.7230786
6. Radivilova T, Kirichenko L, Ageyev D, Tawalbeh M, Bulakh V, Zinchenko P (2019) Intrusion detection based on machine learning using fractal properties of traffic realizations. In: 2019 IEEE international conference on advanced trends in information theory (ATIT). IEEE, Kyiv, Ukraine, pp 218–221. https://doi.org/10.1109/ATIT49449.2019.9030452
7. Barabash O, Shevchenko G, Dakhno N et al (2017) Information technology of targeting: optimization of decision making process in a competitive environment. Int J Intell Syst Appl 9(12):1–9. https://doi.org/10.5815/ijisa.2017.12.01
8. Kalmykov AV, Smidovich LS (2010) Adaptatsiya notatsii BPMN k biznes-protsessam telekommunikatsionnogo predpriyatiya [Adaptation of BPMN notation to business processes of a telecommunication enterprise]. System technologies. Regional intercollegiate collection of scientific papers. Dnepropetrovsk, vol 5, pp 106–118
9. Leszek AM (2002) Analiz trebovaniy i proyektirovaniye sistem s ispol'zovaniyem UML [Requirements analysis and system design using UML]. Williams, Moscow
10. Galkin G (2004) Pokazateli effektivnosti biznes-protsessov [Indicators of the effectiveness of business processes]. Intell Enterp Corp Syst 21(107):50–59

11. TMF eTOM model. http://www.tmforum.org/BestPracticesStandars/BusinessProcessFramework/6637/Home.html
12. Laguna M, Marklund J (2018) Business process modeling, simulation and design. Chapman and Hall/CRC, New York
13. Nerush VB, Kurdecha VV (2012) Imitatsiyne modelyuvannya system ta protsesiv [Simulation of systems and processes]. National Technical University of Ukraine "KPI", p 115
14. Scheer A-W (2012) ARIS—business process modeling. Springer, Berlin, Heidelberg
15. Kryvinska N (2010) Converged network service architecture: a platform for integrated services delivery and interworking. Electronic business series, vol 2. International Academic Publishers, Peter Lang Publishing Group
16. Mukhin V, Kuchuk N, Kosenko N, Artiukh R, Yelizyeva A, Maleyeva O, Kuchuk H (2020) Decomposition method for synthesizing the computer system architecture. In: Advances in intelligent systems and computing, vol 938. Springer, Cham, pp 298–300. https://doi.org/10.1007/978-3-030-16621-2_27
17. Kosenko V, Persiyanova E, Belotskyy O, Maleyeva O (2017) Methods of managing traffic distribution in information and communication networks of critical infrastructure systems. Innov Technol Sci Solut Ind 2(2):48–55. https://doi.org/10.30837/2522-9818.2017.2.048
18. Kulik IYu, Smidovich LS, Maleeva OV (2012) Imitatsionnoye modelirovaniye biz-nes-protsessov telekommunikatsionnoy kompanii [Simulation modeling of a business of a telecommunication company]. Radioelectron Comput Syst 3:167–172
19. Kosenko V, Maleyeva O, Persiyanova E, Rogovyi A (2017) Analysis of information-telecommunication network risk based on cognitive maps and cause-effect diagram. Adv Inf Syst 1(1):49–56. https://doi.org/10.20998/2522-9052.2017.1.09
20. Davydovsky Yu, Reva O, Malyeyeva O (2018) Method of modeling the parameters of data communication network for its upgrading. Innov Technol Sci Solut Ind 4(6):15–22. https://doi.org/10.30837/2522-9818.2018.6.015
21. Parzhin Yu, Kosenko V, Podorozhniak A, Malyeyeva O, Timofeyev V (2020) Detector neural network vs connectionist ANNs. Neurocomputing 414:191–203. https://doi.org/10.1016/j.neucom.2020.07.025
22. Zhambiu M (1988) Iyerarkhicheskiy klaster-analiz i sootvetstviya [Hierarchical cluster analysis and compliance]. Finance and Statistics, Moscow
23. Kanunnikov G. SPU—Sozdaniye setevykh grafikov [SPU—Creating Networks]. http://motosnz.narod.ru/spu.htm
24. Novitsky NI (2004) Setevoye planirovaniye i upravleniye proizvodstvom [Network planning and production management]. «Novoye znaniye», Moscow

Probabilistic Method Proactive Change Management in Telecommunication Projects

Viktor Morozov

Abstract This article considers the issues of improving the efficiency of projects in the field of telecommunications on the basis of distributed information systems (DIS). Traditional approaches are based on stationary studies of such projects. The reasons for the negative completion of projects of this class are investigated. The author proposes the use of a proactive approach to forecasting the state of implementation of such projects in their interaction with the turbulent external environment. At the same time, effective mechanisms for responding to the controlled positive or negative influence of the external environment on the project parameters are searched for and investigated. The author proposes to consider a three-component system based on the project, product and developer organization (P2O). An example of structural interaction of DIS elements is considered. The model of clustering of interaction with the external dynamic environment is offered. The basis of the studied interaction are three groups of processes—project management, product management, interaction management with the external environment. The mathematical model of the description of forecasting of consequences of influences of changes as reaction to external influences from environment is given. The description of experimental researches which have shown efficiency of the offered approach is given. At the same time, the proposed approach focuses on the possibilities of forming the desired actions for change management through active measures.

Keywords Telecommunication projects · Proactive management · Information impacts · Influences

1 Introduction

Telecommunication systems (TSs) are becoming widespread in the modern world and popular in all spheres of human activity. It's hard to imagine today even a simple business function without their use. The rapid development of information

V. Morozov (✉)
Management Technologies Department, Taras Shevchenko National University of Kyiv, Bohdan Gavrilishin Str., 24, Kiev 01601, Ukraine

D. Ageyev et al. (eds.), *Data-Centric Business and Applications*, Lecture Notes on Data Engineering and Communications Technologies 69,
https://doi.org/10.1007/978-3-030-71892-3_18

technologies adds new opportunities to meet the needs of modern business with high-quality relevant solutions. One of these solutions, which has become the main component of telecommunication systems, are cloud technologies, which greatly expanded the capabilities of these systems, but at the same time made them even more complicated. This complexity manifests itself throughout the life-course of the existence of the TS, from the idea to the implemented product, as well as at the stage of its industrial exploitation. As the reasons, you can distinguish the main of them:

- technologies used for the development of TSs are in constant development;
- the range of tasks that face such systems is constantly;
- security, accessibility, stability and scalability issues continue to gain importance;
- the use of distributed capacities and human resources creates ambiguous connections causing problems and incidents;
- the need to integrate conceptually/technically/program different elements, which requires the universality of technological solutions;
- a large number of ambiguous cross-effects from the external turbulent environment increase uncertainty and risks in the design, implementation and operation of the TSs.

Modern telecommunication systems provide for long distance transmission of various types of information and NGN (Next Generation Network) technologies. A heterogeneous multiservice packet switching network allows obtaining for almost unlimited possibilities for such systems. At the same time, NGN uses hardware and software tools focused on the TCP/IP protocol stack as hardware. Today, companies are actively implementing or upgrading their telecommunication systems to ensure business development.

Creating a new or modernizing an existing system requires taking into account the many requirements and parameters that determine the desired result, and can be presented as two components. The first concerns all technological aspects, including available software and hardware, the existing network architecture, modern technologies and ways of integrating them. The second is all that is connected with the processes of creation, implementation and maintenance of the TSs. In particular, it is necessary to take into account the response to deviations in the configuration of the environment and the environmental influences that interfere with the steady state of processes and lead to flows of complex sometimes uncontrolled changes.

Thus, we have a sufficiently large range of problems in the process of creating and implementing modern telecommunication systems that can be represented by the following groups:

- reactive approach to problem and incident management;
- lack of methodologies that would provide effective support to the process of creation and implementation of TS, as well as their further exploitation;
- low ability to recognize and manage the external environment;
- the lack of an integrated approach to product design management processes (TSs), project management processes and stakeholder management processes.

The processes of creation, implementation and support of distributed information systems (DIS) [1, 2] that are built on the basis of modern telecommunication technologies and the latest developments in the IT field are considered in the article. The urgency of the use of DIS and the complexity of these problem groups indicate the need to find and develop modern tools and methods capable of solving such problems and providing the effective management of the processes for creating and implementing DIS.

Actuality of issues of forecasting the environmental impacts on business in general and on individual projects, including on the development and implementation of DIS projects, continues to grow under the influence of globalization, integration and scientific and technological development processes. The presence of poorly predicted and complex processes in the external environment of the projects, their increasing speed and interconnectivity lead to the need to find ways to prevent possible threats and uncontrolled consequences for both the organization and individual projects [3, 4].

This applies to all areas of human activity, including the rapidly evolving IT industry and forms trends in various fields. At the same time IT projects, in the conditions of modern development tendencies and under the influence of turbulent environment, acquire new features—complexity, scale, distribution, speed, riskiness, uncertainty, etc. [5]. It is necessary to develop and use new methods and approaches to manage such projects, one of which is based on the use of proactive management methods.

2 Related Work

The use of a proactive approach to project management and organizational development was discussed in works [3, 5, 6]. Studies of the peculiarities of the implementation of complex IT projects have been studied in publications of Ukrainian and foreign scientists, such as: Bushuieva [6], Teslia [7], Ageyev [8], Hohunskyi [9], Kryvinska [10], Vayno et al. [11, 12], Itchenko [13], Taleb [14], Ervin Laszlo [15], Warrilow [16] and others. The analysis shows that the method of project management is common and well studied. But to manage complex projects there is a problem of choosing the optimal number of managed elements of the system (project). In addition, the effects of a turbulent project environment must be taken into account. In view of these problems, it can be assumed that such experts have not yet been sufficiently studied and need further research.

Considering the problems of influences and interaction with the turbulent external environment, it is safe to say that such an environment has a dynamic nature and is constantly changing and, accordingly, changing the conditions of such interaction with the elements of the project. The results of such interaction form appropriate reactions from the project, which leads to the need for frequent changes in management actions. This in turn complicates the control over the parameters of the project, which was discussed in the works [17, 18].

As shown by the analysis of publications to obtain results and products in complex IT projects, there are a number of unresolved issues, which in turn requires a proactive approach with forecasting and further analysis of the proposed actions. This will be a promising study for further development in this area.

Summing up the preliminary results, it can be noted that the effects of the external turbulent environment are dynamic. This leads to the need to make frequent changes in the process of creating innovative IT products. The consequences of such changes can lead to the cancellation of projects and their premature closure without obtaining results, which was investigated in [19–22]. This in turn points to a number of unresolved issues, and the application of a proactive approach will significantly improve the management of such complex projects and improve the results of project activities.

The purpose of the article is to conduct a study on the interaction between four categories of processes: project management processes, processes for managing the creation of an IT product, stakeholder management processes and processes for managing interactions with the external environment. It will be possible to formalize the method of interaction on the basis of earlier developed models by the authors of such interaction and proceed to experimental modelling. And also, the study is the meaning of proactive management of impacts on a project and organization, as a mechanism that can prevent potential problems or allow the use of favorable opportunities.

3 Materials and Methods

The above description shows a sufficient number of examples of the development of DIS. The authors emphasize the project use of a proactive approach, which has repeatedly shown its effectiveness.

One of them is shown in Fig. 1, in which the authors took direct participation.

To further consider the proposed approach, we need to clarify some assumptions. Because project teams (DIS) and project management implementations involve project teams, the system itself is integrated with product development and management processes. It is advisable to consider it from the standpoint of "Project-Product-Organization" (P2O). Thus it is possible to pass to the model which description was given in [23] and which consists of the following elements:

- product subsystem, which includes elements and processes of creating a complex IT product;
- project management subsystem, which includes project management processes;
- subsystem of organizational management, which includes the processes of production and corporate organizational management and provides product creation.

As noted earlier, the system under consideration is under constant dynamic influences. In [20, 24] it was shown that such systems can be in a state of "rest", which

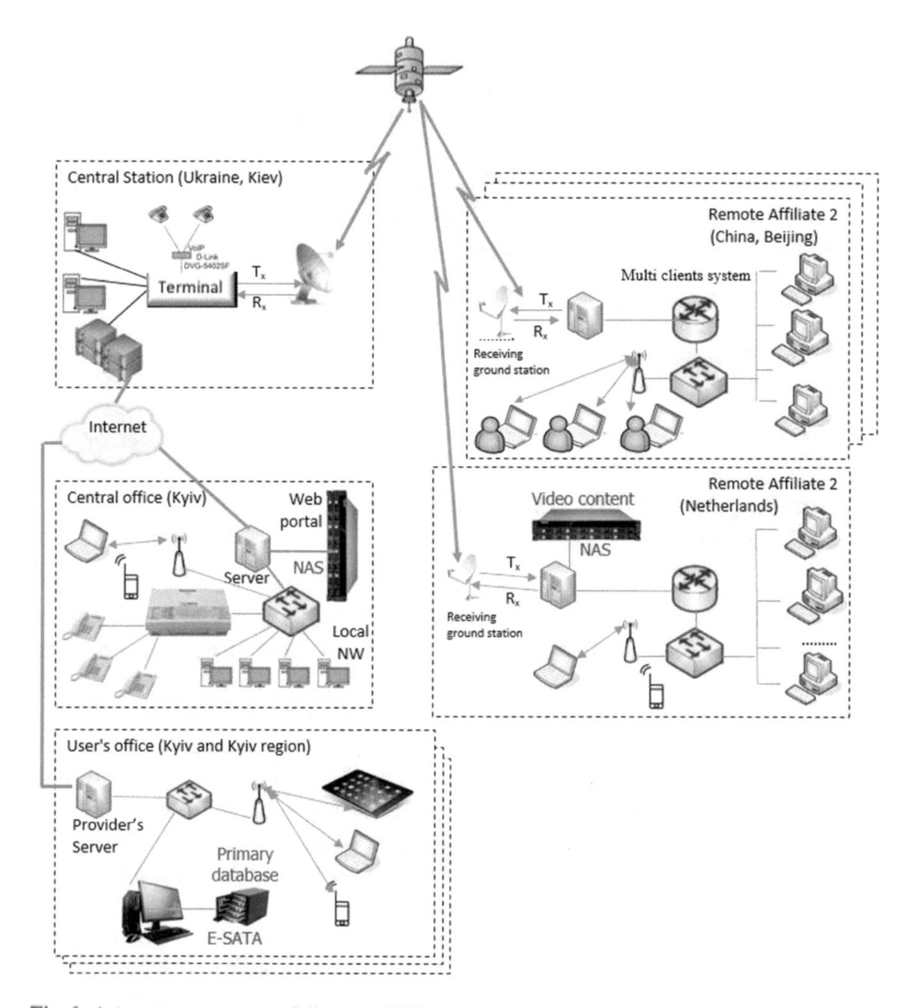

Fig. 1 Interaction structure of discussed DIS

is characterized as a state of normal functioning. This state is fixed as long as the system is not influenced from the outside. That is, the system does not receive any information.

The system can come out of a state of "rest" when it receives information from the outside, changes the values of its parameters, properties and behavior. In this way, it responds to external influences. In order to understand what processes and how they will be involved in this case when interacting with the external environment, we need to divide all processes into three groups:

- Processes that focus on managing the interaction with the external environment of the organization where the projects are implemented;

- Processes that are focused on project management, in particular on project stakeholder management. Such processes consist [25] of five groups of processes: initialization, planning, implementation, monitoring and completion of projects.
- Processes that appear with the management of production processes for the creation of innovative IT products [26]. Such processes also consist of six groups of SaaS processes [27]: strategy development and service management, service design and development, service implementation, service improvement.

Creating such a categorization of processes is appropriate in terms of taking into account the effects of a dynamic external environment and understanding the interaction of the organization with projects, as well as the reaction of the organization and projects to these influences.

3.1 Investigation of the External Environment Impact of Projects

Turning to the concept of "external environment" and its effects on the elements and parameters of projects, it should be noted that it consists of two environments: neighboring internal - organizational environment of the company implementing the project, and remote - external environment in relation to the organization.

In turn, considering the concept of "external environment" in relation to the organization, it can be noted. That it also consists of two levels: direct and indirect influence. The environment of direct influence includes subcontractors of parts of the project product and suppliers, users and investors, competitors, intermediaries and shareholders, etc. Such an environment will further serve as a basis for the formation of the concept of "stakeholders" (IPs) of the project.

The remote environment in our approach will be the economic and political, technological and legal factors of the country, where the project is implemented. This also includes cultural and natural, environmental infrastructure factors and more. These datasets constitute the "background environment" of the project.

Thus, it can be noted that the concept of "environment" from a conceptual point of view can be a multilayer model (Fig. 2), in which processes can have the following characteristics:

- Complexity—a large number of parameters (data sets) that are part of the model;
- Uncertainty—for a large number of parameters it is difficult to determine the numerical characteristics and they must be established experimentally on the basis of experiments;
- Mobility—high fluidity and variability of information over time, which gives the impact a dynamic character;
- Interconnectedness—depending on the changes of some factors, other factors may change, chains of change can be formed, etc.

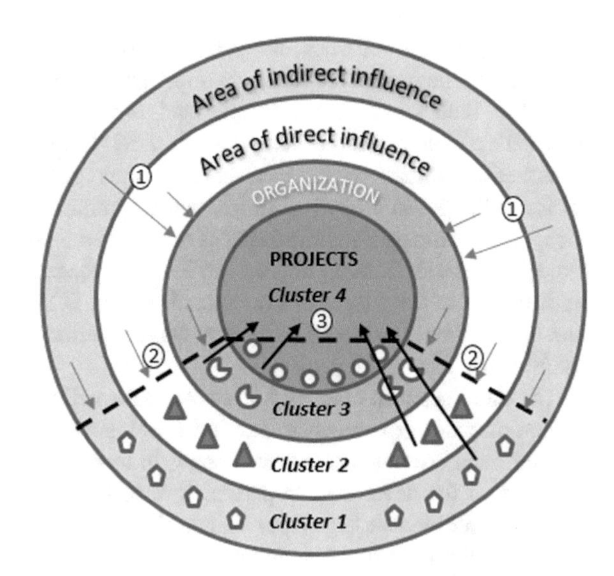

Fig. 2 Model of cauterization of environmental influences

Four clusters of interaction can be identified, when considering the model of environmental impacts (Fig. 2):

- Cluster 1, 2—remote and approximate external environment in relation to the company implementing the project;
- Cluster 1–3—is the external environment of the project, which may include clusters 1 and 2;
- Cluster 3—close project environment;
- Cluster 4—the internal environment of the project, including management processes, project teams and other elements.

Analyzing the possible effects from the external environment, you can mark the first focus group of impacts 1 (in a circle). Then, the influences perceived by the system form groups of clusters. This is a group of influences 2 (in a circle). The third group of impacts is associated with clusters of impacts on the project itself. In this case, the interested parties (IPs) are factors of direct influence of the environment and have very significant impacts on the project [28]. Therefore, it is necessary to allocate them into a separate category and to consider the specifics of interaction with the interested parties and possible proactive actions more carefully.

Based on the global international standard for project management PMBoK (Project Management Body of Knowledge) [29], it can be noted that the internal parties of the project can be in clusters 3 and 4. Then the external stakeholders should be in cluster 2. Based on this, we note that stakeholder processes determine their impacts on the project, which in turn leads to certain reactions to these impacts. This in turn has effects on the external environment to which they belong (complexity, uncertainty, mobility, interconnectedness).

Looking at the statistics of poor completion of projects in the IT industry, it should be borne in mind that DIS creation projects are complex projects. Therefore, the risk of their unsuccessful completion increases significantly. On the other hand, the creation of such systems has a long duration [5], which in turn becomes the basis of numerical influences due to changes in standards, requirements, customer preferences, changes in user capabilities, market conditions and more. Not surprisingly, such global standards include the field of knowledge in knowledge management and project stakeholders. Such impacts are based on the effects of changes in the external environment of the customer's organizations and DIS developer organizations. This situation will be considered as the first class of influences on the process of creating complex products in IT projects.

The next (second) class of influences is formed in the information environment of the companies of the customer and developers. These factors are technological in nature and are not formed by changes in technological processes, usually in the company of the developer. In particular, also in the customer's company, substantiating the problems of choosing standards (technologies) for the management and implementation of project activities.

Turning to the consideration of the third class of influences, we can note that it can be formed in two environments (1, 2), which we can now add to the design environment. These impacts are based on the functional requirements for the product of the IT project, i.e. the information system itself. Of course, such a system can also change over time and such changes can be permanent (continuous improvement).

Considering the fourth class of influences, it can be noted that it is formed on the relationship between the companies of the customer and the developer, which has a certain organizational structure and can act as the main contractor. Such relations are formally fixed in the form of a development contract, in which, too, from time to time certain changes are made. Such changes are often the result of other changes in the external environment of companies, but complicate such relationships.

3.2 Construction of a Mathematical Model

Based on the materials demonstrated above, it can be noted that the processes of managing the creation of complex information systems have a direct impact on the product processes of creating the information product, on production processes. They, in turn, interact and create a cumulative impact on management processes in the form of feedback. All these processes also interact with the information and are subject to the mutual influence of the management processes of groups of enterprises of the customer and the general developer. These interactions could be modeled in stationary conditions without change.

Given the requirements of systems analysis, we must approach the real conditions for creating an information system, including: the processes of design management, development and production of parts of complex IT products. At the same time, we are forced to take into account the contextual content of such systems. Hereinafter, we will call such systems the "external information environment" (EIE) system. We have already noted that such systems are characterized by a large number, power, distribution over time of various influences on all these groups of processes, the consequences of which are difficult to predict. The presence of this factor significantly complicates the modeling process, requires new techniques based on forecasting the state of projects and the production system as a whole.

Based on the above and for further consideration and formal description, the author developed and proposed a conceptual model to take into account the interaction of three groups of processes and clustering, which form the basis of a proactive approach [5, 10]. This model can be represented in Fig. 3.

So, the flow of input information in time t in the form of requirements (problems) in the management of business development is represented as $X(t)$ in Fig. 3; $Y(t)$—initial information about the state of business development in time t, which is determined by the values of the indicators of the organization's development. As $X'(t)$ is the flow of input information in time t in the form of requirements, constraints and assumptions about project management; $Y'(t)$ is the initial information about the state of the project in time t, which is determined by the values of the project implementation indicators.

As $X''(t)$ is the flow of input information in time t as project product requirements (DIS); $Y''(t)$ is output information about the state of preparedness of the project product in time t. The set of flows of input and output information can be represented as (1) and (2):

$$X(t) = \{x_i(t) | i = \overline{1, n}\}, n \in N \tag{1}$$

$$X'(t) = \left\{x'_{i_1}(t) | i_1 = \overline{1, n_1}\right\}, n_1 \in N_1;$$

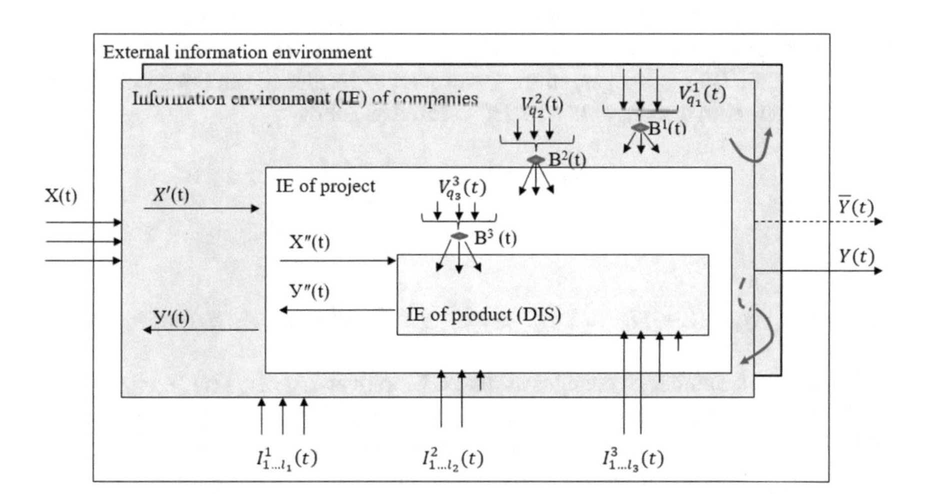

Fig. 3 Conceptual model of proactive management of changes in DIS projects

$$X''(t) = \left\{x''_{i_2}(t)|i_2 = \overline{1, n_2}\right\}, n_2 \in N_2;$$
$$Y(t) = \{y_j(t)|j = \overline{1, m}\}, m \in M,$$
$$Y'(t) = \left\{y'_{j_1}(t)|j_1 = \overline{1, m_1}\right\}, m_1 \in M_1;$$
$$Y''(t) = \left\{y''_{j_2}(t)|j_2 = \overline{1, m_2}\right\}, m_2 \in M_2;$$
$$Y(t) = Y'(t) \cup Y''(t), \quad (2)$$

where N, N_1, N_2—the total number of input characteristics of the system, organizations and the project, respectively.

M, M_1, M_2—the total number of output parameters of the system, the participating organizations of the project and the project, respectively.

In turn, $I^1(t)$ are varieties of set of flows of influences of the environment on the activities of the enterprises of customer and developers, as well as the indirect impact of the external environment on the state of the project (projects) $I^2(t)$ and on the state and functional of the product $I^3(t)$ in the form of its components and elements. In addition there is a category of influences between the customer and the contractor within the project $I^4(t)$. The set of flows of influences can be represented as follows:

$$I(t) = I^1_{k_1}(t), I^2_{k_2}(t), I^3_{k_3}(t), I^4_{k_4}(t), \quad (3)$$

where $k_1 = \overline{1, l_1}; k_2 = \overline{1, l_2}; k_3 = \overline{1, l_3}; k_4 = \overline{1, l_4};\ for\ \forall(l_1, l_2, l_3, l_4)\exists(l, k) \in L;$

L—is a total number of influences at time t.

You can also define the set of information flow changes as signals for external influences, such as those, which have *proactive nature*. Such changes have a focus on the activities of participating organizations $V^1(t)$, on the project $V^2(t)$, or on product $V^3(t)$. It is clear that $V^4(t)$ also reflects the effects of changes in the relations between the project participants and the formation of changes, at least in contractual relations between them. These changes, in turn, can lead to further development of system reactions. The above changes can be represented as follows:

$$V(t) = \left\{V^1_{q_1}, V^2_{q_2}, V^3_{q_3}, V^4_{q_4}\right\}, \quad (4)$$

where:

$$q_1 = \overline{1, d_1}; q_2 = \overline{1, d_2}; q_3 = \overline{1, d_3}; q_4 = \overline{1, d_4};\ for\forall(d_1, d_2, d_3, d_4)\exists(d, q) \in D,$$

D—the total number of changes at time t. Moreover $L \geq D;\ I(t) \geq V(t)$.

The use of a proactive approach in this case allows you to model the parameters of the project and predict the consequences of reactions to change. This also applies to the end result of the project activity. It also provides an opportunity to assess the projected changes and their impact on the project implementation. In turn, this will facilitate decision-making processes to include these changes in the project or leave them without consideration.

It should be noted that all datasets of changes, as well as all project parameters depend on a specific point in time, which is discrete. In this case, each of the moments of time t∈T, where T is the total period of project implementation to create the necessary IT product in full.

To include a certain type of changes in the current model (Fig. 3), sets of switches ("a traffic light") B(t) are used. They characterize the sets of corresponding states of other communicators and their description will be provided below. So:

$$B(t) = \{B^1(t), B^2(t), B^3(t)\}, \tag{5}$$

where

$B^1(t)$ switches that determine the ability to implement changes at the organization level;
$B^2(t)$ switches that determine the ability to implement changes at the project level;
$B^3(t)$ switches that determine the ability to implement changes at the product level

In this case, the states of each of the switches, prohibiting or permitting changes are defined as follows:

$$B(t) = \begin{cases} 0, & \text{information about changes is not perceived by the system} \\ 1, & \text{all suggested changes are allowed} \end{cases}$$

To determine the status of each of the switches (traffic lights), an expert method is used, which involves three groups of experts who analyze the degree of critical impact of changes on project parameters. Depending on the indicator Y (t), multicriteria dependences are currently modeled, indicators of the future state of the system are obtained, and the state of the corresponding switches is determined.

Moving to a more formal consideration of constituents of the conceptual model, we can point out that the reaction of the system to the effects of the EIE can be calculated on the basis of the use of a mathematical apparatus [7], as the value of a compatible conditional probability. If necessary, we will also make changes to certain formulas.

Calculation of the definiteness of system reactions to the impacts:

$$d(Y_j) = \begin{cases} 0,5 \times \sqrt{\frac{p(Y_j)}{1-p(Y_j)} + \frac{1-p(Y_j)}{p(Y_j)} - 2}, & for\ p(Y_j) \geq 0 \\ -0,5 \times \sqrt{\frac{p(Y_j)}{1-p(Y_j)} + \frac{1-p(Y_j)}{p(Y_j)}} - 2, & for\ p(Y_j) < 0 \end{cases} \tag{6}$$

Calculation of the magnitude of the effect (determination of the reaction) of the input information on the reaction of the system, which can be determined on the basis of the deviation of the conditional probability of choosing a reaction from the unconditional:

$$d\left(\frac{Y_j}{X_i}\right) = \begin{cases} 0,5 \times \sqrt{\frac{p\left(\frac{Y_j}{X_i}\right)}{1-p\left(\frac{Y_j}{X_i}\right)} + \frac{1-p\left(\frac{Y_j}{X_i}\right)}{p\left(\frac{Y_j}{X_i}\right)} - 2}, & for\ p\left(\frac{Y_j}{X_i}\right) \geq p(Y_j) \\ -0,5 \times \sqrt{\frac{p\left(\frac{Y_j}{X_i}\right)}{1-p\left(\frac{Y_j}{X_i}\right)} + \frac{1-p\left(\frac{Y_j}{X_i}\right)}{p\left(\frac{Y_j}{X_i}\right)} - 2}, & for\ p\left(\frac{Y_j}{X_i}\right) < p(Y_j) \end{cases} \tag{7}$$

where

X_i input information,

Y_j reaction in the form of flow of output information, $i \in N, j \in M$;

$p\left(\frac{Y_j}{X_i}\right)$ conditional probability of choosing the reaction Y_j under the influence of the input information X_i

$p(Y_j)$ is the unconditional probability of choosing the reaction Y_j

It should be noted that the magnitude of the effects of the flow of calls from the EIE generate a list of changes in each relevant group in the system and are calculated as follows:

$$d\left(\frac{Y_j}{X_i}\right) = \begin{cases} 0,5 \times \sqrt{\frac{p\left(\frac{V_q}{I_k}\right)}{1-p\left(\frac{V_q}{I_k}\right)} + \frac{1-p\left(\frac{V_q}{I_k}\right)}{p\left(\frac{V_q}{I_k}\right)} - 2}, & for\ p\left(\frac{V_q}{I_k}\right) \geq p(V_q) \\ -0,5 \times \sqrt{\frac{p\left(\frac{V_q}{I_k}\right)}{1-p\left(\frac{V_q}{I_k}\right)} + \frac{1-p\left(\frac{V_q}{I_k}\right)}{p\left(\frac{V_q}{I_k}\right)} - 2}, & for\ p\left(\frac{V_q}{I_k}\right) < p(Y_j) \end{cases} \tag{8}$$

where $\left(\frac{V_q}{I_k}\right)$—is the probability of forming V_q changes taking into account the effects of a particular information medium $I_k, k \in L, q \in M$.

The magnitude of the impact of changes (in the organization, in the product, or in the project on the effects of the EIE) on the overall system response (including filters—switches that allow/forbid changes) is calculated as follows:

$$d\left(\frac{Y_j}{V_q}\right) = \begin{cases} B(t) \times 0,5 \times \sqrt{\frac{p\left(\frac{Y_j}{V_q}\right)}{1-p\left(\frac{Y_j}{V_q}\right)} + \frac{1-p\left(\frac{Y_j}{V_q}\right)}{p\left(\frac{Y_j}{V_q}\right)} - 2}, & for\ p\left(\frac{Y_j}{V_q}\right) \geq p(Y_j) \\ B(t) \times (-0,5) \times \sqrt{\frac{p\left(\frac{Y_j}{V_q}\right)}{1-p\left(\frac{Y_j}{V_q}\right)} + \frac{1-p\left(\frac{Y_j}{V_q}\right)}{p\left(\frac{Y_j}{V_q}\right)} - 2}, & for\ p\left(\frac{Y_j}{V_q}\right) < p(Y_j) \end{cases} \tag{9}$$

where $\frac{Y_j}{V_q}$—is the probability of choosing the reaction Y_j taking into account the changes V_q, caused by the influences of a particular information environment $q \in D$.

Then, in a similar way, you should calculate:

$$d\left(\frac{Y_j}{X_i I_k}\right), d\left(\frac{Y_j}{X_i V_q}\right), d\left(\frac{Y_j}{I_k V_q}\right), d\left(\frac{Y_j}{X_i I_k V_q}\right)$$

In addition to the definite reaction to the effects, one should calculate the awareness of the system with respect to the reaction Y_j:

$$\mu(Y_j) = \sqrt{d^2(Y_j) + 1} \tag{10}$$

The system's awareness of the reaction Y_j when the input parameters X_i: are changed:

$$\mu\left(\frac{Y_j}{X_i}\right) = \sqrt{d^2\left(\frac{Y_j}{X_i}\right) + 1} \tag{11}$$

Similarly, by formula (11) we expect the system to be aware of the reaction Y_j with other influences:

$$\mu'\left(\frac{Y_j}{I_k}\right), \mu'\left(\frac{Y_j}{V_q}\right), \mu'\left(\frac{Y_j}{X_i I_k}\right), \mu'\left(\frac{Y_j}{X_i V_q}\right), \mu'\left(\frac{Y_j}{I_k V_q}\right), \mu'\left(\frac{Y_j}{X_i I_k V_q}\right)$$

Additional certainty of the system, which is formed as a result of the effects of influence, we calculate according to the formula:

$$\Delta d\left(\frac{Y_j}{X_i}\right) = d\left(\frac{Y_j}{X_i}\right) \times i(Y_j) - d(Y_j) \times i\left(\frac{Y_j}{X_i}\right) \tag{12}$$

Similarly, we make a calculation:

$$\Delta d\left(\frac{Y_j}{I_k}\right), \Delta d\left(\frac{Y_j}{V_q}\right), \Delta d\left(\frac{Y_j}{X_i I_k}\right), \Delta d\left(\frac{Y_j}{X_i V_q}\right), \Delta d\left(\frac{Y_j}{I_k V_q}\right), \Delta d\left(\frac{Y_j}{X_i I_k V_q}\right)$$

The total effect on the changes in the reactions of the system can be shown as follows:

$$\begin{aligned}\Delta d_{\Sigma}(\overline{Y_j}) = {} & \sum\nolimits_i \Delta d\left(\frac{Y_j}{X_k}\right) + \sum\nolimits_k \Delta d\left(\frac{Y_j}{I_q}\right) + \sum\nolimits_q \Delta d\left(\frac{Y_j}{V_q}\right) \\ & + \sum\nolimits_i \sum\nolimits_k \Delta d\left(\frac{Y_j}{X_i, I_k}\right) + \sum\nolimits_k \sum\nolimits_q \Delta d\left(\frac{Y_j}{I_k, V_q}\right) \\ & + \sum\nolimits_i \sum\nolimits_q \Delta d\left(\frac{Y_j}{X_i, V_q}\right) + \sum\nolimits_i \sum\nolimits_k \sum\nolimits_q \Delta d\left(\frac{Y_j}{X_i, I_k, V_q}\right)\end{aligned} \tag{13}$$

We can form the corresponding algorithm to reflect the processes of interaction of elements of the system of creating complex IT products in the perception of information influences by the system and the formation of reactions to them, which is shown in Fig. 4, which supplements the description of the analysis.

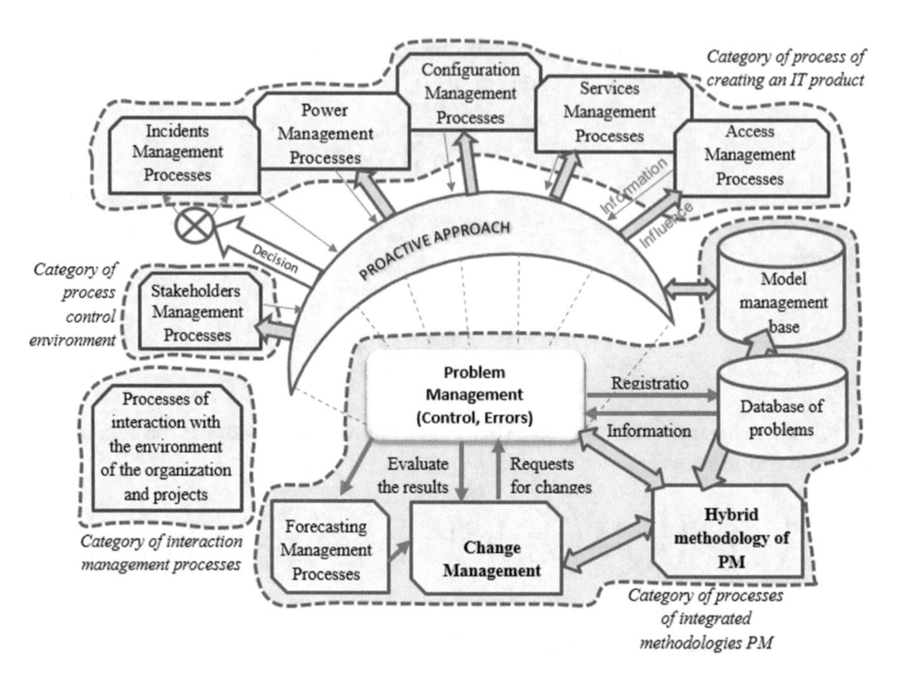

Fig. 4 The model of integrated control processes based on the concept of a proactive approach

3.3 Creating a Model of Interaction Processes

The concept of proactive management can become an integral part of the P2O system. It is advisable to implement proactive processes at the organizational level that seem to be the intermediary between the project and the external environment and, unlike the project, is not a temporary phenomenon. It is necessary to identify the main processes that have a significant impact on product creation, stakeholder interaction and change management in order to understand the key features of the proposed concept.

As can be seen from Fig. 4, visually, the model of information interaction of the components of antisipative management forms a “parachute” with the union of all reactions of the predicted effects. This approach for projects of creation of distributed information systems shows the close mutual information connection of the processes of creation of such systems and processes of their management.

The focus of this model was the process of managing problems to identify the causes of incidents. This process sends information to the input of the change management process, if this is necessary and receives feedback in the form of made corrections. In this case, the proactive component of the problem management process can be implemented for the main processes of creating and implementing an IT project (Fig. 4).

Taking into account the above principles on which the proposed concept is based, the focus of attention shifts from the problem management process to the processes outlined in the concept (Fig. 4). This greatly expands the understanding of the proactive approach to managing impacts during project implementation.

The proposed concept can be presented as a distributed system. Then the distributed subsystem of the product creation project consists of a finite set of independent processes $\{P_1^s, P_2^s, \ldots, P_n^s\}$. Distributed project consists of a finite set of m processes $\{P_1^p, P_2^p, \ldots, P_m^p\}$. Distributed environment can be represented by the finite sets k of independent processes of the organization $\{P_1^o, P_2^o, \ldots, P_k^o\}$ and f of independent processes of interaction with the environment $\left\{P_1^e, P_2^e, \ldots, P_f^e\right\}$. In this case, the sequence of events $E_i^0, E_i^1, \ldots, E_i^x, E_i^{x+1}$ is the execution of the process P_i of the P2O system.

We need to determine the state of the production system (project) at any time, based on the analysis of events that occur as a reaction to external influences. In this case, the target function should be the function of minimizing the cost of implementing changes. In addition, the optimal system formed and the minimization of damage from the environment should be taken into account.

Turning to the formalization and construction of the optimization criteria for the above processes, we introduce the following characteristics of projects [13, 30].

In this case, the input parameters of the project model can also be presented in the form $X = \{x_{i_1} | i_1 = 1, 2, \ldots, N_1\}$, where N_1—number of areas of knowledge of the model. Then the planned cost of the project will be:

$$C_p = \sum_{i_1=1}^{N_1} \sum_{j_1=1}^{T_p} \sum_{i_2=1}^{E} \left(C_1\left(x_{i_1}, t_{j_1}\right) + C_2\left(h_{i_2}\right)\right), \tag{14}$$

when $\forall\left(x_{i_1} \in X\right) \cup \left(q_{i_2} \in Q\right) \exists t_{j_1} \in T_p;\ T_p \geq 0$, and $C_p \leq C_{b,} C_b \geq 0$,

where C_1—the function of the cost of elements with $\{X\}$ on the moment of time $t_{j_1} \in T_p$, C_2—the function of the cost of communication channels between the elements of the model with $\{X\}$, C_b—the budget cost of the project.

Given the impact of the external environment and the project of stakeholders that lead to changes and deviations from the given project parameters, it is possible to determine the actual cost of the project upon its completion $\left(C_j\right)$ and the actual completion time of the project $\left(T_j\right)$:

$$T_f = T_p \pm (f_1(I) + f_2(U) + f_3(V)), \tag{15}$$

$$C_f = C_p \pm (C_3(I) + C_4(U) + C_5(V)), \tag{16}$$

where C_3, C_4, C_5—is the actual cost of making changes due to the many impacts on the project, monitoring the multitude of IT projects and the many executed actions that operate accordingly; f_1, f_2, f_3—the functions of measuring the time intervals of many impacts on the project, monitoring the set of states of the IT project and the set of executed actions that operate accordingly.

In this case, the target functions of the management model of the IT project can be represented as follows:

$$C_f - C_p = \pm \Delta C \rightarrow \min, \tag{17}$$

$$T_f - T_p = \pm \Delta T \rightarrow \min, \tag{18}$$

4 Experimentation

The indicated dependencies at first glance have a certain contradiction and can counteract each other, which can lead to conflicts in the project management process. However, during the study (4–9), we can use the well-known Earned Value method [29], where the parameters ΔT and ΔC co-operate within the limits of this method.

Thus, the traditional dependence of these parameters—the graph of the cost of the project in time is shown by a solid curve of blue in Fig. 5a. This creates the base cost curve of the project. In this case, the prediction of the state of the project on the value, for example, from the 5th time interval, is determined by a simple transfer of the base curve to the point of the current execution. We have a forecast for the cost to complete the project—dashed curve (Fig. 5a). Figure 5b is constructed similarly, but it shows more clearly how deviations are formed by ΔT and ΔC. All this is calculated by standard software, such as Microsoft Project.

The results of one of the studies of the effects of changes on the project cost graph are shown in Fig. 5c. Herewith, the so-called corridor is created taking into account the minimum and maximum changes, which greatly complicates the results of forecasting by traditional methods. The given results show that when taking into account the numerical influences of the external environment on the main parameters of the project we get a toothed graph of execution in terms of cost (Fig. 5c, red curve). Forecasting by traditional methods (assuming that impacts will no longer be available), cost indicators from the point of current performance of works are facing difficulties, forming a certain static corridor. However, the use of predictive functions makes it possible, even taking into account future turbulent influences.

The results of digital experiments of expression research (4–9) are shown in Fig. 6. In this case, the red curve shows a minimum of 50 variants of the experiment, and blue is the maximum distribution of the cost of the project. It is clear that the other 48 curves for the sake of clarity in the demonstration of results are not shown. But you can see the calculated averaged result (gray dotted curve). Also, this figure shows an orange curve, which is the result of the application of a certain set of changes and the reaction in this system.

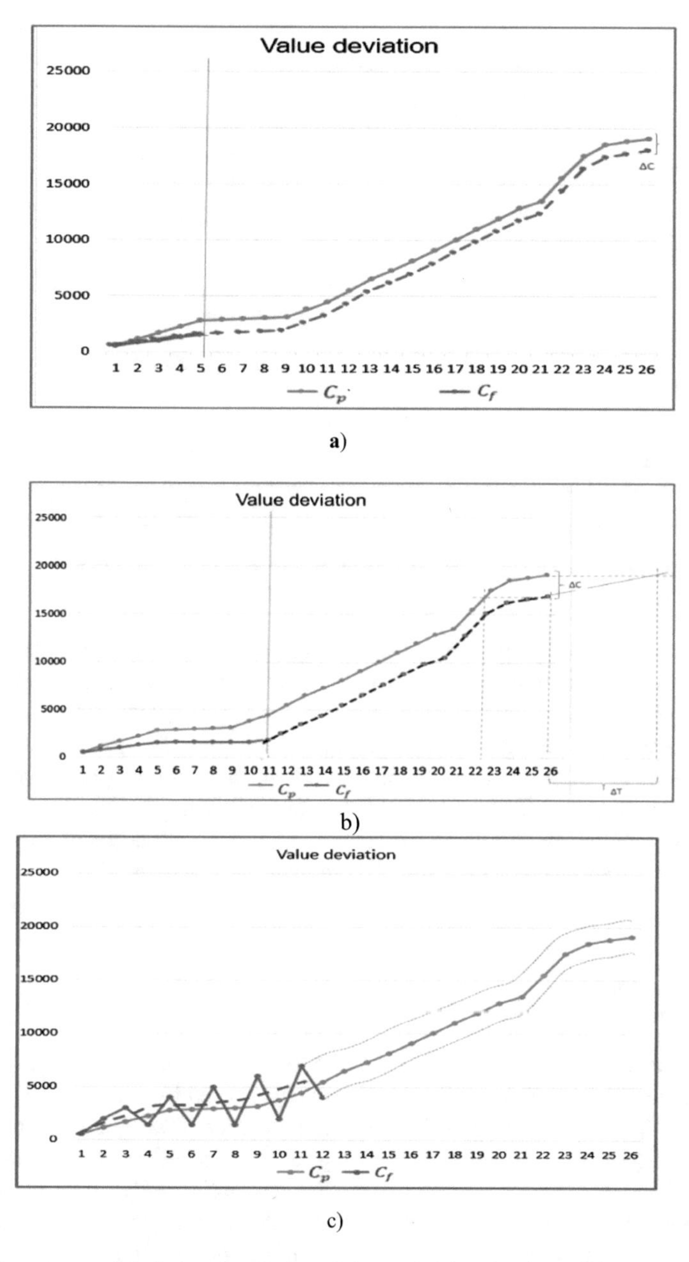

Fig. 5 Management of deviations (prediction of changes) a) in value by traditional methods at the beginning of the project; b) on cost and time by traditional methods in the middle of the project; c) example of changes in real projects

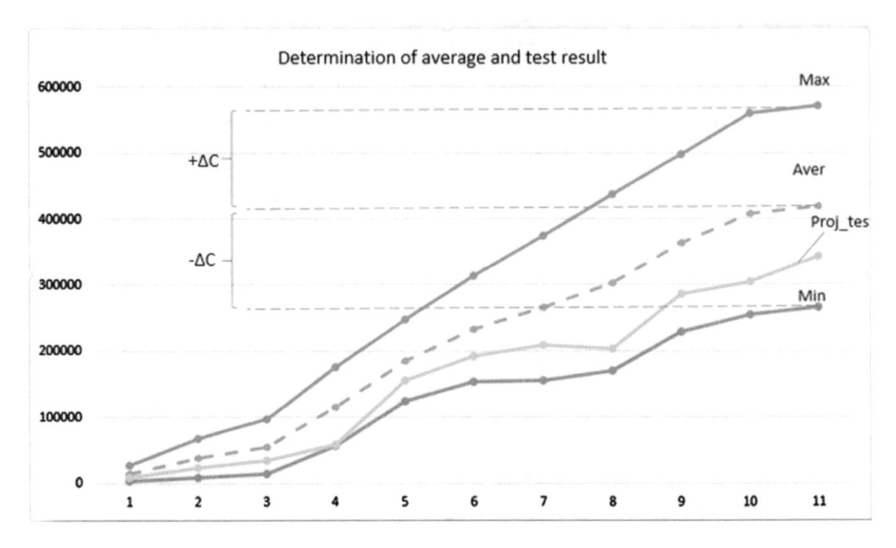

Fig. 6 Results of simulation of changes with deviations and obtaining the test result

In this case, the test graph is between averaged and minimal curves. This indicates a fairly good reaction of the system and the adequacy of the proposed models, investigated as a result of experiments. Thus, it is possible to calculate the actual parameters for each variation of changes during the project implementation, taking into account the effects of the external environment, as well as calculate the predictive function of the system response from the current state of the project.

Such functions are implemented in the interaction of problem management processes, forecast management and change management (Fig. 4).

5 Conclusions

The conducted analysis of the processes of the "product-project-organization" system and its interaction with the turbulent environment allowed formulating the principles within which the functioning of the system takes place. Based on these principles, the assumption is made about the need to expand the standard proactive management approach by separating certain processes that can significantly affect the system's ability to withstand powerful, dynamic and complex impacts. The concept of proactive management has been formulated in the framework of the study and identified proactive management processes

The proposed concept significantly expands the capabilities of proactive management contrary to the accepted understanding of the proactive approach, which is to monitor the environment, detect weak signals, analyze them and develop scenarios for future events. These principles and processes reveal the deep essence of the tasks of proactive management and show the spectrum of its capabilities.

References

1. Cloud terminology—key definitions, Available: https://www.getfilecloud.com/cloud-terminology-glossary. Last accessed 21 Feb 2019
2. Cloud computing and emerging IT platforms: vision, hype, and reality for delivering computing as the 5th utility, Available: http://www.sciencedirect.com/science/article/pii/S0167739X08001957. Last accessed 21 Feb 2019
3. Morozov V, Kalnichenko O, Bronin S (2018) Development of the model of the proactive approach in creation of distributed information systems. Eastern-Eur J Enterp Technol 43/2(94):6–15
4. Bushuyev S, Burkov V (2015) Resources management for distributed projects and programs. In: Torubara VV, Monograph, Nikolayev, 338 p
5. Proactive project management. Available: http://www.itexpert.ru/rus/ITEMS/200810062247/. Last accessed 12 Mar 2019
6. Bushueva N (2007) Models and methods for proactive management of organizational development programs. Monograph. Scientific World, 199 p
7. Teslia (2010) Vvedeniye v informatiku prirody. In: Monografiya K (ed) Maklaut, 255 p
8. Ageyev D et al (2018) Classification of existing virtualization methods used in telecommunication networks. In: Proceedings of the 2018 IEEE 9th international conference on dependable systems, services and technologies (DESSERT), pp 83–86
9. Gogunskiy V, Kolesnikova K, Lukianov D (2016) Lifelong learning is a new paradigm of personnel training in enterprises. Eastern-Eur J Enterprise Technol 4/2(82):4–10
10. Kryvinska N, Zinterhof P, van Thanh D (2007) New-Emerging service-support model for converged multi-service network and its practical validation. In: First international conference on complex, intelligent and software intensive systems (CISIS'07), pp 100–110. IEEE. https://doi.org/10.1109/cisis.2007.40
11. Vajno A, Kobiakov A, Saraev V, Anticipatory management. Electronic data. Access mode: http://webcache.Googleusercontent.com/search?q=cache, https://istina.msu.ru/media/Publications/articles/4f3/98f/4274680/UPREZhDAYuSchEE_UPRAVLENIE_-5_pravka.doc&gws_rd=cr&ei=BpxKWdupN_Pb6QSF2qn4Dg
12. Vaino AE, Kobyakov AA, Saraev VN (2011) Proactive control of complex systems. Vestn Econ integration 11:7–21
13. Itchenko DM (2015) Analysis of approaches to proactive management in the context of their application in the implementation of projects and programs agro-industrial complex. Bulletin of the National Technical University "KhPI". Collection of scientific works. Series: Strategic Management, Portfolio Management, Programs and Projects. Kharkiv: KhPI, no 2(1111), pp 141–148
14. Taleb NN (2017) Black swan under the sign of unpredictability. In: The table NN, 2nd edn. Moscow: [Kolibri], 735 p
15. Laszlo E (1991) The Age of bifurcation. understanding the changing world. Series: World futures general evolution studies (Book 3), 126 p
16. Warrilow S (2019) Change management: the horror of it all, Project Smart. (Online). Available: https://www.projectsmart.co.uk/change-management-the-horror-of-it-all.php. Last accessed 21 Feb 2019
17. Morozov V, Kalnichenko O, Kolomiiets A (2019) Research of the impact of changes based on external influences in complex it projects. In: Proceedings of the IEEE international conference on advanced trends in information theory, ATIT'2019, 481–488. IEEE. https://doi.org/10.1109/ATIT49449.2019.9030441
18. Danchenko OB (2008) Modern metodology of change project management. In: Danchenko OB, Mykhailuta SL (eds) Courier ChDTU, no 3, pp 130–132
19. Proactive project management. (Online). Available: http://www.itexpert.ru/rus/ITEMS/200810062247/. Last accessed 02 Jan 2019

20. Dombrowski MZ, Sachenko AO (2017) The proactive management model of strategic development project on the energy supply companies in a turbulent environment. Bulletin of NTU "KhPI" 2(1224):41–45
21. Prigogine I, Strangers I (1999) Time, chaos, quantum: Per. from english—M.: Publishing group "Progress", 268 p
22. Radivilova T, Kirichenko L, Vitalii B (2019) Comparative analysis of machine learning classification of time series with fractal properties. In: Proceedings of the international conference on advanced optoelectronics and lasers, CAOL, Sozopol, Bulgaria: IEEE, pp 557–560. https://doi.org/10.1109/CAOL46282.2019.9019416
23. Chernyak OI, Zakharchenko PV, Klebanov-Berdiansk TS (2019) The theory of chaos in the economy: under the arm. In: Tkachuk OV (ed), 288 p
24. Barabash OV, Open'ko PV, Kopiika OV, Shevchenko HV, Dakhno NB (2019) Target programming with multicriterial restrictions application to the defense budget optimization. Adv Mil Technol 14(2):213–229. https://doi.org/10.3849/aimt.01291
25. Morozov V, Kalnichenko O, Mezentseva O (2020) The method of interaction modeling on basis of deep learning of neural networks in complex it-projects. Int J Comput 19(1):88–96
26. Komashinsky, Smirnov DA (2003) Neural networks and their use in control and communication systems. Hotline-Telecom 94
27. Teslia Y (2016) Control of informational Impacts on project management. In: Teslia Y, Khlevnyi A, Khlevna I (eds) Proceedings of the 1th IEEE international conference on data stream mining & processing. Lviv, pp 378–391
28. Garaedagi D (2011) System thinking. How to manage chaos and complex processes. Platform for modeling business architecture, Grevtsov Buks (Grevtsov Publicher), 480 p
29. A guide to the project management body of knowledge (PMBOK®) (2017). 6th edn. Delaware, Pennsylvania, Newton Square 19073–3299, Project Management Institute Four Campus Boulevard, USA, 762 p
30. Free ITIL, v.3, Available: http://www.wikiitil.ru/books/2015_Free_ITIL.pdf. Last accessed 10 Mar 2019

Printed in the United States
by Baker & Taylor Publisher Services